Convergence of Cloud Computing, AI, and Agricultural Science

Avinash Kumar Sharma
ABES Institute of Technology, India

Nitin Chanderwal
University of Cincinnati, USA

Rijwan Khan
ABES Institute of Technology, India

IGI Global
PUBLISHER of TIMELY KNOWLEDGE

A volume in the Advances in Environmental Engineering and Green Technologies (AEEGT) Book Series

Published in the United States of America by
IGI Global
Engineering Science Reference (an imprint of IGI Global)
701 E. Chocolate Avenue
Hershey PA, USA 17033
Tel: 717-533-8845
Fax: 717-533-8661
E-mail: cust@igi-global.com
Web site: http://www.igi-global.com

Library of Congress Cataloging-in-Publication Data

Names: Sharma, Avinash Kumar, 1982- editor. | Chanderwal, Nitin, 1978- editor. | Khan, Rijwan, 1981- editor.
Title: Convergence of cloud computing, AI, and agricultural science / edited by Avinash Kumar Sharma, Nitin Chanderwal, and Rijwan Khan.
Description: Hershey, PA : Engineering Science Reference, [2023] | Includes bibliographical references and index. | Summary: "With its academic tone and in-depth exploration of cloud computing in smart agriculture, this book serves as an essential resource for researchers, academics, and professionals in the fields of agriculture, computer science, and environmental science. By examining the convergence of cloud computing, AI, and agricultural science, it provides a roadmap for harnessing technology to revolutionize farming practices and ensure sustainable agri-food systems in the digital era"-- Provided by publisher.
Identifiers: LCCN 2023025888 (print) | LCCN 2023025889 (ebook) | ISBN 9798369302002 (hardcover) | ISBN 9798369302019 (paperback) | ISBN 9798369302026 (ebook)
Subjects: LCSH: Agriculture--Data processing. | Artificial intelligence--Agricultural applications. | Cloud computing.
Classification: LCC S494.5.D3 C683 2023 (print) | LCC S494.5.D3 (ebook) | DDC 338.10285--dc23/eng/20230624
LC record available at https://lccn.loc.gov/2023025888
LC ebook record available at https://lccn.loc.gov/2023025889

This book is published in the IGI Global book series Advances in Environmental Engineering and Green Technologies (AEEGT) (ISSN: 2326-9162; eISSN: 2326-9170)

British Cataloguing in Publication Data
A Cataloguing in Publication record for this book is available from the British Library.

For electronic access to this publication, please contact: eresources@igi-global.com.

Advances in Environmental Engineering and Green Technologies (AEEGT) Book Series

Sang-Bing Tsai
Zhongshan Institute, University of Electronic Science and Technology of China, China & Wuyi University, China
Ming-Lang Tseng
Lunghwa University of Science and Technology, Taiwan
Yuchi Wang
University of Electronic Science and Technology of China Zhongshan Institute, China

ISSN:2326-9162
EISSN:2326-9170

Mission

Growing awareness and an increased focus on environmental issues such as climate change, energy use, and loss of non-renewable resources have brought about a greater need for research that provides potential solutions to these problems. Research in environmental science and engineering continues to play a vital role in uncovering new opportunities for a "green" future.

The **Advances in Environmental Engineering and Green Technologies (AEEGT)** book series is a mouthpiece for research in all aspects of environmental science, earth science, and green initiatives. This series supports the ongoing research in this field through publishing books that discuss topics within environmental engineering or that deal with the interdisciplinary field of green technologies.

Coverage

- Waste Management
- Renewable Energy
- Air Quality
- Cleantech
- Contaminated Site Remediation
- Pollution Management
- Alternative Power Sources
- Water Supply and Treatment
- Radioactive Waste Treatment
- Electric Vehicles

IGI Global is currently accepting manuscripts for publication within this series. To submit a proposal for a volume in this series, please contact our Acquisition Editors at Acquisitions@igi-global.com or visit: http://www.igi-global.com/publish/.

The Advances in Environmental Engineering and Green Technologies (AEEGT) Book Series (ISSN 2326-9162) is published by IGI Global, 701 E. Chocolate Avenue, Hershey, PA 17033-1240, USA, www.igi-global.com. This series is composed of titles available for purchase individually; each title is edited to be contextually exclusive from any other title within the series. For pricing and ordering information please visit http://www.igi-global.com/book-series/advances-environmental-engineering-green-technologies/73679. Postmaster: Send all address changes to above address.

Titles in this Series

For a list of additional titles in this series, please visit: www.igi-global.com/book-series/advances-environmental-engineering-green-technologies/73679

Handbook of Research on AI-Equipped IoT Applications in High-Tech Agriculture
Alex Khang (Global Research Institute of Technology and Engineering, USA)
Engineering Science Reference • © 2023 • 473pp • H/C (ISBN: 9781668492314) • US $335.00

Contemporary Developments in Agricultural Cyber-Physical Systems
G.S. Karthick (Department of Software Systems, PSG College of Arts and Science, India)
Engineering Science Reference • © 2023 • 293pp • H/C (ISBN: 9781668478790) • US $240.00

Perspectives on Ecological Degradation and Technological Progress
Veli Yilanci (Faculty of Political Sciences, Canakkale Onsekiz Mart University, Turkey)
Engineering Science Reference • © 2023 • 342pp • H/C (ISBN: 9781668467275) • US $225.00

Nanopriming Approach to Sustainable Agriculture
Abhishek Singh (Faculty of Biology, Yerevan State University, Armenia) Vishnu D. Rajput (Academy of Biology and Biotechnology, Southern Federal University, Rostov-on-Don, Russia) Karen Ghazaryan (Faculty of Biology, Yerevan State University, Armenia) Santosh Kumar Gupta (National Institute of Plant Genome Research, India) and Tatiana Minkina (Academy of Biology and Biotechnology, Southern Federal University, Rostov-on-Don, Russia)
Engineering Science Reference • © 2023 • 300pp • H/C (ISBN: 9781668472323) • US $245.00

Nematode-Plant Interactions and Controlling Infection
Waleed Fouad Abobatta (Horticulture Research Institute, Agriculture Research Center, Egypt) and Rehab Yasin Ghareeb (Plant Protection and Biomolecular diagnosis Dep., Arid Lands Cultivation Research Institute, City of Scientific Research and Technological Applications, Egypt)
Engineering Science Reference • © 2023 • 300pp • H/C (ISBN: 9781668480830) • US $240.00

Perspectives on Global Biodiversity Scenarios and Environmental Services in the 21st Century
Naveen Kumar Chourasia (Government College, Bichhua, India) and Kavita Chahal (Government College, Bichhua, Chhindwara, India)
Engineering Science Reference • © 2023 • 300pp • H/C (ISBN: 9781668490341) • US $250.00

Artificial Intelligence Applications in Water Treatment and Water Resource Management
Victor Shikuku (Kaimosi Friends University College, Kenya)
Engineering Science Reference • © 2023 • 300pp • H/C (ISBN: 9781668467916) • US $225.00

IGI Global
PUBLISHER of TIMELY KNOWLEDGE

701 East Chocolate Avenue, Hershey, PA 17033, USA
Tel: 717-533-8845 x100 • Fax: 717-533-8661
E-Mail: cust@igi-global.com • www.igi-global.com

Table of Contents

Detailed Table of Contents

The development and widespread use of cloud technologies promises a number of economic prospects at the local, regional, and global levels. Cloud technologies have numerous advantages in terms of management, storage, and exchange of information resources in the virtual space for the public and private sector, citizens, and general users. The chapter provides an overview of the research conducted in the field of cloud technologies in Azerbaijan. The results show that the studies are mostly devoted to the infrastructure issues of cloud technologies, to the research of the conceptual bases of the creation of the government cloud, to the issues of security and personal data protection, as well as to the implementation of measures for the development of the economy of cloud technologies. The international experience in the field of government cloud was studied, and the realization of this concept and the results to be achieved were analysed. Necessary measures for the formation and development of the cloud technologies economy at the government level have been indicated.

The amount of food produced in fields is directly proportional to the Earth's growing population. Taking into account the deteriorating climatic conditions and ineffective agricultural productivity, we must modernize our agricultural methods immediately. India is second in terms of agricultural performance. The main concerns that farmers have are crop rotation, fertilizer use, and irrigation. Due to technological advancements, farmers nowadays are unable to produce their crops to their maximum potential. The main focus is on utilizing contemporary technology in agriculture and using it to boost yield. We can now better estimate crop output thanks to the deployment of IoT in agriculture and cloud computing. Implementing IoT and cloud computing strategies is a way to overcome challenges in agriculture. This chapter gives a brief summary of the integration of cloud and IoT (internet of things) technologies into agricultural practices, which will make agricultural practices smarter.

Agriculture has been the backbone of human civilization. For ages, humans have been using traditional farming techniques to produce high-quality crops, but with the era of digitalization and the challenges of these methods, there has been a shift from traditional farming techniques to smarter farming techniques, which utilise the latest technologies to reduce labour and maximise the yield in a given scenario. Smart agriculture produces a large amount of data from sensors and IoT devices; this is where cloud computing comes into play. Cloud computing-based technology provides storage and analysis facilities to deal with huge amounts of data and produces real-time insights that can help the farmer make better decisions. Cloud computing also helps in reducing the cost and resources incurred in smart agriculture techniques, thus

In this research, resource allocation in machine learning is used to analyze how cloud computing is being applied in smart agriculture. This chapter goes over the advantages of cloud computing for farming and how machine learning can enhance resource allocation for higher agricultural yields and less negative environmental impact. The chapter also examines implementation difficulties for cloud-based agricultural solutions and speculates on potential fixes. Insights for researchers and practitioners in the area are provided by the research, which demonstrates the potential for merging cloud computing and machine learning in smart agriculture to increase productivity and sustainability. The research also assessed the efficacy of the ML-based techniques using a variety of performance indicators, including reaction time and throughput. The management of cloud workloads has shown considerable promise when using machine learning-based methods. This chapter offers a thorough overview of current developments in ML-based cloud workload management and identifies areas for further research.

In order to optimize resource use, having information that is both up to date and accurate is vital. Agriculture utilizes sensors to determine the nutrients, moisture, organic matter, and clay present in the soil. To connect sensors that are positioned in a variety of locations, several technologies are used. Even in the absence of an internet connection, its data will automatically be uploaded to the cloud. Sensors transmit data to adjacent local base stations at varied distances using WiFi, LPWAN, LoRa, and Bluetooth, among other technologies, before transferring it to a distant central base station (CBS). Maintaining the integrity of the signal requires line of sight, coding, and modulation. The data collected by CBS is uploaded to the cloud so that it may be analyzed for trends and visualized. Farmers will now be able to get frequent and up-to-date information without needing to physically be present at the event.

Agriculture plays a crucial role in addressing global challenges such as food security, resource scarcity, and climate change. As the world population continues to grow, the need for sustainable and efficient farming practices becomes paramount. This chapter aims to explore the applications of artificial intelligence (AI) in agricultural sustainability, focusing on how AI technologies can enhance efficiency and resilience in agricultural practices. The chapter provides an overview of the potential benefits and challenges of integrating AI into agriculture, highlighting specific use cases, and discussing the implications for sustainable farming systems. It also examines the role of AI in addressing key sustainability issues including resource management, yield optimization, pest control, and adaptation to environmental conditions. By examining current trends and advancements in AI for agriculture, this chapter contributes to a comprehensive understanding of how AI can revolutionize farming practices and promote sustainable agricultural systems.

This chapter explores the mechanized functioning of tomato and lettuce greenhouses through the integration of internet of things (IoT) and machine learning techniques. It highlights the transformative impact of these technologies on the agricultural industry, enabling enhanced productivity, optimized resource management, and improved crop yield. By leveraging IoT, greenhouses become interconnected systems that allow real-time monitoring and control of environmental parameters. Machine learning (ML) algorithms examine data collected from sensors and historical records to predict crop growth patterns, disease outbreaks, and pest infestations. This enables proactive decision-making and efficient resource allocation. The adoption of IoT in greenhouse farming not only automates operations but also promotes sustainable practices, reducing labor costs and optimizing resource utilization. Overall, this chapter showcases the potential of IoT and ML technologies in revolutionizing greenhouse farming for increased efficiency, sustainability, and food security of tomato and lettuce crops.

Chapter 14
Machine Learning Time Series Models for Tea Pest Looper Infestation in Assam, India 280
Dwijendranath Dwivedi, Krakow University of Economics, Poland
Pradish N. Kapur, PES University, India
Nipun N. Kapur, Indian Institute of Technology, Roorkee, India

In the agriculture industry, pest infestation is a significant challenge that is complicated by the nonlinear relationship with environmental factors. Given the effectiveness of machine learning models in simulating such complex nonlinear phenomena, the authors opted to employ them in the modelling of the life cycle of tea pests, which impact several other crops as well. Accordingly, multiple machine learning models were developed to forecast the occurrence of tea pest looper infestations. They utilized data for just two readily available parameters—temperature and rainfall—to investigate whether predictive models of good quality can be created even with limited data, particularly for small tea growers. After analyzing the various models generated, they discovered that neural network models can produce accurate predictions even with a restricted data set. Therefore, they are optimistic that new age technologies such as machine learning can benefit many small farmers in India who lack access to various technologies and, as a result, have limited data.

Chapter 15
Crop Disease Prediction Using Deep Learning Algorithms 290
Pancham Singh, Ajay Kumar Garg Engineering College, Ghaziabad, India
Mrignainy Kansal, Ajay Kumar Garg Engineering College, Ghaziabad, India
Manoj Kumar Singh, IMS Engineering College, Ghaziabad, India
Sachin Kumar, Galgotias College of Engineering and Technology, Greater Noida, India
Anupam Dwivedi, Ajay Kumar Garg Engineering College, Ghaziabad, India

Good crops yield good food which in turn nourishes the human body and mind. But these crops across the globe face the threat of various diseases that remain unidentified, leading to poorer quality and quantity of crops. But, in recent times, the increasing adoption of smartphones worldwide and current developments in image processing of computers enabled by deep models of learning have made smartphone-based disease detection possible. In this chapter, the authors train a deep convolutional neural network (CNN) model to recognize 18 crop species and 28 diseases by feeding it a pre-available dataset of 70,296 photographs of unhealthy and healthy crop leaves taken under control. On a sustained test set, the trained model shows up to 99% accuracy, proving the feasibility of the method.

Preface

Dear readers,

In the fast-evolving landscape of technology and agriculture, the convergence of Cloud Computing, Artificial Intelligence (AI), and Agricultural Science has opened up remarkable possibilities for the transformation of farming practices. The integration of these cutting-edge technologies has paved the way for smarter, more efficient, and sustainable agricultural systems.

As editors of this reference book, titled *Convergence of Cloud Computing, AI, and Agricultural Science*, we are delighted to present a collection of insightful chapters authored by experts and researchers in the field. This compilation aims to shed light on the foundation and applications of cloud computing and AI in the context of Smart Agriculture, showcasing their potential to revolutionize the agricultural sector and improve development and services.

The first part of this book introduces a comprehensive cloud-based system designed to remotely monitor agriculture farms using the Internet of Things (IoTs). This smart system enables real-time data analysis by specialists and farmers, utilizing diverse data types such as audio, video, images, text, and digital maps. Leveraging AI-based machine learning models, such as Support Vector Machine (SVM), allows for accurate data classification. The processed information becomes readily available to end-users through virtual machines and datacenters, empowering farmers to enhance their skills and prepare for smart agri-food disaster recovery. Additionally, the book explores how these advancements can provide valuable information on international markets relevant to various crops.

The subsequent chapters delve into specific applications and implementations of Cloud Computing and AI in the field of Smart Agriculture. From implementing conversational AI for farmer advisory and communication to predicting the impact of global warming on agriculture using intelligent fuzzy logic controllers, the contributions in this book cover a wide array of topics. These include AI-aided phenomics in high-throughput stress phenotyping of plants, disease detection through hybrid deep learning architecture, greenhouse monitoring systems, rainfall clustering techniques, cloud-based surveillance solutions, IoT integration for smart agriculture, app-based information systems for rural farmers, and much more.

Our intended audience for this book comprises research scholars, graduate and undergraduate students of Computer Science and Agriculture, professors, and professionals working in this dynamic domain. We believe that the diverse content within these pages will not only inspire further research and innovation but also serve as a valuable resource for individuals seeking to explore the possibilities and challenges presented by the convergence of Cloud Computing, AI, and Agricultural Science.

We extend our heartfelt gratitude to all the authors who have contributed their expertise and knowledge to make this book possible. We hope that this compilation will contribute significantly to the advancement of Smart Agriculture and foster a deeper understanding of the transformative potential of Cloud Computing and AI in this crucial sector.

CHAPTER OVERVIEW

Chapter 1: A Systematic Study of Services and Security Model in Cloud Computing – A Brief Overview

In this chapter, the authors, Mrignainy Kansal, Pancham Singh, Manoj Singh, and Sanskar Varshney, present a comprehensive overview of cloud computing as a technical model with on-demand network access, offering a wide range of services and resources, including memory storage, applications, networking, databases, and servers. The authors emphasize the critical requirement for internet connectivity as a condition for accessing these resources. The primary objective of their research is to delve into the security and services aspects of cloud computing environments.

The chapter delves into the intricate relationship between service delivery and security, exploring the challenges and potential solutions to ensure that cloud services are both trustworthy and efficient while maintaining a high level of safety. With the rapid advancement of technology and the increased interconnectedness of systems, the authors address the growing concerns regarding potential threats and vulnerabilities in cloud-based environments.

Furthermore, the chapter provides insights into the various types of services available in the cloud and how each service option offers unique benefits such as efficiency, accessibility, cost savings, flexibility, opportunities for innovation, and more. By presenting a comprehensive overview of cloud computing, the authors aim to equip readers with a deeper understanding of the services and security considerations crucial for harnessing the full potential of cloud-based solutions.

Chapter 2: Toxicological Parameters and Risk Assessment of Some Novel Insecticides – An In-Depth Analysis

Authored by Akanksha Chand, this chapter delves into the critical subject of pesticides, their usage in agriculture, horticulture, and public health for crop protection and pest control. While pesticides play a crucial role in enhancing agricultural productivity, improper or excessive use can lead to significant risks for human health and the environment.

The chapter focuses on understanding pesticide toxicity, which is often measured by lethal dose (LD) or lethal concentration (LC) values. These values provide estimates of the amount of a pesticide that can cause death in a specified population. The analysis encompasses a review of LD and LC values, shedding light on various factors influencing pesticide toxicity, including the mode of action and formulation.

Furthermore, the chapter presents toxicity ranges for selected pesticides commonly employed in agriculture and public health practices. By exploring the health effects associated with acute and chronic exposure to these chemicals, readers gain valuable insights into the potential risks posed by pesticide use.

In conclusion, the chapter emphasizes the importance of regulatory frameworks governing pesticide use and advocates for sustainable and integrated pest management practices. By addressing these crucial aspects, the author highlights the significance of informed decision-making to strike a balance between the benefits of pesticides in safeguarding crops and the imperative of safeguarding human health and the environment. This in-depth analysis serves as a valuable resource for researchers, professionals, and policymakers involved in the domain of agricultural and public health practices.

Chapter 3: Cloud in Digital Government – Problems and Perspectives in the Case of Azerbaijan

Authored by Farhad Yusifov, this chapter explores the vast potential of cloud technologies and their economic implications at local, regional, and global levels. The widespread adoption of cloud technologies offers numerous advantages, particularly in terms of managing, storing, and exchanging information resources in virtual space. These benefits extend to the public and private sectors, citizens, and general users alike.

The chapter provides a comprehensive overview of the research conducted in the specific context of Azerbaijan, focusing on cloud technologies. The results of the research highlight several key areas of focus, including infrastructure issues related to cloud technologies, conceptual bases for creating a government cloud, security and personal data protection concerns, and measures to foster the economic development of cloud technologies.

In addition to examining the domestic landscape, the chapter delves into international experiences with Government cloud implementations, studying the concepts behind these initiatives and the outcomes they seek to achieve. By drawing on global best practices, the chapter aims to enrich the understanding of cloud technologies' potential in the context of digital government.

Furthermore, the chapter outlines essential measures required for the formation and advancement of the cloud technologies economy at the government level. These recommendations serve as valuable guidelines for policymakers and stakeholders seeking to leverage cloud technologies to enhance government services and promote socio-economic development.

In conclusion, this chapter contributes to the growing body of knowledge in the domain of cloud technologies and digital government. The insights provided by Farhad Yusifov shed light on the challenges and possibilities inherent in adopting cloud technologies within the Azerbaijani context and offer valuable perspectives for optimizing the potential benefits of cloud-based solutions in the government sector.

Chapter 4: A Comprehensive Study on Smart Farming for Transforming Agriculture Through Cloud and IoT – Exploring the Smart Farming Landscape With Cloud and IoT

Authored by Ajay Poonia, Lakshmi D., Tanmay Garg, and Vishnuvarthanan G. from VIT Bhopal University, this chapter addresses the pressing need to modernize agricultural methods in response to the increasing global population and the challenges posed by deteriorating climatic conditions and ineffective agricultural productivity.

India, being the second-largest agricultural performer, faces critical concerns related to crop rotation, fertilizer utilization, and irrigation practices. To unlock the full potential of agricultural productivity, contemporary technological solutions are indispensable. Farmers can significantly enhance crop yields by adopting modern technologies in their farming practices.

The chapter focuses on the integration of Internet of Things (IoT) and Cloud Computing technologies into agricultural practices, leading to the emergence of smart farming. By leveraging IoT devices and sensors in the fields, farmers can monitor and optimize various factors such as soil moisture, temperature, and crop health in real-time. This enables them to make data-driven decisions, leading to improved crop output and resource management.

Cloud Computing plays a vital role in smart farming by serving as the centralized platform for collecting, storing, and processing the vast amounts of data generated by IoT devices. The use of Cloud-based solutions ensures scalability, accessibility, and cost-effectiveness, making it an ideal choice for implementing IoT strategies in agriculture.

The chapter presents a comprehensive study on the transformational potential of smart farming, emphasizing how the integration of Cloud and IoT technologies can revolutionize agricultural practices. By embracing smart farming practices, farmers can overcome existing challenges, optimize resource utilization, and increase agricultural productivity sustainably.

In conclusion, the chapter sheds light on the synergy between Cloud and IoT technologies and their role in making agricultural practices smarter and more efficient. The insights provided by the authors serve as valuable knowledge for researchers, practitioners, and policymakers in the agriculture sector, offering a pathway towards a more sustainable and technologically advanced agricultural landscape.

Chapter 5: Analyzing the Applications of Cloud Computing in Smart Agriculture

Authored by Ishan Jain and Brijendra Singh from Vellore Institute of Technology, India, this chapter delves into the transformative impact of cloud computing in the domain of smart agriculture. Agriculture has been the foundation of human civilization, relying on traditional farming techniques for centuries. However, in the digital age, traditional methods face challenges that necessitate a shift towards smarter farming techniques, integrating the latest technologies to optimize labor and maximize crop yield.

Smart agriculture generates a substantial volume of data from various sources, including sensors and Internet of Things (IoT) devices. Herein lies the significance of cloud computing. Cloud-based technology offers storage and analytical capabilities to efficiently manage and process vast amounts of agricultural data, providing real-time insights that enable farmers to make informed decisions.

By leveraging cloud computing, farmers can access valuable information and recommendations based on data analysis, thereby enhancing their decision-making process. This technology also contributes to reducing costs and resource consumption in smart agriculture, ultimately making the model more efficient and beneficial for farmers.

The chapter systematically analyzes diverse applications where cloud computing can be integrated into smart agriculture. It explores how innovative cloud technologies can improve various aspects of farming, such as crop management, resource allocation, and monitoring systems. The authors also emphasize the need to consider potential drawbacks and challenges when adopting the cloud model in agriculture, including data security and privacy concerns.

In conclusion, this chapter presents a comprehensive examination of cloud computing's applications and implications in the context of smart agriculture. The insights offered by Ishan Jain and Brijendra Singh contribute to the growing body of knowledge in agricultural technology and provide a valuable resource for researchers, practitioners, and policymakers seeking to harness the full potential of cloud computing to drive sustainable and efficient farming practices.

Chapter 6: Cloud Computing-Based Smart Agriculture

Authored by Kaushal Kishor and Raj Verma from ABES Institute of Technology, India, this chapter explores the profound impact of cloud computing and the Internet of Things (IoT) on agriculture. With the advent of digital technologies, agriculture has undergone significant transformation, particularly in countries like India, China, and others, where it serves as a vital source of income and livelihood for a large population.

Cloud computing and IoT solutions offer unprecedented benefits to the agricultural sector, enhancing production cost management, performance tracking, and maintenance practices. By leveraging smart drones and instant data collection, farmers can improve crop management and optimize agricultural processes. The integration of computing technologies facilitates the collection, analysis, and utilization of agricultural data, benefiting not only individual farmers but also the nation's agricultural sector as a whole.

The chapter emphasizes the importance of modernizing Indian agriculture through technological advancements to enhance productivity, wealth distribution, and cost management. Precision gardening practices have been successful in other countries, and the adoption of technology-driven solutions can further boost productivity in Indian agriculture.

In this article, the authors propose a comprehensive smart agriculture paradigm that incorporates several technological advances, including sensors, IoT, cloud computing, mobile computing, and big data analytics. By integrating these cutting-edge technologies, farmers can make informed decisions, optimize resource utilization, and enhance overall agricultural productivity.

In conclusion, this chapter advocates for the adoption of cloud computing and IoT-based solutions in agriculture, presenting a vision of smart agriculture that benefits farmers, the economy, and society at large. The insights provided by Kaushal Kishor and Raj Verma serve as a valuable guide for embracing technological advancements to ensure sustainable and efficient agricultural practices in India and beyond.

Chapter 7: IoT-Based Smart Agriculture

Authored by Sheelesh Kumar Sharma, Avinash Sharma, Sushant Sharma, Kabir Shukla, and Ishaan Ishaan from ABESIT Engineering College and ABES Institute of Technology in India, this chapter delves into the transformative potential of IoT technology in the agricultural industry. With the rapid advancement of IoT devices, sensors, and data analytics, the agricultural sector is experiencing a disruptive shift, enabling the monitoring, automation, and optimization of various elements of agricultural operations.

The chapter sheds light on the challenges faced by conventional agricultural practices, including issues like water scarcity, soil deterioration, and insect control. In response to these challenges, the authors stress the need for creative solutions to revolutionize traditional farming methods.

The primary focus of the chapter lies in exploring the transformative impact of IoT technology on agriculture. It delves into the diverse applications of IoT in smart agriculture, encompassing precision farming, livestock monitoring, smart irrigation systems, and crop health monitoring. By leveraging IoT devices such as sensors, actuators, and drones, farmers gain access to real-time data collection and analysis, empowering them to make well-informed decisions.

IoT-based smart agriculture has emerged as a powerful approach with numerous benefits, including higher yields, optimal resource utilization, reduced environmental impact, and increased profitability for farmers. By adopting IoT solutions, agriculture can become more sustainable, efficient, and profitable, overcoming traditional limitations and embracing data-driven approaches to enhance productivity.

In conclusion, the chapter presents a comprehensive exploration of IoT-based smart agriculture, emphasizing the transformative capabilities of IoT technology in addressing agricultural challenges and unlocking new opportunities for the industry. The insights shared by the authors serve as a valuable resource for researchers, practitioners, and policymakers seeking to leverage IoT technology to drive sustainable and innovative practices in agriculture.

Chapter 8: Smart Agriculture Resource Allocation and Cost Optimization Using ML in Cloud Computing Environments

Authored by Pancham Singh, Mrignainy Kansal, Mili Srivastava, and Muskan Gupta from Ajay Kumar Garg Engineering College, India, this chapter delves into the integration of cloud computing and machine learning in the context of smart agriculture. The research explores resource allocation through machine learning techniques and analyzes how cloud computing is being harnessed to enhance agricultural practices.

The chapter highlights the numerous advantages of cloud computing in farming, emphasizing how machine learning can further optimize resource allocation to achieve higher agricultural yields while minimizing negative environmental impacts. By combining cloud computing and machine learning, smart agriculture can benefit from data-driven decision-making and efficient resource utilization.

Moreover, the chapter addresses the challenges and implementation difficulties encountered in deploying cloud-based agricultural solutions and speculates on potential solutions. The insights provided by the research offer valuable guidance to researchers and practitioners in the field, showcasing the potential of merging cloud computing and machine learning to elevate agricultural productivity and sustainability.

The research conducted also evaluates the efficacy of machine learning-based techniques, utilizing various performance indicators, including reaction time and throughput. The management of cloud workloads has shown promising results with the application of machine learning methods, indicating the potential for further advancements in the field.

In conclusion, this chapter presents a comprehensive overview of current developments in machine learning-based cloud workload management in smart agriculture. The authors' work serves as a valuable resource for researchers, practitioners, and policymakers interested in leveraging machine learning and cloud computing to optimize resource allocation and enhance agricultural practices, leading to a more sustainable and productive agricultural landscape.

Chapter 9: Intelligent Farming Utilizing a Soil Tracking Device

Authored by Sabyasachi Pramanik from Haldia Institute of Technology, India, this chapter delves into the significance of up-to-date and accurate information in optimizing resource utilization in agriculture. The chapter focuses on the integration of soil tracking devices equipped with sensors to monitor essential soil parameters such as nutrients, moisture, organic matter, and clay content.

To ensure efficient data collection from various locations, the chapter explores the use of different technologies to connect the sensors. These technologies include WiFi, LPWAN, LoRa, and Bluetooth, allowing seamless transmission of data even in the absence of an internet connection. The collected data is automatically uploaded to the cloud, ensuring real-time availability and analysis.

The data transmission process involves sensors sending information to nearby local base stations, which then relay the data to a distant central base station (CBS). Maintaining signal integrity requires considerations such as line of sight, coding, and modulation techniques. Once the data reaches the CBS, it is uploaded to the cloud, where it can be analyzed for trends and visualized.

By implementing soil tracking devices and cloud-based data analysis, farmers gain access to frequent and up-to-date information without the need for physical presence at the farming site. This empowers farmers to make informed decisions about resource management and agricultural practices, optimizing yields and resource utilization.

In conclusion, this chapter highlights the potential of intelligent farming utilizing soil tracking devices and cloud computing. The integration of sensors, data transmission technologies, and cloud-based analysis offers a powerful solution for enhancing agricultural practices and improving overall farm productivity. The insights shared by Sabyasachi Pramanik serve as a valuable resource for researchers, practitioners, and farmers seeking to harness the benefits of modern technologies in agriculture.

Chapter 10: Artificial Intelligence Applications in Agricultural Sustainability – Enhancing Efficiency and Resilience

Authored by Imed Othmeni, Rim Ben Abdallah, Hiba Ben Aribi, Eya Ben Amara, and Sami Fattouch from the National Institute of Applied Sciences and Technology, Tunisia, this chapter explores the pivotal role of Artificial Intelligence (AI) in addressing global challenges faced by agriculture, such as food security, resource scarcity, and climate change. As the world's population continues to grow, the need for sustainable and efficient farming practices becomes increasingly critical.

The chapter aims to shed light on the applications of AI in promoting agricultural sustainability, with a particular focus on how AI technologies can enhance efficiency and resilience in farming practices. It provides an insightful overview of the potential benefits and challenges associated with integrating AI into agriculture. Specific use cases are examined, showcasing how AI can be applied to revolutionize farming systems and contribute to sustainable practices.

One of the key areas explored is how AI can play a significant role in resource management, yield optimization, pest control, and adapting agricultural practices to varying environmental conditions. By harnessing AI technologies, farmers can make informed decisions, optimize resource utilization, and address sustainability challenges effectively.

Moreover, the chapter delves into current trends and advancements in AI for agriculture, offering a comprehensive understanding of how AI can transform farming practices and promote sustainable agricultural systems.

In conclusion, this chapter makes a valuable contribution to the literature on agricultural sustainability by exploring the potential of AI technologies. By examining the applications, benefits, and challenges, the authors provide researchers, practitioners, and policymakers with valuable insights to leverage AI's potential in promoting sustainable and efficient farming practices, thereby addressing the global challenges faced by agriculture today.

Chapter 11: Empowering Agriculture With Conversational AI – An Application for Farmer Advisory and Communication

Authored by Mrignainy Kansal, Pancham Singh, Mili Srivastava, and Prateek Chaurasia from Ajay Kumar Garg Engineering College, India, this chapter explores the transformative potential of Conversational AI in the agricultural sector. Conversational AI, which includes chatbots, virtual assistants, and voice-based interfaces, has emerged as a trending technology with significant benefits for farmers and agricultural workers.

The chapter highlights how Conversational AI can address challenges in accessing timely and accurate information, providing valuable insights to farmers. It enables early detection of crop diseases, offers real-time market prices, weather forecasts, and aids in soil health monitoring. By streamlining access to government schemes and fostering knowledge sharing among farmers, Conversational AI plays a crucial role in supporting agricultural activities.

Moreover, the integration of Machine Learning (ML), data analytics, and Internet of Things (IoT) devices enhances the accuracy and insights derived from the collected data. Conversational AI's transformative potential contributes to the sustainable growth of agriculture, ensuring food security, and empowering farmers for a prosperous future.

By leveraging this technology, farmers can make informed decisions, optimize resource utilization, and enhance agricultural productivity. Additionally, Conversational AI can support precision agriculture practices through data collection and analysis, further enhancing efficiency and sustainability in farming.

In conclusion, this chapter sheds light on the significant impact of Conversational AI in agriculture, emphasizing its role in empowering farmers and promoting sustainable growth. The insights provided by the authors serve as a valuable resource for researchers, practitioners, and policymakers seeking to harness the benefits of Conversational AI in advancing agricultural practices, ensuring food security, and supporting the livelihoods of farmers.

Chapter 12: Impact of Mandatory Food and Agriculture Laws on Food Manufacturing Industry – Review of 75 Years of Challenges and Changes

Authored by Shipra Saraswat, Rijwan Khan, Ashok Kumar, and Akshay Singh from ABES Institute of Technology, India, this chapter explores the significant impact of mandatory food and agriculture laws on the food manufacturing industry. Food safety and quality are paramount in ensuring good food governance and the public's access to nutritious, uncontaminated food.

The chapter delves into the diverse food safety regulations, legislation, and standards that vary from one nation to another, highlighting the importance of ensuring food purity and safety. Additionally, the authors discuss the comprehensive scope of agricultural law, encompassing all aspects of the agricultural industry. Supporting and accommodating farmers is a crucial aspect of agricultural law, with a focus on mitigating losses sustained by farmers and addressing critical farming issues, such as land ownership, environmental and social equity, and climate change.

An objective and credible analysis of the economic impact of required food and agricultural legislation on the businesses that provide them is essential. Throughout the chapter, the authors provide a detailed discussion of the challenges and changes faced by the food manufacturing industry over the past 75 years due to mandatory food and agricultural laws.

By examining the historical trajectory and experiences of the food manufacturing industry, the chapter offers valuable insights into the evolution of food governance and its influence on the business landscape. This retrospective analysis is vital for understanding the implications of food and agricultural laws on the industry's development, sustainability, and compliance.

In conclusion, this chapter serves as a comprehensive review of the complex relationship between mandatory food and agriculture laws and the food manufacturing industry. The insights provided by Shipra Saraswat, Rijwan Khan, Ashok Kumar, and Akshay Singh contribute to the understanding of food governance, agricultural regulations, and their profound implications on the food manufacturing sector. This research is of significant importance for policymakers, researchers, and industry stakeholders seeking to foster a safe, sustainable, and resilient food supply chain.

Chapter 13: Mechanized Functioning of Tomato and Lettuce Greenhouses Using IoT and Machine Learning Techniques

Authored by Shipra Saraswat, Rijwan Khan, Ayush Gupta, and Abhay Mehta from ABES Institute of Technology, India, this chapter delves into the transformative impact of integrating Internet of Things (IoT) and machine learning techniques in the mechanized functioning of tomato and lettuce greenhouses. By leveraging these advanced technologies, greenhouse farming experiences significant improvements in productivity, resource management, and crop yield.

The chapter highlights the interconnected nature of IoT-enabled greenhouses, allowing real-time monitoring and control of crucial environmental parameters. By analyzing data collected from sensors and historical records through machine learning algorithms, greenhouse operators can predict crop growth patterns, disease outbreaks, and pest infestations. This proactive approach to decision-making enables efficient resource allocation and precise agricultural practices.

The adoption of IoT in greenhouse farming not only automates operations but also promotes sustainable practices by reducing labor costs and optimizing resource utilization. This results in increased efficiency and improved crop yields for tomato and lettuce crops, ensuring food security and contributing to a sustainable agricultural system.

Overall, this chapter showcases the potential of IoT and machine learning technologies in revolutionizing greenhouse farming practices. The insights provided by the authors serve as a valuable resource for researchers, practitioners, and policymakers seeking to embrace innovative technologies to enhance greenhouse farming's efficiency, sustainability, and contribution to food security.

Chapter 14: Machine Learning Time Series Models for Tea Pest Looper Infestation in Assam, India

Authored by Dwijendranath Dwivedi from Krakow University of Economics, Poland, Pradish Kapur from PES University, India, and Nipun Kapur from IIT Roorkee, India, this chapter addresses the significant challenge of pest infestation in the agriculture industry, with a focus on the tea pest Looper in Assam, India. Pest infestation is a complex phenomenon influenced by nonlinear relationships with environmental factors.

The authors employ machine learning models to model the life cycle of tea pests, considering its impact on various other crops as well. The study aims to forecast the occurrence of tea pest Looper infestations using machine learning techniques. Despite limited data availability, the researchers explore the effectiveness of predictive models in simulating complex nonlinear phenomena in pest infestations.

Using only two readily available parameters, namely temperature and rainfall data, the study investigates the creation of predictive models with good quality, especially for small tea growers who may have limited data resources. Various machine learning models are analyzed to identify the most accurate forecasting approach.

The results reveal that Neural Network models exhibit accuracy even with restricted data sets, indicating the potential of machine learning technologies to benefit small farmers in India. As technology accessibility can be a challenge for many farmers, particularly those with limited data, the application of machine learning provides a promising solution.

In conclusion, this chapter showcases the utility of machine learning time series models in forecasting tea pest Looper infestations in Assam, India. The insights presented by the authors contribute to the understanding of how new age technologies, such as machine learning, can empower small farmers by enabling accurate predictions and supporting pest management practices. This research is particularly relevant for the agricultural community in India, providing valuable knowledge for sustainable crop protection and enhancing agricultural productivity.

Chapter 15: Crop Disease Prediction by Using Deep Learning Algorithms

Authored by Pancham Singh, Mrignainy Kansal, Manoj Singh, Sachin Kumar, and Anupam Dwivedi from Ajay Kumar Garg Engineering College, India, this chapter focuses on the critical issue of crop diseases and their impact on global food production. Crop diseases pose a threat to crop quality and quantity, leading to potential food scarcity and agricultural challenges worldwide.

The chapter explores the potential of deep learning algorithms, specifically deep convolutional neural network (CNN) models, in predicting and recognizing crop diseases. By utilizing a pre-available dataset of 70,296 photographs of both unhealthy and healthy crop leaves taken under controlled conditions, the authors train the CNN model to recognize 18 crop species and 28 diseases.

The results show promising accuracy, with the trained model achieving up to 99% accuracy on a sustained test set. This demonstrates the feasibility and effectiveness of using deep learning algorithms for crop disease prediction and detection.

The chapter emphasizes the significance of current developments in image processing and the increasing adoption of smartphones worldwide. These advancements have enabled smartphone-based disease detection, providing an accessible and efficient means of monitoring crop health and mitigating disease-related risks.

Overall, the study showcases the potential of deep learning algorithms in addressing crop disease challenges and improving global food security. By utilizing ever-larger and publicly available image datasets for training deep learning models, researchers and practitioners can further enhance the accuracy and effectiveness of crop disease prediction and management strategies.

In conclusion, this chapter highlights the transformative impact of deep learning algorithms in crop disease prediction and detection. The insights provided by the authors contribute to the understanding of how technological advancements in image processing and deep learning can play a crucial role in safeguarding crop health, optimizing agricultural practices, and ensuring a sustainable and abundant food supply for the global population.

Chapter 16: Identification of Tomato Plant Disease Using Faster R-CNN and RetinaNet

Authored by Indrajeet Kumar from Graphic Era Hill University, India, Aman Aman Bisht from DIT University, India, and Jyoti Rawat from VertexPlus Technologies Limited, India, this chapter addresses the significant issue of plant diseases affecting tomato crops and their impact on productivity and agricultural economy.

The chapter focuses on the identification of tomato plant diseases at an early stage to mitigate losses and enhance crop yield and quality. Identifying plant diseases accurately can greatly contribute to producing a higher quantity of good-quality tomatoes. However, disease identification can be challenging due to factors such as lack of knowledge, multiple diseases affecting the plants, or incorrect disease predictions.

The chapter presents implementations of two classification algorithms, Faster R-CNN and RetinaNet, for the identification of tomato plant diseases. Both algorithms are initiated from transfer learning, allowing for efficient training and testing with different hyperparameters to achieve improved results.

The authors report impressive results with average precision (AP)-50 for Faster R-CNN at 93.11% and RetinaNet at 95.54%. These results demonstrate the effectiveness of the proposed approach in accurately identifying tomato plant diseases.

The preliminary outcomes of the study are promising, suggesting that the implementation of Faster R-CNN and RetinaNet can be valuable for enhancing harvest quality and precision agriculture. By leveraging these advanced algorithms, farmers can detect and manage plant diseases promptly, leading to better crop yields and reduced losses.

In conclusion, this chapter highlights the significance of employing advanced classification algorithms for the early identification of tomato plant diseases. The insights provided by the authors contribute to the advancement of agricultural practices, offering valuable tools to improve crop health and productivity. The research holds potential benefits for farmers, agricultural practitioners, and policymakers, contributing to sustainable and efficient tomato cultivation and bolstering the agricultural economy.

IN SUMMARY

In this edited reference book, *Convergence of Cloud Computing, AI, and Agricultural Science*, we have explored the transformative potential of cutting-edge technologies in the agricultural sector. The convergence of cloud computing, artificial intelligence (AI), and agricultural science has paved the way for novel solutions, revolutionizing farming practices and addressing key challenges in the industry.

Throughout the book, esteemed researchers and experts have contributed valuable chapters, each focusing on specific applications and advancements in this dynamic field. From the development of smart cloud-based systems for remote monitoring of agriculture farms using IoTs to the application of conversational AI for farmer advisory and communication, the range of topics covered exemplifies the wide-ranging impact of technology in agriculture.

We have witnessed how cloud computing and IoT have empowered precision agriculture, providing real-time data and insights to farmers for better decision-making. The integration of AI in agricultural sustainability has enhanced efficiency, resilience, and resource management, fostering the sustainable growth of the sector.

Furthermore, deep learning models and machine learning algorithms have been employed to predict crop diseases, enabling early detection and effective pest management strategies. The use of AI-based phenomics has revolutionized high-throughput stress phenotyping of plants, supporting the development of more resilient crop varieties.

The chapters on smart farming and greenhouse monitoring demonstrate how IoT and cloud computing technologies are creating a paradigm shift in agricultural practices, leading to optimized resource allocation and higher yields. Moreover, cloud-based architectures have facilitated effective surveillance in agriculture, ensuring enhanced security and monitoring capabilities.

The exploration of smart agriculture applications in different regions, such as Azerbaijan and India, highlights the global relevance and potential of these technologies in addressing agricultural challenges worldwide.

The editors and authors of this book extend our gratitude to all the contributors for their invaluable research, insights, and expertise. We believe that this comprehensive collection of chapters will serve as a valuable resource for researchers, academics, practitioners, and policymakers interested in leveraging technology to advance the agricultural sector.

As editors, we are hopeful that this book will inspire further research, innovation, and collaboration in the convergence of cloud computing, AI, and agricultural science. The future holds tremendous possibilities for enhancing food security, sustainability, and efficiency in agriculture through the continuous integration of advanced technologies.

In conclusion, *Convergence of Cloud Computing, AI, and Agricultural Science* sheds light on the pivotal role of technology in shaping the future of agriculture. It is our sincere hope that this book will contribute to the betterment of the agricultural industry and ultimately lead to a more resilient, sustainable, and prosperous future for farmers and societies around the world.

Regards,

Avinash Kumar Sharma
ABES Institute of Technology, India

Nitin Chanderwal
University of Cincinnati, USA

Rijwan Khan
ABES Institute of Technology, India

Chapter 1
A Systematic Study of Services and Security Model in Cloud Computing:
A Brief Overview

Mrignainy Kansal
Ajay Kumar Garg Engineering College, Ghaziabad, India

Pancham Singh
Ajay Kumar Garg Engineering College, Ghaziabad, India

Manoj Kumar Singh
IMS Engineering College, Ghaziabad, India

Sanskar Varshney
Ajay Kumar Garg Engineering College, Ghaziabad, India

ABSTRACT

Cloud computing is a technical model with on-demand network access, delivering various types of services, providing resources such as memory storage, applications, networking, databases, servers, etc. All these resources always keep with us but with one condition: "the internet." The major purpose of this chapter is to know the security and services requirements and their solutions in a cloud computing environment. This rigorous research extensively explores the complex relationship between service delivery and security. It also looks at the issues and potential fixes for providing services that are trustworthy, efficient, and safe. Because of the accelerated pace of technological advancement, the networked nature of systems, and the rising reliance on internet services, the potential of threats and vulnerabilities has increased significantly. After that, the authors focus on what types of services are given in the cloud where each type of service provides different options like efficiency, accessibility, savings, flexibility, opportunity, innovation, and many more.

DOI: 10.4018/979-8-3693-0200-2.ch001

1. INTRODUCTION

The Term Cloud means a virtual memory space on the Internet. Cloud Computing is a technology model that uses services of servers for Storing, Managing and Accessing Data over the internet. Documents, E-mails, Chats, Videos, Blogs, Training Spreadsheets, Pictures, Address Books, Contact No, Presentations etc. All these are present in the cloud and we can access them through any device like mobiles, laptops, tablets, and desktops (Li et al., 2012).

Based on extensive research and corroborating sources, cloud computing is poised to become one of the leading technologies in the upcoming years. Prominent industry players like Sun Microsystems, Google, IBM, Amazon, and Microsoft have undertaken the establishment of new data centers in various global locations. This strategic expansion aims to ensure redundancy, fault tolerance, and consistent availability of cloud computing services, mitigating the risks of website outages or failures.

The greatest option nowadays for individuals searching for quick implementation methods is the cloud. Cloud computing involves the dissemination of applications, including both hardware and software components, across virtual data centres via the Internet. It represents a flexible, parallel, distributed, virtual, and configurable system that offers versatility and scalability. Customers can customise cloud computing services, and they pay for resources and services according to how much they use them (Dudiki et al., 2022).

Here, we explained the whole overviews of the paper that there is not only three main cloud services there are so many services that are available for cloud computing and also there features which would be helpful for user to understand it and also we collected all the information about the cloud security, their importance and problem arised due to cloud.

The structure of the paper is as follows: Section 2 provides an overview of the background of cloud computing. Section 3 contains a survey of year-wise data on the cloud The paper describes the primary cloud service models in section 4. A brief introduction to cloud security is covered in section 5. The importance of cloud security is discussed in Section 5.1. The issues with cloud security are covered in Section 6.

2. BACKGROUND

2.1 Characteristics of Cloud Computing

Here are alternative descriptions of the characteristics of Cloud Computing (see Figure 1):

1. **On-demand self-services:** It means we access any email, chats, and messages from anywhere but we have a valid user id and password and also the Internet (N et al., 2022).
2. **Broad network access:** Here we can access cloud servers from any device whether it is a mobile, tablet, or laptop.
3. **Resource pooling:** The numerous IT resources that are shared and sporadically utilised by the various applications include networks, servers, storage, apps, and services.
4. **Rapid elasticity:** IT resources had to be adaptable enough to scale quickly as needed to supply computing services. Every time the service provider needed something, they gave it to him.

5. **Measured services:** Cloud providers charge users based on their usage of computing resources, often on a pay-as-you-go basis (Harish Naik & Kannan, 2023).
6. **Security:** Cloud provides heavily in Security because to protect their user's data and ensure them to not leak sensitive data or information.

Figure 1. Characteristics of cloud computing

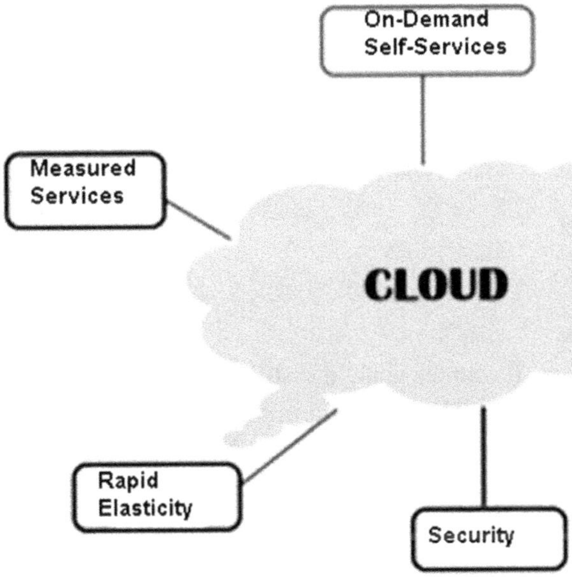

2.2 Types of Cloud

Public clouds, Private clouds, and Hybrids clouds are the three forms of Cloud Computing. All deployment models are based on locations, to find which order is suitable for our organisation or company (See in Figure 2). Each deployment model has their unique properties, benefits and uses (Butt et al., 2020).

2.2.1 Public Clouds

Third-party providers offer public cloud services to the general public through online platforms, featuring flexible pay-as-you-go pricing options. These public cloud services are accessible to a wide range of users. Users of public cloud services can access infrastructure that is owned and maintained by service providers as well as other parties without having to buy their own hardware, software, or auxiliary components. By hosting the cloud infrastructure, a service provider opens up public clouds to all users (Jouini & Rabai, 2017). They become a practical option for managing the infrastructure and provide solutions for lowering the cost of IT infrastructure. On the public cloud, multiple users, not just one, can access the services. It is location independent. The fundamental characteristics are Multitenancy.

Examples of public clouds are Google, Amazon, Microsoft, oracle cloud, AWS etc. and all of them are required the Internet.

Figure 2. Types of cloud

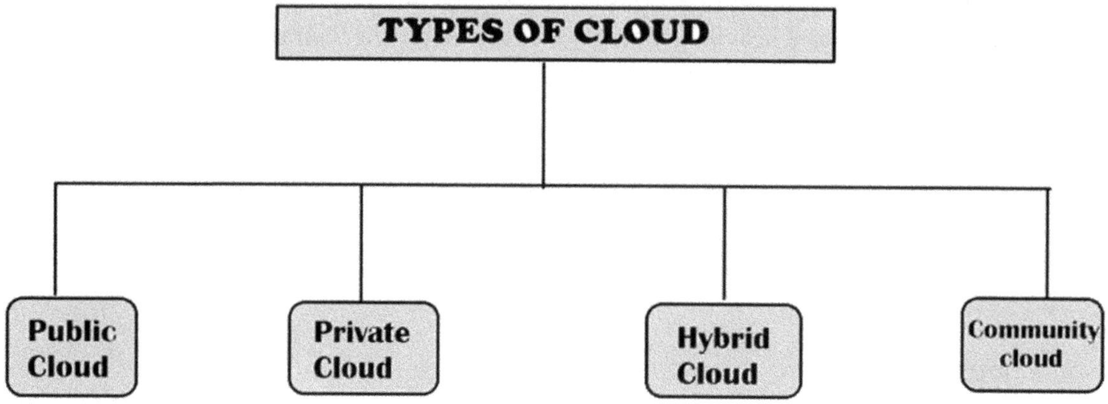

2.2.2 Private Cloud

Customers can dynamically deploy computer resources using a distributed system known as a private cloud that uses private infrastructure. The primary objective of a private cloud is to leverage cloud architecture while retaining ownership and control over the data center, rather than offering cloud services to external parties in an "as-a-service" model (Chavan et al., 2022). Although private clouds are more expensive than public clouds, they also provide more security.

2.2.3 Hybrid Clouds

Two or more private, shared, or public clouds can make up a hybrid cloud. While each cloud is distinct, they all share connections and gain from various deployment techniques. Hybrid cloud deployment allows for a combination of public and private clouds, providing varying degrees of control over private cloud service providers. This enables flexibility and utilization of both public and private cloud resources (Wei & Zhang, 2018).

2.2.4 Community Clouds

A distributed system known as a community cloud is created by combining the services of many clouds to cater to the specific needs of a certain industry, community, or business sector. However, assigning work to different organisations is difficult. The Community Cloud's infrastructure is shared by organisations with comparable objectives and duties. The cloud might be managed by your organisation or a different part (Yu, 2018).

3. SYSTEMATIC SURVEYS OF DIFFERENT CLOUD COMPUTING SECURITY

See Table 1.

Table 1. Systematic surveys of different cloud computing security

S.No	Focused Topic	Description
1.	Cloud Privacy Regulation (Ziani & Medouri, 2019)	This paper states about privacy issues and security issues, privacy regulation.
2.	Secure Cloud Architecture (Ghosh & Dolui, 2020)	This Paper tells the threats, vulnerabilities, architecture of cloud.
3.	Balancing In Virtual Machines for Cloud Computing (Shei et al., 2016)	The article suggests an approach for dynamically allocating resources by optimizing the load on virtual machines (VMs) and achieving load balancing.
4.	Security Implementation (N et al., 2022)	It Implements the security models to save new technologies from attackers
5.	Plagiarism Detection (Xia, 2020)	This Paper States that how to get plagiarism check or detection on cloud.
6.	Controlled Multimedia Cloud Architecture (Mishra et al., 2022)	Multimedia content management and Processing of cloud.
7.	A Study on cloud Computing Services (Mehra et al., 2018)	This paper proposes the services model of cloud.
8.	Data Integrity for shared data (Chen & Liu, 2022)	Paper Concludes User can easily modified data in the cloud either shared and stored.
9.	Privacy in Cloud ERP System (Arulkumaran et al., 2023)	This article explores the difficulties and resolutions associated with ensuring the security of cloud-based Enterprise Resource Planning (ERP) systems.
10.	Data Security and privacy information challenges in cloud computing (Xiao & Chen, 2015)	In this paper it gives details about data security and privacy preservation in cloud storage system.
11.	Grid and cloud Computing security (Xiong et al., 2022)	This paper tells, the major purpose of this article is to know the security requirements and their solution in grid and cloud computing environment.
12.	Cloud Computing and its role in the Information Technology (Yu, 2018)	This paper focused mainly on major cloud paradigm security threats.
13.	Research paper on cloud computing (Chavan et al., 2022)	The paper proposes about all the history, types and services, and uses of the cloud.
14.	Threats and Mitigation Strategy (Xiong et al., 2022)	Told About cloud model decryption, encryption, malicious behavior, and secured communication.
15.	Review and Classification of Cloud Computing (Schiering & Kretschmer, 2012)	According to this research, community cloud computing technologies provide excellent opportunities for higher education.
16.	Cloud computing security: A survey of service-based models (Trikha, 2021)	This study sought to summarise the state of the field's research and provide a taxonomy that links vulnerabilities to effective responses.

4. SERVICES OF CLOUD COMPUTING

Nowadays Cloud Computing is a most on-demand service, with high & new infrastructure, among all the technologies that come into the market, As Compare to other Traditional service providers it has many advantages such as availability is high, cloud automation, investment savings, scalability, fault tolerance capability, no redundancy, High performance and so on. That's why IT companies such as Google, Amazon, Microsoft, and Salesforce has been followed these services (Mahlaule et al., 2022).

There are a variety of applications and infrastructures like data processing, data storage, CRM and so on that are used by Cloud Computing. All those Services provided by cloud computing are used by governments, organisations, companies and third persons also. Cloud computing has a number of characteristics including virtualization, multi-tenancy, elasticity, network access, on-demand networks, etc.

The most common and frequently used cloud computing services offered to Companies and third persons or parties are Software-as-a-service(SAAS), Platform-as-a-service(PAAS), and Infrastructure-as-a-service (IAAS).

4.1 Software-as-a-Service (SaaS)

Software as a Service (SaaS), formerly known as an on-demand service, has reached a significant level of prominence in the present times. As we can say, these service provides software to a third party or customer either as a call-for-service or via subscription (Blancaflor et al., 2023), charged according to the usage of software etc. Big companies like Microsoft, and Google provided the applications to the third or end user. In the coming future that Saas will become common use by every company, organisation, and end user. This service is dependent on the internet without the internet it cannot be usable. Distinctive Features of SaaS are:

a. The "one to many" model has been delivered by Software.
b. Users do not need to improve the software.
c. It needs less hardware (X. Yan et al., 2012).
d. Do not require a special hardware version or software.
e. The application programming interface provides connectivity between software components that can be integrated.

Companies like Microsoft offic365, dropbox, Cloud9anlytoics, salesforce are providing Saas services.

4.2 Platform-as-a-Service (PaaS)

Platform as a Service (PaaS) facilitates the quick and easy creation of web applications in a computing environment, eliminating the need to invest in costly software and hardware maintenance. PaaS, similar to SaaS, provides a platform for software development that is delivered over the internet, rather than offering the software itself. In this case, the service provided by the company includes guidance on fundamental development standards and effective distribution methods . the Operating System, the database, the setting in which the programming language can be used, and the web servers all fall under the categories of the environment in this case. On the same platform that their end users execute their programme, this enabled developers to create, test and implement their software (Schiering & Kretschmer, 2012).

Google App Engine and Microsoft Azure are two instances of PaaS. Some basic PaaS features include:

1. Software's Built-in scalability which includes load balancing and failover.
2. It is easier to develop. Only requires a PC and the Internet.
3. Using Common standards, integration with web services and databases.

4.3 Infrastructure-as-a-Service (IaaS)

Servers, storage, networks and operating system all Belongs to the cloud computing infrastructure, which is distributed as an on-demand service using Infrastructure-as-a-service(IaaS). Clients opt to acquire resources such as servers, software, data centers, and networking infrastructure as fully outsourced, on-demand services rather than making direct purchases.

According to the Internet Engineering Task Force (IETF), the Infrastructure as a Service (IaaS) paradigm is considered the foundational service model. In this model, users have the ability to utilize virtual machines that are specifically allocated to them, enabling access to infrastructure resources. With the growing number of providers offering development tools as part of the Infrastructure as a Service (IaaS) model, including the option to install different types of cloud environments, the distinction between Platform as a Service (PaaS) and IaaS is becoming less defined (Sanger & Johari, 2022). Another term used for this concept is Hardware as a Service.

Companies like Amazon web services, IBM, Bluestick, and Vmware are providing Infrastructure-as-a-service.

4.4 Big-Data-as-a-Service (BDaaS)

An all-inclusive big data solution is offered to businesses through a cloud-based platform called BDaaS. This organizational structure incorporates components such as Hadoop as a Service (HaaS), Data as a Service (DaaS), and Data Analytics as a Service (DAaaS). The exponential expansion of data is a key driver in this domain. Forecasts indicate a substantial rise in the global big data as a service (BDaaS) market, projected to grow from $1.8 billion in 2015 to $7 billion by 2020 (Oh et al., 2010).

BDaaS is divided into 3 parts.

4.4.1 Hadoop-as-a-Service (HaaS)

Hadoop serves as the underlying infrastructure for numerous big data initiatives and applications. It is a robust data management and storage platform that facilitates tasks such as data sharing, analysis, and various other operations. As social media and online communication have become more commonplace, firms ranging from Facebook to Yahoo are using Hadoop more frequently. According to Research and Markets, from 2014 to 2019, the market for Hadoop-as-a-service is expected to grow by approximately 85% annually. HaaS (Hadoop as a Service) has emerged as a viable substitute for on-premises Hadoop deployments. The following points help distinguish between the different HaaS options:

- Data scientists and data centre administrators must have access to what they require.
- Data at Rest must be kept in HDFS by HaaS.
- HaaS must offer flexibility.

Companies who offer Hadoop as a service offer it “full-baked.” Usually, their own Hadoop version closely follows Apache Hadoop’s initial release. The advantages of HaaS are listed below:

- Fault-tolerance
- Ease of use

- No hardware
- Performance

4.4.2 Data-as-a-Service (DaaS)

With cutting-edge technologies and groundbreaking software platforms, SaaS continues to rule the digital industries. Its identical twin brothers, DaaS, assist in providing high-quality data for your company's marketing initiatives. Daas technology solution offers real-time customer and market data. No matter the location or platform, DaaS isolates data from connected applications to provide consumers with useful informationDaaS can be used to collect data from various heterogeneous sources such as enterprise resource planning (ERP) systems, data warehouses, transactional databases, and customer relationship management (CRM) tools (L. Yan et al., 2009). The DaaS approach brings the following benefits:

- Data moved from one platform to another easily.
- Compatibility across platforms.
- Worldwide ease of access.

4.4.3 Data-Analytics-as-a-Service (DAaaS)

Analytical as a Service (DAaaS) and Data Analytics as a Service (DAaaS) are cloud-based delivery methods that provide a range of analytical tools for analyzing different types of data. Users of the service can upload corporate data and get analytical data by using the cloud, which is composed of analytical algorithms and machine learning principles. Big data and cloud, two of her trends, are combined in DAaaS. It is a platform-based strategy that offers cloud-based analytics with a wide range of functionality, from data gathering through end-user visualisation, reporting, and interactivity. Knowledge workers can have individualised access to information records thanks to analytics as a service (Yadlapati et al., 2023). For instance, business analysts and data scientists may interact with the information and quickly provide more insightful results as a result. The benefit of DAaaS is that it lowers the entry barrier to more sophisticated analytical skills without forcing users to commit a significant amount of internal resources to the project, such as employees and equipment. Instead of complex custom projects, customers follow simpler steps:

- Our company's data scientist searches the AppStore for a suitable analytics app.
- They rent an analytics application for a certain period of Date and data volume.
- They Modify the analytic skills application to suit its requirement.
- After that, the Internal system feeds data into the analytics application.
- The consequences are offered for any other purposes (J. S. Park et al., 2011).

4.5 Database-as-a-Service (DBaaS)

Cloud computing includes services like SaaS, PaaS, and IaaS, and DBaaS (sometimes referred to as Managed Database Service) fits into this category. This makes it possible for users involved in database activities to access and utilise a cloud database system without having to pay for it.

Software as a Service (SaaS), whose demand is increasing at such a rapid rate, includes DBaaS and cloud databases. Database as a Service (DBaaS), in its simplest form, is self-service, on-demand database consumption combined with operational automation. DBaaS is built on the same pay-as-you-go payment structure as cloud computing services, thus as is well knowledge, you only pay for the services you utilise. The capability offered by this DBaaS is equivalent to that of conventional classical and relational database models. By avoiding database configuration, management, upgrading, and security, DBaaS enables businesses to save time and money. An information manager component of DBaaS utilises an API to handle all instances of fundamental information. The user can access this API using a management console, which is often an online tool the user can use to provide and de-provision instances of information as well as manage and aggregate information (Deepika et al., 2022).

4.6 Information-as-a-Service (INaaS)

Information as a Service (INaaS) is a service that offers a secure and standardized approach to effectively communicate, generate, and extract valuable information from the abundance of available data, ensuring it is delivered at the appropriate time and in the suitable format . By having access to this, any software may use any type of information that is housed remotely and can access any type of information via the API (Application Programming Interface). Businesses utilise data from one or more apps that combine data from several more sources. The core idea behind INaaS is " decoupling data consumers from the existence of additional of network infrastructure" (Yeo & Park, 2013). With no regard to the kind or location of the data, INaaS is a method that is used to turn data into information and provides actionable business insights in an organised style and with a streamlined flow.

4.7 Integration-as-a-Service (INTaaS)

"Integration as a Service" (INTaaS) is a cloud-based offering that focuses on providing integration services. It involves defining clear interfaces, web services, and interactions between software and data sources. This integration-as-a-service concept offers a comprehensive solution that represents seamless connectivity with backend systems, resources, files, and operational applications. Reputable companies like MuleSoft, Dell Boomi, CloudSwitch, and others offer integration as a service (Weihua & Shibing, 2013).

4.8 Security-as-a-Service (SECaaS)

Because of how dangerously they use and utilise the Internet, users are vulnerable to security problems. SECaaS now becomes relevant in this situation. It serves as protection against the most persistent cyberattacks. SECaaS, which stands for Security as a Service, is built upon the Software as a Service (SaaS) model. However, it should be noted that SaaS is only suitable for specific information security services within this outsourcing approach to security management. SECaaS, or security controls as a service, refers to products like online-sold antivirus software.

The SECaaS is a subscription-based security service that offers security services for information systems and is hosted by Cloud providers to meet corporate objectives and safeguard the Confidentiality, Integrity, and Availability (CIA) of information or essential data.

Security-as-a-Service offers many benefits including:

- Regular updates of virus definitions independent of user compliance.
- Faster provisioning of user accounts.
- Better security know-how than is normally available from administrative tasks within the organization.
- Products for cloud security also assist in reducing the demand for expensive security analysts and professionals.

4.9 Testing-as-a-Service (TaaS)

Cloud computing enables the creation of Testing as a Service (TaaS) for cloud, where cloud services and cloud applications can be effectively developed and tested. Under the TaaS model, any testing done in connection with a company's business operations can be completely under the control of a third person that specialises in offering proper test conditions in line with User specifications. It can either assign the testing area to outside service providers or hire specialists and analysts to assist and counsel their staff. New business prospects, problems, needs for creative service models, and QoS standards and requirements all result from this (Nwobodo, 2016). The process of moving current application requirements to the cloud is driven by four major factors. They provide the necessary foundation or tools to facilitate cloud testing within an organization, including the elements of people, processes, governance, and infrastructure.

4.10 Anything-as-a-Service (XaaS)

"X as a service," "anything as a service" is just a few of the portmanteaux words that make up the phrase "XaaS." A wide and large category of services connected to remote access & cloud computing are referred to as XaaS. The most modern type of cloud computing service is called XaaS. It gives users the option of combining personal, private cloud services with online public cloud service delivery. Indicated how widespread cloud-based IT resources are given that, how many different types of services are currently made available in this manner, the term "XaaS" is somewhat paradoxical when used to describe their pervasiveness. XaaS (Everything as a Service) encompasses the widely recognized models of Software as a Service (SaaS), Infrastructure as a Service (IaaS), and Platform as a Service (PaaS). These models provide users with convenient access to software applications, infrastructure resources, and development platforms without the need for extensive in-house management and maintenance (Chavan et al., 2022).

Various service models are available, such as Network as a Service, Desktop as a Service, Disaster Recovery as a Service, Monitoring as a Service, and even Marketing as a Service and Healthcare as a Service, among others. These service models offer organizations the flexibility to access specific functionalities and resources without having to manage them internally.

5. OVERVIEW ON CLOUD SECURITY

As businesses attempt to incorporate more digital tools and services into their infrastructure, it is imperative to consider security concerns for these new technologies. This is what is meant by "cloud security." By definition, it is a collection of procedures and resources developed to protect users of web-based applications from both internal and external threats. There are several reasons why a company would

Figure 3. Cycle of cloud computing

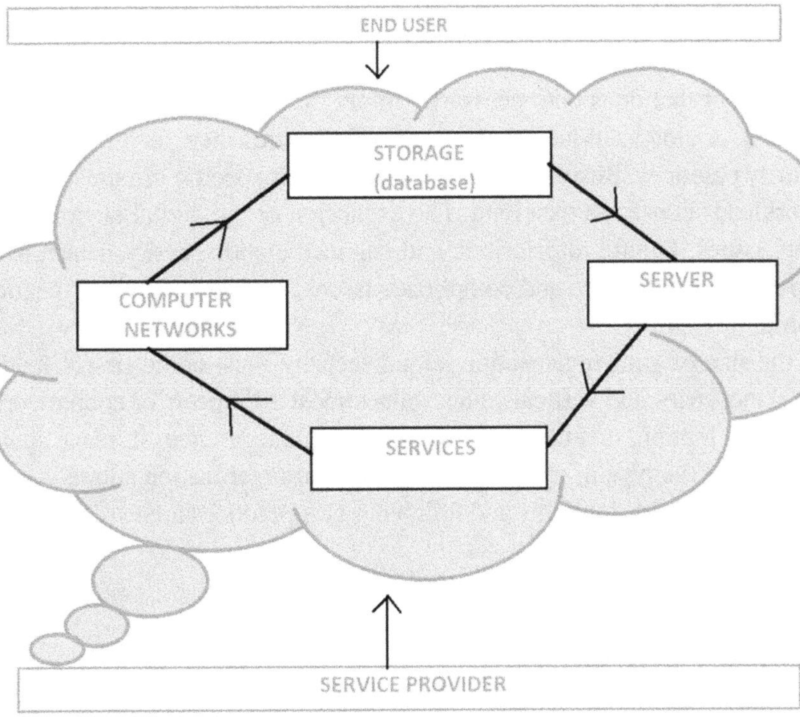

decide to carry out a digital transformation strategy (Mehra et al., 2018). Maintaining competition may be vital in some circumstances. Other times, it could be carried out to raise consumer or employee satisfaction. Whatever the justification, moving to the cloud shouldn't compromise security.

A large migration to cloud settings can have several negative repercussions if done improperly, even though more current technologies are enabling organisations to expand potentiality beyond the constraints of on-premises infrastructure (Raj et al., 2021). (See in Figure 3)Understanding how connected cloud technology can help modern businesses while implementing cloud security best practices is crucial for striking the right balance.

5.1 Why Is Cloud Security Important?

Cloud computing might make businesses function more effectively. Among other things, this details enhancing data collection and storage to deliver better customer service, to increase mobility through remote working and rapid scaling, to improve comfort through connected systems with rapid file and data sharing, and much more. Any company's cloud infrastructure must be safe for it to function, due to the dangers of improper configuration and the constant threat from hackers. Security in the cloud is important here. You may better safeguard your digital assets and limit the chance of human mistakes with cloud security, which reduces the possibility that a breach would have been halted and resulted in a large loss for your firm (Kaur & Kaushal, 2011). The adoption of cloud environments and various computing models such as IaaS, PaaS, or SaaS has witnessed a significant surge in today's business landscape. Despite providing ample resources to their departments, organizations often encounter challenges associated with the dynamic nature of infrastructure management, particularly when integrating new applications and services. To streamline and optimize their IT operations, enterprises can leverage as-a-service models, which allow them to outsource time-consuming IT tasks.

Although the administration of this infrastructure may be delegated to independent cloud computing service providers, but that does not necessarily imply that the security and accountability of data assets follow. The bulk of cloud service providers actively secures their servers' security by defaulting to recognised security practices. Businesses must consider their specific circumstances when securing data, apps, and workloads housed in the cloud. The expansion of the digital environment has led to an increase in security issues. Failure to prioritize and enhance cloud security measures can expose organizations to substantial governance and compliance issues, irrespective of the location where client information is stored.

Regardless of the size of your organisation, cloud security must come first. Cloud infrastructure, which spans several industries and verticals, underpins almost all aspects of contemporary computing. For a successful cloud adoption, it is necessary to have adequate defences in place against modern cyberattacks. Irrespective of the type of cloud environment your organization adopts—public, private, or hybrid—ensuring cloud security solutions and following best practices are essential to sustain uninterrupted business operations.

6. CLOUD SECURITY PROBLEMS

There is no doubt the many advantages of cloud computing, but there are some security issues as well. The following is a list of several security concerns with cloud computing (Che Fauzi et al., 2012).

1. **Data loss:** Data loss is One of the problems with cloud computing. This is also known as a data leak. We are conscious of the fact that some of the sensitive data in your database are not entirely within our control. So it is possible that if a cloud service's security is compromised, hackers will gain access to our personal information or sensitive data.
2. **Interference of hackers and insecure APIs:** From our perspective, the Internet is also taken into consideration while discussing the cloud and its services. We also understand that the simplest approach to connect to the cloud is by using an API. Ensuring the security of APIs and interfaces utilized by external users is paramount. While public accessibility to services in cloud computing is limited, the vulnerability lies in the possibility of unauthorized access to these services. Safeguarding against such risks is essential to mitigate potential threats and maintain a secure environment. Therefore, hackers might easily steal or destroy personal data utilising these sites.
3. **Changing service provider:** A Vendor lock-in is a major security issue with cloud computing. Switching vendors will present a number of challenges for many corporations. For instance, switching from Amazon Web providers (AWS) to Google Cloud Services (GCS) may be challenging for a corporation owing to the need to move all of their data and issues resulting from the fact that the two cloud providers employ several different features and approaches. Additionally, it's likely that Google Cloud and other providers have costs that are different from those of AWS (Peng, 2022).
4. **Lack of skill:** Inexperienced workers in an IT company often face a range of challenges such as workplace difficulties, transitioning to a new service provider, the requirement for an extra function, and struggling with understanding certain features. These issues frequently arise and can pose sig-

nificant obstacles to productivity and performance. Therefore, using cloud computing necessitates having the proper credentials.

5. **Denial of service (DoS) attack:** A Denial of Service (DoS) attack is a malicious act where an individual or group intentionally overloads a network, server, website, or application with an excessive amount of data or traffic, aiming to disrupt its regular operations and make it inaccessible to legitimate users. Major corporations within industries such as banking and the public sector face the brunt of DoS attacks, being the primary targets. In the unfortunate event of a DoS attack, data loss becomes a critical consequence. Consequently, the process of data recovery becomes a demanding and time-consuming endeavor, requiring substantial investments of time and effort.
6. **Multitenancy:** Multitenancy is commonly used in cloud computing environments, where multiple users or organizations access the same cloud infrastructure to run their applications or services, it can also be used in many software like computer relationship management (CRM) and ERP Systems, where multiple departments of a large organisation to access the same system while maintain their own data.

7. CONCLUSION

In this chapter, it investigate into Challenging and complex process of Cloud Services Selecting. There exist so many Cloud Providers who offer a wide range of Cloud Services. The objective is that there exist many Cloud Services not only the three main Services (IAAS, PAAS, SAAS) Here we tell that we use all these services when needed by the users and not only depend on the main services of the cloud computing . In the rapidly expanding global cloud computing market, valued at $100 billion, competition between cloud providers and outsourcing companies continues to intensify. As developers anticipate this trend, it becomes crucial for them to consider the adoption of cloud computing. Regardless of the specific approach taken by a cloud provider, it is essential to prioritize horizontal scalability of virtualized resources, such as computing, storage, and networking, over the performance of individual nodes . Alternatively, The researcher might also investigate the overview of Cloud Security, their importance and the problem faced by users, So that users could understand the security breaches while using Cloud. Overall, cloud computing services have become an essential part of the modern IT landscape, and are expected to continue to grow in popularity and importance in the coming years.

REFERENCES

Arulkumaran, G., Jayagopalan, S. L, T. M., & Balamurugan, P. (2023). An Effective Analysis of Proficient Two Level Security Contraptions For Loading Data In Cloud. *2023 1st International Conference on Innovations in High Speed Communication and Signal Processing (IHCSP),* 536–540. 10.1109/IHCSP56702.2023.10127217

Blancaflor, E., Saunar, B. Y. P., Darrel, C., Bilbao, T., Villarias, I. H. B., Paula, V., & Mapue, I. (2023). The Use of Cloud Computing and its Security Risks in a Philippine Education System: A Literature Review. *2023 11th International Conference on Information and Education Technology (ICIET),* 66–70. 10.1109/ICIET56899.2023.10111146

Butt, U. A., Mehmood, M., Shah, S. B. H., Amin, R., Shaukat, M. W., Raza, S. M., Suh, D. Y., & Piran, M. J. (2020). A Review of Machine Learning Algorithms for Cloud Computing Security. *Electronics (Basel), 9*(9), 9. Advance online publication. doi:10.3390/electronics9091379

Chavan, J., Patil, R., Patil, S., Gutte, V., & Karande, S. (2022). A Survey on Security Threats in Cloud Computing Service Models. *2022 6th International Conference on Intelligent Computing and Control Systems (ICICCS),* 574–580. 10.1109/ICICCS53718.2022.9788148

Che Fauzi, A. A., Noraziah, A., Herawan, T., & Mohd. Zin, N. (2012). On Cloud Computing Security Issues. In J.-S. Pan, S.-M. Chen, & N. T. Nguyen (Eds.), *Intelligent Information and Database Systems* (pp. 560–569). Springer. doi:10.1007/978-3-642-28490-8_58

Chen, T., & Liu, H. (2022). Discussion on Network Information Security Based on Cloud Computing Environment. *2022 8th Annual International Conference on Network and Information Systems for Computers (ICNISC),* 65–68. 10.1109/ICNISC57059.2022.00023

Deepika, K., R., & Dalip. (2022). Security Enabled Framework to Access Information in Cloud Environment. *2022 International Conference on Machine Learning, Big Data, Cloud and Parallel Computing (COM-IT-CON), 1,* 578–582. 10.1109/COM-IT-CON54601.2022.9850906

Dudiki, N., Sangeetha, S., Manna, A., Pokhariyal, R., Mohanaprakash, T. A., & Srivastava, A. P. (2022). A Hybrid Cryptography Algorithm to Improve Cloud Computing Security. *2022 5th International Conference on Contemporary Computing and Informatics (IC3I),* 1020–1023. 10.1109/IC3I56241.2022.10072437

Ghosh, U., & Dolui, S. (2020). Cloud Computing: It's Security and Introduction to Amazon Web Services. In S. Dawn, V. E. Balas, A. Esposito, & S. Gope (Eds.), *Intelligent Techniques and Applications in Science and Technology* (pp. 942–951). Springer International Publishing. doi:10.1007/978-3-030-42363-6_108

Harish Naik, B. M., & Kannan, J. (2023). A Research on Various Security Aware Mechanisms in Multi-Cloud Environment for Improving Data Security. *2023 International Conference on Distributed Computing and Electrical Circuits and Electronics (ICDCECE),* 1–5. 10.1109/ICDCECE57866.2023.10151135

Jouini, M., & Rabai, L. B. A. (2017). A Security Risk Management Model for Cloud Computing Systems: Infrastructure as a Service. In G. Wang, M. Atiquzzaman, Z. Yan, & K.-K. R. Choo (Eds.), *Security, Privacy, and Anonymity in Computation, Communication, and Storage* (pp. 594–608). Springer International Publishing. doi:10.1007/978-3-319-72389-1_47

Kaur, P. J., & Kaushal, S. (2011). Security Concerns in Cloud Computing. In A. Mantri, S. Nandi, G. Kumar, & S. Kumar (Eds.), *High Performance Architecture and Grid Computing* (pp. 103–112). Springer. doi:10.1007/978-3-642-22577-2_14

Li, H., Tian, X., Wei, W., & Sun, C. (2012). A Deep Understanding of Cloud Computing Security. In J. Lei, F. L. Wang, M. Li, & Y. Luo (Eds.), *Network Computing and Information Security* (pp. 98–105). Springer. doi:10.1007/978-3-642-35211-9_13

Mahlaule, S., van der Poll, J. A., & Ochola, E. O. (2022). Towards an Enhanced Framework to Facilitate Data Security in Cloud Computing. In M. Saraswat, H. Sharma, K. Balachandran, J. H. Kim, & J. C. Bansal (Eds.), *Congress on Intelligent Systems* (pp. 227–242). Springer Nature. doi:10.1007/978-981-16-9113-3_18

Mehra, N., Aggarwal, S., Shokeen, A., & Bura, D. (2018). Analyzing Cloud Computing Security Issues and Challenges. In P. K. Pattnaik, S. S. Rautaray, H. Das, & J. Nayak (Eds.), *Progress in Computing, Analytics and Networking* (pp. 193–202). Springer. doi:10.1007/978-981-10-7871-2_19

Mishra, S., Kumar, M., Singh, N., & Dwivedi, S. (2022). A Survey on AWS Cloud Computing Security Challenges & Solutions. *2022 6th International Conference on Intelligent Computing and Control Systems (ICICCS),* 614–617. 10.1109/ICICCS53718.2022.9788254

N, B., K, P., S, M., S, H., M, V. K., & MRM, V. (2022). A Novel Framework for Cyber Security Attacks on Cloud-Based Services. *2022 Fourth International Conference on Cognitive Computing and Information Processing (CCIP),* 1–4. doi:10.1109/CCIP57447.2022.10058673

Nwobodo, I. (2016). Cloud Computing: Models, Services, Utility, Advantages, Security Issues, and Prototype. In Q.-A. Zeng (Ed.), *Wireless Communications, Networking and Applications* (pp. 1207–1222). Springer India. doi:10.1007/978-81-322-2580-5_110

Oh, T. H., Lim, S., Choi, B. Y., Park, K.-R., Lee, H., & Choi, H. (2010). State of the Art of Network Security Perspectives in Cloud Computing. In T. Kim, A. Stoica, & R.-S. Chang (Eds.), Security-Enriched Urban Computing and Smart Grid (pp. 629–637). Springer. doi:10.1007/978-3-642-16444-6_79

Park, J. S., Yi, K. J., & Park, J. H. (2011). SSP-MCloud: A Study on Security Service Protocol for Smartphone Centric Mobile Cloud Computing. In J. J. Park, H. Arabnia, H.-B. Chang, & T. Shon (Eds.), *IT Convergence and Services* (pp. 165–172). Springer Netherlands. doi:10.1007/978-94-007-2598-0_18

Peng, J. (2022). Research on E-government Information Security Based on Cloud Computing. *2022 IEEE 10th Joint International Information Technology and Artificial Intelligence Conference (ITAIC), 10*, 312–316. 10.1109/ITAIC54216.2022.9836548

Raj, A., Jain, N., & Chauhan, S. S. (2021). Mapping of Security Issues and Concerns in Cloud Computing with Compromised Security Attributes. In R. Agrawal, G. Sanyal, K. Curran, V. E. Balas, & M. S. Gaur (Eds.), *Cybersecurity in Emerging Digital Era* (pp. 24–40). Springer International Publishing. doi:10.1007/978-3-030-84842-2_2

Sanger, A. K. S., & Johari, R. (2022). Survey of Security Issues in Cloud. *2022 International Mobile and Embedded Technology Conference (MECON)*, 490–493. 10.1109/MECON53876.2022.9751959

Schiering, I., & Kretschmer, J. (2012). The Infrastructure Level of Cloud Computing as a Basis for Privacy and Security of Software Services. In J. Camenisch, B. Crispo, S. Fischer-Hübner, R. Leenes, & G. Russello (Eds.), *Privacy and Identity Management for Life* (pp. 88–101). Springer. doi:10.1007/978-3-642-31668-5_7

Shei, S., Kalloniatis, C., Mouratidis, H., & Delaney, A. (2016). Modelling Secure Cloud Computing Systems from a Security Requirements Perspective. In S. Katsikas, C. Lambrinoudakis, & S. Furnell (Eds.), *Trust, Privacy and Security in Digital Business* (pp. 48–62). Springer International Publishing. doi:10.1007/978-3-319-44341-6_4

Trikha, P. (2021). A Critical Review on Security Issues in Cloud Computing. In D. Goyal, V. E. Bălaş, A. Mukherjee, V. Hugo C. de Albuquerque, & A. K. Gupta (Eds.), Information Management and Machine Intelligence (pp. 643–652). Springer. doi:10.1007/978-981-15-4936-6_70

Wei, Y., & Zhang, Y. (2018). Cloud Computing Data Security Protection Strategy. In X. Sun, Z. Pan, & E. Bertino (Eds.), *Cloud Computing and Security* (pp. 376–386). Springer International Publishing. doi:10.1007/978-3-030-00006-6_34

Weihua, J., & Shibing, S. (2013). Research on the Security Issues of Cloud Computing. In Z. Du (Ed.), *Intelligence Computation and Evolutionary Computation* (pp. 845–848). Springer. doi:10.1007/978-3-642-31656-2_115

Xia, P. (2020). Data Security Risk and Preventive Measures of Virtual Cloud Server Based on Cloud Computing. In Z. Xu, R. M. Parizi, M. Hammoudeh, & O. Loyola-González (Eds.), *Cyber Security Intelligence and Analytics* (pp. 40–45). Springer International Publishing. doi:10.1007/978-3-030-43306-2_6

Xiao, Z., & Chen, J. (2015). Cloud Computing Security Issues and Countermeasures. In W. E. Wong (Ed.), *Proceedings of the 4th International Conference on Computer Engineering and Networks* (pp. 731–737). Springer International Publishing. 10.1007/978-3-319-11104-9_85

Xiong, H., Cheng, J., Wang, Y., Yang, X., & Wang, K. (2022). Research on Cloud Computing Security Technology of Power System. In Y. Tian, T. Ma, M. K. Khan, V. S. Sheng, & Z. Pan (Eds.), Big Data and Security (pp. 269–283). Springer. doi:10.1007/978-981-19-0852-1_21

Yadlapati, D., Siddhartha, N., Seelamneni, M., Nali, A. Y., Sangaraju, H. R., & Sridhar, P. S. V. S. (2023). Security Management Approaches Over the Cloud. *2023 International Conference on Sustainable Computing and Data Communication Systems (ICSCDS)*, 1277–1282. 10.1109/ICSCDS56580.2023.10105026

Yan, L., Rong, C., & Zhao, G. (2009). Strengthen Cloud Computing Security with Federal Identity Management Using Hierarchical Identity-Based Cryptography. In M. G. Jaatun, G. Zhao, & C. Rong (Eds.), *Cloud Computing* (pp. 167–177). Springer. doi:10.1007/978-3-642-10665-1_15

Yan, X., Zhang, X., Chen, T., Zhao, H., & Li, X. (2012). The Research and Design of Cloud Computing Security Framework. In Y. Wu (Ed.), *Advances in Computer, Communication, Control and Automation* (pp. 757–763). Springer. doi:10.1007/978-3-642-25541-0_95

Yeo, S.-S., & Park, J. H. (2013). Security Considerations in Cloud Computing Virtualization Environment. In Grid and Pervasive Computing (pp. 208–215). Springer. doi:10.1007/978-3-642-38027-3_22

Yu, S. (2018). Towards the Cloud Computing from Services to Securities and Challenges. In X. Sun, Z. Pan, & E. Bertino (Eds.), *Cloud Computing and Security* (pp. 334–343). Springer International Publishing. doi:10.1007/978-3-030-00009-7_31

Ziani, A., & Medouri, A. (2019). Cloud Computing: Security and Privacy Issues. In M. Ben Ahmed, A. A. Boudhir, & A. Younes (Eds.), *Innovations in Smart Cities Applications* (2nd ed., pp. 668–677). Springer International Publishing. doi:10.1007/978-3-030-11196-0_55

Chapter 2
Toxicological Parameters and Risk Assessment of Some Novel Insecticides: An In-Depth Analysis

Akanksha Chand
State Niti Aayog, India

ABSTRACT

Pesticides are widely used in agriculture, horticulture, and public health to protect crops and control pests. However, pesticides can also pose a significant risk to human health and the environment, especially when used improperly or in excessive amounts. The toxicity of a pesticide is often expressed in terms of its lethal dose (LD) or lethal concentration (LC) values, which provide an estimate of the amount of the pesticide that can cause death in a specified population. In this chapter, the authors review the LD and LC values and explore the different factors that influence pesticide toxicity, such as the mode of action and formulation. Additionally, they present the toxicity ranges of selected pesticides commonly used in agriculture and public health and discuss the health effects associated with acute and chronic exposure to these chemicals. The chapter will conclude with a discussion of the regulatory frameworks governing pesticide use and the need for sustainable and integrated pest management practices.

INTRODUCTION

As the development of new insecticides continues to advance, it becomes increasingly important to evaluate their toxicological parameters and assess the associated risks. Novel insecticides offer promising solutions for pest control, but their potential impacts on human health and the environment must be thoroughly understood. Toxicological parameters encompass a range of factors, including acute and chronic toxicity, effects on non-target organisms, and persistence in the environment. Risk assessment plays a crucial role in determining the safety and suitability of these insecticides for use.

DOI: 10.4018/979-8-3693-0200-2.ch002

Figure 1. Chemical structure depiction of Buprofezin

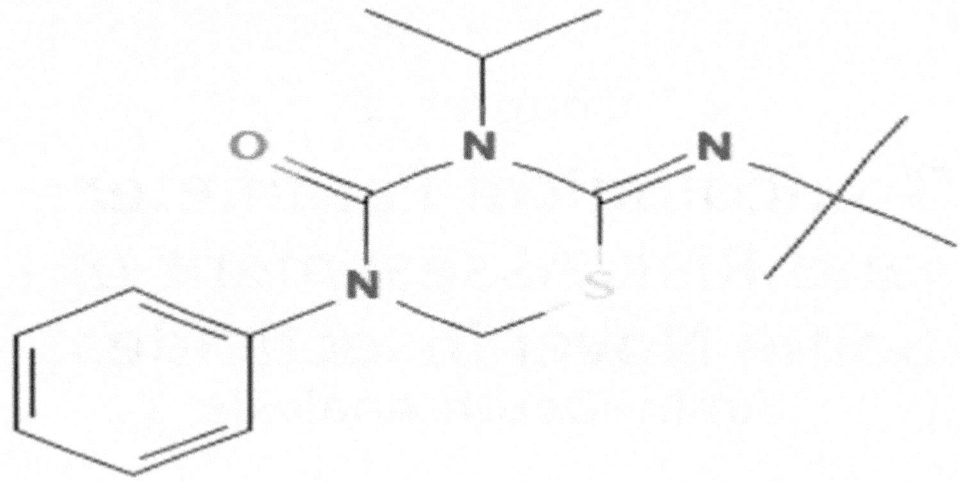

Toxicological parameters form the foundation for understanding the potential hazards posed by novel insecticides. Acute toxicity studies investigate the immediate effects of exposure to determine the lethal dose (LD50) or concentration (LC50) that causes adverse effects in test organisms. Chronic toxicity studies, on the other hand, focus on the long-term effects of repeated or continuous exposure to insecticides, providing insights into potential reproductive, developmental, and carcinogenic effects. These parameters enable scientists and regulators to assess the risks associated with human exposure and determine safe levels of application.

Furthermore, the effects of novel insecticides on non-target organisms such as beneficial insects, pollinators, aquatic organisms, birds, and mammals are a significant concern. The persistence of insecticides in the environment also warrants attention.

In conclusion, the evaluation of toxicological parameters and the rigorous risk assessment of novel insecticides are fundamental steps in ensuring the responsible and sustainable use of these chemical tools for pest control. Hence, the toxicological parameters of some insecticides are discussed below.

BUPROFEZIN

Buprofezin is a non-systemic narrow-spectrum foliar insecticide in the thiazine class belonging to IRAC Group 16 (chitin biosynthesis inhibitor-type1). It does not exhibit cross-resistance to other pesticide classes. It acts as a molting inhibitor against some Coleoptera, Hemiptera, Acarina and Homoptera such as plant hoppers, whiteflies and scale insects. It is widely used in more than 80 countries. Insect growth regulators (IGRs) have lower toxicity against vertebrates than most conventional insecticides.

Specifications

MF: $C_{16}H_{23}N_3OS$
MW: 305.4g/mol

IUPAC Name: 2-tert-butylimino-5-phenyl-3-propan-2-yl-1,3,5-thiadiazinan-4-one
Isomeric SMILES:
CC(C)N1C(=NC(C)(C)C)SCN(C1=O)C2=CC=CC=C2

Formulation Type

It is marketed under the trade names Applaud, Maestro, Podium, Profezon, Sunprofezinm, and Viappla and is available in DP, GR, SC, and WP forms.

Dose

Buprofezin is an insecticide with persistent larvicidal action against Homopterans, some coleopterans and also Acarina. It is found to be effective against Cicadellidae, Delphinidae, and Deltocephalinae in rice when applied at the rate of 50-250g a.i./ha; mealybug in grapes, Whiteflies in citrus, cotton and vegetables@ 250 -750ga.i./ha; scale insects and mealybugs (Pseudococcidae) in citrus and top fruit @ 25-50g a.i./ha and yellow mite in chillies @ 75-150ga.i./ha.

Mode of Action

Buprofezin is a persistent insecticide and acaricide which acts as a chitin synthesis inhibitor (type-1). It interferes with cuticle deposition and also affects hormone levels of nymphs associated with molting and prostaglandin synthesis, eventually leading to death Additionally, it prevents adults from ovipositing. It does not translocate in plants and serves as a contact and stomach poison.

Toxicological Profile

- **Mammalian toxicity:** It was found to be moderately toxic to mammals with acute oral LD_{50} of >2000mg/kg and LD_{50} skin and eye of >5000 against the rats. It had an ADI/RfD(JMPR) of about 0.01m/kgb.w.
- **Phytotoxicity:** It was found slightly phytotoxic to Chinese cabbage and Japanese persimmon in glasshouse conditions.
- **Toxicity to bees and natural enemies:** It had chronic LD_{50} value of >163.5μg/bee for the honey bee which shows that it is non-toxic to honeybees and also it does not have any detrimental effect on any predator such as C*yrtorhinus lividipennis* and parasites such as *Encarsia Formosa*.
- **Birds:** The acute oral LD_{50} for birds was >2000mg/kg which indicates that it was moderately toxic to birds.
- **Fishes:** The acute oral LC_{50} for carp fish was 0.57mg/l.
- **Health effects:** Acute exposure to buprofezin can cause skin and eye irritation, as well as respiratory irritation and nausea if the chemical is inhaled or ingested. Symptoms of acute toxicity can vary depending on the dose and duration of exposure, and may include dizziness, headaches, and stomach upset. Chronic exposure to pesticides can cause neurological effects and blood disorders (S., 2009). A study on Nile tilapia found that buprofezin can cause tissue damage mediated by oxidative stress and inflammatory response (Rehab A. Azouz, 2020). A study on fish exposed to

sub-acute doses of fipronil and buprofezin revealed abnormal changes in their erythrocyte morphology such as necrosis, micronuclear formation, and hyperchromatosis (Qureshiet al, 2016).

CLOFENTEZINE (TETRAZINE)

Clofentezine is a tetrazine that functions as a mite growth regulator and organochlorine acaricide and. It is a member of monochlorobenzene. It is an acaricide used for the residual control of mites. it predominantly functions as an ovicide, with some activity in early motile stages. Used on a wide range of crops – apples, pears, stone fruit, nuts, ornamentals, almonds etc.

Figure 2. Chemical structure depiction of Clofentezine

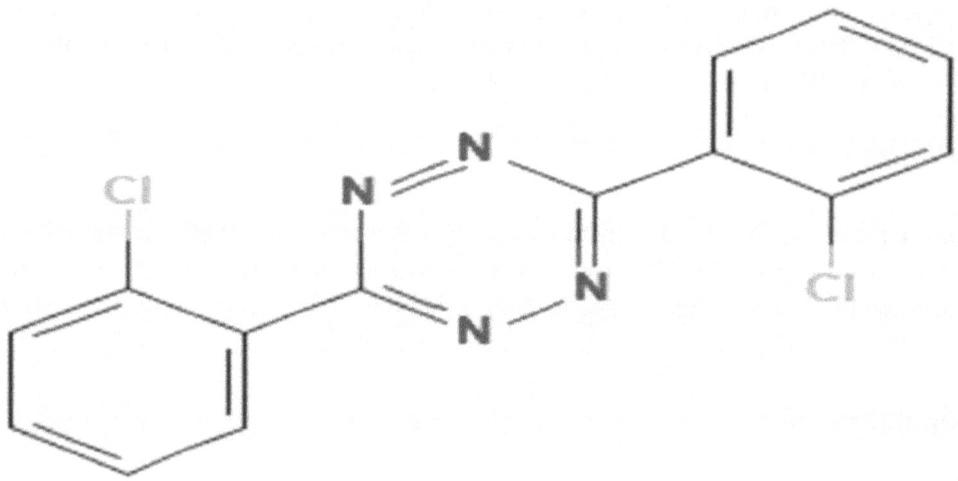

Specification

MF: $C_2H_2N_4$
MW: 82.06g/mol
IUPAC Name: tetrazine
Isomeric SMILES: C1=CN=NN=N1

Formulation Type

Clofentezine is an acaricide belonging to the unique group, the tetrazine. it is marked as Apollo®, Acaristop®. It is mainly available as SC formulation. In many nations around the world, these formulations are registered and sold as either 200 or 500 g/l. Clofentezine may be combined with bifenthrin as Torant CL and Percut, containing 200 g/l clofentezine + 40 g/l bifenthrin. Clofentezine is currently registered for use in 25 countries, including Australia, New Zealand, France, Germany, Israel, Spain, Switzerland and the United Kingdom. Trade names include Apollo, Apolo, Acaristop, Cara or Niagara.

Mode of Action

Clofentezine act as an acaricide/ovicide. It primarily functions as an ovicide because it inhibits cell growth and differentiation during the last stage of embryonic and early larval development. It has a unique contact mode of action with prolonged lingering effect. It is used against spider mites on ornamentals, food and non-food crops grown outdoors, indoors and in orchards.

Crops

Clofentezine is used as an acaricide/ovicide for deciduous fruits (apples, pears, stone fruit and citrus), nuts, almonds, cotton, cucurbits, vines and ornamentals.

Pest Range

It is mainly used as an acaricide against spider mites.

Toxicological Profile

- **Mammalian toxicity:** As evaluated by the US EPA (USEPA 1999), toxicity of clofentezine was classified (Makhteshim 2005) as follows:
    - 1-Acute oral, dermal and inhalation toxicity: Category III (Moderately toxic).
    - 2-Acute eye and skin irritation: Category IV (Slightly toxic).
- **Toxicity to humans:** Clofentezine do not pose any major toxicity risk to humans, like it does to other mammals. Its ADI range is 0 – 0.02 mg/kg/day.
- **Acute toxicity limits:** It poses a low acute oral toxicity with LD_{50} 2100 mg/kg and inhalation LC_{50} for rats as 0.89 mg/l.
- **Toxicity to bees and natural enemies:** It is moderately toxic to honeybees with acute LD50 84.5 µg a.i./bee (48h). In field experiments, does not impact any predatory mites or beneficial insect species.
- **Birds:** It possess low toxicity to birds with acute LD_{50} of 7500 mg/kg for bobwhite quail.
- **Fish and Aquatic Invertebrates:** It was found to be highly toxic to aquatic invertebrates having acute EC_{50} 0.0008 mg/l (48h) and fishes as well with acute LC_{50} of 0.015 mg/l (96h) in rainbow trout. However, it was moderately toxic to algae with acute EC_{50} 0.32 mg/l (72h).

The main toxicological effects seen in studies in rats and dogs were hepatotoxicity and alterations to the thyroid, particularly follicular hyperplasia, Humans are not at high risk of poisoning from clofentezine.

- **Health effects:** Regarding the health implications of acute or ongoing exposure to clofentazine, there is no precise evidence available. The search results do not contain any specific information regarding the health effects of acute or chronic exposure to clofentazine. However, a study on chlordane, another pesticide, found that acute exposure to chlordane can cause adverse health effects such as nausea, irritability, headaches, and neurological effects due to the parent compound and its metabolites5. Chronic exposure to chlordane can cause blood disorders and permanent alterations of nervous system function. (Agency for toxic substance and diseade registry, 2014).

Figure 3. Chemical structure depiction of Flonicamid

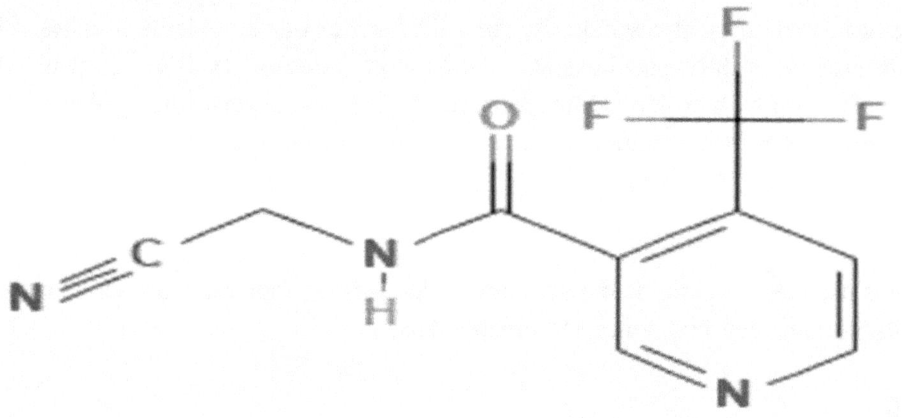

FLONICAMID

Flonicamid is a pyridinecarboxamide compound in which the nicotinamide is altered by the addition of a trifluoromethyl group at position 4 on the pyridine ring and a cyanomethyl group at the carbamoyl nitrogen atom. It has a role as a xenobiotic, an environmental contaminant and an insecticide. It is functionally related to a nicotinamide. Flonicamid is a systemic insecticide discovered by ISK(Ishihara Sangyo Kaisha, Ltd). ISK has developed flonicamid on a global basis since the late 1990s. Flonicamid is the exclusive insecticide within IRAC Group 29, with no other members in this class

Specification

MF: $C_9H_6F_3N_3O$
MW: 229.16g/mol
IUPAC Name: N-(cyanomethyl)-4-(trifluoromethyl)pyridine-3-carboxamide
Isomeric SMILES: C1=CN=CC(=C1C(F)(F)F)C(=O)NCC#N

Formulations

Flonicamid is offered in various formulations such as 10%DF, 50%DF, 50%SG, and 1%G, and it is commercially available under different trade names, including Ulala, Teppeki, Carbine, Beleaf, Aria, and Turbine.

Registered Countries

Flonicamid is primarily used as a foliar application and is registered in more than forty countries across the Americas, Asia, Europe, and Africa:

- Asia-China, India, Japan, South Korea, Taiwan, etc.

- Europe- Belgium, Czech Republic, France, Germany, Hungary Romania, Italy, Netherlands, Poland, Spain, Switzerland, UK, etc.
- America-Canada, Brazil, Mexico, USA, etc.

Dose

When applied as a foliar spray at a rate of 50-100 g a.i./ha, flonicamid demonstrates outstanding insecticidal activity against aphids and other sucking pests. such as thrips, whiteflies, leafhoppers, planthoppers, plant bugs, and mealybugs in fruit trees (apples and peaches), cereals (wheat and rice), potatoes, cucurbits (EPA), cotton, leafy vegetables and ornamentals. Flonicamid is also used as an oil application for control of the foliar pests.

Crops

Potatoes, Fruit trees, Cereals, Cotton, Vegetables, Ornamentals, etc.

Insects

Aphid, Whitefly Planthopper Leafhopper Psyllid Plant bug Thrips.

Mode of Action

Flonicamid possesses a unique mode of action that sets it apart from conventional insecticides, ensuring no cross-resistance. Moreover, it exhibits remarkable translaminar and systemic activity by utilizing xylem vessels. It offers good residual control for about 2-3 weeks.

Toxicological Profile

Flonicamid boasts a favorable toxicological, environmental, and eco-toxicological profile.

- **Mammalian toxicity:** It was found to be moderately toxic to rat having acute LD_{50} oral as 884 mg/kg and LD_{50} dermal as > 5,000mg/kg b.w. The ADI/RfD (FSC) was about 0.073mg/kg b.w.
- **Toxicity to Bees:** It is reported to cause no adverse effects on pollinating insects and beneficial insects when applied @50-100 g a.i./ha. It had an acute oral LD_{50} of > 60.5 µg/bee (48 h) and acute contact LD_{50} of > 100 µg/bee (48 h) for honeybees.
- **Birds:** It was found to be moderately toxic to birds with acute oral LD_{50} of > 2,000 mg/kg for male and female quail.
- **Fish and Aquatic Invertebrates:** It possessed very low aquatic toxicity having an LC_{50} value of about > 100 mg/L for trout and bluegill, 96 h. For *Daphnia magna,* it had an EC_{50} of > 100 mg/L (48 h).
- **Health effects:** Regarding the health implications of acute or chronic exposure to flonicamid, there is no precise evidence available. However, Health Hazard Assessment by the California Office of Environmental Health Hazard Assessment states that flonicamid is a nicotinoid insecticide used on cotton, alfalfa, and a variety of fruits, and exposures may occur to agricultural workers (Office

Figure 4. Chemical structure depiction of Oxadiazine

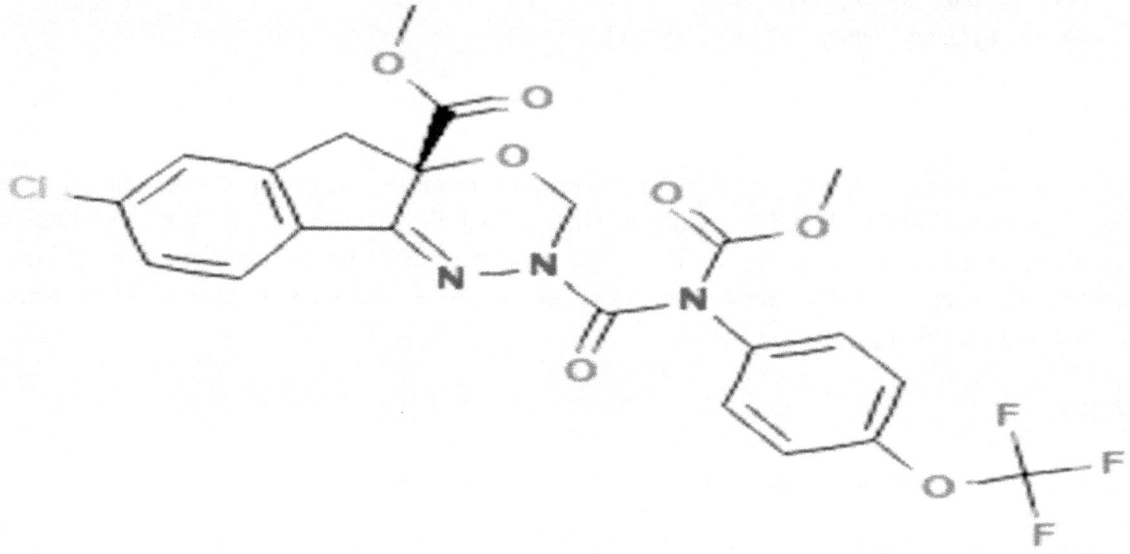

of Environmental Health Hazard Assessment, 2011). A study found that both males and females in the 20 mg/kg group and the females in the 50 mg/kg group demonstrated treatment-related signs of vomiting, and ataxia was also noted (Pesticide Directory). Another study on acute inhalation toxicity found that clinical signs of toxicity consisted of exaggerated breathing up until day 2 and brown staining around the snout (Lambré, 2015).

OXADIAZINE (INDOXACARB)

Indoxacarb is an ISO (International Organization for Standardization) approved name for the new oxadiazine insecticide. It is an organochlorine insecticide and a methyl ester. It acta as a Voltage-gated sodium channel blocker. It is an oxadiazine pesticide that acts against lepidopteran larvae.

Specification

MF: C22H17ClF3N3O7
MW: 527.8g/mol
IUPAC Name: methyl (4aS)-7-chloro-2-[methoxycarbonyl-[4-(trifluoromethoxy)phenyl]carbamoyl]-3,5-dihydroindeno[1,2-e][1,3,4]oxadiazine-4a-carboxylate
Isomeric SMILES:
COC(=O)[C@]12CC3=C(C1=NN(CO2)C(=O)N(C4=CC=C(C=C4)OC(F)(F)F)C(=O)OC)C=CC(=C3)Cl

Formulation Type

Produced by DuPont, Flonicamid is marketed as Steward and Avaunt, intended for agricultural use. Additionally, it is available as Technical Indoxacarb for agriculture purposes and as Advion Fire Ant Bait, designed for use by commercial applicators. Indoxacarb is available as EC, SC, WG, TB and broadcast granule. At the time of initial registration, indoxacarb was formulated as a 30% WG for use in fruits and vegetables. It was also formulated as a 15%SC liquid for use on cotton only.

Registration

In October 2000, the U.S. Environmental Protection Agency (U.S. EPA) granted its initial registration to Flonicamid, classifying it as a reduced-risk pesticide (U.S. EPA, 2000).

Mode of Action

Indoxacarb exhibits dual capabilities, functioning as both a larvicide and an ovicide. In its larvicidal role, it acts as both a stomach and contact poison. Upon ingestion or contact, feeding ceases within 2 to 8 hours. Indoxacarb acts by binding to a site on the sodium channels and blocking the flow of sodium ions into nerve cells resulting in impaired nerve function, feeding cessation, paralysis, and death (Brugger, 1997).

Crops

Indoxacarb, a wide-ranging foliar insecticide, is registered for application on diverse crops, including fruits such as apples, pears, and tomatoes, as well as vegetables like bok choy, Brussels sprouts, broccoli, cabbage, Chinese cabbage, peppers, and cauliflower, corn, eggplant, kohlrabi, peas and lettuce, soybeans, alfalfa and cotton.

Dose

It is advisable to apply indoxacarb at a dose of 30-75g a.i./ha.

Insect Range

Indoxacarb is an effective, non-systemic, synthetic organophosphate alternative insecticide, offering excellent field activity against several lepidopteran pests. Notably, it targets beet armyworm, corn earworm, cabbage looper, fall armyworm, diamondback moth, imported cabbageworm, southern armyworm, tomato pinworm, and tomato fruit worm (DuPont, 2006; California Department of Pesticide Regulation, 2006) as well as certain homopteran and coleopteran pests. Indoxacarb is categorized as a pro-insecticide. Upon ingestion by insects, the active enantiomer DPX-KN128 experiences rapid degradation in the insect gut, leading to its conversion into more toxic form called IN-JT333. This transformation process takes place swiftly in lepidopterans, as observed in studies conducted by Wing et al. in 1998. However, in sucking insect pests, its conversion to IN-JT333 occurs at slower rates but still holds toxicological significance (Wing et al., 2000).

Selectivity of Indoxacarb

Indoxacarb acts by inhibiting sodium channels and demonstrating reversibility of the blocking effects in mammalian neurons. This phenomenon accounts for the notable selective toxicity of indoxacarb, effectively distinguishing between target insects and non-target organisms. (Tsurubuchi and Kono, 2003; Zhao et al., 2003; Silver and Soderlund, 2005). In insects (e.g., lepidopterans) the metabolism of the parent to IN-JT333 is very rapid. However, in mammals (e.g., rats) conversion of the indoxacarb to IN-JT333 is a minor pathway, and IN-JT333 is extensively metabolized and eliminated in the urine (Brugger, 1997; Hetrick et al., 2005).

Toxicological Profile

- **Mammalian toxicity:** Mammalian toxicity studies shows that DPX-MP062 enantiomer possesses moderate acute toxicity against mammals. The acute oral LD_{50} value of indoxacarb was about 843mg/kg b.w. The ADI/RfD was reported to be about 0.01mg/kg b.w. (JMPR) and 0.06 mg/kg b.w (EC).
- **Toxicity to Bees:** In bees, indoxacarb has high contact toxicity (acute contact LD_{50}= 0.18ug/bee) and nearly harmless when consumed through food(U.S. EPA, 2000). A study by DuPont states that after the spray has dried, indoxacarb has a low impact on honeybees. For eg., according to one experiment honeybees remained unaffected even after three hours post-application when exposed to residues of indoxacarb at the maximum proposed single-use rate (0.11 a.i./acre) in alfalfa (Brugger, 1997). Although, additional field data show that when DPX-MP062 is sprayed at a rate of 133 g/hectare, it resulted in considerable mortality after a 24-hour exposure period (Hetrick et al., 2005).
- **Birds:** Indoxacarb exhibits moderate toxicity to avian species both on a sub-acute dietary basis (LD50=808 mg/kg diet) and an acute single oral dose basis (LD50=98 mg/kg for bobwhite quail) (Hetrick et al., 2005). In contrast, the metabolite IN-JT333 demonstrates slight toxicity to avian species on an acute oral basis (U.S. EPA, 2000).
- **Fish and Other Aquatic Invertebrates:** The DPX-KN127, and associated isomer has LC_{50} ranging from 0.024 to 2.94 mg/L and EC_{50} ranging from 0.0542 to >0.37 mg/L are found to be moderate to highly toxic to freshwater fish and invertebrates and estuarine fish and invertebrates, respectively. For freshwater fish and invertebrates, the chronic toxicity levels ranged from 0.0036 to 0.25 mg/L, whereas for estuarine fish and invertebrates, the range was from 0.017 to 0.042 mg/L (Mathews, 2001)
- **Health Effects:** Chronic exposure to indoxacarb can deplete blood-forming elements in the bone marrow and lymphoid organs in mice and rats due to the formation of metabolite (Kaur Sandeep, 2016). Indoxacarb is also toxic to honey bees (Pesticide and you, 2012). Some studies have shown sublethal effects associated with indoxacarb exposure, such as a significant reduction in the survival of larvae, pupae, and offspring (Lemes et al., 2021). According to some studies, acute exposure to indoxacarb can cause loss of reflexes, breathing difficulties, and tremors (Merec, 2022). Acute toxicity studies suggest that indoxacarb can also cause skin sensitization and neurotoxicity in rats, such as a decrease in motor activity (Pesticide and you, 2012). In addition, acute inhalation studies with indoxacarb mixtures have shown lung noise, which indicates the development of acute lung injury and high permeability pulmonary edema (Pesticide and you, 2012).

Figure 5. Chemical structure depiction of Spirodiclofen

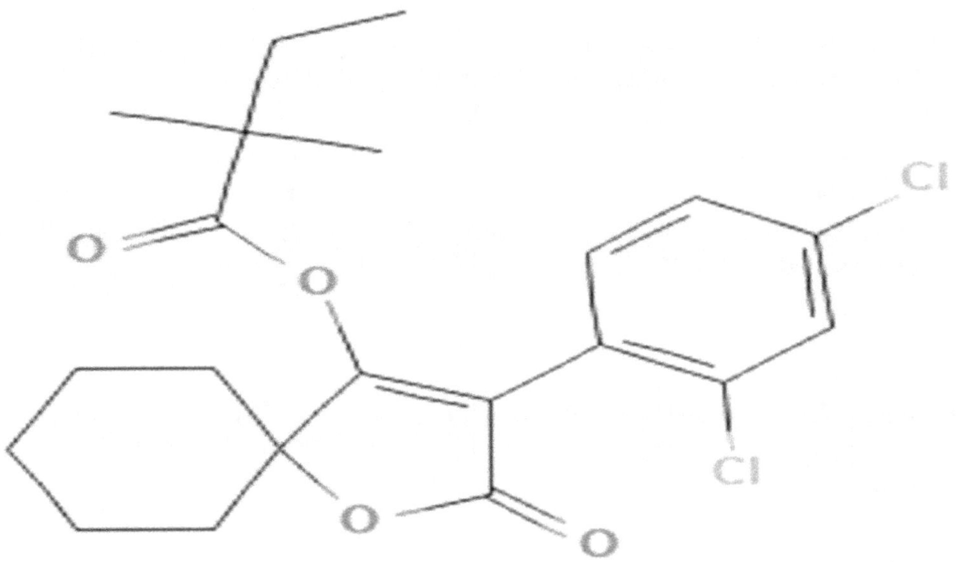

SPIRODICLOFEN

Spirodiclofen belongs to the organochlorine acaricides, characterized as a dichlorobenzene and an oxaspiro compound, as well as a gamma-lactone. It shares a functional relationship with a 1,3-dichlorobenzene.

Specification

MF: C21H24Cl2O4
MW: 411.3g/mol
IUPAC Name: [3-(2,4-dichlorophenyl)-2-oxo-1-oxaspiro[4.5]dec-3-en-4-yl] 2,2-dimethylbutanoate
Isomeric SMILES: CCC(C)(C)C(=O)OC1=C(C(=O)OC12CCCCC2)C3=C(C=C(C=C3)Cl)Cl

Formulation

Bayer Crop Science formulates the end-products of Spirodiclofen as suspension concentrate {SC 322.5 and 240 g ai/L (Envidor 240 SC)}, wettable powder (WP 360 g ai/kg), or water dispersible granule (WG 380 g ai/kg). It functions as both an insecticide and an acaricide.

Mode of Action

Spirodiclofen is categorized as an Insect Growth Regulator (IGR) based on its mode of action. It is a selective, non-systemic foliar insecticide and acaricide classified under the chemical class of ketoenols or Medtronicc acids. Its pesticide mode of action involves inhibiting lipid synthesis.

Crops

Spirodeclofen controls San Jose scale and mites in citrus (orange [sweet and sour], lemon, lime, citrus hybrids [includes chironja, tangelo, and tangor], kumquat, grapefruit, calamondin, citrus citron, mandarin [tangerine], pummelo, satsuma mandarin), plum [includes Chickasaw plum, damson plum, and Japanese plum], plumcot, prune),pome fruit (apple, crabapple, loquat, mayhaw, pear, Oriental pear, quince), grapes, stone fruit (apricot, cherry [sweet and tart], nectarine, peach, and tree nut crops (almond, Brazil nut, butternut, cashew, beechnut, chestnut, Chinquapin, pecan, pistachio, filbert, hickory nut, Macadamia nut, and walnut.

Dose

Spriodeclofen is applied as foliar spray application (ground only) @0.16 to 0.53 lbs a.i./acre.

Insect Range

The foliar-applied acaricide is active against a range of sucking insects on fruit and nuts for example-mites; Pear sucker; Scale insects; Earwigs; Aphids; Whitefly.

Toxicological Profile

- **Mammalian toxicity:** Spirodiclofen exhibits low acute toxicity to rats when administered orally, dermally, or through inhalation. It has low acute toxicity of LD50: >2500 mg/kg bw/d to mammals. The Spirodiclofen toxicity exposure ratio (TER) for small herbivorous mammals (as indicator species) in orchards is estimated to be >147. The NOEC (No Observed Effect Concentration) during the chronic toxicity test was about 29.6 mg/kg bw/d i. e. 29.6 mg/kg bw/d is the highest concentration of the substance that did not cause any observed toxic effects when chronically tested. TER is estimated to be 6.1. Spirodiclofen was not found to be genotoxic (in vitro and in vivo tests) in mammalian or microbial systems.
- **If farm animals fed on fodder and forage crops:** A study showed that no residues were found in milk and tissues when lactating cows were dosed at 1.29 ppm dry feed. Due to this, no residues are expected to be present in tissues and milk at the mean and maximum calculated dietary burden of 0.74 ppm. As a result, an estimated maximum residue level (MRL) for spirodiclofen of 0.004() mg/kg for milk, 0.01() mg/kg for meat from mammals other than marine mammals, and 0.05(*) mg/kg for mammalian edible offal is considered valid.
- **Toxicity to Bees:** Spirodiclofen demonstrates low toxicity to bees both through contact (LD50: >200 μg/bee) and oral exposure (LD50: >196 μg/bee). Hazard quotients for oral exposure (Qho) and contact (Qhc) are estimated to be <0.73 and <0.72, respectively.
- **Non-target arthropods:** In a laboratory acute contact toxicity study, Envidor showed 100% effects on predatory mites. Extended lab studies showed effects above the trigger effect level of 50%. The Hazard Quotient based on the lowest LR50 for Typhlodromos pyri was estimated to be 60 (in-field). In two separate field studies, a substantial decrease in the abundance of Typhlodromus pyri was observed after 27 and 89 days, respectively.

- **Earthworms:** It shows low acute toxicity to earthworms (LC50: >1000 mg/kg d.w. soil). TER is estimated to be 10417. This value does not exceed the trigger (<10).
- **Birds:** It shows low acute toxicity towards the birds with LD50: >2000 mg/kg bw/d and its TER (toxicity exposure ratio; TER = Toxicity / Exposure) is estimated as >257 for the small insectivorous bird in orchards (as the indicator species). Similarly, the moderate dietary toxicity is LC50: >1061 mg/kg bw/d) and TER for the indicator species in orchards is estimated as >244. The NOEC (No Observed Effect Concentration) during the chronic toxicity test is about 51 mg/kg bw/d.
- **Fish and Aquatic Invertebrates:** It shows moderate acute toxicity to Rainbow trout with LC50: >58.3 mg a.s./L) at 96HAE. All TER calculations for spirodiclofen pass the EU triggers (Acute: 100, Chronic: 10) with a buffer zone of 20 meters.
- **Invertebrates:** It is low acute toxicity to Daphnia magna (48h EC50: >100 mg as./L). All TER calculations for spirodiclofen pass the EU triggers (Acute: 100, Chronic: 10) with a buffer zone of 30 meters.

Environmental Fate

- **Degradation in soil:** Spirodiclofen exhibits a moderate to high degradation rate, with a DT50 (half-life) ranging from 1 to 13 days and a geometric mean DT50 of 5.4 days. The DT90 (time required for 90% degradation) ranges from 3.8 to 44 days, with a geometric mean of 18.4 days.
- **Degradation in water:** At 20 °C, the hydrolysis of spirodiclofen in sterile aqueous buffer solutions yields different DT50 values: 120 days at pH 4, 52 days at pH 7, and 2.5 days at pH 9. The hydrolytic stability of spirodiclofen decreases with an increase in pH. However, photolysis is not a significant degradation pathway for spirodiclofen.

In aerobic water-sediment studies with a sediment content of 2.1-2.6%, the substance demonstrates low CO2 formation rates, indicating that it is not readily biodegradable.

- **Fate in the air:** Under field conditions, the volatilization of spirodiclofen from both soil and leaf surfaces was found to be negligible, with a DT50air (half-life in the air) of approximately 2.7 hours.
- **Health effects:** It is not an eye or dermal irritant, but it is a potential skin sensitizer (U S Environmental Protection Agency, 2005). Chronic exposure to spirodeclofen is very toxic to aquatic life with long-lasting effects (National Institute for Public Health and the Environment, 2015).

Regulatory Frameworks Governing Pesticide Use and the Need for Sustainable and Integrated Pest Management Practices

Pesticides play a crucial role in modern agriculture by protecting crops from pests and diseases. The use of pesticides in agriculture has brought many benefits to the society, but the inappropriate use of agrochemicals raises concerns about potential adverse effects on human health, the environment, and non-target organisms. For this reason, many countries have established monitoring programs and legal regulations to control the use of pesticides on crops. Additionally, there is an increasing recognition of the need for sustainable and integrated pest management (IPM) practices to minimize reliance on pesticides.

Regulatory frameworks governing pesticide use serve as guidelines and rules to ensure that pesticides are used safely and responsibly. These frameworks vary across countries but generally involve the registration and licensing of pesticides, setting maximum residue limits (MRLs) in food and feed, establishing safe use practices, and monitoring and enforcement of compliance. These regulations aim to protect human health, prevent environmental contamination, and ensure the sustainable use of pesticides. Despite regulatory efforts, concerns remain about the potential risks associated with pesticide use harming beneficial insects, birds, and other non-target organisms. Pesticide residues in food and water have also raised concerns regarding food safety. In response to these concerns, there has been a growing emphasis on sustainable and integrated pest management practices. It entails embracing a comprehensive approach that takes into account the entire agro-ecosystem, encompassing crop rotation, implementation of resistant crop varieties, habitat manipulation, and utilization of biological control methods. By promoting biodiversity, enhancing natural pest control mechanisms, and minimizing pesticide applications, sustainable pest management practices help mitigate the negative impacts associated with conventional pesticide use.

By adopting these practices, farmers can reduce pesticide dependency, preserve beneficial insects and natural enemies, protect biodiversity, and maintain the health and productivity of agro-ecosystems. Furthermore, sustainable pest management contributes to the long-term sustainability and resilience of agricultural systems, helping to address.

DEFINITION

1. An active ingredient is a biologically active component of a pesticide formulation responsible for its toxicity (phytotoxicity for herbicides) or it is the ability to control the target pest.
2. The acceptable daily intake (ADI) is an estimate of the amount of a substance in food or drinking water that can be consumed daily over a lifetime without presenting an appreciable risk to health.
3. Insect growth regulators (IGRs) are insecticides that mimic hormones in young insects. They disrupt how insects grow and reproduce. IGRs can control many types of insects including fleas, cockroaches, and mosquitos.
4. LD50 (Lethal Dose 50%) is a statistically derived dose which is expected to kill 50% of test animals when administered as a single exposure. For inhalation toxicity, LC50 (Lethal Concentration 50%) is used, it is the amount of a substance suspended in the air required to kills 50% of a test animals:
    a. LD50: mg/kg bw. mg/kg bw/d stands for mg of substance per kg of body weight administered per day.
    b. LC50: mg/L. The estimated air concentration of a substance administered via inhalation route is typically expressed as mg/L. Alternatively, ppm may be used as a substitute for mg/L in some cases.
5. A maximum residue level (MRL) The maximum concentration of a pesticide residue permitted in or on food or feed of plant and animal origin, which is legally acceptable when a plant protection product (PPP) is used correctly in accordance with good agricultural practice.
6. Pesticides are chemical compounds that are used to kill pests, including insects, rodents, fungi and unwanted plants (weeds).

7. Acute toxicity of a pesticide refers to the effects from a single dose or repeated exposure over a short time.
8. Chronic toxicity refers to the effects of long-term or repeated lower level exposures to a toxic substance over a long period of time.

REFERENCES

Agency for Toxic Substance and Disease Registry. (2014). *Medical Management Guidelines for Chlordane.* Retrieved from https://wwwn.cdc.gov/TSP/MMG/MMGDetails.aspx?mmgid=349&toxid=62

Amanda, A. F. (2021, February). Acute and Chronic Toxicity of Indoxacarb in Two Populations of *Plutella xylostella* (Lepidoptera: Plutellidae). *Journal of Economic Entomology, 114*(1), 298–306. doi:10.1093/jee/toaa260 PMID:33179745

Kaur Sandeep, C. S. (2016). Indoxacarb interaction alters immunotoxic and genotoxic potential of endotoxin. *Jpurnal of Pesticide Science, 41*(3), 65–70. doi:10.1584/jpestics.D16-012

Lambré, K. L. (2015). *FLONICAMID.* JMPR. WHO. Retrieved from https://apps.who.int/pesticide-residues-jmpr-database/Document/242

Matthews, G. (2001). The Pesticide Manual. Crop Protection -. *Crop Protection (Guildford, Surrey), 20*(6), 539–539. doi:10.1016/S0261-2194(01)00019-9

Merec. (2022). *safety data sheet Indoxacarb.* Retrieved from https://www.merck.com/docs/product/safety-data-sheets/ah-sds /Indoxacarb%20Formulation_AH_CA_EN.pdf

National Center for Biotechnology Information. (2023). *PubChem Compound Summary for CID 107720, Indoxacarb.* Retrieved June 14, 2023 from https://pubchem.ncbi.nlm.nih.gov/compound/Indoxacarb

National Institute for Public Health and the Environment. (2015). *CLH report for spirodiclofen.* Retrieved from https://echa.europa.eu/documents/10162/41b26907-3ca7-4059-84f2-4694826f3598

Office of Environmental Health Hazard Assessment. (2011). *CIC Consultation: Flonicamid.* Retrieved from https://oehha.ca.gov/media/downloads/crnr/101211flonicamidcic.pdf

Pesticide and You. (2012). *Chemical Watch Factsheet.* Retrieved 2023, from https://www.beyondpesticides.org/assets/media/documents/pesticides/factsheets/Indoxacarb.pdf

Pesticide Directory. (n.d.). *Adverse Effects Flonicamid.* Retrieved from https://www.fluoridealert.org/wp-content/pesticides/epage.flonicamid.effects.htm

Qureshi, I.Z., Bibi, A., Shahid, S., & Ghazanfar, M. (2016). Exposure to sub-acute doses of fipronil and buprofezin in combination or alone induces biochemical, hematological, histopathological and genotoxic damage in common carp (Cyprinus carpio L.). *Aquat Toxicol., 179*, 103-14. doi:10.1016/j.aquatox.2016.08.012

Rehab, A., & Azouz, H. O.-Z. (2020). Buprofezin toxication implicates health hazards in Nile tilapia (Oreochromis niloticus). *Aquaculture Research*. doi:10.1111/are.14884

S., L. E. (2009). *Potential Health Effects of pesticides*. Pesticide Education Program in Penn State's College of Agricultural Sciences. Retrieved 2023, from https://extension.psu.edu/potential-health-effects-of-pesticides

US Environmental Protection Agency. (2005). *Pesticide fact sheet*. Retrieved from https://www3.epa.gov/pesticides/chem_search/reg_actions/registration/fs_PC-124871_11-Aug-05.pdf

Chapter 3
Cloud in Digital Government:
Problems and Perspectives in the Case of Azerbaijan

Farhad Yusifov
https://orcid.org/0000-0002-9114-9972
Institute of Information Technology, Baku, Azerbaijan

ABSTRACT

The development and widespread use of cloud technologies promises a number of economic prospects at the local, regional, and global levels. Cloud technologies have numerous advantages in terms of management, storage, and exchange of information resources in the virtual space for the public and private sector, citizens, and general users. The chapter provides an overview of the research conducted in the field of cloud technologies in Azerbaijan. The results show that the studies are mostly devoted to the infrastructure issues of cloud technologies, to the research of the conceptual bases of the creation of the government cloud, to the issues of security and personal data protection, as well as to the implementation of measures for the development of the economy of cloud technologies. The international experience in the field of government cloud was studied, and the realization of this concept and the results to be achieved were analysed. Necessary measures for the formation and development of the cloud technologies economy at the government level have been indicated.

INTRODUCTION

The rapid development of the Internet has led to the emergence of new services. The Internet creates an environment of interpersonal communication and satisfies people's information needs (Ali et al., 2021). Recently, the Internet has begun to meet people's computing needs. Due to the technology called "cloud computing", access to software, databases, and other services is obtained through various devices. Cloud computing is able to further expand the capabilities of information technologies, eliminating a number of technical and economic barriers. Eliminating those barriers will enable millions of specialists to make

DOI: 10.4018/979-8-3693-0200-2.ch003

new contributions to the information technology market, and billions of users to benefit from these opportunities (Gourisaria et al., 2020).

According to experts' predictions, cloud technologies will have a significant impact on society and the economy as one of the most important turning technologies in the next 20 years. The rapid growth of data processing, storage, and transmission capabilities lays the foundation for the economics of cloud technologies.

The development and widespread use of cloud technologies promises a number of economic prospects at the macro and micro levels. Therefore, cloud services are considered one of the main, prospective segments of the information economy and its leading direction, the Internet economy. That is why a number of leading countries of the world allocate a special place to cloud technologies in their national strategies and programs for the development of the information society, information and communications technology (ICT), information economy, or the Internet (Mahmudov, 2014; Government Cloud Concept, 2019; Ali et al., 2021).

Azerbaijan's perspective of the telecommunications sector until 2025 consists of strengthening the country's telecommunications and information technology infrastructure, expanding the services provided in this field, and encouraging all sides to use these services. As a continuation of this vision, one of the main priorities is to become a leading technology-based country in the region as a result of the wide application of services used in everyday life in the period after 2025. For this purpose, it is planned to develop high-speed broadband infrastructure with wide coverage in the country. To achieve this goal, it is planned to achieve the following three strategic goals (ICT Roadmap, 2016):

- Improvement of management structures and strengthening of ICT.
- Increasing productivity and operational efficiency in the activities of business entities.
- Digitization of government and social environment.

At the same time, the regulatory and legal framework will be improved in order to develop communication links, form a stable telecommunication infrastructure, expand the use of the Internet and encourage the participation of the private sector. For this purpose, the regulator will manage issues such as transparent and efficient national frequency allocation, effective management of the national frequency spectrum, effective organization of relations between market subjects and further increase of the competitive environment, stimulation of investments in the telecommunications sector, control of service quality and protection of consumers' interests. the establishment of the institution will be considered.

ICT is considered an important component of a developed economy with high performance. The ICT sector includes a wide range of manufacturing and service sectors, including the telecommunications industry. By creating conditions for communication and information exchange between business subjects and consumers, ICT has a supporting role in the development of society, government and economy.

The large-scale application of ICT has a significant impact on the creation of new jobs and productivity growth. Especially in developing countries, the application of ICT has become an important factor in driving growth in GDP (ICT Roadmap, 2016). ICT has a multiplier effect in creating new jobs. International experience shows that each new job created in ICT stimulates the creation of 2-4 new jobs in other sectors of the economy. Wages for new jobs created in this sector are higher than the average wage. It should be noted that there is a noticeable increase in productivity in industries where ICT is applied, and SMEs that use the most advanced technologies develop faster.

Governments can use the possibilities of ICT in many areas. In this regard, various initiatives have been taken by a number of states in order to create new sources of income, improve the efficiency of services, the safety and social welfare of citizens. For example, the use of effective electronic governance mechanisms can reduce the "shadow economy" and tax evasion. Online services can increase the efficiency and speed of transactions, and eliminating the human factor in simple transactions can reduce corruption. By establishing an effective monitoring system, the level of national security can be increased through activities such as increasing traffic safety and reducing crime. ICT initiatives also lead to increased social and financial inclusion.

Telecommunication infrastructure and technology, which are important components of the ICT sector, are of particular importance for business activity. Thus, the telecommunications industry offers a platform for mobile transactions, as well as electronic payments and other electronic transactions. At the same time, the telecommunications infrastructure and industry allow to improve economic activity indicators through new business models. Given that the ICT industry is constantly renewing and innovating, it can be noted that this sector has great potential to further increase its benefits.

Innovations in the ICT sector include the following three trends: data abundance, ease of use, and increased processing capabilities. More than 6 trillion gigabytes (GB). The "Internet of Things", big data, and social networks with wide coverage allow increasing the abundance of information even more. The increase in the level of interaction with users, the introduction of innovative "form-factors" and the "consumer-oriented" IT system makes the use of information even easier. Advances in "cloud" computing and "machine learning" are increasing the possibilities of data processing.

The telecommunications sector consists of four main actors with different roles: service providers, consumers, operators and the government. As the telecommunications industry is an important component of the ICT sector architecture, innovations and trends in communication technologies also affect the dynamics and key players of this sector.

One of the main factors is the supporting tools that expand the application of communication technologies (for example, increasing the level of accessibility of products and services). Over the past decade, there has been a rapid increase in internet usage, as well as the level of internet access through mobile devices. This trend is expected to continue in the future.

In the context of business-to-business interactions, services that enhance Internet connectivity are more accepted. Such value-added services include machine-to-machine communications, cloud services, enterprise mobility, unified communications platforms, and other services. Digital innovations are constantly changing purchasing patterns and preferred services in the context of the relationship between companies and consumers. Thus, multifunctional services are widespread and have begun to replace traditional mobile services. For example, increased interest in the use of IP telephony ("VoIP") communication services has begun to reduce consumer demand for traditional data and SMS services and, consequently, the revenues of this sector.

These trends also directly affect the participants in the telecommunications sector, and as a result of declining revenues, the fixed income of telecommunications operators is declining globally. As for fixed revenues, these revenues have surpassed voice service revenues and this trend is expected to continue (ICT Roadmap, 2016).

Taking into account the increasing volume of information in our daily life activities, there is a need to find technological solutions along with new methods for the management of Big data and their useful use. All this is of great importance against the background of the increase of data not only in terms of volume but also in terms of variety (for example, audio and video).

More advanced analytics enables the analysis and processing of Big data in real-time, and this is considered a significant step forward, differentiating the data processing process from the current model based on the analysis of past events. In the future, predictive analysis will become more important, which will enable more accurate and future-oriented decision-making in the public and private sectors. All this will create more opportunities to take preventive measures based on the forecast.

As one of the main participants in the telecommunications sector, government plays a major role in the development of communication technologies and in providing important supporting factors for the successful application of technologies, with different levels of participation. These factors include infrastructure, talent and skills, products and services measures that drive the development of the telecommunications industry. Governments use various financing instruments that facilitate infrastructure development, either by directly investing in shares or indirectly by applying incentive mechanisms for private enterprises. To support infrastructure development, states are focusing on improving access to fixed broadband and expanding mobile network coverage. Ensuring the social benefits of broadband has become an important factor in countries' strategies, and this infrastructure is constantly being expanded. For this purpose, different alternatives for broadband network implementation are considered. These alternatives include infrastructure-based competition, the appointment of a publicly funded operator, and the creation of a joint venture with a national fiber-optic communications company. In this sense, states have expanded the deployment of 4G technology through spectrum allocation and coverage commitments for broadband. In addition, measures have been taken to improve the telecommunications infrastructure and technology in order to improve the operational base, which is of particular importance for electronic payments and business activities (ICT Roadmap, 2016).

BACKGROUND

In international practice, the use of "cloud computing" technology is considered a radical change in the activities of organizations, leading to the deployment of applications and data in the "cloud" from the traditional storage of data on a personal computer desk or physical server (Gourisaria et al. 2020; Ali et al., 2020).

Since 2000, the research conducted in the field of information technology service or equipment purchase, information security, energy consumption and reduction of personnel costs has led to the use of computer resources from a single source of grid computing technology (grid computing) and introduced hosting services. However, since the use of these technologies is not sufficiently effective in meeting the growing needs of organizations, the possibility of outsourcing computing services and the provision of services by data centers on an industrial scale has created the foundation for the emergence of cloud technology.

Cloud technology provides remote access to multiple computers and servers, applications hosted on them, software platforms or infrastructure from any type of device.

According to the calculations of the European Commission, as a result of the participation of interested parties from all sectors in "cloud" solutions, the cost allocated by organizations to ICT will decrease to 10-20 percent. At the same time, "cloud" technology provides access to mobile work mode, rapid productivity, international standardization, new management and business opportunities, and additional markets for the public and private sectors (Government Cloud Concept, 2019).

"Cloud computing" technology has numerous advantages in terms of management, storage and exchange of information resources in the virtual space for the public and private sector, citizens, general users. Most importantly, it frees users from technological and technical issues, as well as the necessary requirements and costs of retaining professionals. "Cloud computing" technologies enable service providers to attract as much technical capacity as needed to their information systems and resources, to the electronic services provided through them, without incurring additional costs or wasting time (Government Cloud Concept, 2019).

As a result of the activities carried out in the country in the field of e-government infrastructure formation, the use of information systems and resources of government institutions is expanding, information exchange is ensured between these systems, and the provision of electronic services based on them is increasing. In order to meet the demand, to develop the e-government infrastructure, it is necessary to obtain operational, uninterrupted access to the information resources of government institutions, as well as to optimize the costs required for the creation and provision of electronic services. Such factors determine the formation of the Government Cloud (G-Cloud), the establishment of a stable, secure and efficient ICT infrastructure in the country and the exploitation of information systems of government bodies through it. In the case of Azerbaijan, the top-level plan for the development (scaling) and migration of the G-Cloud and the transition plan is as follows (as shown in Figure 1).

Figure 1. Development of g-cloud and top-level planning

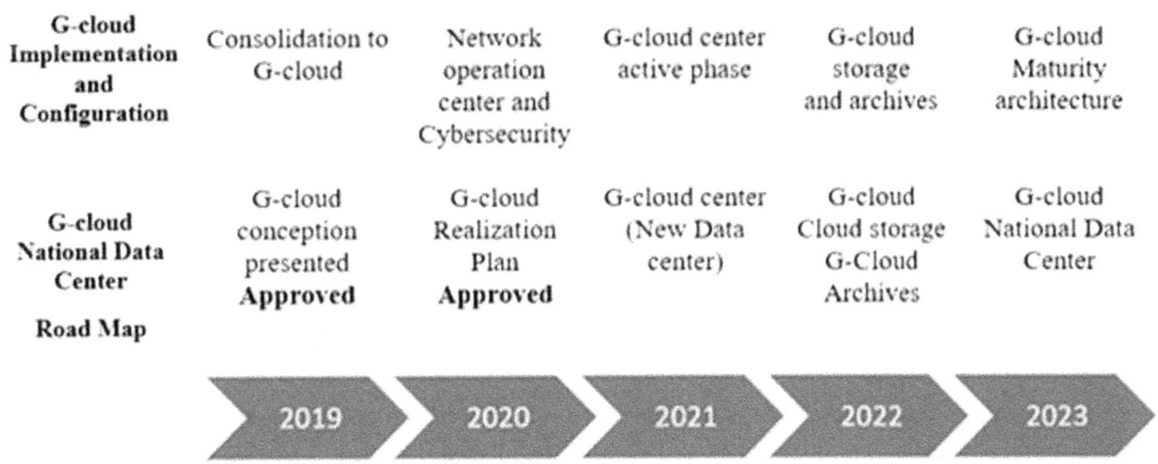

G-Cloud provides the basis for stable, flexible, secure and optimized cost-effective operation of the information systems owned by government bodies in a single data center infrastructure that meets international standards, ensures the use of cloud services and allows information to be obtained from any location is cloud technology.

With the application of this technology, the government information systems are kept safe, formed and managed centrally in a sustainable infrastructure, and the data center that meets international standards is the national information space of the government. G-Cloud in terms of more efficient use of ICT and its advantages in public administration, improvement of existing infrastructure management, and increase of the quality and security of services provided through it:

Table 1. SWOT analysis of telecommunications and information technologies

Strengths	Weaknesses
● non-residential areas for deployment of telecommunication devices and equipment, telephone sewerage facilities for laying fiber-optic cables, power supply, etc. availability of basic infrastructure; ● availability of fiber-optic cable infrastructure laid to homes by cable television and other internet providers in Baku and other big cities; ● availability of broadband services provided by fiber-optic cable and modern corporate networks within the institution, such as the offices of government bodies and large business institutions; ● favorable geographic location for East-West international transit traffic.	● High determination of tariffs for the use of public communication network resources by telecommunication operators to private providers operating in the field of ICT; ● organization of broadband internet services mainly with old copper communication lines and low service quality. The average internet speed of a user is about two times less than the world average; ● the fact that the software for the used equipment is the private property of the manufacturer creates dependence on network security issues; ● availability of physically and morally obsolete communication equipment; ● non-effective use of frequency resources; ● lack of sufficient ICT literacy.
Opportunities	**Threats**
● large number of young people in the age range who can be potential users of telecommunication products; ● proximity to large regional markets, including a tendency to global markets; ● Existence of the necessary regulatory and legal framework in the field of ICT.	● presence of strong competition in access to foreign financial markets; ● growing challenges of network security and information protection in modern technology.

- Ensuring the operation of flexible ICT systems that support the government policy and strategy and meet the necessary requirements;
- Development of e-government and transition to “digital government”;
- Ensuring a high level of quality during the operation of government information systems and the provision of electronic services, increasing stability, and increasing user satisfaction;
- Acceleration of mutual integration of government information systems and resources;
- Collecting data on public services and conducting analyzes for the purpose of improving the quality of services;
- Organization of services and applications based on “open data” (Open Data);
- Support of innovative solutions based on “Artificial Intelligence” (AI) and robot technologies;
- Effective provision of activity needs;
- Application of advanced and high technologies in public administration;
- Cost prevention during the creation and use of information systems;
- Will contribute to the optimization of the demand for personnel resources and the improvement of the qualification level of personnel.

Given the limited resources, public administration reforms should focus on effective centralization of functions through consolidation and standardization of ICT capabilities. SWOT analysis of telecommunications and information technologies in Azerbaijan are shown in Table 1.

The long-term vision in the ICT sector for the period up to 2025 is to contribute to the development of the country by continuing innovation-oriented reforms in this sector, increasing the range and volume of modern services, expanding the scope of the fast broadband network, ensuring the application of standards that meet important international requirements in the telecommunications sector, and the technical development of society. is to achieve an increase in the level of literacy and the formation of a digital economy. As part of this goal, the implementation of mechanisms related to the construction of strong infrastructure in Azerbaijan by 2025 will be supported.

In this technology-based development process, standards will be improved in Azerbaijan, and conditions will be created to attract large-scale investments to the telecommunications industry while maintaining a more coordinated function of the government. Until this period, the competitive environment in the country's telecommunication sector, the level of public participation will be further increased, and these goals will be supported by an improved regulatory and legal framework. Increasing the integration of information services to key sectors and creating a favorable environment for optimized operations will increase the interest of both local and foreign investors in Azerbaijan.

It is planned to continue this development process in Azerbaijan in the period after 2025. In the long-term perspective, in order to ensure the implementation of the target vision of the digital economy, the level of innovation-oriented technological literacy will be increased, special attention will be paid to the training of talented students in vocational education institutions and universities. In order to increase technological literacy and develop an innovative way of thinking, lessons on programming will be included in the educational programs of universities and relevant vocational schools.

The emergence of the information society will determine additional innovative goals. In order to stimulate the development of startups and innovative enterprises, the creation of clusters and high-tech parks will be considered. In this case, the relevant investments of the state and other interested parties will play the role of the main supporting force. In order to attract more innovation-promoting start-ups to such parks and innovation centers, their application will be implemented in a wider scope (regions). This approach will enable the enterprises in those parks to become the main driving force for Azerbaijan's transition to the digital economy. Also, the information society will be encouraged and instructed to use electronic services more through the implementation of relevant plans. Thus, the development of the general business environment will be achieved as a result of technological literacy, an increase in the application of high-tech tools and access to a quality broadband network.

In the period after 2025, the integration of advanced digital apps into daily life will expand in Azerbaijan. At a time when most government documents and processes are already digitized, the transition to paperless documentation in the country will be encouraged in the long term. This target vision envisages a digital banking system where transactions are paperless and electronic services are widespread. In Azerbaijan, it is planned to create a complete digital banking experience in which various stakeholders interact, including banks, digital payment systems, mobile wallets, online shopping platforms and even social networks. A strong banking and communication infrastructure will strengthen the possibility of e-commerce in various fields from retail to services during this period. Also, all the processes in the country's healthcare sector will be digitized on the basis of the e-health services established up to this period, and the database of patients' medical records will be expanded.

During this period, a transition to the use of more powerful digital solutions will be ensured in the country's internal operations, and for this purpose, the digitization trend in the format of large-scale data will be used as the main tool for the optimization of operations conducted in both the public and private sectors. Companies will invest in this concept to better understand customer requirements and use the results in targeted marketing activities. Also, these enterprises will use "cloud" computing technologies for data storage and processing. Wireless communication operators will offer solutions based on cloud computing technologies for their customers.

In the Strategic Roadmap are the following measures to be taken in the field of developing Cloud Technologies (ICT Roadmap, 2016):

- *Supporting local infrastructure development*

Will determine the appropriate support mechanisms for the formation and development of the necessary infrastructure related to the transition to the digital economy. These types of support mechanisms include giving certain incentives to the private sector for developing infrastructure and software, developing a national cloud computing system or localizing key software resources, encouraging the use of those software, and using ICT on a large scale. acceleration of use and stimulation of recruitment of IT specialists will be relevant. When the support mechanisms are prepared, there will be close cooperation with high technology parks, their infrastructure model will be used, including the successful experiences already achieved in this direction in Azerbaijan.

- *Creation of G-cloud*

Through the creation of G-Cloud, actions on the operation of information systems of government bodies, construction of infrastructure and formation of standards will be coordinated.

- *Ensuring mutual recognition of electronic signature, including "Asan Imza" (eng. easy signature) at the international level*

Reducing costs and increasing coordination in public administration with the creation of the G-cloud.

- *Improvement of government IT systems*

Possibilities of creating a centralized infrastructure for government bodies will be investigated and implemented. Necessary steps will be taken for government bodies and private institutions to use the technical capabilities of the existing Data Center, including proposals for regulatory and legal acts. Application of Cloud technologies and interaction of information systems will be ensured.

RELATED WORKS

In recent years, cloud computing and its implementing different platforms the core of information economy and innovations, has emerged as a new computing paradigm with significant benefits (Liang et al., 2019). In the study provided a systematic literature review of data governance and cloud data governance (Al-Ruithe et al., 2019). The systematic literature review covered published contributions from both academia and industry. State of the art of data governance for cloud and non-cloud have been discuses in this study. In research provided a systematic literature review of the adoption of Cloud Computing in electronic government (e-government) (Abied et al., 2022). In the research concentrates on the adoption of cloud computing in governments, the benefits, models, and methodologies utilized, and the analysis techniques. This study has critically peer-reviewed articles that concentrate on cloud computing for e-government. It describes a systematic evaluation of the empirical studies focusing on cloud adoption studies in e-governments. One of the most recent surveys was on the effect of cloud computing on the sustainability of government services (Mohammed et al., 2020). In this study, the authors identify the gap in incorporating cloud computing as a platform for establishing sustainable services. Results show that research on the role of cloud computing in ensuring provision of sustainable e-government services still scarce. The study provided a systematic review of the benefits and risks of cloud computing in e-government tasks (Assaf et al., 2021). The study is devoted to analyzing the efficiency and optimization of government service resource allocation in a cloud computing environment (Guo et al., 2023). The study analyzed the patterns and evolutionary trends of government service resource allocation efficiency in each region. The studies observed that the G-Cloud cannot be implemented without the support of government service resources (e.g., infrastructure construction, support software, application systems, and information resources) and cloud computing resources (Wyld, 2010; Ivanus & Iovan, 2014; Liang et al., 2017). In study proposed a hybrid model for adopting cloud computing in e-government (Ali et al., 2018). This research demonstrates e-government's obstacles and cloud computing features. The proposed model determines how the entity connects to each of three (Local Governmental, Regional Governmental and Wide Governmental) clouds and what the relation between them is. In study identifies and explores the critical factors associated with information security requirements of cloud services within the Australian regional local government context (Ali et al., 2020). There is proposed a conceptual cloud computing security requirements model, in order to promote a balanced view on cloud security for governments. Using this model, governments can work together to demand uniform security requirements for adopting cloud services. This study investigates the implementation of cloud computing in both a practical setting and from an organisational user perspective via three UK local government authorities (Jones et al., 2019). Government organizations have turned to cloud-based services to reduce their total investment in IT infrastructure and resources (e.g. data centres) and take advantage of the various benefits of cloud computing technology. This study seeks to investigate the cloud computing opportunities, challenges, and service models for e-government in the Middle East (Almutairi & Thuwaini, 2015). As shown in the study, many governments in Middle East have identified the significance of IT in improving and sustaining their economy. The facts gathered are analyzed and a comparison is done between Europe and Middle East with regards to the use of cloud computing for e-government. The study assesses the outcomes related to Korean government policies to foster cloud industry growth in the public sector and efforts to mitigate associated risks related to privacy and security (Lee et al., 2020). In research proposed a strategic framework to adopt cloud-based solutions in the public sector to improve e-government processes in developing countries (Al-Maatouk et al., 2018). The framework was formulated based on the collected data analysis and the conclusions from experts' interviews. This study provides detailed guidelines to a successful launch and implementation of cloud-based solutions for e-government initiatives in the public sector. The results of the literature review show that different aspects of cloud technologies and their implementation in the public sector were analysed, studies concentrate on the adoption of cloud computing in governments, conceptual approaches, models, and methodologies, and the analysis techniques. In Table 2. summarizes the details of the related work by providing a comparative survey that seeks to key contributions and outlines some of the features or issues.

In international practice, there are different approaches to G-cloud in different countries.

In Great Britain, G-Cloud is applied in the "pay as you use" model, and the information resources of public institutions in transport, education, health, internal affairs, defense and other fields are placed in a single data center infrastructure through "cloud computing" technology, as well as performed provision of public procurement (Government Cloud Strategy, 2011; G-Cloud framework, 2013). In the ICT strategy prepared by the Cabinet Ministers of the United Kingdom in 2011, the reduction of public expenditure and better provision of electronic services in e-government were considered as the main goals. In that strategy, the government's view of this field, including the implementation of the G-Cloud was defined

Table 2. A comparative survey of the related work

References	Key Contribution						
	Literature Review	**Concept Approaches**	**G-Cloud**	**E-Services**	**Models**	**Case Study**	**Empirical Study**
Zwattendorfer & Tauber (2013)		+	+		+	+	+
Ivanus & Iovan (2014)		+	+		+	+	
Liang et al. (2017)		+	+		+	+	+
Ali et al. (2018)		+	+	+	+		
Jones et al. (2019)	+	+	+		+	+	+
Liang et al. (2019)		+	+			+	+
Assaf et al. (2021)	+	+	+		+		
Gourisaria et al. (2020)	+	+			+		
Al-Ruithe et al. (2019)	+	+	+				+
Abied et al. (2022)	+	+	+		+		+
Guo et al. (2023)		+	+	+	+		+
Mohammed et al. (2020)	+	+	+	+			
Lee et al. (2020)		+	+	+		+	+
Al-Maatouk et al. (2018)		+	+				+
Ali et al. (2021)	+	+	+	+	+		+

as one of the main directions, and the road map covering the years 2011-2016 was approved (Government Cloud Strategy, 2011). As a result of a joint project between the Cabinet Ministers of the United Kingdom and the Ark Data Center, the government's Crown Hosting Data Center has started to operate. In this infrastructure, public services are provided on the basis of cloud technology. For example, the Office for National Statistics has left its data center and moved its entire infrastructure to the Crown Hosting Data Center. In accordance with the agreement between the government and service providers within the framework of the G-Cloud (G-cloud framework), the services of the G-Cloud are currently being used in the IaaS, PaaS and SaaS service models (Government Cloud Concept, 2019).

In Ireland, public services are delivered through the G-Cloud, enabling government agencies to deliver public services to all user groups by reducing costs and increasing efficiency, simplifying inter-agency communication and cross-system integration across agencies, sharing information in a "whole of government" (WoG) approach, and new digital government services are provided to citizens and business entities in accordance with the law (Cloud Computing Advice Note, 2019). In the "Government ICT strategy" presented to the public by the Irish government in 2015 and the "Government services reform plan" and "Government services renewal plan" which are part of it, the digitization of "government-citizen-business" relations, improvement of government ICT management, provision of target working groups with knowledge and skills on information technologies applied in public services is planned (Public Service ICT Strategy, 2015).

In the United States of America, by using the full potential of the single data center infrastructure, it is planned to direct the real income to be obtained from this area to the welfare of citizens, as well as to reduce the costs of equipment and software tools by applying cloud technology, and to increase the information security of government bodies. In February 2010, the federal government launched the

Federal Data Center Consolidation Initiative (FDCCI, 2011) and developed a roadmap. In the guide program, it is planned to inventory the assets of retail data centers belonging to state organizations, to place the assets in the centralized infrastructure during 2010, and to include them in the budget proposals for the following year 2012. In this regard, the development of the "cloud" strategy and the migration of the services of state institutions and traditional systems to the "cloud" by 2015 have been set as a goal.

Evaluating the migration to the "cloud" as an initial step, the federal government decided to consider the potential costs of the government bodies related to the transition to the "cloud", to appoint authorized persons from each of them within 6 months, established the formation of a working group and the installation of a dashboard to monitor the migration as a support mechanism for the consolidation plan in the migration process of systems and software. It was also determined that government institutions and private organizations use only up to 30 percent of their server capacities in their data centers. In 2011, the government approved the "Federal Cloud Computing Strategy", prepared a transition plan to the G-Cloud, eliminated 800 retail data centers belonging to government agencies as a result of migration, and reduced the state's IT costs by a quarter (Government Cloud Concept, 2019; Federal Cloud Computing Strategy, 2019).

In Norway, the "cloud" service is distinguished from traditional IT service models, it is determined that the "cloud" service can reduce the amount of capital investments necessary for innovation, start-up activities, it is determined that there is no need for large investments for equipment or software licenses, as well as services provided by companies that establish their activities through the Internet. in terms of the increase in the volume of customer requests, this is considered an innovative solution. In 2016, the government developed a Cloud Computing Strategy for Norway (Cloud Computing strategy for Norway, 2016). In the strategy, in order to achieve economic efficiency, the use of "cloud" services as an alternative during the operation of ICT systems for government bodies and private companies, the use of cloud services in order to achieve cost optimization in the provision of digital services and state ICT systems was considered important. ICT principles such as compliance with technical standards defined by the Norwegian Agency for Public Administration and E-Government, flexibility in government and business processes, and scalability in operations from the point of view of adaptation to changes, have also been applied to the cloud (Cloud Computing strategy for Norway (2016); Government Cloud Concept, 2019).

In Estonia, the application of modern ICT solutions has been recognized as a suitable option for increasing efficiency in public administration, improving the welfare of citizens, as well as "cloud" technology for ensuring digital continuity. "Estonia - Digital Agenda for 2020", prepared by the government in 2013, mentioned cloud technology as one of the trends affecting the provision of continuous integration of government information systems and government portals (Digital Agenda 2020 For Estonia, 2013). Ensuring the stability of the state information systems of the Republic of Estonia in the event of natural disasters, cyber-attacks and even war and proactive protection of state information were considered fundamental in this agenda. One of the measures envisaged in this Agenda for 7 years until 2020 is to store and protect digital copies of all information and state information systems and resources that are critical for the daily and future activities of the government in "data embassies" located in other countries. For this purpose, it is planned to open the first "data embassy" of Estonia in the data center located in Betzdorf, Luxembourg (Yusifov & Farajova, 2021). Continuing the work started by the consortium in the form of public-private cooperation in Estonia, the G-Cloud managed by the State Information Communication Fund under the Ministry of Communications and Economic Affairs currently provides cloud services for public institutions (Government Cloud Concept, 2019; Yusifov & Farajova, 2021). As a continuation of the Agenda 2020, Estonia's Digital Agenda 2030 includes a vision and an action plan concerning the development of the Estonian economy, state and society with the help of digital technology in the next decade (Estonia's Digital Agenda 2030, 2011).

In Turkey, as a result of merging the data centers of government institutions, a single data center is being built, and IT infrastructure is being developed for the provision of e-government services.

Establishment of a single data center as a result of merging the data centers of goverment institutions by the Supreme Council of Science and Technology of the Republic of Turkey and "Cloud Computing" document published in 2013 provides for the creation of a "cloud" information platform in the goals to be fulfilled by 2023 related to "cloud" solutions.

The eleventh development plan (2019-2023) adopted in 2019 envisages the creation of industrial cloud platforms, the expansion of cloud services, the use of cloud computing in the public sector, and the development of technological infrastructure. Also, one of the main priorities in the National Artificial Intelligence Strategy (2021-2025) document is the development of cloud technologies by the Digital Transformation Office of the Presidency of the Republic of Turkey and the Ministry of Industry and Technology. One of the main goals is to create at least 10 sectoral cloud platforms for data exchange.

According to the information of the European Union Agency for Information Security and Network (ENISA) in 2015, the countries of the European Union are currently implementing various projects on the application of "cloud computing" technology in e- government" (Government Cloud Concept, 2019).

RELATED WORKS IN THE FIELD OF CLOUD TECHNOLOGIES IN AZERBAIJAN

Today, the rapid implementation of various innovations in the field of information technology continues. One of them is cloud technologies. The role of cloud technologies in providing services to the public and private sector is seriously increasing due to the significant expansion of capabilities in the field of data processing, transmission and storage (Government Cloud Concept, 2019).

According to the definition proposed jointly by the International Telecommunication Union and the International Organization for Standardization in 2013, cloud technologies are a paradigm that provides access to physical or virtual resources that can be used and adjusted independently, if necessary, network availability to an expanding and flexible resource base (Cloud Computing, ITU). According to experts' predictions, cloud technologies will have a significant impact on society and economy as one of the most important turning technologies in the next 20 years (Information Economy Report 2013). The rapid growth of data processing, storage and transmission capabilities lays the foundation for the economics of cloud technologies.

The development and widespread use of cloud technologies promises a number of economic prospects at the macro and micro levels. Therefore, cloud services are considered one of the main, prospective segments of the information economy and its leading direction, the Internet economy. That is why a number of leading countries of the world allocate a special place to cloud technologies in their national strategies and programs for the development of the information society, ICT, information economy, or the Internet.

However, in addition to the advantages of cloud technologies, there are also certain problems. These problems are primarily related to the high cost of cloud technologies, information security and legal regulation issues. In addition, the development, application and expansion of the scope of cloud technologies necessitates the definition and implementation of a scientifically based state policy in this area. The research work shows the nature, emergence and development stages of cloud technologies (Mahmudov, 2014). The classification and forms of cloud services are explained. Statistics and forecasts related to the pace of development of cloud technologies are analysed. The economic advantages and disadvantages of the relevant technologies and the problems of legal regulation are investigated. Proposals are made regarding the goals of the state policy related to the development of the cloud technologies economy.

Trends in the development of cloud technologies are currently accompanied by a radical change in the interaction between telecommunications, business and the public sphere. Due to these changes, great opportunities are created in the field of information processing, storage and transmission. For example, Intel's current 22-nanometer processors are 4,000 times faster than those released in 1971. Between 1986 and 2007, global "technological storage" doubled every three years. In 1993, when the Internet browser was created, the maximum speed of Internet access was 56 Kbit/s, but in 2013, Internet services with a speed of 2 Gbit/s are already offered on the consumer market. This is 36 thousand times more than the indicator in 1993 (Information Economy Report 2013; Mahmudov, 2014). Currently, large multinational companies that provide cloud services have hundreds of thousands of servers with huge databases in different parts of the world. Cloud technologies affect most segments of the ICT sector to one degree or another. The demand for broadband Internet services is driving the growth of telecommunication services revenues. However, revenues from telephone communication services may decrease due to the increase of users working on the basis of cloud technologies with the help of voice Internet communication protocols (Mahmudov, 2014). Currently, the cloud technology economy is dominated by several large cloud service providers, all of which are American companies (IBM, Microsoft, Google, HP, AT&T, etc.). Those companies are the first to enter the cloud services market and gain an advantage. The amount of investment necessary to create a large cloud service platform is quite large: the creation of a cluster of data centers requires more than half a billion dollars. A number of features of cloud technologies are quite attractive from an economic point of view. Whether they are start-ups or small firms, or large business structures, the economic opportunities of cloud technologies attract them. According to estimates, companies' costs related to the creation and operation of their data centers make up about half of their total costs (Yakusheva, 2012). Therefore, it is more profitable for business organizations to use the services provided by these technologies. In the research work (Reese, 2011; Yakusheva, 2012; Williams, 2012; OECD Internet Economy Outlook 2012; Information Economy Report 2013; Mahmudov, 2014), the following features of the economic opportunities of cloud technologies are indicated.

Expansion (Scaling) (Mahmudov, 2014). The introduction of new products and services, the expansion of the sales channel and the increase in the number of customers require the organization's information system to keep up with the increasing load and process a large amount of data. Flexible and reliable operation increases customer satisfaction. Scalable software add-ons allow you to sustain a large load due to the increase in the number of copies released simultaneously. As is well known, the simultaneous release of a large number of copies uses typical equipment, which lowers the overall cost of the corresponding costs and facilitates the accompanying infrastructure.

1. *Flexibility*: Flexible response to changes in business conditions is one of the main conditions of successful business. For example, changing market conditions and competitor activity may require rapid introduction of new products or services. At this time, it is necessary to implement the full cycle of planning, designing and development of the information system. Flexibility allows you to increase the capacity of the infrastructure without the need for an initial investment in hardware and software. Flexibility is due to expandable software add-ons. Thus, it solves the issue of instantaneous change of the number of computing resources intended for the operation of the information system.

2. *Multitenancy:* Multi-tenancy – to different user groups, organizations and user categories, etc. it is one of the methods of reducing costs due to the maximum use of common resources for providing services. Multitenancy can be more attractive to software companies. Because this method allows you to save the costs of paying for cloud platform resources and to use the available computing resources to the maximum (Mahmudov, 2014).
3. *Pay per use:* Paying for the resources used is one of the features of cloud technologies that allow to direct part of the capital costs to operational costs. As soon as the necessary resources are obtained, the costs related to the operation of the organization's information system can be optimized. By applying the multitenancy method, resources can be divided between different users, reducing costs even more. Flexibility allows to flexibly increase or decrease the amount of resources, that is, to adapt the costs to the actual needs of the organization.
4. *Self-service:* In modern conditions, it is required to expand or modify information systems in order to bring a new product or service to the market flexibly. Traditionally, the selection, purchase and installation of equipment for changes in information systems were solved in advance. Solving these cases took a long time. Self-service allows the user to obtain the required resources in a few minutes (Reese, 2011; Yakusheva, 2012; Williams, 2012; OECD Internet Economy Outlook 2012; Information Economy Report 2013).
5. *Instant access to resources:* According to Frank Gens, vice president of the well-known US company International Data Corporation (IDC), which is engaged in researching the IT market, these companies eagerly apply cloud technologies not only because of their economic advantages, but also because of their convenience. While IT operations take days and weeks in traditional data centers, these tasks are performed in a few minutes thanks to cloud technologies. For example, let's say that a company's new pilot project requires the commissioning of several new servers. Previously, it was necessary to buy equipment and place it in a data center, which took weeks and months. When using cloud technologies, it is possible to simply activate the required number of virtual servers and start working (Yakusheva, 2012; Williams, 2012; OECD Internet Economy Outlook 2012; Information Economy Report 2013).
6. *Resource-intensive software tools:* Before the advent of cloud technologies, resource-intensive software tools were mainly available to large companies. But with the advent of cloud technologies, even small firms have the opportunity to solve resource-intensive computing problems. Of course, the prospects for corporate websites are not so important, but when it comes to business analytics systems that require the processing of large data sets, the commercial prospects of cloud technologies should be taken more seriously. In addition, cloud technologies allow to "democratize" the processing of large multimedia information arrays (Information Economy Report 2013).
7. *Removal of obstacles related to creativity of programmers:* Although cloud technologies are more of a commercial rather than a technical nature, it is precisely here that greater opportunities for programmers to innovate arise. Thus, cloud technologies allow us to get rid of many tedious tasks that took a lot of time and created problems in the past. For example, every programmer knows how many difficulties it is necessary to overcome to move the software from the experimental laboratory to production. In the conditions of cloud technologies, the configuration of virtual machines is the same when it is developed and applied in the production process. This allows you to avoid problems related to differences in hardware and software. In such conditions, programmers can focus all their attention on creative issues. This allows for more productive and efficient work in less time and with less resources (Williams, 2012; OECD Internet Economy Outlook 2012).

In the study, the current situation, problems and security issues of the cloud-based e-government system were investigated (Alakbarov, 2021). In recent times, citizens and government institutions widely use the services of the e-government system. This requires the establishment of the e-government system of the government on the basis of new information technologies. On the other hand, the prices of information and telecommunication equipment used by organizations to expand their network infrastructure are rising and demand for additional costs is created. At the same time, the rapid increase in the number of electronic services (e-service) in the e-government system leads to the expansion and complexity of information exchange between citizens and government institutions. The e-government model on the basis of cloud technologies creates great opportunities for the country's economic and socio-political development, raising the standard of living and the development of democracy. In connection with the mentioned, the study proposed to use cloud technologies for effective management of the e-government system, providing users with quality service and increasing the efficiency of mixed communication between citizen-government structures. The evaluation of the e-government system, the purpose of the e-government, problems arising in the use of the existing e-government, the concept, models and services of cloud technologies were analyzed and studied in the research work. At the same time, the advantages, problems and security issues created by the cloud-based e-government system for citizens and government institutions have been widely analyzed (Alakbarov, 2021).

Cloud technologies can also be used to increase collaboration between different organizations within the government. The distribution of cloud technologies among central authorities and organizations leads to a reduction in infrastructure costs. With the widespread use of cloud technologies, it is possible to achieve transparency in the activities of state bodies (Danielsen et al. 2019; Alakbarov, 2021). In this sector, cloud technologies have great potential, not only for the government, but also for millions of people. The application of cloud technologies in the e-government system helps to perform the following tasks in a better quality (Mohammed et al., 2017):

- improving and increasing the number of provided government services;
- obtaining information and establishing contact with the appropriate organizations of the government, expanding the opportunities of citizens;
- achieving more transparent and broad accountability of the government;
- ensure a higher level of data security.

The application of cloud technologies creates benefits such as convenience of service, cost reduction, integration of services, appropriate security and reliability. With these features, its use in e-government can be the best alternative. The cloud-based e-government system created on the basis of cloud computing technologies has the following advantages (Rastogi, 2010; Takabi et al. 2010; Carlin & Curran, 2011; Ttripathi & Parihar, 2011; Vats, Sharma & Rathee, 2012; Alessandro, Barbara & Alberto, 2017; Mohammed et al., 2017):

- **Accessibility.** the ability to use e-services for all users from any place with Internet, using any computer. Users can switch from regular PCs and laptops to more compact and convenient netbooks. When data is stored on cloud servers, users can access it anytime and anywhere. If there is an Internet, there is no such thing as a forgotten file anymore, they are always in memory. When data is stored on cloud servers, their copy is automatically stored on other servers, sometimes located on different continents. When a personal computer is stolen, the user does not lose the information he needs, he can get it again because it is stored on any other cloud server. Thus, these mentioned ones play an important role in the success of citizens in using e-government systems. E-government system created on the basis of cloud technologies provides 24/7 services to citizens. It is always available online and citizens have the opportunity to use services anytime and anywhere (Alakbarov, 2021).

- **Cost savings.** In a cloud-based e-government system, public organizations and government agencies do not need to purchase and install ICT equipment, unlike in a traditional e-government system, and they can obtain these resources by renting them from the cloud. This frees organizations from additional costs. The cost of ICT services for public organizations and citizens is reduced in a cloud-based e-government system, as they rent ICT resources and services. Cloud-based e-government system reduces government costs and frees organizations from purchasing very expensive systems and hiring professional staff by changing investment costs to operating costs (Mohammed, et al. 2017).
- **Efficiency.** Ensuring the efficient use of e-services for citizens and enterprises is one of the main features of the e-government system. The cloud-based e-government system facilitates the delivery of e-services to citizens. Also, in such a system, it is possible to create new solutions that are technically and economically efficient without using cloud services. Service models of cloud computing technologies provide efficient service to companies and government institutions (Mohammed, et al. 2017; Danielsen et al. 2019; Alakbarov, 2021).
- **Fast scaling**. Users can automatically access any amount of cloud computing resources at any time. The virtual resources of the system are unlimited, scalable and provide users with the required amount of resources. Citizens are one of the main stakeholders in the e-government system. Over time, more citizens are involved in the use of e-government services, and as a result, the gap between supply and demand increases. To bridge the gap between supply and demand, the system should be more flexible and scalable. Cloud computing technologies are considered to be a scalable technology because they can provide the necessary resources in a short time according to the dynamically growing number of users.
- **Conducting an audit.** The cloud-based system has the ability to quickly process and analyze Big data, which allows you to detect all kinds of fraud. Employees of government organizations can use IT to carry out corruption through e-services. To ensure the security of the government against corruption, the audit process should be carried out from time to time. This system allows the creation of new applications that provide protection mechanisms to help increase security and transparency (Mohammed, et al. 2017).
- **Accident recovery.** Natural disasters such as floods, earthquakes, war and civil unrest can cause e-government system malfunctions and major problems. Keeping these in mind, full backups and recovery solutions of system data should be available in different geographical locations. Applications and data need to be replicated in multiple data centers, so that data can be moved from one data center to another data center as needed, so that damaged data can be recovered as quickly as possible. A cloud-based e-government system allows you to back up and restore data in the cloud. The mentioned system performs the migration of applications more conveniently than in a traditional data center. Cloud-based systems restore organizations' data faster and more effectively than the traditional model during a disaster.
- **Policy management.** Using the e-government system will support the correct establishment of political relations between the government and citizens. Cloud technologies help government and

citizens to connect with each other more easily. This, in turn, supports the government to conduct the policy correctly.

- **Green technologies.** The use of numerous ICT equipment and data processing centers in the public sector has a negative impact on the ecosystem. Thus, the operation of these systems requires high energy consumption, which in itself causes the release of high amounts of harmful gases (carbon dioxide, etc.) into the air. Cloud computing systems reduce energy consumption by 80% compared to other technologies (Zwattendorfer & Tauber, 2013). Thus, in a cloud-based e-government system, as citizens use virtual resources, energy consumption is reduced, which is a more favorable option for protecting the ecosystem.

Ensuring data security in the cloud-based e-government system is considered one of the urgent issues. Security issues have been investigated in a number of research studies. Let's look at the issues of data security in the cloud-based e-government system. Illegal access to customer's personal data by staff of cloud providers is considered one of the potential threats to cloud data. The cloud user must ensure that all their data stored in the cloud is protected. Providers should create confidence in users to use the e-government system. Let's look at the problems and security issues that arise in the cloud-based e-government system (Mohammed, et al. 2017).

- **Privacy.** When placing citizens' data in the cloud, it should be guaranteed that access to that data will be limited to only those authorized by the provider. Unauthorized access to user data by cloud personnel is a potential risk to individuals' cloud data. Customers should be assured and appropriate regulations should be applied. Privacy policies and procedures should ensure the security of users' data on cloud servers. Cloud users must be sure that the data stored in the cloud is securely protected. In order to protect the confidentiality of users, providers should offer procedures and schemes that ensure the implementation of user control and security. In fact, third parties are responsible for the storage and use of data in their systems. As they store data on servers located in different locations, they are responsible for protecting the privacy of citizens' personal data (Mohammed, et al. 2017).
- **Lack of user control.** Users trust the system less when they don't have control over their data. In cloud-based e-government, user data is stored in third-party data centers where citizens cannot control it. This system allows cloud computing providers to access private information of citizens. In such a situation, the protection of intellectual property and personal information is very important (Alakbarov, 2021).
- **System rejections.** Malfunctions in the system affect the efficient use of e-services, which reduces users' confidence in the reliability of the system. Some e-services should be available 24/7 for citizens, but sometimes these services are not available for citizens. Also, data loss or data security breach cannot be compensated by the service provider.
- **Security.** Security ensures availability and completeness of data in cloud computing systems. Organization of the security system at a high level plays an important role in trusting cloud-based e-government systems. In the context of cloud-based e-government, data security must be ensured at a high level (Mohammed, et al. 2017; Danielsen et al. 2019; Alakbarov, 2021).
- **Self-service on demand.** Self-service on demand is one of the main features of the cloud computing system. This feature is accessed through a virtual environment or management. On-demand self-service is an interface available to all cloud service users. Users can assign and simultaneous-

ly change the virtual machines they need according to demand without using the provider's interaction service. Providers must protect the security of this cloud service from unauthorized access, and this access must be accessible only to a small number of authorized persons (Mohammed, et al. 2017).

- **Data leakage.** Data leakage also affects citizens' and public's trust in cloud-based e-government system. The e-government system stores confidential information about users and enterprises, therefore ensuring data security is one of the important conditions. Data leakage creates distrust of citizens in using the cloud-based government system.

In addition to the problems mentioned above, the dependence of the user's data storage on cloud service providers, the reliability of communication channels, failure to develop standards guaranteeing security, the fact that the user's computer is constantly connected to the Internet, the speed of the communication channel is high, etc. such problems affect the effective use of cloud-based e-government system (Alakbarov, 2021).

In the study, the security problems of cloud technologies were analyzed (Aliguliyev & Abdullayeva, 2013). In the process of conducting the research, the reasons for the emergence of cloud technologies, the main concepts, characteristics, service models, and application models were studied. Various security issues such as identity management, web applications, virtualization, trust management relevant for the security of cloud technologies have been investigated, the main problems in this field have been identified and a number of recommendations for their solution have been given.

In the conducted research, the security issues of cloud-based SCADA (Supervisory Control and Data Acquisition) systems, which are widely used in the monitoring and management of the oil and gas industry, were considered (Alakbarov & Hashimov, 2020). Traditional SCADA systems are very expensive, inflexible, difficult to scale, so there are many problems in data collection, transmission and processing. Moving SCADA system applications to the cloud provides opportunities to reduce costs, improve scalability, manage the system efficiently, increase reliability, and solve resource-related problems. The research shows the existing loopholes that can hinder the security of cloud-based SCADA systems. Taking security measures and assessing risks in cloud-based SCADA systems is one of the urgent issues. For this purpose, the security problems and risks arising in the use of cloud-based SCADA systems have been analyzed and suggestions have been made to help solve them to some extent.

The study analyzed the problems (resource scarcity and energy consumption) in solving problems that require large computing and memory resources in mobile devices, and the widespread use of mobile computing cloud (MCC) technologies was noted to overcome these problems (Alakbarov, 2021). Different architectural-technological principles of mobile clouds are analyzed and their advantages are shown. The models, services, features and advantages of computing clouds are shown. The current state of the MCC, opportunities, emerging problems and suggestions and recommendations for their solution are given. At the same time, security and privacy issues were analyzed in MCC. It was proposed to use cloudlet-based MCC, which provides opportunities for mobile users to solve problems that require large computing resources. At the same time, the architectural-technological principles of cloudlet-based hierarchically structured MCC were developed and its advantages were shown. It is noted that the proposed model satisfies the demand for computing and memory resources of mobile devices, increases the transmission capacity of communication channels in the network, eliminates delays and ensures security. The research work is devoted to the issues of using cloud technologies in the e-government system (Aliguliyev et al., 2015).

The methods of using cloud technologies in the electronic state and the prospects of the population using the services of the e-government system created on the basis of this technology more efficiently were analyzed. The benefits of cloud technologies service such as convenience, cost efficiency, integration of services, appropriate security and reliability factors are mentioned to be the best alternative for its use in e-government.

According to international expert assessments, over the next 10 years, as a result of the development of digital technologies and their application to various economic sectors, 7 economic sectors of a completely new quality will gain greater importance and importance and attract more investments.

These seven economic and high-tech areas are (Abbasov & Ibrahimov, 2019; Digital Transformation of Industries):

1. Production of "smart" things and systems based on artificial intelligence technologies;
2. Autonomous unmanned vehicles and other moving devices (unmanned boats, military vehicles, drones, etc.);
3. Large-scale data (BigData) and cloud technologies;
4. Industry of personal orders based on individual characteristics and 3D printing technologies;
5. Internet of Things and 5G technologies;
6. Development and mass production of robots and drones;
7. Development of social media, social networks and communication platforms industry.

In the study, by examining the existing technical and logical capabilities of the national e-signature infrastructure related to electronic signature (e-signature), the optimal management of the load on the system and the investigation of problems were determined (Imamverdiyev & Abbasov, 2021). Since e-signature provides the role of an identity card for identification of citizens in the electronic environment, providing access to informative and interactive electronic services formed in the e-government ecosystem and safe use of these services is carried out by means of e-signature. In the study, the problems of improving the national e-signature infrastructure in terms of new technological challenges, increasing mobility opportunities, ensuring high productivity in devices with limited resources, and high security requirements for wide use of e-services are analyzed and actual scientific and practical issues are determined. Research directions were analyzed for modeling the work of the centers for the assessment of the security of the national infrastructure components and the trust chain by examining the international experience in the direction of solving the identified problems. In the study, the effectiveness of the national e-signature infrastructure as a security component in the service areas of the centers is determined by identifying its own security problems (Imamverdiyev & Abbasov, 2021).

The conducted research is devoted to the analysis of the problems that arise during the transition of the oil and gas industry to Industry 4.0 technologies (Imamverdiyev, 2021). Oil and gas companies rely on the application of Industry 4.0 technologies to maintain their competitive advantage in the face of the deterioration of the structure of hydrocarbon resources, the volatility and sharp decrease in the price of energy resources. Oil and gas companies have been using technologies such as the Internet of Things, cloud computing, and artificial intelligence to optimize various processes for a long time. The advantage of Industry 4.0 is that it also integrates existing technologies in one system and provides higher synergy based on this. The "Oil and Gas 4.0" concept, which aims to form such a unified system based on the Industry 4.0 approach, has entered the agenda of companies in the last 2-3 years. The formulation of the concept of "Oil and Gas 4.0" is still in its early stages, and while it has attracted a lot of attention, there remains a number of misunderstandings about how it will be used to add value to oil and gas operations. The aim of this work is a systematic analysis of the processes of transition to Industry 4.0 technologies in the oil and gas industry and the problems encountered at this time. The main characteristics of Industry 4.0 and the state of application of its basic technologies in the oil and gas industry are analyzed. The problems of the main components of the "Oil and Gas 4.0" concept are analyzed and appropriate recommendations are developed. Systematization, generalization, comparative analysis and systematic approach methods were used in the research. The obtained results are expected to have important applications in the development and implementation of the digital transformation strategy based on Industry 4.0 in the oil and gas industry. In the study (Alakbarov, 2019), the architectural and technological principles of mobile computing clouds created on the basis of cloud technologies were analyzed. At the same time, a comparative analysis of mobile computing cloud architectures was conducted. The advantages of mobile computing clouds with a hierarchical structure created on the basis of cloudlets are shown. It is noted that the proposed architecture more efficiently satisfies the demand for computing and memory resources of mobile equipment and eliminates network delays.

In the study, the concept of Serverless computing technology (Function as a Service - FaaS), which is the latest trend in cloud technologies was investigated (Alekbarov & Hasanli, 2020). Widespread services of cloud technologies were reported and the demand for these services was analyzed. When users use computing cloud services (IaaS, PaaS, SaaS), selection of virtual machines based on their technical capabilities, operating system startup, balanced distribution of resources, etc. are directly involved in solving many issues such as In order to eliminate the mentioned shortcomings, the use of serverless computing service was evaluated as an urgent issue. In the research, serverless and server-based architectures were also compared, platforms offering serverless computing services were studied, and various programming languages supported by them were investigated. The advantages and disadvantages of FaaS service are shown. In the study, the economic efficiency and energy-saving issues of using cloud computing were investigated (Alekbarov & Hasanli, 2020). Cloud computing combines two important aspects of a green IT approach: energy saving and efficient use of resources. In terms of energy conservation, cloud computing uses server virtualization to minimize server resources, resulting in less electricity consumption. Green cloud computing is considered an approach that companies can use to minimize energy consumption and overall carbon emissions more efficiently than they already are. According to estimates, organizations that have moved their workloads to efficient cloud solutions can save between 30% and 40% compared to those that use traditional IT infrastructures (Salimov, 2022).

CLOUD IN DIGITAL GOVERNMENT: IN THE CASE OF AZERBAIJAN

The transition to "digital government" involves the use of data analysis, smart devices, intelligent systems, automation of workplaces, and increasing efficiency through data integration for the organization of public services. Using the infrastructure capabilities of the G-Cloud in our country to ensure the transition to the "digital government" will allow achieving the desired result in this area.

Thus, advances in the field of "cloud computing" will increase the possibilities of data processing and transmission. In accordance with global development trends, "machine learning", "machine-to-machine communication" (M2M), "Artificial Intelligence" (AI), "Big Data" and "Internet of Things" (IoT) will become important elements in the life of society in Azerbaijan. These technologies will act as a basis for the organization of quality public services in the country, they will also be used as new sources of information. For example, in the long term, it is planned to replace all utility meters in Azerbaijan with digital meters, which will allow for improvements in consumption control and payment of utility fees, as well as in distribution through the analysis of the obtained data.

The implementation of the G-Cloud in the "digital government" will increase the quality of real-time access to information on government services and create a favorable environment for citizens and government institutions. Thus, the transition from "document-based" service to "data-based" service will be ensured, which makes the presence of advanced infrastructure related to the speed and quality of information exchange inevitable during the proactive organization of services on the "non-stop shop" principle. At the same time, obtaining the information required by the state only once and ensuring the "once-only" principle during work with citizens requires close integration between state information systems and resources.

Migration of existing and newly created state information systems and resources to the G-Cloud infrastructure, improvement and modernization in the migration process will play a special role in order to accelerate the process of transition to digital government.

"Cloud computing" technology can provide high-quality coordination of public services in obtaining information in a secure and authorized manner. This infrastructure can be used not only for information exchange, but also for conducting information-request operations for organizations' own activities, supporting decision-making, and applying analytical systems with proactive and predictive capabilities. Great importance is attached to data analysis and making data-based decisions in the organization of public services in digital government.

Digital government implies the use of technologies and relations resulting from the development of ICT in the organization of qualitatively new types of public services. For example, currently, blockchain technology allows data or databases in the information infrastructure to be exchanged in a shared form, not centralized, and stored in the form of blocks in a secure mode based on cryptographic technology, as well as the history of each transaction sent or confirmed over the network as a component of those transactions. It is developing as a new technology that allows keeping information together. This technology makes it possible to apply different approaches in the provision of electronic services and to organize services with higher speed and quality. With the availability of computing capabilities, cloud technology is a favorable environment for the application of this new technology.

Also, cloud technology will be applied in the formation, storage, management and use of national spatial data, formation of metadata and provision of services. These steps to be taken will serve the formation and development of the necessary infrastructure related to the transition to digital government in the country, cloud technologies will have a significant impact on solving these issues.

Approaches to G-Cloud. The general view of the G-Cloud in the Republic of Azerbaijan is based on international experience.

The formation of the G-Cloud will serve to coordinate measures for the operation of information systems of government bodies, the construction of infrastructure and the formation of standards, the reduction of the costs of creating and maintaining information systems in state administration, and the creation of a centralized, secure, stable infrastructure for government bodies.

Necessary steps will be taken for government bodies, as well as private institutions to use the technical capabilities of the existing Data Center, including relevant proposals on normative legal acts will be prepared.

First of all, effective organization of effective use of ICT resources available in the public sector and their maintenance, increasing the productivity of these systems by reducing costs for this activity, expanding their use and introducing new services will be ensured.

As a result of the conditions and infrastructure that will be created, government bodies will be able to obtain services in accordance with demand, at reasonable prices, with reduced operating risks based on the rules of state procurement. This, in turn, will encourage the development of e-government, increase the efficiency of using ICT in public administration, strengthen electronic relations between the state, citizens and business, and the transition to digital government as the next stage in the development of e-government.

Thus, the ICT infrastructure that can support the growing demand will be developed, cloud services will be provided and improved, integration and information exchange between state information systems and resources will be strengthened.

The current Concept will play the role of a support mechanism for achieving the strategy until 2020, long-term until 2025 and after 2025 target visions for the development of the G-Cloud:

In the strategic vision until 2020, the operation of information systems applied in state administration, economic and social processes will be gradually organized on the basis of “cloud” technology, taking into account the necessity, and the transition to the G-Cloud infrastructure will be ensured;

In the long term until 2025, innovative reforms will be continued to expand the application of ICT and increase the efficiency of use, the range and volume of electronic services will be increased, the application of standards that meet international requirements will be ensured, work will be done in the field of forming a digital economy, and in the network of regional data centers Government infrastructure of the cloud and its potential will be strengthened and will serve to meet this demand. In these times, the transition to the use of digital solutions in internal operations in the country will gradually be more widely ensured, and for this purpose, the technical capabilities of the Data Center will be the main tool for processing Big data related to the optimization of operations, electronic governance and monitoring systems, both in the public sector and in the private sector;

In the target vision after 2025, in the long-term perspective, in the field of ICT, the creation of a strong potential ICT industry in the region and specialization in internationally important information transit, services, product production, including the provision of regional services based on cloud technologies will be ensured.

Also, cloud computing technology will be widely used for data storage and processing. Telecommunications operators will also offer solutions based on cloud computing technology for their customers. Despite the fact that service in the cloud is a new type of activity, it has already become a profitable field. By providing state support to the Cloud technology-based development process, conditions will be created to attract investments to the ICT industry in the Republic of Azerbaijan, which, in turn, will increase the interest of local and foreign investors in the field. The studies of the European Commission show that by 2020, as a result of the implementation of the “open cloud” in Europe, a value of 250 billion euros in GDP will be generated, at the same time, during 2015-2020, a total of 600 billion euros worth of income will be generated and 2.5 million additional jobs will be created.

The operator of G-Cloud will ensure the organization of the activity of G-Cloud. The operator will take measures related to the creation of the G-Cloud, the construction of its infrastructure, the organization of services, the transition of the information systems of state bodies to this platform, the effective organization of the storage of state information resources and systems in the Data Center, as well as the inventory and expertise of state information systems.

In order to organize the activities of the G-Cloud, the infrastructure and cloud services of the TIER III level Data Center commissioned in the country will be used. For this, meeting the needs of state institutions, localization of the main software resources, consolidation (centralization) of various data centers that do not meet the standards are envisaged.

In addition, cloud services will be promoted in the country by using technological solutions developed by private companies in public-private partnerships. The widespread use of "Cloud" services by business entities will be one of the main factors that create the basis for increasing efficiency and productivity in their business operations, and will also serve to increase the competitiveness of small and medium-sized businesses (SMEs) by increasing the level of use of digital technologies.

During the realization of Cloud services, the general approach available in international practice will be applied. The operator of G-Cloud will provide services based on infrastructure (IaaS), platform (PaaS), application software (SaaS) models. For their implementation, a technological environment suitable for open (public), private (private) and hybrid (hybrid) deployment models will be provided. Accordingly, the service models have the following characteristics (Government Cloud Concept, 2019):

- infrastructure service model (IaaS) – the data center provides technical, network and electrical supply, virtualization is applied to physical servers, that is, the use of virtual servers;
- platform service model (PaaS) – technical, network and electrical supply of the data center, virtual servers as well as operating systems, database services are applied (for example Oracle, MySQL);
- application software service model (SaaS) – services are offered through ready-to-use application programs along with technical, network and power supply of the data center, virtual servers, operating systems, database services (for example office programs (MS Office), Enterprise Resources Planning (ERP), electronic document circulation (DMS)).

Deployment models differ in the following features:

- open deployment model (public) – the provided services are open for public use and privacy restrictions are not applied. This model is suitable for open data provided by information systems;
- private deployment model (private) – privacy restrictions are applied to the provided services and designated for use by authorized users. This model is a suitable choice for confidential, backup or archived, critical data and inter-system confidential data exchange;
- hybrid deployment model (hybrid) – is used to provide both open and confidential information during the provision of services;
- public deployment model (community) – the provided services are presented to user groups of organizations operating in different fields and cooperating with each other for information sharing.

During the realization of Cloud services, the following existing principles will be observed:

- self-service - unilateral use of the "cloud" service by the user without interacting with any service provider, without dealing with server work and technical support issues;
- access to a wide network - providing the user with the use of services from any location and device (computer, laptop, mobile phone, tablet, terminal, etc.);
- resource deployment - deployment of information systems and resources in accordance with the user's needs;

- speed and flexibility – making relevant changes automatically and flexibly at any time according to the user's needs;
- monitoring of services - monitoring of the services provided in the "cloud", monitoring the system and providing accountability.

The G-Cloud Concept will serve the following purposes (Government Cloud Concept, 2019):

1. Selection of a unified approach for the ICT infrastructure of government information systems and organization of efficient management and use, cost optimization with the application of cloud technology;
2. Use of government information resources (information, work and services) from any location and efficiency of public-private ICT service provision;
3. Efficient use of the infrastructure of the Data Center, effective organization of government information systems and deployment in a sustainable infrastructure;
4. Provision of infrastructure and service needs of state institutions in the G-Cloud in a fast and flexible form and provision of services according to demand;
5. Expanding the use of cloud technology in the country, increasing efficiency in activities, implementing public-private cooperation;
6. Application of prices and service qualities in accordance with the principles of transparent public procurement in the provision of "cloud" services;
7. Increasing the quality of electronic services and increasing citizen satisfaction with services;
8. Continuous, safe and prompt exchange of information between state information systems and resources;
9. Concentration of information centers existing in state institutions that do not meet international standards;
10. As a result of the use of centralized services by state institutions, reducing the costs of creating their own server rooms to zero and reducing current operation and personnel costs;
11. Ensuring sustainable development, competitiveness and economic diversification in the national economy and economic sectors due to the renewal and improvement of the government's information technology infrastructure in the G-Cloud;
12. Implementation of innovations in public administration by providing public services on the "cloud computing" platform.

Since the organization of information technology infrastructure in the G-Cloud will lead to radical changes in the maintenance and management of information systems of state institutions, the transition from a traditional service to a new service based on modern cloud technology requires the preparation of state institutions for this process.

In order to ensure the transition to the G-Cloud, the operator of the G-Cloud will initially carry out the inventory and expertise of the state information systems, an assessment will be carried out together with the State Agency for Citizen Services and Social Innovations under the President of the Republic of Azerbaijan, and according to the results of the assessment, the state information systems will be implemented in place a review will be conducted and institutions' readiness for the transition will be evaluated.

Depreciation status of equipment and application programs of state institutions to be placed in the G-Cloud in the infrastructure of the Data Center, assessment of state information systems, inclusion of the obtained results in the Transition Plan, provision of communication with state institutions acting as service users, and operating costs of migration to the cloud is one of the determining factors in its implementation in stages and sequentially. In addition, the main conditions for preparation for G-Cloud are as follows:

- **security measures:** the operator ensures the security and cyber security of the G-Cloud together with the Special Communications and Information Security State Agency of the Special State Protection Service of the Republic of Azerbaijan;
- **amortization status:** review of the amortization period of the state information systems that will be transitioned to the G-Cloud in the evaluation process of the state information systems to be conducted by the State Agency for Citizen Service and Social Innovations under the President of the Republic of Azerbaijan together with the operator, based on the results obtained, information about the end of the amortization period according to the Transition Plan at the initial stage migration of systems to the G-Cloud from scratch, migration or colocation of information systems that have not expired at the next stage to the G-Cloud;
- **data migration:** ensuring data integrity when migrating outdated systems and applications to the G-Cloud in platform-as-a-service (PaaS), application-as-a-service (SaaS) models;
- **network infrastructure:** improvement of the existing communication infrastructure by the operator for their connection to the G-Cloud and Provision of connection of state institutions with telecommunication channels and internet network by the Special State Security Service of the Republic of Azerbaijan Special Communication and Information Security State Agency;
- **life cycle of equipment and software:** regular adaptation of the equipment and application programs that will ensure the functioning of the G-Cloud to the requirements of the time;
- **maintenance costs:** taking into account in the state budget the costs required for the maintenance and operation of the information systems of state institutions in the G-Cloud.

At the same time, attention will be paid to the following points to ensure the reliability and safe use of data in the "cloud" environment and to minimize migration risks as shown in Table 3.

The Strategic Roadmap for the development of telecommunications and information technologies in the Republic of Azerbaijan is aimed at increasing the development potential of information and communication technologies in the country, improving the ICT infrastructure for the digitization of the economy, effectively meeting the demand for ICT services at the expense of internal resources, and was prepared taking into account the new challenges and opportunities facing the telecommunications sector in order to further increase the efficiency with the application of ICT in the activities of the public sector.

The Strategic Road Map includes a strategic vision and action plan until 2020, a long-term vision for the period until 2025, and a target vision for the period after 2025, covering short, medium and long-term periods. In the Strategic Road Map, 3 strategic goals and 10 priorities have been defined to achieve the goals set by 2020. The Strategic Road Map envisages the creation and development of effective mechanisms for expanding the application of innovations. Important steps have been taken in the direction of faster integration of Azerbaijan into the global electronic space, formation of new forms of social and economic activity, creation of the information and knowledge market, increase of efficiency in various fields of the economy, improvement of the quality of products and services, and development of ICT in general.

Table 3. Reliability and safe use of data in the "cloud" environment

Awareness	One of the main problems is that institutions that move to the cloud without understanding the cloud face risks. Operational and architectural problems arise if the team of specialists of the joining institution is not familiar with the characteristics of "cloud" technologies and the principles of expansion of "cloud" applications. In order to minimize these problems, measures will be taken to carry out extensive information work, to increase the qualifications of employees working in institutions, and to provide methodical support of the G-Cloud operator to institutions during the transition period and later;
Systems analysis	Due to the fact that the shortcomings of the systems to be placed in the "cloud" environment do not become a problem for the overall system, appropriate measures will be taken to investigate and reveal the problems by scanning the systems. This includes both security issues and system architecture compatibility issues related to the transition to the "cloud". In connection with these issues, measures will be taken for the G-Cloud operator to provide methodical support to institutions during the transition period and later. At the same time, measures will be taken to promptly eliminate threats and respond to possible risks in time;
Authentication	vulnerabilities in the technologies used are quite risks for the cloud. In this regard, measures will be taken to implement multi-factor authentication mechanisms, use intrusion detection systems, divide the network into segments, and apply the principle of minimum usage permissions.

The long-term vision for the period up to 2025 in the sector of information and communication technologies contributes to the country's development by continuing innovation-oriented reforms in this sector, increasing the range and volume of modern services, expanding the scope of the fast broadband network, and ensuring the implementation of standards that meet important international requirements in the telecommunications sector, to achieve the development of the country, increase the level of technical literacy of the society and the formation of the digital economy.

In addition to the above, in the period until 2025, a large part of document circulation and administrative processes in Azerbaijan (for example, state documents, provision of state services) will be carried out electronically, and as a result, by ensuring the automation and efficiency of such operations, the favorable business environment in the country will further improve.

In this technology-based development process, standards will be improved in Azerbaijan, and conditions will be created to attract large-scale investments to the telecommunications industry while maintaining a more coordinated function of the state. Until this period, the competitive environment in the country's telecommunication sector, the level of public participation will be further increased, and these goals will be supported by an improved regulatory and legal framework. Increasing the integration of information services to key sectors and creating a favorable environment for optimized operations will increase the interest of both local and foreign investors in Azerbaijan.

As expected results of the implementation of the G-cloud Concept are mainly the following:

- Establishment of G-Cloud in Azerbaijan in a TIER III level Data Center, which is a centralized infrastructure;
- Construction of the National Data Center, expansion and strengthening of the infrastructure of the G-Cloud;
- transition and migration of government information systems and resources to the G-Cloud;
- Increasing the level of digitalization of e-services;
- Using the possibilities of G-Cloud in digital government;
- making electronic and mobile services of government institutions accessible to citizens;
- organization of services in G-Cloud for analysis of government big data based on cloud computing;
- organization of open data services and mobile applications in the G-Cloud;

- Preparation of the technological environment for the construction and commissioning of service platforms in the G-Cloud with the application of artificial intelligence, distributed registry technology and other new technologies;
- Preparation of national standards for information security, service level agreement and data usage in the G-Cloud;
- Transformation of the Data Center into the country's national information space and the G-Cloud into the country's regional information center.

It should be noted that a number of public and private companies in Azerbaijan currently offer cloud services. "AZCLOUD" is the cloud provider that created the first Data Center with TIER III, ISO 20000, ISO 22301 and ISO 27001 certificates in Azerbaijan and the South Caucasus region (AzCloud computing). It was commissioned by "AzInTelecom" LLC in December 2016. "AzInTelecom" LLC, operating under the Ministry of Digital Development and Transport of the Republic of Azerbaijan, has 4 main service areas. "AzInTelecom" LLC, which aims to provide safe, reliable and adaptable services to those who prefer quality in the digital world, has formed its company strategy around fundamental values such as human value, reliability, innovation, flexibility, transparency and privacy protection. AzInTelecom Data Centers have the largest availability zone in the South Caucasus region, including Azerbaijan. Baku Data Center owned by the company provides large-scale, multifaceted cloud such as "Cloud Infrastructure", "Colocation", "AZCLOUD Drive", "Archiving", "Backup", "Microsoft" and "Oracle" products, "Virtual IP Telephony" to the public and private sector. provides its services. AzScienceNet scientific computer network provides institutes and organizations of the Azerbaijan National Academy of Sciences with modern network services necessary for the implementation of scientific research, scientific-practical and educational issues. The creation of the AzScienceNet network integrates the scientific research and educational environment of Azerbaijan into the European scientific research and educational space, allowing the use of technologies and services in accordance with world standards in our country, and the more efficient functioning of our researchers in this environment. The AzScienceNet network offers services such as AzCloud, Owncloud and Cloud Storage (AzScienceNet cloud service). Also, the AzScienceNet network plays the role of an experimental base in the direction of personnel training in the field of network technologies and relevant scientific research works conducted by the Institute of Information Technologies. Mobile cloud services provided by Azercell company can be shown (Azercell). "MicroTech" LLC provides services such as the application of new technologies, cloud solutions, development of various business solutions, automation of production and various types of management processes, integration solutions between various technologies, etc. in the ICT market of Azerbaijan. Also, the MicroTech company provides consulting services for the proper implementation and management of IT and telecommunication solutions in business processes to clients of various orientations (MicroTech).

DISCUSSIONS AND CONCLUSION

Currently, providing certain incentives to the private sector for the development of infrastructure and software, developing a national "cloud computing" system, as well as localization of key software resources, promoting the use of such software, and large-scale use of ICT it is intended to accelerate the development and stimulate the recruitment of IT specialists. When developing support mechanisms, close cooperation with high-tech parks, their infrastructure model, as well as successful experiences already achieved in this direction in Azerbaijan are used. In order to improve the electronic services of state institutions, actions on the operation of information systems of state institutions, construction of infrastructure and formation of standards will be coordinated through the creation of G-Cloud. For this purpose, development of G-cloud aims to reduce costs and increase coordination in public administration. State institutions will further develop functional compatibility between different sectors, and it is planned to ensure the functioning of separate information systems as a single system.

In the long-term view until 2025, the range and number of electronic services will be increased, the implementation of standards that meet international requirements will be ensured, and the infrastructure and potential of the G-Cloud will be strengthened in the network of regional data centers.

The overview of the scientific and research works conducted in the field of cloud technologies in Azerbaijan shows that researches are mainly carried out in the following directions:

- Security issues of cloud technologies;
- G-Cloud:
- Use of Cloud technologies in Industry;
- Mobile cloud technologies and security;
- Cloud-based SCADA systems;
- Economics of cloud technologies;
- Cloud technologies in education;
- FaaS platforms;
- Cloud-based digital signature.

The analysis of the scientific research conducted in the field of cloud technologies in Azerbaijan shows that there have been no studies on the factors affecting the use of cloud technologies by IT workers. Researches are mostly devoted to infrastructure issues of cloud technologies, researching the conceptual foundations of creating a G-Cloud, issues of security and personal data protection, as well as the implementation of measures for the development of the economy of cloud technologies.

It is appropriate to implement the following measures for the formation and development of the economy of cloud technologies at the government level (Mahmudov, 2014; Information Economy Report 2013) (as shown in follow):

The main reason for the rapid development of cloud technologies is their economic viability, availability, and ease of use. Business organizations benefiting from the possibilities of these technologies can significantly save their financial resources. In the modern economy, information acts as one of the main production factors for any type of business activity. In this regard, it should be considered that the opportunities provided by cloud technologies are aimed at meeting one of the main needs of modern business structures. However, the application of these technologies is accompanied by a number of problems. There are problems, the solution of which is the direct responsibility of the enterprises applying cloud technologies. These are primarily related to economic, technical and security issues within the enterprise. The tasks assigned to the state should be carried out in the context of national security, legal regulation and macroeconomic policy.

Table 4. Actions for the formation and development of the economy of cloud technologies at the government level

Assessment of the country's readiness to apply cloud technologies.	First of all, the government should analyze the current situation in the country, reveal the weaknesses that need to be eliminated for the successful application of cloud technologies, and determine which issues are more appropriate to solve based on cloud technologies.
Preparation of a national strategy for the development of cloud technologies.	Based on the assessment of the readiness level, a national strategy for the development of cloud technologies can be developed. It can be in the form of a separate document, it can be considered as part of the country's ICT development strategy.
Solving infrastructure issues.	This includes measures to improve the reliability and availability of broadband Internet infrastructure, as well as regular monitoring of the corresponding service quality. Effective regulatory norms of the telecommunications sector play an important role in this matter. Uninterrupted electricity supply is also one of the important issues.
Solving regulatory and legal problems.	Solving regulatory and legal problems related to the application of cloud technologies, ensuring reliable protection of users' interests. This includes key issues such as data mining, electronic transactions and cybercrime. It is also important to take into account advanced international practice in the preparation of relevant legislation.
Analysis of possible directions of development within the framework of the economy of cloud technologies.	It is important to pay attention to three main areas: the development of national data centers, the potential for the integration of services in the base of cloud technologies, and the development of new cloud services.
Human resources supply.	It is important to develop the software necessary for the transition to cloud services and their integration, to carry out the management and organizational work required to reorganize and establish business processes, as well as to train specialists related to legal and logistics issues. Consideration of factors influencing the use of cloud technologies by IT professionals.
Provision of public institutions with cloud services.	Given the role of government agencies in the Internet economy, it is important to form and develop a G-Cloud.

ACKNOWLEDGMENT

This research was conducted within the framework of the cooperation project TR & AZ For CC 120N283. I would like to express my gratitude to my colleague Dr. Sulduz Salimov from the Institute of Economics for his cooperation and support.

REFERENCES

Abbasov, A., & Ibrahimov, O. (2019). *Digitalization of the world economy and social life: Global trends, achievements, and challenges to the national interests and growth perspectives of Azerbaijan.* https://mpra.ub.uni-muenchen.de/105176/

Abied, O., Ibrahim, O., & Kamal, S. N. (2022). Adoption of Cloud Computing in E-Government: A Systematic Literature Review. *Pertanika Journal of Science & Technology, 30*(1), 655–689. Advance online publication. doi:10.47836/pjst.30.1.36

Al-Maatouk, Q., Othman, M., Rana, M., & Al-Rahmi, W. (2018). A cloud based framework for e-government implementation in developing countries. *IACSIT International Journal of Engineering and Technology*, *7*(4), 3018–3021. doi:10.14419/ijet.v7i4.21532

Al-Ruithe, M., Benkhelifa, E., & Hameed, K. (2019). A systematic literature review of data governance and cloud data governance. *Personal and Ubiquitous Computing*, *23*(5-6), 839–859. doi:10.100700779-017-1104-3

Alakbarov, O. (2019). Research of architectural-technological principles of mobile cloud computing. *Problems of Information Technology*, *2*, 92–104.

Alakbarov, R.G. (2021). Cloud-based electronic government system: current situation, problems and security issues. *Information Society Problems*, *1*, 18–31.

Alakbarov, R.G. & Hashimov, M.A. (2020) Security issues in cloud-based SCADA systems. *Information Technology Problems, 2*, 3–12.

Alekbarov, R.G. (2021). Mobile cloud technologies: current situation, problems and security issues. *Information Technology Problems, 1*, 15–32.

Alekbarov, R.G. & Hasanli, A.Sh. (2020). Serverless computing service in cloud technologies. *Information Technology Problems, 2*, 72–79.

Alessandro, A., Barbara, R., & Alberto, P. (2017). E-government and cloud: Security implementation for services. *Proceedings of the 4th International Conference on eDemocracy & eGovernment (ICEDEG)*, 79-85.

Ali, K. E., Mazen, S. A., & Hassanein, E. E. (2018). A proposed hybrid model for adopting cloud computing in e-government. *Future Computing and Informatics Journal*, *3*(2), 286–295. doi:10.1016/j.fcij.2018.09.001

Ali, O., Shrestha, A., Chatfield, A., & Murray, P. (2020). Assessing information security risks in the cloud: A case study of Australian local government authorities. *Government Information Quarterly*, *37*(1), 101419. Advance online publication. doi:10.1016/j.giq.2019.101419

Ali, O., Shrestha, A., Osmanaj, V., & Muhammed, S. (2021). Cloud computing technology adoption: An evaluation of key factors in local governments. *Information Technology & People*, *34*(2), 666–703. doi:10.1108/ITP-03-2019-0119

Aliguliyev, R.M. & Abdullayeva, F.C. (2013). Research and analysis of cloud technology security issues. *Information Technology Issues, 1*(7), 3-14.

Aliguliyev, R.M., Alakbarov, R.G., & Hashimov, M.A. (2015). Issues of application of cloud technologies in the electronic government system. *Problems of Information Society*, *1*, 3-11.

Almutairi, N. N., & Thuwaini, Sh. F. (2015). Cloud Computing Uses for E-Government in the Middle East Region Opportunities and Challenges. *International Journal of Business and Management, 10*(4), 60–69. doi:10.5539/ijbm.v10n4p60

Assaf, A., Hamsir, A.W.I., & Muhammad, M. (2021). Benefits and Risks of Cloud Computing in E-Government Tasks: A Systematic Review. *E3S Web of Conferences, 328.* doi:10.1051/e3sconf/202132804005

AzCloud computing. (n.d.). www.azcloud.az

Azercell. (n.d.). www.azercell.com/az/corporate/cloud-services.html

AzScienceNet cloud service. (n.d.). https://azsciencenet.az/az/azcloud

Carlin, S., & Curran, K. (2011). Cloud computing security. *International Journal of Ambient Computing and Intelligence*, *3*(1), 14–19. doi:10.4018/jaci.2011010102

Cloud ComputingI. T. U. (n.d.). www.itu.int

Cloud Computing strategy for Norway. (2016). www.regjeringen.no

Danielsen, F., Flak, L. S., & Ronzhyn, A. (2019). Cloud Computing in eGovernment: Benefits and Challenges. *Proceedings of the Thirteenth International Conference on Digital Society and eGovernments (ICDS)*, 71-77. www.weforum.org/reports/digital-transformation-of-industries

Estonia's Digital Agenda 2030. (2021). www.mkm.ee

Federal Cloud Computing Strategy. (2019). https://cloud.cio.gov

Federal Data Center Consolidation Initiative. (2011). *Data Center Consolidation Plan & Progress Report.* https://www.dhs.gov/

G-Cloud framework, Guidance Applying to the G-Cloud framework. (2013). *How to apply to sell cloud hosting, software and support to public sector buyers.* https://www.gov.uk/guidance/g-cloud-suppliers-guide

Gourisaria, M. K., Samanta, A., Saha, A., Patra, S. S., & Khilar, P. M. (2020). An Extensive Review on Cloud Computing. Data Engineering and Communication Technology. Advances in Intelligent Systems and Computing, 1079. doi:10.1007/978-981-15-1097-7_6

Government Cloud Concept. (2019). *Decree of the President of the Republic of Azerbaijan on the Concept of "Government Cloud" (G-cloud), creation of "Government Cloud" (G-cloud) and provision of "cloud" services.* https://e-qanun.az/framework/42560

Government Cloud Strategy. (2011). https://service.gov.uk

Guo, Y., Yin, Q., Wang, Y., Xu, J., & Zhu, L. (2023). Efficiency and optimization of government service resource allocation in a cloud computing environment. *Journal of Cloud Computing (Heidelberg, Germany)*, *12*(1), 18. doi:10.118613677-023-00400-2 PMID:36789367

Imamverdiyev, Y.N. (2021). Oil and Gas 4.0 existing solutions, prospects and problems. *Information Society Problems*, *1*, 32–42.

Imamverdiyev, Y.N., & Abbasov, H.H. (2021). Current scientific-research problems of the national e-signature infrastructure. *Information Technology Problems*, *1*, 33–45.

Information Economy Report. (2013). The Cloud Economy and Developing Countries. United Nations Publication UNCTAD.

Internet, O. E. C. D. (2012). *Economy Outlook*. OECD Publishing.

Ivanus, C., & Iovan, S. (2014). Governmental cloud-part of cloud computing. *Informações Econômicas, 18*(4), 91–100. doi:10.12948/issn14531305/18.4.2014.08

Jones, S., Irani, Z., Sivarajah, U., & Love, P. E. D. (2019). Risks and rewards of cloud computing in the UK public sector: A reflection on three Organisational case studies. *Information Systems Frontiers, 21*(2), 359–382. doi:10.100710796-017-9756-0

Lee, S., Choi, Y., & Ra, J. (2020). Impact of public cloud computing service in Korean government organizations. *ICIC Express Letters, 11*(3), 313–318. doi:10.24507/icicelb.11.03.313

Liang, Y., Qi, G., Wei, K., & Chen, J. (2017). Exploring the determinant and influence mechanism of e-government cloud adoption in government agencies in China. *Government Information Quarterly, 34*(3), 481–495. doi:10.1016/j.giq.2017.06.002

Liang, Y., Qi, G., Zhang, X., & Li, G. (2019). The effects of e-Government cloud assimilation on public value creation: An empirical study of China. *Government Information Quarterly, 36*(4), 101397. Advance online publication. doi:10.1016/j.giq.2019.101397

Mahmudov, R. Sh. (2014). Analysis of the current state of the economy of cloud technologies and development prospects. *Problems of the Information Society, 2*, 14-23. www.microtech.az

Mohammed, F., Ibrahim, O., Nilashi, M., & Alzurqa, E. (2017). Cloud computing adoption model for e-government implementation. *Information Development, 33*(3), 303–323. doi:10.1177/0266666916656033

Mohammed, F., Olayah, F., Ali, A., & Gazem, N. A. (2020). The effect of cloud computing adoption on the sustainability of e-government services: A review. *International Journal of Advanced Science and Technology, 29*(5), 2636–2642.

National Artificial Intelligence Strategy 2021-2025. (2021). https://cbddo.gov.tr/en/nais

Public Service ICT Strategy. (2015). https://ictstrategy.per.gov.ie

Rastogi, A. (2010). Application of cloud computing for the better e-governance in developing countries. *Oriental Journal of Computer Science & Technology, 3*(2), 343–348.

Reese, D. (2011). *Cloud Computing*. BHV-Petersburg.

Roadmap, I. C. T. (2016). *Strategic Roadmap for the development of telecommunications and information technologies*. Decree of the President of the Republic of Azerbaijan. https://president.az/az/articles/view/22382

Salimov, S. M. (2022). Comparative analysis and evaluation of ecological indicators of the Republic of Azerbaijan. Baku. "Science and Knowledge".

Takabi, H., Joshi, J. B. D., & Ahn, G. (2010). Security and Privacy Challenges in Cloud Computing Environments. *IEEE Security and Privacy, 8*(6), 24–31. doi:10.1109/MSP.2010.186

Tripathi, A., & Parihar, B. (2011). E-Governance challenges and cloud benefits. *Proceedings of IEEE International Conference on Computer Science and Automation Engineering (CSAE 2011)*, 351-354. 10.1109/CSAE.2011.5953237

Vats, K., Sharma, S., & Rathee, A. (2012). A Review of Cloud Computing and E-Governance. *International Journal of Advanced Research in Computer Science and Software Engineering*, *2*(2). http://www.ijarcsse.com/docs/papers/february2012/volume_2_issue/-V2I2074.pdf

Williams, B. (2012). *The Economics of Cloud Computing*. Cisco Press.

Wyld, D. C. (2010). The cloudy future of government IT: Cloud computing and the public sector around the world. *International Journal of Web & Semantic Technology*, *1*(1), 1–20.

Yakusheva, N.A. (2012). Calculation of economic efficiency of cloud computing. *Vestnik MGTU im. N.E. Baumann. Sir. "Priborostroenie", 5: Informatics and systems*, 224-235.

Yusifov, F. & Farajova, A. (2021). Protection of personal data in electronic governance: "Data embassy" concept. *Information Society Problems, 2*, 119-130.

Zwattendorfer, B., & Tauber, A. (2013). The Public Cloud for E-Government. *International Journal of Distributed Systems and Technologies*, *4*(4), 1–14. doi:10.4018/ijdst.2013100101

KEY TERMS AND DEFINITIONS

Artificial Intelligence (AI): Artificial intelligence is a field, which combines computer science and robust datasets, to enable problem-solving. AI is the simulation of human intelligence processes by machines, especially computer systems.

Big Data: Big data is high volume, high velocity, and or high variety information assets that require new forms of processing to enable enhanced decision making, insight discovery and process optimization.

Cloud Computing: Cloud computing is the on-demand availability of computing resources (including analytics, databases, networking, servers, and storage), as services over the internet.

Data Center: Data Center is a facility composed of networked computers, storage systems and computing infrastructure that organizations use to assemble, process, store and disseminate large amounts of data.

Digital Government: Digital government refers to the comprehensive digital strategies and structures set up by the government to care for the needs and concerns of its citizens.

E-Government: E-government is the use of information and communication technologies (ICTs) to improve the activities of public sector organisations. Some definitions restrict e-government to Internet-enabled applications only, or only to interactions between government and outside groups.

Government Cloud: Government cloud is a government computer system that store data over the Internet rather than a physical data server. The government cloud is cybersecure, Internet-based data storage for public sector entities.

Internet of Things (IoT): Internet of Things (IoT) refers to a network of physical devices, vehicles, appliances, and other physical objects that are embedded with sensors, software and network connectivity that allows them to collect and share data.

Chapter 4
A Comprehensive Study on Smart Farming for Transforming Agriculture Through Cloud and IoT

Ajay Poonia
VIT Bhopal University, India

D. Lakshmi
https://orcid.org/0000-0003-4018-1208
VIT Bhopal University, India

Tanmay Garg
VIT Bhopal University, India

G. Vishnuvarthanan
VIT Bhopal University, India

ABSTRACT

The amount of food produced in fields is directly proportional to the Earth's growing population. Taking into account the deteriorating climatic conditions and ineffective agricultural productivity, we must modernize our agricultural methods immediately. India is second in terms of agricultural performance. The main concerns that farmers have are crop rotation, fertilizer use, and irrigation. Due to technological advancements, farmers nowadays are unable to produce their crops to their maximum potential. The main focus is on utilizing contemporary technology in agriculture and using it to boost yield. We can now better estimate crop output thanks to the deployment of IoT in agriculture and cloud computing. Implementing IoT and cloud computing strategies is a way to overcome challenges in agriculture. This chapter gives a brief summary of the integration of cloud and IoT (internet of things) technologies into agricultural practices, which will make agricultural practices smarter.

DOI: 10.4018/979-8-3693-0200-2.ch004

1. INTRODUCTION

The agricultural production system has undergone seismic change as a result of the development of precision agriculture technology during the last few decades. The quality of the yield increases the opportunity for the export business and contributes a major part to the GDP growth rate. The majority of Asian countries are still in the early stages of developing and adapting agritech. Smart agriculture systems built on the IoT are crucial for the successful utilization of land, water, and energy resources. One approach to achieving this is by adopting IoT and Cloud Computing (CC) strategies. Things in the term "IoT" refer to actual physical things embedded with sensors, computing power, software, and other technologies, that are connected to other systems and devices via the Internet or other networks, for receiving or transferring agronomic information to data centers. CC is a concept that involves integrating cutting-edge technology assistance mechanisms for the backend in order to provide computing facilities for a range of users. Applications for SA can employ a variety of communication protocols with different capabilities. To make an appropriate and ideal choice, it is necessary to analyze the aspects of potential protocols while creating a smart agriculture system. An IoT network is made up of tangible assets, network elements, and data analysis software. CC offers an analytical capability using big data and Deep Learning (DL) models to get maximum profit from the robust computing environment. According to Sadu et al. (2022), agricultural operations produce a large amount of agronomic data, which is swiftly processed and examined. The creation of an automated system makes it possible to water irrigation, sow seeds, and quantify the moisture in the soil. Taylor et al. (2019) state that the primary elements of pest surveillance for agricultural diseases include sensing, evaluating, and treating. Recent developments in image processing have made it possible to analyze agricultural images to classify various plant diseases for pest management purposes.

Figure 1 shows the Taxonomy of Agriculture, in which it classifies smart practices into major categories, such as Agricultural Data Gathering, Edge computing mechanisms, Agricultural data transmission, and Cloud computing system analysis, providing a systematic overview of agricultural activities. It also specifies the flow of information from perception to application through processing, communication, and middleware.

2. CHALLENGES WITH TRADITIONAL AGRICULTURE TECHNIQUES

The eighteenth century saw a significant advance in chemical understanding. To increase crop productivity, several synthetic compounds are produced for agricultural use. With time, there was an increase in the use of these chemicals in agriculture, but along with their advantages came certain drawbacks. Traditional farming focuses on treating the soil and plants with poisonous chemicals, which were formulated artificially in a lab. These elements protect the plants against disease and pests. Chemical use results in building up fatty tissues in our body, and if the concentration is high enough, a disease like cancer may occur. The use of these synthetic compounds has a number of serious drawbacks, one of which is that they kill out all life in the topsoil and subsoil. It requires both microfauna (protozoa, nematodes, and arthropods) and microflora (bacteria, fungi, and actinobacteria) to be valuable. The loss of microbial life in soil has eventually been attributed to the use of powerful chemical agents that kill or inhibit undesirable fungus or fungal spores, kill or inhibit harmful insects and other pests, and manage undesirable plants (herbicides), to the point that it is necessary to use extremely large amounts of fertilizer to nurture

Figure 1. Taxonomy

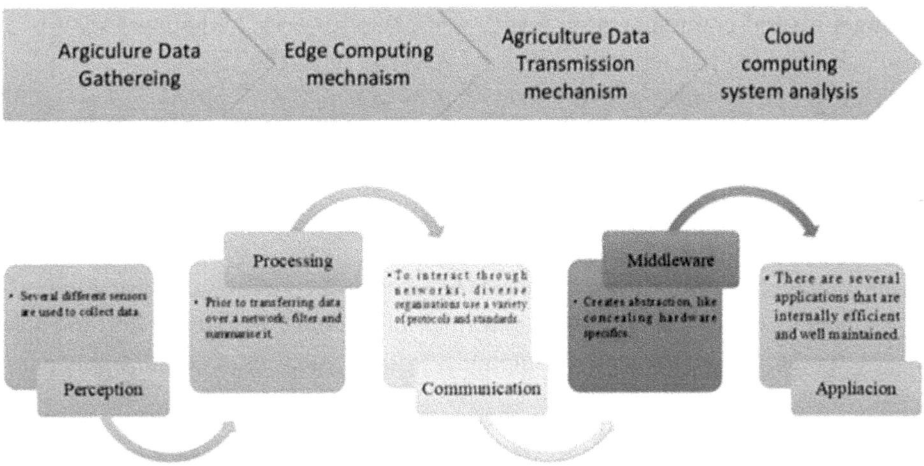

adequate crops. As a result, the ground becomes nearly sterile and unable to produce organic material on its own. In layman's language, whether we are in urban or rural locations, we are ingesting poison rather than nourishment. Resource waste is another limitation of traditional farming. Smart agriculture enables us to reduce these elements.

2.1. Limitations With Modern Methods

2.1.1. Price

IoT technologies, such as sensors and gateways, are used in smart agriculture practices and must work in harsh outdoor environments. The need for hardware to match system requirements has an impact on the cost of these components, including base station infrastructure, sensors, and gateways. The vast amount of data that is produced might cost money. Use CC devices to analyze data and make educated choices, and make sure that data is collected, sent, and stored in a reliable manner. According to Avşar et al. (2022), when considering maximum data rates and allowed download and upload limitations, it is advised to select the optimal communication protocol and subscription plan that ensure dependable and effective data transportation.

2.1.2. Security

IoT systems, which include sensors, base stations, and gateway infrastructure, are related to smart agriculture approaches. These parts can be pricey and are subject to rules governing dust and water resistance. IoT security flaws can result in data loss, alteration, privacy concerns, and decreased client confidence. Physical in-field devices may potentially be open to outside access, which poses a serious risk to the development of crops or plants. Vulnerabilities may arise on hardware with low processing and memory capabilities.

Based on the security concerns in each system tier, it is possible to categorize risks and vulnerabilities in IoT systems. Vulnerabilities can also be the result of bugs and restrictions in hardware and software. Attacks may target several communication levels in ZigBee networks, whereas LPWAN systems are susceptible to packet forging and replay attacks, say Avşar E. et al. (2022).

IoT systems are generally vulnerable because security standards or guidelines are not followed. To provide comprehensive security, all parties in the manufacturing and supply chains must act responsibly. As stated by Avşar E. et al. (2022), cyber incidents targeting the agriculture sector have the potential to disrupt deliveries and compromise sensitive information, leading to adverse consequences for both individuals and the economy.

The smart agriculture ecosystem may make use of blockchain technology to improve cybersecurity and reduce risks. Applications based on blockchain may track and record every step of the food manufacturing process, boosting confidence, guaranteeing compliance with regulations, and addressing possible issues.

2.1.3. Communication Proficiency

In a smart agricultural setup, maintaining seamless communication is vital to ensuring precise and intelligent decision-making. To prevent unfavorable results, it is imperative to consistently collect data from sensors and continuously monitor in-field actuators. Wireless communication is commonly used in Smart Agriculture applications, but it can be influenced by factors such as interference and weather conditions. Employing hardware protection measures improves the quality of communication and reduces maintenance costs. While licensed cellular networks like 5G and NB-IoT present fewer interference issues, unlicensed frequency bands like WiFi, ZigBee, LoRaWAN, and Sigfox face higher risks of interference.

2.1.4. Hardware and System Layout

Investing in dependable equipment for smart agriculture leads to increased costs for both initial investment and ongoing maintenance, thereby decreasing overall profits in the agriculture industry. When designing sensors and gateways for a Smart Agriculture system, factors such as distance, data transfer rate, and power consumption must be meticulously considered. The intricacy of this optimization challenge demands a thorough assessment of multiple constraints.

3. THE RATIONALE OF THE STUDY

In some of the least developed nations, agriculture may account for more than 25% of GDP, playing a significant part in the GDP, which accounts for 4% of the global GDP. In the face of a global food crisis that is driving millions of people into extreme poverty and leaving 205 million people in 45 countries with so little food that their lives are at risk, the old agricultural system and practices are useless. The invasion of Ukraine by Russia has made this issue worse. In order to develop intelligent and automated agriculture techniques and systems, it is now functionally necessary to connect cloud computing and IoT with agriculture research. The chapter focuses on how data collection, analysis, and decision-making utilizing cloud computing and IoT may develop a solid architecture to manage the automation of farming operations to battle the exponentially growing needs for food.

4. SMART AGRICULTURE

4.1. Assisting Technologies for IoT

Rathor S. & Kumari S. (2021) mention that Internet of Things (IoT) technologies incorporate Wireless Sensor Networks (WSNs), Big Data analysis, and Cloud Computing (CC) to distribute system resources and data based on users' demands. This distribution can be achieved through various models like Laas, PaaS, and SaaS, utilizing communication protocols as the backbone of IoT systems. These protocols facilitate network-wide data sharing and application connectivity.

In typical farm monitoring IoT systems, a variety of sensors are deployed, including soil and ambient sensors. The data collected from these sensors, such as rainfall, CO2 levels, humidity, soil moisture (Madushanki A. R. et al., 2019), temperature, and plant health data (S. M. Kamruzzaman et al., 2018), play a crucial role in generating plans to enhance productivity and maximize financial gains. These sensor data are transmitted to cloud servers through gateways using the system's designated connection protocol.

Figure 2 highlights Advanced Technologies used in E-Agriculture.

4.1.1. Wireless Sensor Network (WSN)

Symeonaki et al. (2019) describe it as an assemblage of wireless sensors and nodes strategically placed throughout a physical area to monitor and gather data. These sensors are equipped with various capabilities, such as humidity, light, motion, temperature, and sound sensors, enabling them to monitor and collect environmental data.

WSN is divided into five categories.

- **Terrestrial WSN (TWSN):** These are employed mostly in the field of environmental monitoring. In these networks, hundreds of thousands of sensors are placed at random or in advance in a certain region.
- **Multimedia WSN:** A promising and cutting-edge network of wirelessly linked devices, multimedia wireless sensor networks include video, audio feeds, etc. For data retrieval and compression, nodes in the sensor field are wirelessly linked to one another.
- **Mobile WSN:** By utilizing Mobile Sink (MS), wireless sensor networks (WSNs) may regulate their energy usage and increase their overall lifespan. During data transmission, MS in mobile WSNs (MWSNs) travels in a regulated manner.
- **Underwater WSN:** Nodes are positioned on the water's surface and below it in underwater wireless sensor networks. In the same network, all nodes communicate and exchange data with one another and with base stations. Communication systems use electromagnetic, optical, or acoustic waves to transport data.
- **Underground WSN:** It is a sensor network made up of sensors usually placed in the ground for agriculture or on the walls of a mine to monitor soil changes. At a specific depth, sensors for irrigation and environmental monitoring programs are buried.

Figure 2. Advanced smart agriculture technologies

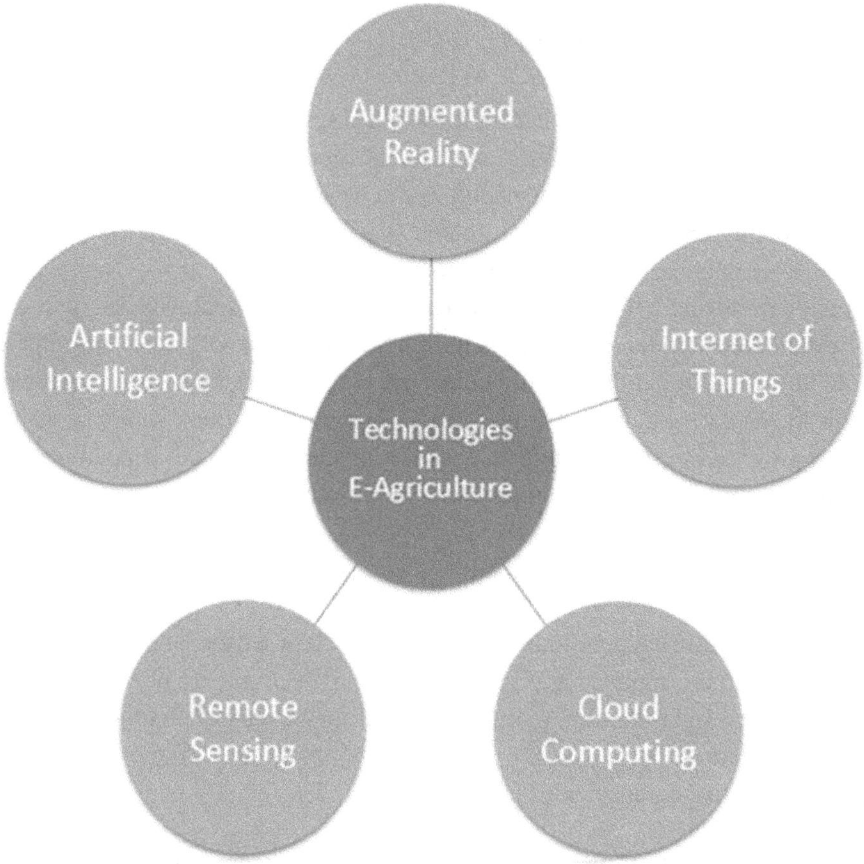

4.1.2. Cloud Computing and IoT

Symeonaki et al. (2019) explain that cloud computing involves the dynamic allocation of system resources and data to users as per their requirements. It can be presented in different forms, such as IaaS, PaaS, and SaaS. The virtually boundless capacities and resources offered by the cloud play a crucial role in overcoming technological limitations within the IoT, such as storage, processing, or energy constraints. By leveraging these cloud capabilities, the IoT can effectively distribute and offer novel services to address diverse real-world situations.

Several services and information management systems employ cloud computing technology. This breakthrough in agricultural cloud computing offers a great opportunity to employ agricultural applications in industry for remote areas sustainable development.

According to Symeonaki et al. (2019), the agriculture sector can benefit from robust computational capabilities through cloud computing. By integrating technologies like Wireless Sensor Networks (WSN) with RFID (Radio Frequency Identification) for data collection and monitoring in growing regions, along with Cloud computing for efficient data transport, storage, and analysis over the network, significant advancements in agricultural automation can be achieved. This research into IoT agricultural applications holds significant theoretical implications and practical importance, as it has the potential to bring about revolutionary developments in the agricultural industry.

According to Kolomvatsos K. & Anagnostopoulos C. (2021), the qualities made possible in agriculture by the use of CC for the IoT are summarized in the list below:

I. Data Gathering And Remote Access Storage: CC applications focused on monitoring soil moisture, temperature, humidity, light, and other factors can be effectively combined with various information-gathering devices available in the market, such as WSN and RFID sensors. In addition to providing ample storage capacity to handle the substantial volume of data related to farming practices, CC also offers additional infrastructure that facilitates decision-support, knowledge sharing, and experience exchange within the global agricultural community.
II. Users are billed based on their actual usage of diverse resources, which are made available upon request, allowing them to access Information and Communication Technology (ICT) resources at a cost-effective rate. This instant access to cloud resources empowers farmers to avoid investing in costly ICT hardware and efficiently obtain the resources they require.
III. Integrating CC into the IoT in agriculture offers farmers a practical solution when they encounter complex issues during various stages of crop production. By seeking agricultural expert guidance through the cloud platforms, farmers can promptly access valuable advice and instructions to address any uncertainties that may arise. This immediate access to expert support enables producers to effectively handle clarification needs and receive precise instructions without delay.
IV. Land Record Automation: The availability of extensive storage capabilities has enabled the global digitization of land records. These storage facilities allow land records to be uploaded and securely stored, along with relevant data specific to each location, such as production history and soil test results. Diligent examination of the data by various private or commercial operators ensures its accuracy.
V. Weather Forecasting: Weather forecasts and analyses can be offered for specific time frames to aid farmers in making informed decisions about their crops.

Figure 3 depicts an IoT cloud architecture for precision agriculture to optimize agricultural practices and improve crop yields, showing the process flow after collecting data from sensor nodes for cloud-based data processing, and real-time analytics.

4.1.3. Big Data Analytics

According to Lutu (2018), various organizational departments like staff, supply chains, financial divisions, and marketing campaigns produce a sizable amount of data every day. The growing trend of utilizing big data analytics enables businesses to gain insights into their customers and convert vast volumes of unstructured data into valuable knowledge. As technology continues to advance rapidly, it highlights the diverse applications of big data analysis in various fields. Here is a concise overview of the different types of big data analysis:

I. According to Kayser et al. (2018), descriptive analytics plays a pivotal role in interpreting historical data and offering valuable perspectives on previous trends, facilitating the recognition of patterns

Figure 3. IoT cloud computing framework for precision agriculture

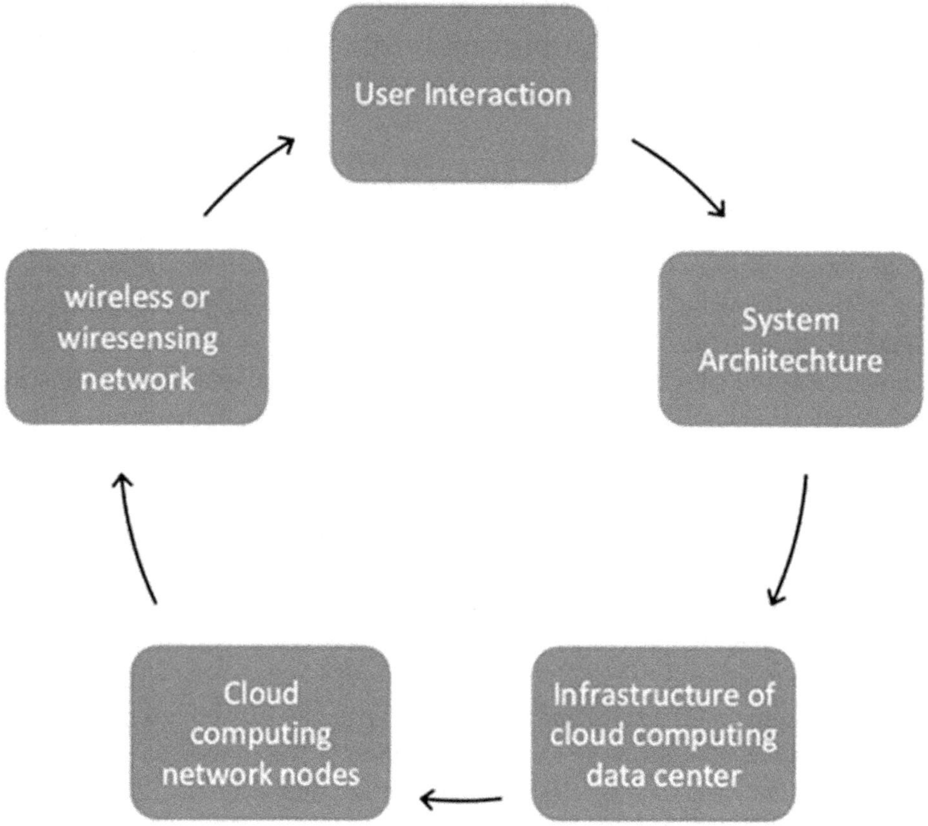

within different customer segments. Illustrative instances of descriptive analytics encompass condensing and grouping data, uncovering association rules, and performing market basket analysis.

II. According to D. Sharma et al. (2021), diagnostic analytics serve as a diagnostic tool that identifies the root causes of problems. Employing methods such as drill-down, data mining, and recovery, it examines the reasons behind customer churn and assesses customer health ratings. This strategy can offer significant benefits for e-commerce enterprises.

III. D. Sharma et al. (2021) explain that predictive analytics involves utilizing historical and current data to forecast future events. It relies on models derived from data mining, Artificial Intelligence (AI), and Machine learning (ML), and is commonly utilized by both businesses and customers. Examples of predictive analytics applications include Churn Risk, Next Best Offers, and Renewal Risk Analysis.

IV. D. Sharma et al. (2021) describe prescriptive analytics as a data-driven approach that utilizes extensive data, statistics, and models to predict future performance, primarily relying on AI and ML techniques. By integrating both descriptive and prescriptive analytics, this methodology offers tailored solutions to specific problems.

A simple explanation of how big data works is as follows:

I. Upadhyay N. (2018) states that businesses collect both structured and unstructured data from various origins, including cloud storage, mobile apps, and IoT sensors. Storing data in the cloud facilitates easy accessibility for analytical devices. Predictive analytics, which relies on AI and ML models, anticipates future performance based on past and current data. Perspective analytics, a fusion of descriptive and predictive analytics, offers targeted solutions to specific problems.
II. Kapoor A. & Attri V. (2021) emphasize the importance of effective data management for precise analytical investigations. To handle larger, unstructured datasets efficiently, they recommend utilizing batch processing and stream processing methods, which reduce the time gap between data collection and analysis. Employing these approaches accelerates processing speed and improves decision-making capabilities.
III. Boonyopakorn P. (2018) highlights the significance of precise data structuring and the removal of irrelevant or redundant elements to achieve improved outcomes. Ensuring data cleaning is carried out is vital to preventing deception and incorrect interpretations.
IV. According to Santur Y. et al. (2016), extracting valuable insights from big data requires time, which is achieved through advanced analytics techniques that integrate diverse datasets. Data mining is employed to discover patterns and relationships within datasets, while predictive analytics uses historical data to make forecasts about potential risks and outcomes. For uncovering patterns in vast and complex datasets, Deep Learning (DL) leverages ML and AI technologies.

Figure 4 differentiates among the different types of big data analysis, such as descriptive, inquisitive, predictive, prescriptive, and pre-emptive analytics.

4.1.4. Communication Protocols

According to Rathor S and Kumari S. (2021), communication protocols serve as the fundamental basis of the IoT system, enabling widespread data sharing and application connectivity. In typical farm monitoring IoT systems, various sensor types, such as soil and ambient sensors, are deployed. These systems employ a wireless network architecture that covers extensive areas, making preparations for rural areas quite intricate. To simulate a rural wireless network with the objective of connecting multiple IoT devices, specific procedures must be followed. This involves selecting a suitable wireless network technology and evaluating its performance under diverse real-world scenarios, as mentioned by Viswanathan et al. (2022).

4.1.5. Embedded System

A computing system that has computing hardware and software embedded to carry out certain functions is known as an embedded system. Embedded systems are created to fulfill a specific set of responsibilities as opposed to general-purpose computing or personal computing, which are capable of doing a variety of tasks. A microcontroller or microprocessor, memory (ROM, cache, RAM), input/output units (keyboard, Monitor, etc.), networking devices (WiFi adapters, Ethernet), and storage (SSD, HDD, etc.) are the main elements of an embedded system. Graphics processors, digital signal processors (DSPs), and processors that are specialized for certain applications are a few examples of the specialized processors found in some embedded systems. Real-Time Operating Systems (RTOS) are examples of embedded operating systems that are used by embedded systems. Digital watches, point-of-sale terminals, vending

Figure 4. Types of big data analysis

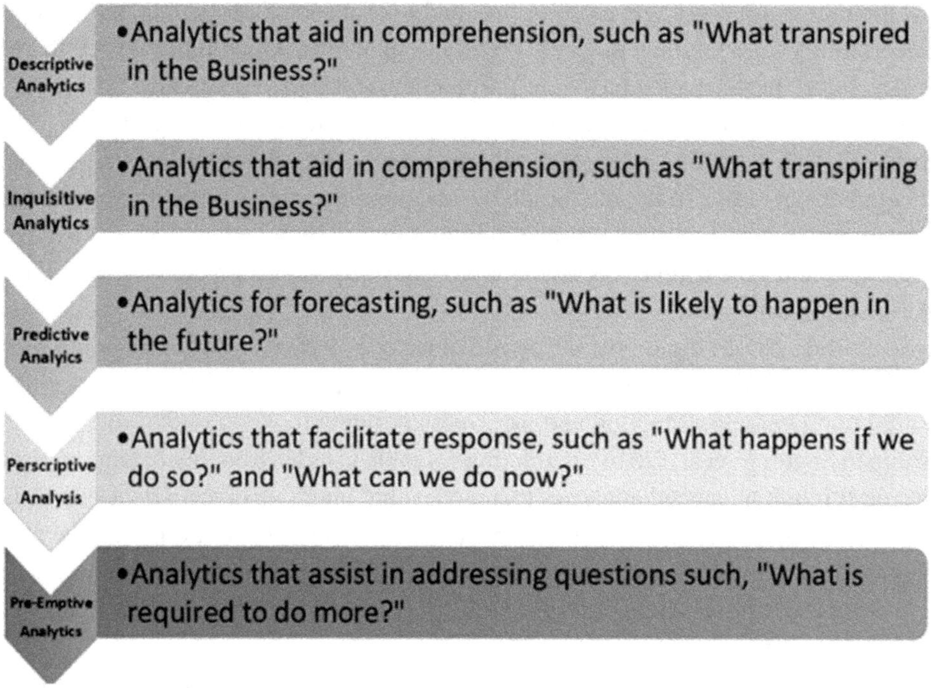

machines, appliances (like washing machines), and other inexpensive miniaturized devices are examples of embedded systems.

4.2. The Architecture of Smart Agriculture

As smart agriculture becomes more robust and complex, IoT architecture plays a major role in gathering useful information about humidity, and soil parameters like fertility, water content, temperature, and precipitation. It also gathers data about plant health through physical devices employed in the physical layer. The data generated by the physical sensors of IoT devices deployed on farms has increased exponentially, to billions, every second. According to the WHO suggested article by B. Marr in 2017, predictions approximate that there will be a sharp 300% growth in just 6 years, i.e., around 55.7 billion connected devices globally, producing 73.1 zettabytes of data (compared to the 18.3 ZB created in 2019).

Moreover, if incorrect user inputs, data breaches, or other problems occur, they can have a cascading effect on the distributed data system. This could result in significant inaccuracies that hinder decision-making in agriculture operations, hinder efforts to enhance productivity, reduce costs, and manage risks. Additionally, it may lead to serious consequences such as legal repercussions, financial losses, identity fraud, or disruptions in the food supply chain. Therefore, it is imperative for these architectures to guarantee connectivity, scalability, intelligence, dynamic adaptability, reliability, privacy, and secure data storage.

1. Connectivity ensures that there is a link among different levels of IoT, i.e., sensors, actuators, and other electronics that allow the automation of various farming tasks, that are connected to each other.

2. Scalability allows an increasing number of devices like actuators, sensors, and drones to be connected to the IoT environment. In order to keep up with the quick developments farmers make according to their farmland, crops, and scalability, the IoT system must be able to handle the vast expansion and appropriately analyze the generated data.
3. Intelligence provides keys to analyze generated data, which helps in extracting knowledge about crop health, soil conditions, and weather patterns in real-time analysis or predictive analysis. The analysis is what helps a farmer forecast crop yields, disease outbreaks, and pest infections, which helps them optimize resource allocation, and plan harvesting schedules.
4. Dynamic nature is an essential component of an IoT architecture. This function guarantees that these intelligent system devices will dynamically adapt to their changing environments, disease outbreaks, equipment failures, and resource constraints and perform the necessary actions.
5. The reliability of IoT architecture is to collect data accurately and consistently and provide accurate measurement, while at the same time maintaining the stability of the operation of smart farming, by developing a robust fault tolerance system that maintains data integrity, system security, and system upgrades. It is defined as the capacity of a product to carry out a necessary function under specified circumstances for a certain amount of time, which can be evaluated in terms of probability.
6. Within the context of IoT, wherein nearly every physical or digital entity or device is assigned a distinctive identification and the ability to independently interact over the Internet or a similar network, the significance of privacy cannot be overstated. Privacy serves as a safeguard, preserving the confidentiality of an individual's information from being exposed. Failure to address privacy concerns may lead to detrimental consequences, including financial data breaches, identity theft, and unauthorized access to or alterations to legal documents, among other potential issues.
7. Data storage holds significance in the context of the IoT data storage within agriculture, as it is subject to legal obligations, industry norms, and compliance regulations aimed at safeguarding integrity, security, and privacy. Through data storage, the validation and auditing of sensor and device data become possible, ensuring the accuracy, coherence, and integrity of the collected data, thereby enabling reliable analysis and decision-making. Auditing capabilities further contribute to accountability and transparency by facilitating the tracing and verification of data sources, usage, and modifications.

All of the aforementioned characteristics collectively contribute to ensuring the Quality of Service (QoS) for IoT architecture. If we address the mentioned features and challenges effectively, it enables us to meet the architecture requirements, such as:

1. Performance, which pertains to the maximum acceptable end-to-end latency for the application.
2. Reliability, ensuring a minimal occurrence of errors, failures, or crashes, and minimizing disruptions.
3. Availability; ensuring the service can be accessed anytime and anywhere as promised to farmers.
4. Operational ability and meeting the operational commitments made to farmers.
5. User experience, encompassing aspects like user-friendly interfaces, responsiveness, and overall user satisfaction.

By carefully addressing these aspects, the IoT architecture can deliver a high-quality service and meet the needs of its users.

To outline the data flow across the layers of IoT architecture, the perception layer is where data is generated, then transmitted (by the network layer) to the middleware layer, which provides device management and data processing and transfer capabilities. Finally, at the application layer, the processed data is offered to the farmer as a service. While there is no standard architecture, we will be reviewing two widely used IoT architectures, i.e., the five-layer architecture and the three-layer architecture.

4.2.1. Three-Layer Architecture

The typical architecture of IoT systems is organized into several layers. According to Avşar, E. et al. (2022), it is widely accepted that there is no universally defined list of layers. However, an application designed for smart cities may consist of various layers, such as network, physical, virtualization, database, data analytics, sensing, cognitive, middleware, and application layers. While selecting layers, we have opted for those commonly found in earlier research, which include the application layer, the network layer, and the perception layer. The figure provided in Figure 5 illustrates this three-layer architecture and offers an overview of the functions performed by each layer.

Figure 5. Three-layer architecture

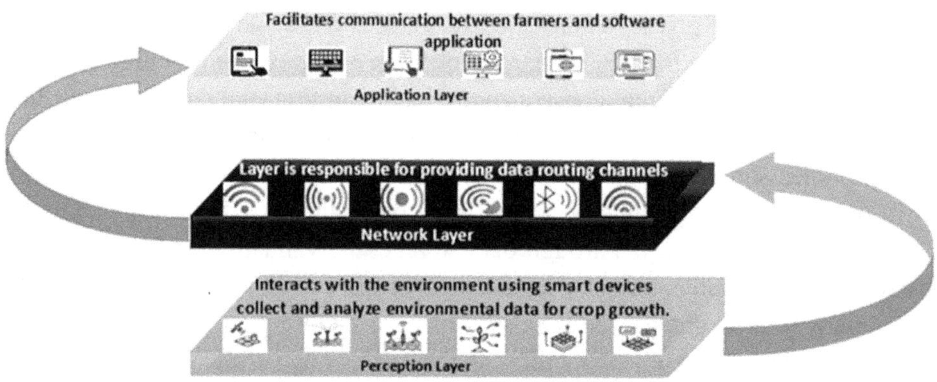

4.2.1.1. Perception Layer

The perception layer—implemented at the base of the IoT ecosystem—is referred to as the sensor layer. It interacts with the surrounding environment using smart devices like actuators, sensors, Global Positioning System (GPS) terminals, and cameras in order to gather and analyse data about soil moisture, temperature, humidity, and crop growth. The key duties of this layer include stream processing, attaching devices to the IoT framework, and sending data to the top layer via its interfaces. Heterogeneous devices are set up at this layer using standardized plug-and-play methods. Big data is produced at this level by the IoT.

4.2.1.2. Network Layer

Also called the communication layer, it is a key part of the IoT ecosystem. In order to send information over the network as packets, it is responsible for providing data transfer channels. This layer controls messaging between several servers, device and cloud routing, and network connectivity. Data is transferred to IoT applications, devices, and servers through network devices. Being a combination of communication channels with the Internet, it helps connect the application layer and the perception layer. IoT gateways, hubs, and switches, as well as numerous communication protocols, including Wi-Fi, Bluetooth, and ZigBee, make up this layer. The deployment of wireless protocols at this layer is essential because of the ease with which high-risk regions can have wireless devices deployed. To communicate data from other applications via interfaces and gateways, several different communication standards and protocols are widely used at this layer. The universality of the IoT concept is realized through different addressing and routing capabilities, which guarantee that varied devices may be swiftly and easily incorporated into a single collaborative framework.

4.2.1.3. Application Layer

It is the top layer of the IoT network, and facilitates communication between farmers and software applications. It is the responsibility of this layer to obtain data forwarded by the network layer and use the information obtained to provide insights on crops or make smart decisions on their own for better crop yield. It is the duty of this layer to provide a storage space where a database backup of the acquired data may be kept. The layer further anticipates the status of sensors and actuators in the future by examining the type of information that was gathered.

4.2.2. Five-layer Architecture

4.2.2.1. Business Layer

This layer assists in creating a business graph, architecture, flowchart, etc. from the output of this layer for various business models, which helps in choosing the subsequent business approaches.

4.2.2.2. Application Layer

Similar to "three-layer architecture", it enables connections between applications and farms. This layer is accountable for obtaining data received from the network layer and using the data to provide insights into crops or independently make wise judgments for increased crop yield. The application layer is responsible for providing a storage location where a database backup of the obtained data may be preserved. By examining the sort of information that was received, this layer makes further predictions about the condition of physical objects in the future.

4.2.2.3. Processing Layer

This layer is responsible for analyzing, cleaning, filtering, and storing data in big data processing modules, databases, and cloud storage. The processing layer manages the unstructured and structured data by sending them to their proper data storage types, i.e., unstructured data is sent to data lakes, while structured data is forwarded to data warehouses because of the space they consume. The structured and unstructured data are always saved in a format readable by the software.

4.2.2.4. Transmission Layer

Also known as the "Network Layer," it is in charge of operating by forwarding data securely to the processing unit, i.e., the perception layer. This layer uses Bluetooth, 3G, Zigbee, and Wi-Fi technologies as the transmission standard for sending data securely over wired and wireless networks. The information is thus transmitted to the middleware layer.

4.2.2.5. Perception Layer

Similar to the perception layer of "three-layer architecture", this layer is in charge of collecting data from the physical layer of the system. Data is collected from sensors like an Ion-Selective Electrode (ISE), a light detection and Ranging (LiDAR) sensor, an Ion-Selective Field Effect Transistor (ISFET), and a tensiometer, while the actuators bring changes in the physical environment by adjusting the moisture in the soil, chemical nutrients in the soil, spraying pesticides, and more. Of course, some key duties of this layer are the same, such as stream processing, attaching devices to the IoT framework, and sending data to the application layer through network channels.

Figure 6 depicts a five-layer design and gives a description of how each layer operates.

Figure 6. Five-layer architecture

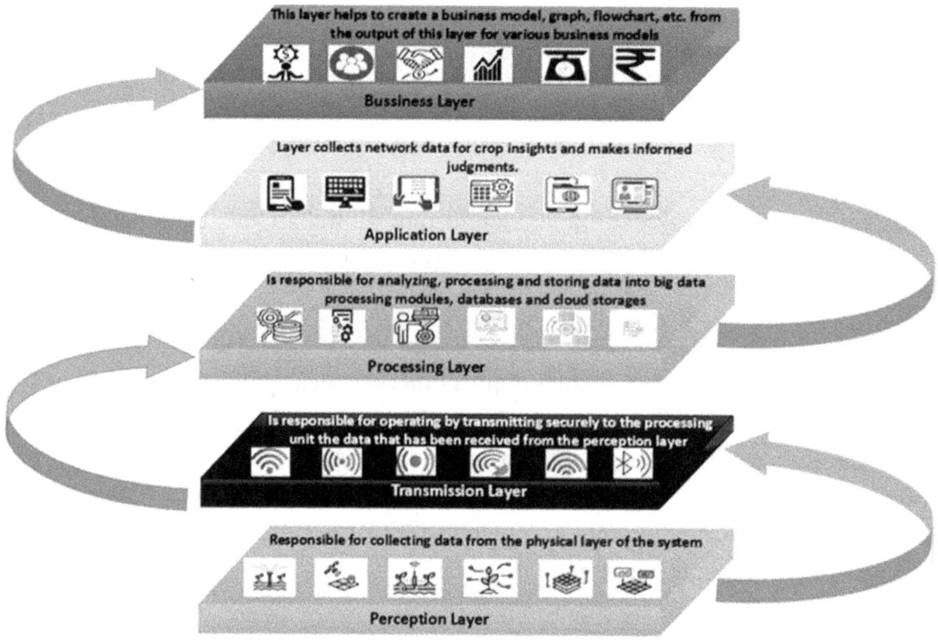

4.3. Smart Agriculture Wireless Communication Protocols

Wireless communication protocols can be divided into two primary types: long-range and short-range, depending on the coverage area. The system's conditions and requirements can be taken into account when choosing how to implement these protocols.

There are several long-range network protocols that can transport data transmission across a long distance between devices, including NB-IoT (Nair et al. (2019)), Sigfox (Lauridsen et al. (2017)), and LoRaWAN (Nair et al. (2019)). These are utilized by SA systems that need the connection of components across extremely broad areas. Conversely, short-range network protocols like ZigBee (Xiaojing et al. (2012)) and Wi-Fi enable information to be exchanged in closer proximity and are better suited for systems that are smaller in scale. Automated irrigation systems use wireless communication protocols to save time and water. gathering environmental information for weather forecasting. Monitoring soil quality to maximize yield Cost-saving crop monitoring. Disease prevention through pest management Utilizing renewable energy sources as smart agriculture applications to reduce costs and promote sustainability.

4.3.1. ZigBee

Considering the 802.15.4 standard, The ZigBee Alliance developed it. It is a low-cost Wireless Personal Area Network (WPAN) communication technology that is ideal for networking and home automation. Its characteristics are peer-to-peer, cluster tree, and star topologies; low power consumption; security; cost-effectiveness; and dependability, says Zhihua S. (2016). It also supports 16-bit and 64 -bit addressing modes. Several applications, including low-power WSN topologies with several sensors and routers, show the utility of the ZigBee protocol in SA. DL techniques are used by ZigBee-based WSN designs to track the moisture, temperature, and humidity content of the soil. On single-board computers & microcontroller boards, rapid prototyping is accomplished using widely accessible hardware.

An issue with ZigBee is that it does not provide a sleep mode, which might result in wastage of power. Programming and hardware may be used to switch between active and sleep modes. Examples include water-saving networks with mesh topologies for monitoring greenhouses and transmitting gathered soil parameter data to an aggregate node for processing. According to Avşar et al. (2022), ZigBee has problems in Smart Agriculture because of its poor power usage, limited range, and slow data rates. It is better suited for small-scale experimental trial areas than large agricultural land or long sensor node distances.

4.3.2. Wi-Fi

In 1997, Wi-Fi first appeared as a GPS substitute for indoor location-based and context-aware applications. According to Santur, Y et al. (2016), with a transmission range of up to one kilometer and the lowest data throughput of 150 KB/s, it is generally used in Wireless Local Area Networks (WLANs) with 2.4 GHz or 5 GHz frequency bands.

For IoT applications like information gathering, cloud connectivity, and weather forecasting, Wi-Fi is a great option since it is widely accessible in both indoor and outdoor areas. In hybrid monitoring designs, SA can be integrated with other protocols like ZigBee for low-data transmission channels and Wi-Fi for monitoring and streaming video. IoT systems have a high Wi-Fi power consumption, which makes it challenging to operate in outdoor settings. Photovoltaic solar panels can be used to solve the power consumption problem. A tree-structured WSN was built to transmit gathered environmental and soil data to a main fertilization unit. The power consumption was decreased by modifying the intervals for data gathering and transmission. Weather predictions and measurements of soil moisture can be applied to develop a smart watering system using Wi-Fi. The system integrates weather forecast data from an open Application Programming Interface (API) with sensor data for effective irrigation, connecting sensors to Unmanned Aerial Vehicles (UAV) for water spraying, say Avşar E. et al. (2022). The Wi-Fi protocol is difficult to use due to high power consumption restrictions when gear cannot be connected to a main power line. Potential alternatives like solar panels must take installation and upkeep costs into account.

4.3.3. Sigfox

The IoT and M2M systems, as well as rural regions, can benefit from the low-data-rate wireless communication technology known as Sigfox, which uses ultra-narrowband cellular networks. At first, it was applied to summertime alpine grasslands to locate cattle, say Lavric A. et al. (2019).

It can split networks into layers through which end-to-end communication facilitates interaction and offers smart agriculture application services. These services can keep an eye on the temperature in grain storage facilities, stop cattle theft, safeguard farmhouses, enhance colony health, and keep an eye on influencing food temperatures along the whole cold chain. The ultra-narrow band and low-duty cycle design of Sigfox are used in a water quality monitoring system in Smart Agriculture, resulting in little power usage. The solution ensures off-grid monitoring by using a solar panel to charge a Li-Po battery. Significant aspects of Sigfox are its bidirectional control method and enhanced security. With excellent coverage throughout the majority of European nations and no extra infrastructure, Sigfox is a flexible SA application. According to Fernández-Ahumada et al. (2019), both Sigfox and LoRaWAN were employed in an irrigation management system, with sensors connecting with a local gateway and data being sent to a cloud server through LoRaWAN. In agricultural applications, such as irrigation management applications, effective data transfer may be hampered by Sigfox's low power consumption and unlicensed ISM band frequencies. Situations of over- or under-irrigation may be caused by signal interpretations and faulty transmission. It could be more appropriate for agricultural applications to use licensed frequencies.

4.3.4. NB-IoT (Narrow Band IoT)

The Third Generation Partnership Project (3GPP) created the radio access technology known as NB-IoT specifically for IoT devices. It coexists with GSM and LTE and provides vast coverage, extended battery life, and strong network security. It provides great network security at a reasonable cost with a 200 kHz bandwidth, say Wang et al. (2017).

In Smart Agriculture, NB-IoT has demonstrated success, especially when used to track cattle and monitor environmental conditions in greenhouses. For geo-fencing applications, the livestock tracker system's rapid reporting of cattle position and battery levels is helpful. The greenhouse monitoring system uses a variety of sensors to gather and keep track of data on water pH, light intensity, and energy usage. For firmware and software updates, the system has a two-way communication mechanism and operates on batteries. Enhanced security, half-duplex communication, expanded coverage, and variable power consumption are some of the characteristics of NB-IoT. In the SA design, NB-IoT is utilized to monitor paddy storage and relay humidity data to a gateway. The battery can power the system for 48 hours at 10,000 mAh. However, NB-IoT comes with membership fees and hardware modifications for current LTE infrastructure while providing faster data rates than Sigfox and LoRaWAN, say Foubert et al. (2020).

4.3.5. LoRaWAN

The LoRa Alliance focused on the IoT paradigm while developing the network layer protocol known as LoRaWAN. It employs a star topology in which nodes collect sensor data and transfer it to gateways for IP translation and transmission to remote access application servers. The front-end and back-end components of the LoRaWAN architecture are handled by gateways and end-node devices, which also manage security, data storage, and query scheduling, state Mahmoud M. S. & Mohamad A. A. (2016).

Although LoRaWAN is frequently employed in smart city applications because of its broad network coverage, its applicability in social contexts is growing. In order to forecast the weather, manage irrigation, and monitor farms, governments and corporate operators have invested in LoRaWAN infrastructure. LoRaWAN's low power consumption makes it perfect for SA systems that use solar energy harvesting because it triples battery life. According to Avşar et al. (2022), smart systems, among other applications and computing situations, can manage up to 1500 messages per minute of LoRaWAN protocol traffic thanks to the IoT architecture's dynamic layers. A middleware layer is included in the LoRaWAN-based SA design for scalability, module attachment, and simplicity of deployment. Three-month field tests have shown reliable data transmission. Similar to Sigfox, LoRaWAN uses unlicensed frequencies, has security difficulties, and performs poorly in rural locations with rough terrain.

4.4. Introduction to Cloud Computing

A user does not need to go to a specific location to access data thanks to cloud computing, which offers remote access over the internet. Cloud computing provides a wide range of services, including the ability to quickly move sizable IoT-produced packages over the internet, secure files, store any type of information, etc.

- Figure 7 demonstrates the data collection process on a cloud server, where sensor nodes equipped with various sensors capture environmental parameters such as soil moisture, plant health, temperature, and weather forecasting in the agricultural field.
- The collected data is wirelessly transmitted to the cloud server through communication protocols like Wi-Fi, LoRaWAN, or other connectivity technologies. This enables data processing, analysis, and presentation, empowering farmers with valuable insights for informed decision-making in precision agriculture.

The following are some services that cloud computing offers:

- SaaS (Software as a Service)

It is a cloud-based software distribution model that enables users to access internet software without having to buy and install it on their computers. Users may pay membership fees on a monthly, yearly, or pay-per-use basis.

- PaaS (Platform as a Service)

It enables web application development and deployment, with end users controlling backend scalability via CSPs. It covers systems for database administration, middleware, storage, and server infrastructure. The host web server provides the required services when the user specifies the room types and specifications.

- IaaS (Infrastructure as a Service)

Figure 7. Data is collected from sensors and sent to a cloud server through WiFi, ethernet, etc.

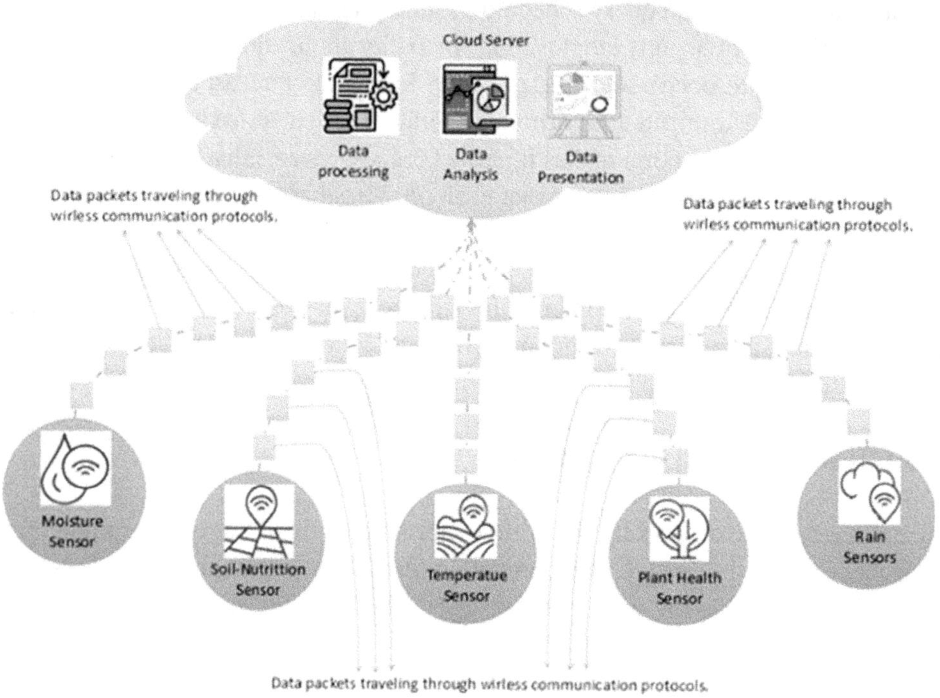

It provides computing resources on a pay-as-you-go basis, freeing customers from worries about infrastructure administration, equipment acquisition, and available space.

Types of Cloud Computing:

- Public Cloud

It is a service based on the internet that anybody may use, either for free or at a monthly cost. It reduces resource waste, enables multiple users to use storage, and is under corporate and organizational control. It is not location-specific.

- Private Cloud

Internal or corporate clouds are also referred to as private clouds, and they enable user protection, privacy, and security of sensitive data. Although it offers more performance, it is more expensive and has a smaller operating space. Among them is Ubuntu.

- Hybrid Cloud

The goal of Hybrid clouds is to create an automated and well-managed environment by combining public and private clouds. It combines fault-finding and non-fault-finding tasks, with the private cloud handling the former and the public cloud handling the latter. Google and Cisco are two such examples.

While private clouds like Ubuntu offer superior performance but come at a higher cost and have more restricted use cases, hybrid clouds are adaptable, economical, and secure.

4.5. Smart Agriculture Uses IoT Sensors

Processing the information provided by these sensors, such as rain, CO2, humidity, temperature, plant health data (Kamruzzaman S. M. et al. (2019)), and soil moisture (Madushanki A. R. et al. (2019)), which may aid in the creation of plans for increasing productivity and maximizing financial gain. After collection, data is transferred to the cloud through gateways using the system's connection protocol.

4.5.1. Sensors That Are Based on Location

It employs GPS satellite signals to pinpoint altitude, longitude, and latitude to a few feet. It takes three fixed-location sensors to triangulate the position.

4.5.2. Optical Sensors

Organic matter, moisture, and clay levels in the soil are measured with it. Usually, drones have these sensors mounted on them.

4.5.3. Electrochemical Sensors

The crucial data on pH and soil nutrient levels is provided by electrochemical sensors.

4.5.4. Mechanical Sensors

The mechanical resistances of the soil are discovered using mechanical sensors.

4.5.5. Dielectric Soil Moisture Sensors

A dielectric soil moisture sensor uses the dielectric constant to determine the soil's moisture content.

4.5.6. Parrot Sensors

The temperature, moisture, and soil salinity of the plant are all monitored by parrot sensors. The farmers' cell phones get the information.

4.5.7. Spruce Sensor

Spruce is a sensor gadget for controlling irrigation. The user has access to the data whenever and wherever they choose because it is kept on a cloud server.

4.5.8. Koubachi

Water is sprinkled onto the garden plants using Koubachi. It functions as a node that gathers information from several sensors, including air temperature, soil moisture, sunlight, and others.

4.5.9. Other Sensors

An overview of the sensors used for soil analysis, leaf disease analysis, and weather forecasts is provided in Table 1. It describes the specific sensors used in each category to gather the relevant data. The soil sensors are used to acquire soil parameters like moisture levels and nutrient content that can be measured. Leaf sensors are used to assess plant health and optimize irrigation by measuring leaf water potential, biotic stress, and photosynthetic rates. The weather forecast sensors predict atmospheric conditions, such as temperature, humidity, and precipitation. That ensures informed decision-making in agriculture, crop health monitoring, and weather forecasting applications.

Table 1. Sensors used for different applications

The Role of Sensors in Soil Analysis			
Hydra probe II soil sensor	Pogo portable soil sensor	ECH2O soil moisture sensor	MP406 Soil moisture sensor
EC sensor (EC250)	ECRN-50 low-REC rain gauge	ECRN-100 high-REC rain gauge	Tipping bucket rain gage
107-L temperature	Sensor (BetaTherm 100K6A1B Thermistor)		
The Role of Sensors in Leaf / Plant Health Monitoring			
237 leaf wetness sensor	the SenseH2TM hydrogen sensor	Leaf wetness sensor	YSI 6025 chlorophyll sensor
Field scout CM1000TM	TT4 multi-sensor thermocouple	LT-2 M (leaf temperature sensor)	TPS-2 portable photosynthesis
PTM-48A photosynthesis monitor	Cl-340 hand-held photosynthesis	107-L temperature	Sensor (BetaTherm 100K6A1B thermistor)
The Role of Sensors in Weather Forecasting			
CM-100 compact	Met station one (MSO)	XFAM-115KPASR	HMP45C
SHT71	SHT75	Cl-340	hand-held photosynthesis
107-L temperature			

5. SMART AGRICULTURE IOT AND CLOUD APPLICATIONS

5.1. Climate Conditions Monitoring

Sunlight and precipitation are necessary for photosynthesis and plant growth. The measurement and monitoring of weather-related data are critical in agriculture. Data analytics techniques and weather forecasting enable farmers to plan ahead for anticipated problems like pesticide poisoning or pre-rain irrigation. Farmers have the option to refrain from initiating irrigation if the soil already contains sufficient moisture and rain is expected shortly.

Examples of forecasting techniques used in smart agriculture are as follows.

Bright Keswani et al. (2018), proposed a valve control method that predicts soil moisture content using an optimization algorithm based on Variable Learning Rate Gradient Descent (VLRGD). A comparative investigation shows that VLRGD delivers lower MSE and RMSE values than a Gradient Descent based Neural Network prediction system. The integration of Structural Similarity Index (SSIM)-based methods with bilinear interpolation results in a comprehensive soil moisture requirement map. Subsequently, irrigation valve control orders are generated using fuzzy logic weather models.

Amarendra Goap et al. (2018) proposed smart irrigation systems for soil analysis and weather forecasting. This paper proposes an IoT-based smart irrigation architecture with a hybrid ML method to forecast soil moisture. The system leverages data from sensors and weather forecasts to reduce the error rate and increase accuracy. The technology is affordable and adaptable to many agricultural purposes. The suggested technique can be extended with numerous nodes and is intended to be used in future research to do water-conservation assessments.

In Gautham Rajagopal et al.'s (2014) study, the authors propose building a farm automation system for about USD 16 using parts that are readily available on the market and an open hardware architecture. Farmers' phones' alphanumeric inputs are used in the interface. Even in rural India, the system's low-cost intelligence enables energy and water maximization and maximizes crop production.

5.2. Drones for Agriculture

In the agricultural sector, drones serve various purposes, including crop spraying, crop monitoring, irrigation management, soil and field analysis, and planting tasks. Drones represent one of the most significant applications of IoT technology in agriculture. They provide aerial views and graphical maps of crops, empowering farmers to identify areas that require immediate attention and intervention. Moreover, drones perform evaluations of crop health, irrigation systems, progress tracking, spraying, and planting tasks. They prove to be valuable in reducing labor and saving time. Drones come in two main types: airborne drones and ground-based drones. Prior to using drones on a field, farmers need to determine the optimal flight height or ground resolution. By capturing images of the crops, drones enable farmers to promptly address the needs of specific crops. In comparison, satellite images can be affected by weather and lighting conditions, while UAVs offer a superior solution for image data collection. The advancements in UAV technologies are cost-effective, time-efficient, and enable the capture of high-resolution images in a non-destructive manner.

DB A. K. et al. (2023) have developed a technology that utilizes a drone for remote and instantaneous measurement of soil moisture content. The proposed approach involves gathering data from a soil moisture sensor and an Arduino FNO to monitor real-time soil moisture. This data is transmitted to the drone, which then sends it to the cloud for immediate monitoring and analysis using MATLAB through the Thingspeak IoT analytics platform. The proposed work consists of two main sections: the transmitting unit, comprising an Arduino UNO, a nRF24L01 radio transceiver, and a 9V battery, which is responsible for soil moisture measurement and data transmission to the UAV drone's receiving unit. The receiving unit includes a node MCU ESP8266-12e board, a nRF24L01 radio transceiver, a 5V USB step-up module, a 3.7 volt rechargeable battery, and a 4G Data Wi-Fi USB Stick, which acquires and transmits soil moisture data to the Thingspeak cloud in real-time.

5.3. Ingenious Greenhouses

The greenhouse industry has earned significant recognition in the agricultural community due to its ability to produce fresh agricultural products rapidly. However, the production costs of greenhouses have increased by approximately half due to labor and energy consumption expenses. To enhance food production, greenhouses maintain controlled conditions and optimum parameter settings, making even minor adjustments that could potentially lead to significant production increases. Conventional greenhouse farming is inadequate to meet the needs of the rapidly growing global population. The advancement of IoT-based technologies now enables the modernization of agriculture, which can effectively address the demands of an expanding population. The IoT-based smart greenhouse has become an essential component of agricultural information technology. According to Jamil F. et al. (2022), the Greenhouse Monitoring System (GMS) is being utilized in smart farming to manage greenhouse machinery and monitor it remotely. Additionally, a smart greenhouse can implement the GMS based on scientific management practices to boost output, enhance quality, and safeguard plants from adverse climate conditions.

In their study, Jamil et al. (2022) introduce an advanced smart contract designed for autonomous greenhouse conditions, emphasizing prediction, optimization, and control components. This approach leads to a significant increase in crop yield while reducing energy consumption by 41% compared to the baseline, and by 19% compared to the optimization strategy. Moreover, the smart contract demonstrates improved forecast accuracy with RMSEs of 1.32, 0.079, and 0.066 for CO2, humidity, and temperature, respectively. The effectiveness of the perfect smart contract is verified using the Hyperledger caliper, which confirms its efficiency.

5.4. Livestock Observation

Water is essential for the survival, development, and reproductive processes of plants. By employing sensors to measure soil moisture and water levels, farmers can ascertain the specific water needs of plants. By monitoring the water content and utilizing actuators for control, farmers can determine if any action is necessary to ensure adequate hydration. Additionally, various minerals like potassium, nitrogen, iron, sulfur, and calcium have an impact on the pH level of the soil. An ideal pH range of 5.5 to 7.5 promotes optimal growth conditions for plants.

ZigBee and LoRaWAN are widely utilized protocols for collecting environmental data and monitoring soil moisture levels. As an alternative option for monitoring soil characteristics and transferring data across short distances, Bluetooth is also commonly used.

The research by J.M. Domnguez-Nino et al. (2020) demonstrates the viability of implementing automated sensor-based irrigation scheduling in orchards. The proposed system provides precise irrigation amounts throughout the growing season by taking into account weather conditions and crop vegetation cycles. This irrigation system is built on the principles of water balance and local data collected from sensors. Automated feedback is provided using capacitance sensors, ensuring consistent interpretation and integration of decision-making processes. The algorithm used in this system requires less manual intervention compared to the involvement of professional technicians, while still aligning with ET (evapotranspiration) as measured by a weighted lysimeter. This system is well-suited for horticulture applications and can also adjust irrigation frequency based on different levels of tree vigor, resulting in less frequent watering for smaller tree patches.

D. Clarke et al. (2019) present a comprehensive framework for fully integrated data analytics, which facilitates the transfer and storage of crucial sensor data from agricultural microcontrollers. This framework is employed in a prototype plant monitoring system, where a Raspberry Pi is used for data processing and an IoT Cloud system is utilized for real-time applications. The microcontroller's prototype version is equipped with a capacitive moisture sensor, a humidity sensor, and three temperature sensors. The architecture of the system utilizes a publish-subscribe approach for communication between a standalone ESP32 microcontroller and an MQTT Broker. Users can access sensor data through a web application or by subscribing to the MQTT broker. To manage users and enable access to the gathered sensor data, the ESPlantMonitoring web application was developed.

A. Na et al. (2016) introduced a novel concept that holds promise for transforming agricultural practices and enhancing productivity: the remote monitoring of soil parameters. The crucial variables that aid in characterizing the soil and making informed decisions regarding crop selection and fertilizer applications include soil temperature, pH value, and soil moisture content. To monitor pH, this study utilizes antimony electrodes. For determining soil moisture content, the researchers leverage the inverse relationship between soil resistance and moisture levels and construct the necessary circuitry. Soil temperature is calculated using the Dallas one-wire protocol in conjunction with the DS18B20 sensor. The system incorporates Bluetooth technology to facilitate data transmission to nearby mobile devices. The STM32 Nucleo platform forms the foundational basis for the entire system.

In their proposed study, Mishra et al. (2023) concentrate on the utilization of IoT and Digital Twins in Livestock Farming (DTLF) to enhance the health and welfare of farm animals. They have put forth a four-layer design for IoT applications. The study demonstrates the application of IoT sensors, machine learning algorithms, monitoring the growth of dairy cattle, simulating cattle emotions, early detection of serious livestock lumpy skin disease, minimizing animal feed wastage, and increasing animal productivity. While these approaches present challenges, they hold great potential for effectively monitoring and assessing animal wellbeing. The incorporation of the Digital Twin (DT) mechanism is expected to improve disease analysis prediction for individual farms by an impressive 92%, similar to previous groundbreaking advancements. The efficacy of DTLF is expected to reach 92% for analyzing heartbeat, 94% for analyzing temperature range, and 94% for assessing humidity range as a whole.

Joshi et al. (2023) proposed the implementation of an IoT-based system to automate and improve crop monitoring, moisture monitoring, and water harvesting for precision agriculture. The system consists of four main components that collaborate to collect, process, and analyze data from various sensors. Using the Arduino IDE, Wi-Fi, several sensors and GSM modules, an LCD display, and a DC motor, the suggested system collects and analyzes data on soil moisture, crop growth rate, water level, and animal intervention. In case of any abnormal situation, the system immediately sends SMS alerts to the farmer. Farmers can access and analyze the stored data in the cloud to make informed decisions.

The design, implementation, and evaluation of the proposed IoT-based Water Harvesting, Moisture Monitoring, and Crop Monitoring Systems for Precision Agriculture are detailed in the study. The system has shown promising results by efficiently managing crop monitoring, moisture monitoring, and water harvesting and storage. The monitoring of animal intervention ensures crop safety by preventing unwelcome animal intrusions. Farmers can receive timely alerts through SMS, email, or app notifications. The adoption of a cloud-based data management and storage system allows for effective data processing and analysis, thereby enhancing decision-making and increasing agricultural yields. Overall, this technology provides a viable solution for precision farming and promotes environmentally friendly agricultural practices.

5.5. Crop Water Management

The possibility of implementing automated sensor-based irrigation scheduling in orchards is viable and can accurately administer appropriate water amounts by relying on local feedback and maintaining water balance. This approach ensures optimal water utilization for food production by adapting to weather patterns and crop growth cycles. Farmers frequently use excessive amounts of water to avoid yield reduction due to water stress, but this behavior can have a negative impact on production and increase the risk of water-related illnesses. The Smart Water Management Platform (SWAMP) project has been dedicated to achieving three crucial objectives simultaneously: enhancing agricultural output, reducing costs, and promoting environmental sustainability.

C. Kamienski et al. (2018) explain that even though IoT, intelligent water management systems, and related technologies seem like obvious choices, their practicality must be demonstrated through deployment in real-world settings. Moreover, the platforms used to create IoT applications need to be highly adaptable to accommodate different crops, climatic conditions, and countries. To enhance smart crop water management within the context of smart irrigation, the SWAMP project devised IoT-based technologies and approaches and conducted trials in Brazil, Italy, and Spain. This article outlines the pilots, architecture, and scenario-based development methodologies employed from the SWAMP project's perspective.

L. Doron et al. (2017) are working on a cost-effective precision irrigation platform as a component of the European research project FIGARO. The primary goal of FIGARO is to enhance water productivity in water-intensive crops. The project is creating a well-organized platform that incorporates a Decision Support System module designed for crop-specific management. By implementing this platform, the project aims to reduce the consumption of fresh water in agricultural settings. This initiative is essential for enabling agriculture to adapt to climate change and promote the sustainable use of natural resources. Funding for this project is provided in part by the 7th Framework Programme for Research and Technological Development, co-funded by the European Commission.

T.A. Khoa et al. (2019) present a practical and cost-effective smart agricultural system in Vietnam, utilizing LoRa module transmission to regulate environmental conditions in farming. Compared to similar precision options, this system is more affordable while maintaining comparable accuracy. The system records measurements and continuously monitors them, sending the data to consumers through a mobile application and network. Additionally, it can issue warnings and activate reduction tools when necessary. In a trial conducted in the Cho Lach area of Ben Tre province, Vietnam, two IoT nodes were used to sense temperature, humidity, and moisture levels. The wireless communication module and integrated electronics worked seamlessly. The results indicate that this technology offers Vietnamese farmers an excellent approach to smart agricultural control. However, it should be noted that real-time data generated from IoT nodes requires time for storage and analysis before finding the optimal solution.

5.6. The Best Use of Fertilizer

Fertilizer refers to the substance that supplies vital nutrients required for the growth of plants. These nutrients play a crucial role in the development of certain plants, such as potassium (K) aiding in water circulation through stems, phosphorus (P) being essential for flowers, fruits, and roots, and nitrogen contributing to leaf growth. To maintain soil nutrient levels and prevent uncontrolled usage, which can lead to climate change and deforestation, it is crucial to apply the appropriate fertilizers in the correct proportions. Effective fertilization necessitates identifying the lacking nutrients, providing the right quantity, and applying them in the appropriate locations. To ensure these conditions are met, innovative fertilization techniques based on the Internet of Things (IoT) have been introduced.

G. Lavanya et al. (2019) propose an IoT system to address the challenges of agricultural soil nutrient assessment, which is often hindered by manual laboratory testing, leading to improper fertilizer application by farmers. The system incorporates a fuzzy rule-based approach and an NPK sensor, which detects nitrogen, phosphorus, and potassium levels in crops. Through SMS alerts, the system notifies farmers when their crops are deficient in these essential nutrients. Experimental simulations demonstrate the system's cost-effectiveness, accuracy, and intelligent functioning, ensuring timely and appropriate fertilizer application. This low-cost, precise, and intelligent IoT device serves as a valuable tool for farmers engaged in agriculture.

A. Giri et al. (2016) have created a system that assesses the soil's nitrogen (N), phosphorus (P), and potassium (K) concentrations, along with its overall nutrient content. Based on these measurements, the system will dispense the required amount of fertilizer through an irrigation system. The user will be informed of all the data updates through email.

P. Benincasa et al. (2018) conducted a study to assess the early nitrogen (N) status and predict the winter wheat yield using in-field data along with a remote sensing Normalized Difference Vegetation Index (NDVI). Six different N rates were applied, with half of the rates applied during shooting and the other half during tillering, in ascending order. The study observed a minimal 20 kg N difference between consecutive N treatments. Various indicators of crop vegetative and N status, including plant N% concentration, crop N absorption, ratio of transmitted to incident photosynthetically active radiation, leaf SPAD values, and NDVI derived from SAT (Satellite) and UAV data, were compared. The results showed significant correlations between in-field variables and NDVI at all crop stages. Strong connections were found between NDVI, in-field data, and grain yield during the shooting stage, suggesting that they can be used interchangeably depending on the specific situation.

5.7. Energy Harvesting

Especially in SA systems, IoT devices and sensors are designed for remote locations where a stable power source is scarce. Traditional batteries are not suitable as they require frequent replacement or recharging. To address this issue, an alternative approach involves harnessing energy from the surrounding environment through specialized circuitry. Establishing these techniques is crucial to ensuring a reliable and efficient power supply for IoT devices. By addressing challenges such as irregular energy harvesting and low panel conversion efficiency, the lifespan of a WSN can be extended. These techniques utilize various elements, including radio frequency, thermal energy, vibration, and solar energy.

As S. Kumari et al. (2019) emphasize, it is crucial to incorporate renewable energy sources in order to reduce the ecological footprint and provide consumers with an economically viable solution. To encourage the widespread adoption of harvesting systems, user-friendly interfaces like applications, websites, or emails can be employed. Additionally, integrating weather models and prediction algorithms into the system can further enhance its efficiency. For instance, in a portable irrigation system, monitoring solar radiation changes enables the adjustment of runtimes, optimizing battery production.

According to E.M. Jung et al. (2019), the introduction of a wideband, omnidirectional energy harvester has opened up numerous possibilities for mass-deployment applications, such as precision agriculture and structural health monitoring. This innovative harvester can collect energy from all nearby FM towers and stands out for its broader bandwidth and superior gain, which enable it to maintain omnidirectionality. Remarkably, it can harvest up to 923 W on a rooftop situated 1.54 km away. The system operates efficiently across a wide band, achieving a 56% efficiency level with a -17 dBm sensitivity.

As highlighted by S. Sadowski et al. (2018), the concept of smart farming has become a reality. Precision agriculture (PA) leverages this technology to monitor crops and deliver specific treatments, leading to reduced stress for farmers and enhanced efficiency. The study introduces an IoT-enabled solar-powered smart agricultural monitoring system, employing prototype solar-powered nodes to observe the environment and relay data to a central base station. This system demonstrates the promising potential of smart farming in agriculture, as exemplified by prolonged device longevity and extended battery life enabled by energy-harvesting equipment.

5.8. Healthcare or Managing Pests and Plant Diseases

Insufficient soil nutrients and water, coupled with the spread of illnesses by pests, significantly impact crop output and quality. This hampers the ability of plants to thrive and produce fruitful yields. To mitigate crop losses, early detection of insects is crucial. IoT systems can play a vital role in this process by utilizing sensor data to aid in the identification of pests.

D. Popescu et al. (2020), the authors emphasized algorithms, methodologies, and tools for smooth interoperability. The article offers a case study for collaborative UAV-WSN operation in PA and includes optimum paths for UAV-enabled field data collection and efficient use of limited ground sensor network resources. In order to balance path length and time spent in cluster heads, the article suggests combining segment and loiter tracking modes. The UAV platform Data provides collection applications and large-scale monitoring that rely less on external communication and processing infrastructure. It is intended to conduct a field evaluation to confirm the system's influence on crop management.

Semios (2021) suggests that the irrigation used in agriculture is the world's largest user of water, necessitating a heavy reliance on technology to maximize water utilization, cut down on energy use, and enhance crop quality. The SWAMP project creates IoT based techniques for precise irrigation and tests them in Brazil, Spain, and Italy. The document presents the project's pilots, architecture, and scenario-based development approach.

According to the study by S.A. Nandhini et al. (2018), they used a Wireless Multimedia Sensor Network (WMSN) camera to take pictures and examine the detection of infected leaves on farms. Subsequently, relevant characteristics were extracted from these images to facilitate categorization. To achieve automated identification and classification of plant leaf diseases, the researchers propose a robust system based on the Internet of Things (IoT) and Compressed Sensing (CS) techniques.

In their suggested approach, the segmented leaf images undergo compression through statistical methods using CS before being transmitted wirelessly. To retrieve features for disease classification, a combination of a Support Vector Machine (SVM) classifier and a Gray level co-occurrence Matrix (GLCM) is employed.

The results obtained from the system are highly promising, with an impressive overall detection accuracy of 98.4% and a classification accuracy of 98.4%, surpassing existing approaches in the literature. The experimental evaluation of the system is performed in real-time, utilizing sensors from a Raspberry Pi 3 and the Thingspeak IoT platform.

The researchers plan to implement this technology in real agricultural fields, capturing images from various locations and storing crop-related information for future use. This innovative approach holds great potential for effective disease detection and monitoring in practical farming scenarios.

6. DISCUSSION

6.1. Challenges

The adoption of IoT technology, such as sensors and gateways, is crucial for implementing smart agriculture practices. However, the high cost and security concerns pose significant obstacles. The cost of the hardware and software, both of which must withstand extreme external conditions, can affect the system's effectiveness. To ensure reliable data processing and storage, robust computer hardware is necessary. To prevent data loss, tampering, privacy breaches, and loss of customer trust, security measures such as dust and water resistance should be employed.

Maintaining uninterrupted communication is vital for accurate decision-making. In this regard, blockchain technology can play a pivotal role in enhancing cybersecurity and reducing risks. By implementing these security measures and leveraging blockchain, smart agriculture can overcome challenges and pave the way for more efficient and secure farming practices.

Figure 8 presents the challenges faced in the implementation of smart agriculture practices, particularly for small-scale farmers, who may struggle with high initial setup and maintenance costs when adopting advanced technologies. Ensuring data privacy and addressing potential cybersecurity vulnerabilities is essential due to the heavy reliance on data collection and analysis in smart agriculture, emphasizing the need for robust security measures to protect sensitive information.

The figure also emphasizes the significance of effective communication in smart agriculture, which involves seamless integration of various devices, agricultural equipment, and central control systems to facilitate real-time data transmission and coordination. To achieve this, efficient communication networks like LoRaWAN, Sigfox, and others enable smooth data transfer from actuators and sensors, providing farmers with timely insights for accurate decision-making and improved resource management.

In addition, the figure highlights the importance of a well-designed Hardware and System layout to address challenges related to interoperability, standardization, and scalability in the hardware setup, aiming to establish an ideal ecosystem for smart agriculture practices.

6.2. Future Trends

6.2.1. Communications Through Satellite and Aerial Systems

In a SA system, uninterrupted communication is essential for accurate, smart decision-making. To avoid undesirable outcomes, it is crucial to gather sensor data consistently and to continuously check on in-field actuators. Weather and interference can have an impact on applications for SA that use wireless communication. Hardware protection enhances communication quality and lowers maintenance expenses. While licensed cellular networks like 5G and NB-IoT offer lesser interference concerns, unlicensed frequency bands like Wi-Fi, ZigBee, LoRaWAN, and Sigfox face higher interference hazards. These solutions, however, call for more infrastructure in terms of hardware and subscription fees.

Figure 8. Challenges in smart agriculture

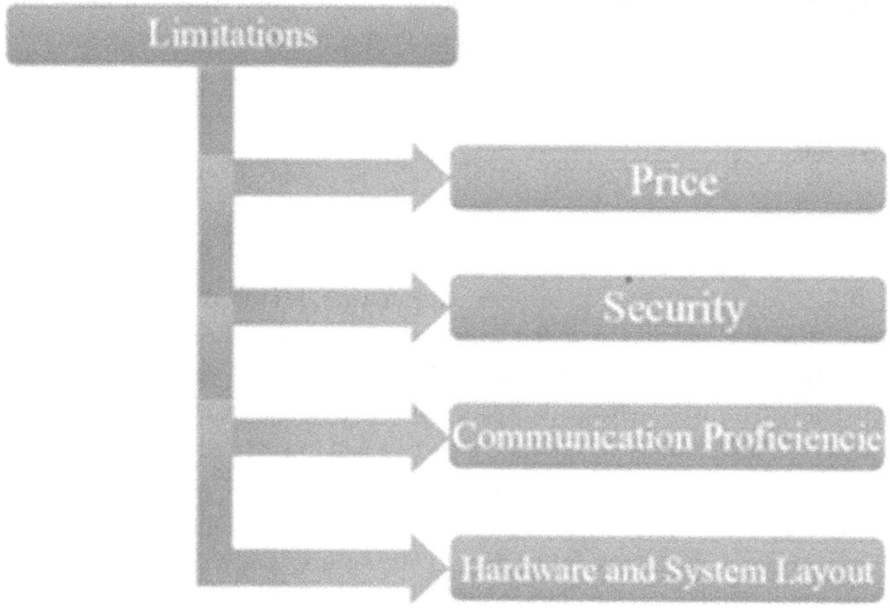

6.2.2. The Effectiveness of Energy Production and Consumption

In remote locations, providing on-grid electricity to all components of autonomous vehicle systems can often be challenging. To address this issue, efforts are being made to reduce the power consumption of IoT devices and harness energy from the surrounding environment. By adjusting the operation frequency and duration of sleep mode and utilizing specialized communication protocols, low power consumption can be achieved without compromising security and reliability.

While renewable energy sources like solar energy are expected to be increasingly utilized, their limitations in supplying electricity during nighttime and overcast conditions may pose constraints. Despite these challenges, ongoing research and advancements aim to improve the energy efficiency and sustainability of autonomous vehicle systems in remote settings.

6.2.3. Applications and Protocols

Currently, most experiments related to SA applications are conducted on a limited scale, focusing on the evaluation of various IoT hardware and communication protocols. However, for future progress, it will be essential to implement large-scale systems to effectively assess their practicality and gather valuable feedback.

Fast-paced technological advancements continuously reshape daily routines, presenting novel possibilities while diminishing the significance of certain aspects. By identifying shared domains and communication protocols through a thorough comprehension of evolving SA trends and application areas, we can gain valuable insights that will shape the future of smart agriculture.

6.2.4. Building Blocks for LPWAN Technologies

Low Power Wide Area Network (LPWAN) protocols, like NB-IoT, offer advantages such as low power consumption, broad coverage, and reliable data transmission for wireless access network systems. While these protocols require infrastructure construction, they are expected to rapidly expand their coverage.

However, LoRaWAN surpasses the limitations of traditional LPWAN technologies, particularly in terms of power consumption and coverage. Its simplified infrastructure makes installation and usage more straightforward, providing scalability and security benefits for smart agriculture (SA) systems. Despite the challenges, it is predicted that LPWAN technologies will continue to progress and evolve in the future, driven by ongoing research and innovation.

7. CONCLUSION

In conclusion, the integration of IoT and AI technologies in agriculture holds immense promise for revolutionizing the industry and driving it towards a more sustainable and efficient future. By harnessing the power of real-time data monitoring and intelligent decision-making, farmers can optimize their resource usage, increase productivity, and minimize labor demands. The seamless communication and information sharing enabled by IoT create a connected agricultural ecosystem, facilitating collaboration among stakeholders and promoting smarter farming practices.

Moreover, the adoption of IoT-based solutions addresses critical challenges such as environmental preservation, hygiene, and local considerations, ensuring a more responsible and ethical approach to agriculture. With the ability to track present conditions and predict future harvests, IoT empowers farmers, schools, and government agencies to make informed decisions and plan for more successful agricultural outcomes.

By embracing IoT-based rural frameworks, we can elevate the quality of rural goods and boost overall agricultural productivity. This transformative advancement not only enhances the livelihood of farmers but also contributes to sustainable development, reducing our dependence on climate change and preserving our valuable natural resources.

As we continue to embrace farm IoT, we usher in a new era of agricultural advancement, paving the way for a smarter, more connected, and environmentally conscious farming industry. The future of agriculture lies in harnessing the potential of IoT and AI, creating a world where innovative technologies and responsible practices work in harmony to feed our growing global population while safeguarding our planet for generations to come.

REFERENCES

Aasha Nandhini, S., Hemalatha, R., Radha, S., & Indumathi, K. (2018). Web enabled plant disease detection system for agricultural applications using WMSN. *Wireless Personal Communications*, *102*(2), 725–740. doi:10.100711277-017-5092-4

Abbasi, A. Z., Islam, N., & Shaikh, Z. A. (2014). A review of wireless sensors and networks' applications in agriculture. *Computer Standards & Interfaces*, *36*(2), 263–270. doi:10.1016/j.csi.2011.03.004

Ari, A. A. A., Gueroui, A., Labraoui, N., & Yenke, B. O. (2015). Concepts and evolution of research in the field of wireless sensor networks. arXiv preprint arXiv:1502.03561.

Augustin, A., Yi, J., Clausen, T., & Townsley, W. M. (2016). A study of LoRa: Long range & low power networks for the internet of things. *Sensors (Basel)*, *16*(9), 1466. doi:10.339016091466 PMID:27618064

Avşar, E., & Mowla, M. N. (2022). Wireless communication protocols in smart agriculture: A review on applications, challenges and future trends. *Ad Hoc Networks*, *136*, 102982. doi:10.1016/j.adhoc.2022.102982

Benincasa, P., Antognelli, S., Brunetti, L., Fabbri, C. A., Natale, A., Sartoretti, V., Modeo, G., Guiducci, M., Tei, F., & Vizzari, M. (2018). Reliability of NDVI derived by high resolution satellite and UAV compared to in-field methods for the evaluation of early crop N status and grain yield in wheat. *Experimental Agriculture*, *54*(4), 604–622. doi:10.1017/S0014479717000278

Boonyopakorn, P. (2018, October). Applying Data Analytics to Findings of User Behaviour Usage in Network Systems. In *2018 International Conference on Information Technology (InCIT)* (pp. 1-6). IEEE. 10.23919/INCIT.2018.8584865

Centenaro, M., Vangelista, L., Zanella, A., & Zorzi, M. (2016). Long-range communications in unlicensed bands: The rising stars in the IoT and smart city scenarios. *IEEE Wireless Communications*, *23*(5), 60–67. doi:10.1109/MWC.2016.7721743

Clarke, D., Al-Aqrabi, H., Hill, R., Mistry, P., & Lane, P. (2019). Implementing a Lightweight Cloud-Based Process Monitoring Solution for Smart Agriculture. *Smart City and Informatization: 7th International Conference, iSCI 2019, Guangzhou, China, November 12–15, 2019 Proceedings*, *7*, 379–391.

DB, A. K., Doddabasappa, N., Bairwa, B., Kumar, A., & Raju, G. (2023, April). IoT-based Water Harvesting, Moisture Monitoring, and Crop Monitoring System for Precision Agriculture. In *2023 International Conference on Distributed Computing and Electrical Circuits and Electronics (ICDCECE)* (pp. 1-6). IEEE.

Domínguez-Niño, J. M., Oliver-Manera, J., Girona, J., & Casadesús, J. (2020). Differential irrigation scheduling by an automated algorithm of water balance tuned by capacitance-type soil moisture sensors. *Agricultural Water Management*, *228*, 105880. doi:10.1016/j.agwat.2019.105880

Doron, L. (2017). Flexible and precise irrigation platform to improve farm scale water productivity. *Impact*, *2017*(1), 77–79. doi:10.21820/23987073.2017.1.77

Fernández-Ahumada, L. M., Ramírez-Faz, J., Torres-Romero, M., & López-Luque, R. (2019). Proposal for the design of monitoring and operating irrigation networks based on IoT, cloud computing and free hardware technologies. *Sensors (Basel)*, *19*(10), 2318. doi:10.339019102318 PMID:31137517

Foubert, B., & Mitton, N. (2020). Long-range wireless radio technologies: A survey. *Future Internet*, *12*(1), 13. doi:10.3390/fi12010013

Giri, A., Dutta, S., & Neogy, S. (2016, October). Enabling agricultural automation to optimize utilization of water, fertilizer and insecticides by implementing Internet of Things (IoT). In *2016 International Conference on Information Technology (InCITe)-The Next Generation IT Summit on the Theme-Internet of Things: Connect your Worlds* (pp. 125-131). IEEE.

Goap, A., Sharma, D., Shukla, A. K., & Krishna, C. R. (2018). An IoT based smart irrigation management system using Machine learning and open source technologies. *Computers and Electronics in Agriculture, 155*, 41–49. doi:10.1016/j.compag.2018.09.040

Jamil, F., Ibrahim, M., Ullah, I., Kim, S., Kahng, H. K., & Kim, D. H. (2022). Optimal smart contract for autonomous greenhouse environment based on IoT blockchain network in agriculture. *Computers and Electronics in Agriculture, 192*, 106573. doi:10.1016/j.compag.2021.106573

Joshi, K., Singh, R., & Kumar, N. (2023, March). Cloud Based Real Time Soil Moisture Content Monitoring Using IoT and Unmanned Aerial Vehicles. In *2023 Sixth International Conference of Women in Data Science at Prince Sultan University (WiDS PSU)* (pp. 206-210). IEEE. 10.1109/WiDS-PSU57071.2023.00050

Jung, E. M., Cui, Y., Lin, T. H., He, X., Eid, A., Hester, J. G., Abowd, G. D., Starner, T. E., Lee, W.-S., & Tentzeris, M. M. (2019). A wideband, quasi-isotropic, kilometer-range FM energy harvester for perpetual IoT. *IEEE Microwave and Wireless Components Letters, 30*(2), 201–204. doi:10.1109/LMWC.2019.2957653

Kamienski C., Soininen J. P., Taumberger M., Fernandes S., Toscano A., Cinotti T. S., ... & Neto, A. T. (2018, June). SWAMP: An IoT-based smart water management platform for precision irrigation in agriculture. In *2018 Global Internet of Things Summit (GIoTS)* (pp. 1-6). IEEE.

Kamruzzaman, S. M., Pavel M. I., Hoque M. A., & Sabuj S. R. (2019). Promoting greenness with IoT-based plant growth system. *Computational Intelligence and Sustainable Systems: Intelligence and Sustainable Computing*, 235-253.

Kapoor, A., & Attri, V. (2021, December). A Review of Tools and Technologies for Big Data Analytics. In *2021 10th International Conference on System Modeling & Advancement in Research Trends (SMART)* (pp. 13-16). IEEE. 10.1109/SMART52563.2021.9676237

Kayser, V., Nehrke, B., & Zubovic, D. (2018). Data science as an innovation challenge: From big data to value proposition. *Technology Innovation Management Review, 8*(3), 16–25. doi:10.22215/timreview/1143

Keswani, B., Mohapatra, A. G., Mohanty, A., Khanna, A., Rodrigues, J. J., Gupta, D., & De Albuquerque, V. H. C. (2019). Adapting weather conditions based IoT enabled smart irrigation technique in precision agriculture mechanisms. *Neural Computing & Applications, 31*(S1), 277–292. doi:10.100700521-018-3737-1

Kolomvatsos, K., & Anagnostopoulos, C. (2021). Proactive, uncertainty-driven queries management at the edge. *Future Generation Computer Systems, 118*, 75–93. doi:10.1016/j.future.2020.12.028

Kumari, S., Modani, V., Oswal, S., Sarma, S. S., & Kulkarni, P. (2019). An Android-Based Solution for Solar Power Harvesting and Irrigation Using IoT. In *International Conference on Intelligent Data Communication Technologies and Internet of Things (ICICI) 2018* (pp. 1136-1147). Springer International Publishing. 10.1007/978-3-030-03146-6_132

Lauridsen, M., Vejlgaard, B., Kovacs, I. Z., Nguyen, H., & Mogensen, P. (2017, March). Interference measurements in the European 868 MHz ISM band with focus on LoRa and SigFox. In *2017 IEEE Wireless Communications and Networking Conference (WCNC)* (pp. 1-6). IEEE. 10.1109/WCNC.2017.7925650

Lavanya, G., & Rani, C., & GaneshKumar, P. (2020). An automated low cost IoT based Fertilizer Intimation System for smart agriculture. *Sustainable Computing: Informatics and Systems, 28*, 100300.

Lavric, A., Petrariu, A. I., & Popa, V. (2019). Long range sigfox communication protocol scalability analysis under large-scale, high-density conditions. *IEEE Access : Practical Innovations, Open Solutions, 7*, 35816–35825. doi:10.1109/ACCESS.2019.2903157

Lutu, P. E. (2018, December). Improving Classification Performance for Minority Classes through the use of Positive-Versus-Negative Ensemble Models. In *2018 International Conference on Intelligent and Innovative Computing Applications (ICONIC)* (pp. 1-6). IEEE. 10.1109/ICONIC.2018.8601260

Madushanki, A. R., Halgamuge, M. N., Wirasagoda, W. S., & Syed, A. (2019). Adoption of the Internet of Things (IoT) in agriculture and smart farming towards urban greening: A review. *International Journal of Advanced Computer Science and Applications, 10*(4), 11–28. doi:10.14569/IJACSA.2019.0100402

Mahmoud M. S., & Mohamad A. A. (2016). *A study of efficient power consumption wireless communication techniques/modules for internet of things (IoT) applications*. Academic Press.

Marr, B. (2017, October). How Much Data Do We Create Every Day? The Mind-Blowing Stats Everyone Should Read. *Forbes*.

Mishra, S., & Sharma, S. K. (2023). Advanced contribution of IoT in agricultural production for the development of smart livestock environments. *Internet of Things, 22*, 100724. doi:10.1016/j.iot.2023.100724

Na, A., Isaac, W., Varshney, S., & Khan, E. (2016, October). An IoT based system for remote monitoring of soil characteristics. In *2016 International conference on information technology (InCITe)-the next generation IT summit on the theme-internet of things: Connect your Worlds* (pp. 316-320). IEEE. 10.1109/INCITE.2016.7857638

Nair K. K., Abu-Mahfouz A. M., & Lefophane S. (2019, March). Analysis of the narrow band internet of things (NB-IoT) technology. In *2019 conference on information communications technology and society (ICTAS)* (pp. 1-6). IEEE.

Popescu, D., Stoican, F., Stamatescu, G., Ichim, L., & Dragana, C. (2020). Advanced UAV–WSN system for intelligent monitoring in precision agriculture. *Sensors (Basel), 20*(3), 817. doi:10.339020030817 PMID:32028736

Rajagopal G., Lodd V. M., Vignesh A., Rajesh, R. & Vijayaraghavan V. (2014, August). Low cost cloud based intelligent farm automation system using Bluetooth low energy. In *2014 IEEE Region 10 Humanitarian Technology Conference (R10 HTC)* (pp. 127-132). IEEE.

Rathor, S., & Kumari, S. (2021, October). Smart agriculture system using iot and cloud computing. In *2021 5th International Conference on Information Systems and Computer Networks (ISCON)* (pp. 1-4). IEEE. 10.1109/ISCON52037.2021.9702345

Reynolds, J. (2003). *Going Wi-Fi: Networks Untethered with 802.11 Wireless Technology*. CRC Press. doi:10.1201/9781482280869

Sadowski, S., & Spachos, P. (2018, November). Solar-powered smart agricultural monitoring system using internet of things devices. In *2018 IEEE 9th Annual Information Technology, Electronics and Mobile Communication Conference (IEMCON)* (pp. 18-23). IEEE. 10.1109/IEMCON.2018.8614981

Sadu, V. B., Budidha, N., Billewar, S., Pimo, S. J., Kumar, D. R., & Kumar, T. C. A. (2022, November). Digitalization of agriculture by using the application of IoT and robotics. In *2022 International Interdisciplinary Humanitarian Conference for Sustainability (IIHC)* (pp. 183-186). IEEE. 10.1109/IIHC55949.2022.10060724

Santur, Y., Karaköse, M., & Akin, E. (2016, September). Improving of personal educational content using big data approach for mooc in higher education. In *2016 15th International Conference on Information Technology Based Higher Education and Training (ITHET)* (pp. 1-4). IEEE. 10.1109/ITHET.2016.7760728

Semios. (n.d.). *Integrated Pest Management*. https://semios.com/ipm/

Sharma D., Singh A., & Singhal S. (2021). The Technological Shift: AI in Big Data and IoT. *The Smart Cyber Ecosystem for Sustainable Development*, 69-90.

Shatnawi, M. Q., Yassein, M. B., Abuein, Q., & Nsuir, L. (2019, December). Big data analytics tools and applications: survey. In *Proceedings of the Second International Conference on Data Science, E-Learning and Information Systems* (pp. 1-4). 10.1145/3368691.3368741

Symeonaki, E. G., Arvanitis, K. G., & Piromalis, D. D. (2019). Cloud computing for IoT applications in climate-smart agriculture: A review on the trends and challenges toward sustainability. In *Innovative Approaches and Applications for Sustainable Rural Development: 8th International Conference, HAICTA 2017, Chania, Crete, Greece, September 21-24, 2017, Selected Papers 8* (pp. 147-167). Springer International Publishing. 10.1007/978-3-030-02312-6_9

Taylor, G. A., Torres, H. B., Ruiz, F., Marin, M. N., Chaves, D. M., Arboleda, L. T., ... Mouazen, A. M. (2019). pH measurement IoT system for precision agriculture applications. *Revista IEEE América Latina*, *17*(05), 823–832. doi:10.1109/TLA.2019.8891951

Upadhyay, N. (2018). *CABology: Value of Cloud, Analytics and Big Data Trio Wave*. Springer. doi:10.1007/978-981-10-8675-5

Viswanatha, V., Kumari, A., & Sathisha, B. M. (2022, October). Implementation of IoT in Agriculture: A Scientific Approach for Smart Irrigation. In *2022 IEEE 2nd Mysore Sub Section International Conference (MysuruCon)* (pp. 1-6). IEEE. 10.21203/rs.3.rs-2392461/v1

Wang, Y. P. E., Lin, X., Adhikary, A., Grovlen, A., Sui, Y., Blankenship, Y., Bergman, J., & Razaghi, H. S. (2017). A primer on 3GPP narrowband Internet of Things. *IEEE Communications Magazine*, *55*(3), 117–123. doi:10.1109/MCOM.2017.1600510CM

Xiaojing, Z., & Yuanguai, L. (2012, September). Zigbee implementation in intelligent agriculture based on internet of things. In *2nd International Conference on Electronic & Mechanical Engineering and Information Technology* (pp. 1842-1846). Atlantis Press.

Zhihua, S. (2016, August). Design of smart home system based on ZigBee. In *2016 International Conference on Robots & Intelligent System (ICRIS)* (pp. 167-170). IEEE. 10.1109/ICRIS.2016.35

Chapter 5
Analysing the Applications of Cloud Computing in Smart Agriculture

Ishan Jain
Vellore Institute of Technology, India

Brijendra Singh
https://orcid.org/0000-0003-2608-3388
Vellore Institute of Technology, India

ABSTRACT

Agriculture has been the backbone of human civilization. For ages, humans have been using traditional farming techniques to produce high-quality crops, but with the era of digitalization and the challenges of these methods, there has been a shift from traditional farming techniques to smarter farming techniques, which utilise the latest technologies to reduce labour and maximise the yield in a given scenario. Smart agriculture produces a large amount of data from sensors and IoT devices; this is where cloud computing comes into play. Cloud computing-based technology provides storage and analysis facilities to deal with huge amounts of data and produces real-time insights that can help the farmer make better decisions. Cloud computing also helps in reducing the cost and resources incurred in smart agriculture techniques, thus making the model more efficient and useful for farmers. This chapter will analyse various applications in which innovative cloud computing technologies can be used in smart agriculture and the drawbacks that should be considered while adopting the cloud model.

1. INTRODUCTION

It is anticipated that the global population will reach 9 billion by the year 2050, which represents an increase of about a quarter in comparison to the current population (World Population Projected to Reach 9.8 Billion in 2050, and 11.2 Billion in 2100 according to United Nations Department of Economic and Social Affairs. Such a high rise in population will certainly require higher demand for food and various

DOI: 10.4018/979-8-3693-0200-2.ch005

agricultural products. Considering the rise in the global population, it is very crucial to be more vigilant regarding the quality of food and the nutrition content derived from it. Food production should be nearly tripled by 2050 in order to satisfy future food demands. To meet the global food demand in 2050, grain crop production must be increased to 3 billion tonnes from the existing from 2.1 billion tonnes and meat production to 470 million tonnes from the existing 200 million tonnes (Tripathi et al., 2019).

In the present scenario, a significant portion of farmers and growers' time is devoted to monitoring and comprehending the condition of crops, which leaves them with limited time for actual farming activities (Sisinni et al., 2018). Therefore, there is a need for precise and advanced technologies to propel the agricultural sector forward. Moreover, in order to achieve sustainable agriculture, it is crucial to minimize the impact of climatic and environmental factors. Additionally, 42% of the population in the world depends on the agricultural sector for its bread and butter, and is considered as crucial for economies of developing nations (Aznar-Sánchez et al., 2019). Hence, there is a need for revolutionizing the sector with the implementation of Industry 4.0 technologies in agriculture will help alleviate the increasing challenges faced by agricultural processes worldwide.

Technological development has paved a new way for the use of electronical systems in the agricultural environment. Technologies like IoT, Augmented Reality, big data analysis, and artificial intelligence have been integrated with traditional farming techniques with the goal to increase sustainability, productivity, and efficiency (Saban et al., 2022). Cloud computing is an essential component of smart agriculture, as it provides storage services to handle the large amount of data generated by agricultural devices and sensors and the required computing power to analyse the data. Cloud computing enables farmers to view real-time information related to field, soil, moisture, crops, weather, insects, etc. i.e., the components that play a crucial role in agriculture. Smart farming is a technique that uses modern technologies and the traditional approach to increase the quantity and quality of products while decreasing human labour and costs required for production.

Before exploring the applications of cloud computing in the agricultural sector, it is essential to understand the fundamentals of cloud computing and smart agriculture, as well as how they are interconnected. By leveraging cloud computing and Industry 4.0 technologies, the agricultural sector can move toward a more sustainable and efficient future, meeting the growing demand for food while reducing the impact on the environment. This chapter will explore the various ways in which cloud computing can be applied to the agricultural sector and the benefits that it can bring to farmers, growers, and the global population as a whole.

1.1 Cloud Computing

Cloud Computing is the availability of IT Resources on demand. In this fast-paced world, we need immediate access to these resources from any part of the world. (Qian et al., 2009) Whether someone is using e-commerce platform or someone is trying to access his/her banking information, it is all part of cloud computing (Soul et al., 2017).

As shown in Figure 1, there are four different types of clouds namely, Public Cloud, Private Cloud, Hybrid Cloud, and Community Cloud (Bhargav & Manhar, 2020):

- Public Cloud: Public Clouds are clouds that are third party managed that provide cloud services over the internet to the public. These services cost according to the usage of resources.

- Private Cloud: Private Clouds are clouds that work on a private infrastructure where resources are allocated according to the fluctuating commands of the users. Here, the cost is the same every month, regardless of the usage of the resources.
- Hybrid Cloud: A type of cloud that combines multiple computation environments.
- Community Cloud:A type of distributed system that is designed by integrating services from multiple clouds to cater to the unique requirements of a user or a community. It may be managed by an organization or third party.

Figure 1. Cloud deployment models

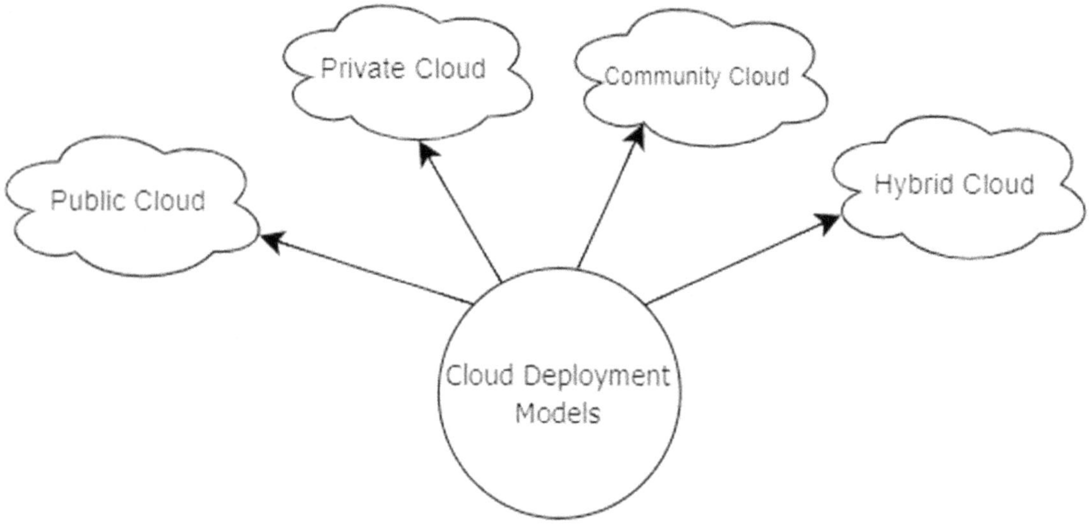

Cloud Service Providers provide services according to different models. The three standard models, as shown in Figure 2, in cloud computing are (Choudhary et al., 2016):

- Infrastructure as a Service (IaaS): Moreover known as Hardware-as-a-Service (HaaS), refers to a computer infrastructure that is managed online. The principal advantage of implementing Infrastructure as a Service (IaaS) is its cost-effectiveness and convenience for users, as it eliminates the need for purchasing and managing physical servers.
- Platform as a Service (IaaS): It enables the delivery of applications composed of services maintained by a third party via the cloud using the Platform-as-a-Service (PAAS) architecture. The technology provides the capacity for flexible expansion of applications, allowing software developers to design internet-based applications and services that can be deployed through a range of options, including public, private, or hybrid.
- Software as a Service (IaaS): The Software-as-a-Service (SaaS) enables the offering of software applications to customers as a service. It refers to a type of computer software that is installed on a server and can be accessed through the internet.

Figure 2. Cloud computing services

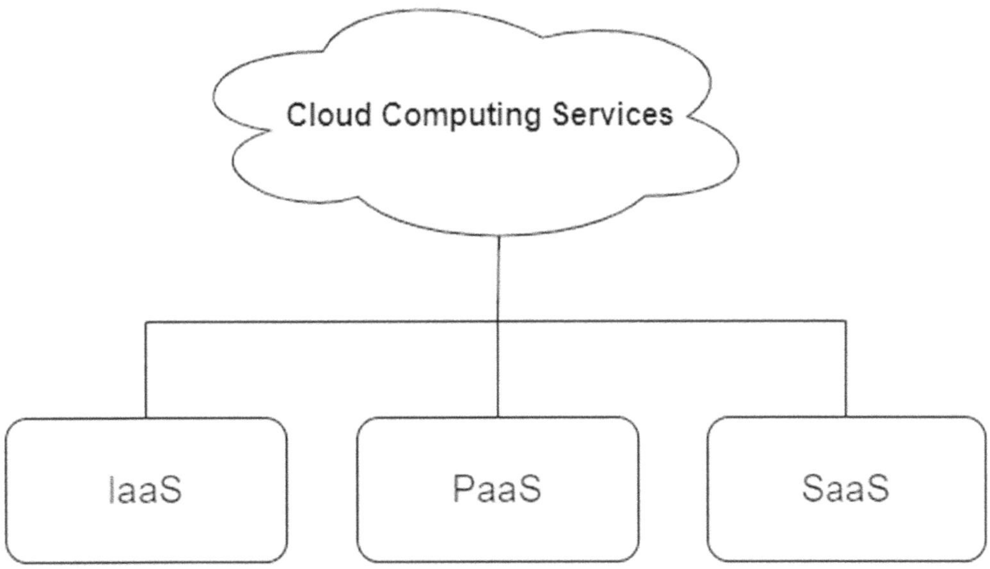

1.2 Smart Agriculture

Smart Agriculture refers to the management of farms using modern information and communication technology to increase the quantity and quality of products while optimizing the human labour required for the produce. (Taylor & Marcus, 2018) Smart agriculture uses IoT devices and sensors along with cloud computing and big data analytics to observe, diagnose, decide, and take action regarding emerging issues and allows farmers to react quickly. (Yang et al., 2021)

Smart agriculture consists of controlling and monitoring various processes of the farming process. It can be the following.

1.2.1. Automation

Agriculture automation is the process of using various technological equipment to enhance and automate agricultural businesses. Day -to-day processes such as irrigation, weeding, and harvesting can be automated using drones, robots, and other smart machinery that can reduce the labour and speed up the process. (Phasinam et al., 2022)

1.2.2. Resource Optimization

Various resources such as land, labour, water, chemicals including fertilizers and pesticides can be monitored using sensors and analysed in order to maximize the efficiency of every resource. Soil moisture and nutrients can be monitored so that it is only irrigated when the moisture content gets low and can alert the farmer when nutrients of the soil get depleted.

1.2.3. Agricultural Production Processes

With today's technology, almost every aspect of a farm can be made smart using devices. The production, process and storage can be monitored, the weather and other geographical features of the farm can be foreseen remotely (Rathod et al., 2022). Crops and animals can also be taken care of using smart cameras and sensors, hence making the overall agricultural production smart and efficient. Figure 3 picturizes the possible uses of technology in agriculture making it smart.

Figure 3. Smart agriculture system

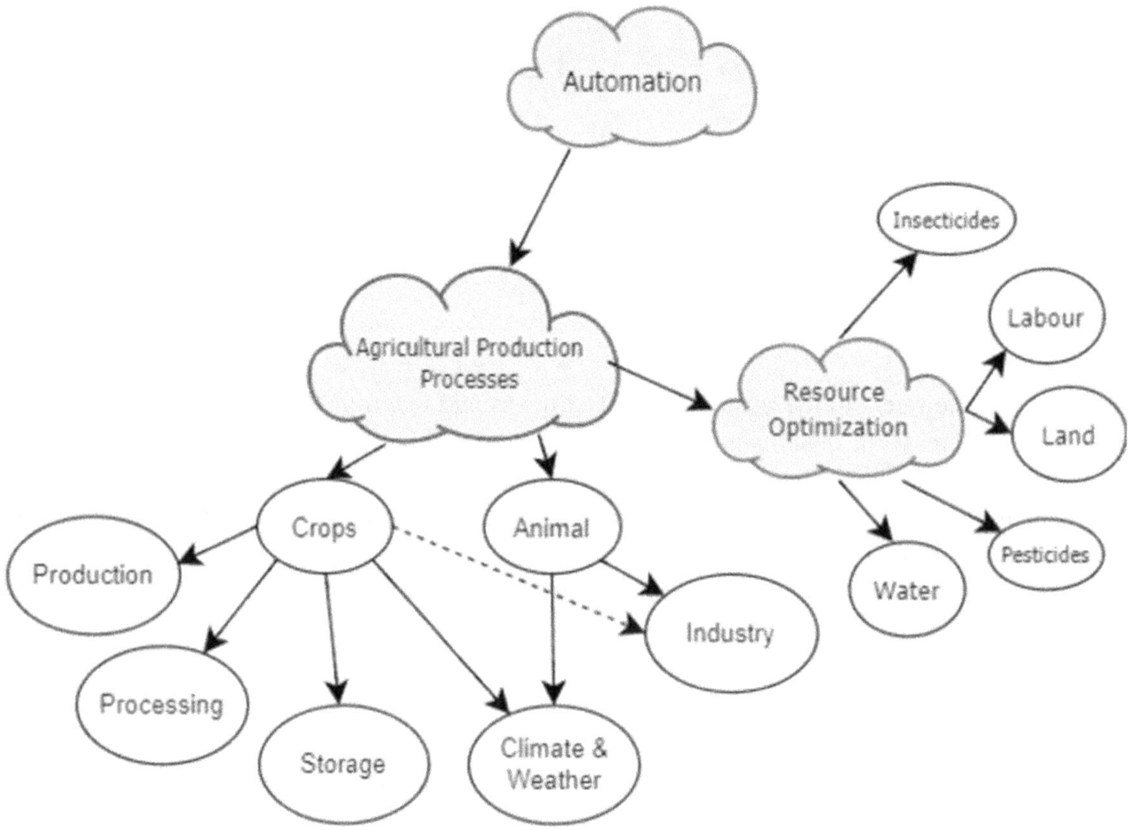

2. CLOUD COMPUTING IN AGRICULTURE

Cloud computing has elastic provisioning of resources which when applied to the field of agriculture can be fruitful for the farmer, researchers, and the market. Cloud can help create a centralized agricultural information data bank that can be available to users from anywhere and at any time. Figure 4 demonstrates the use of cloud computing in smart agriculture. (Hori et al., 2010) Since the cloud is connected to the internet, it can generate meaningful insights based on the farm data, which can help the farmer in decision-making, it can also help enhance the farmer's technical knowledge and solve problems on the go by connecting with experts through cloud technology (Goraya et al., 2015).

Figure 4. Cloud computing in smart agriculture

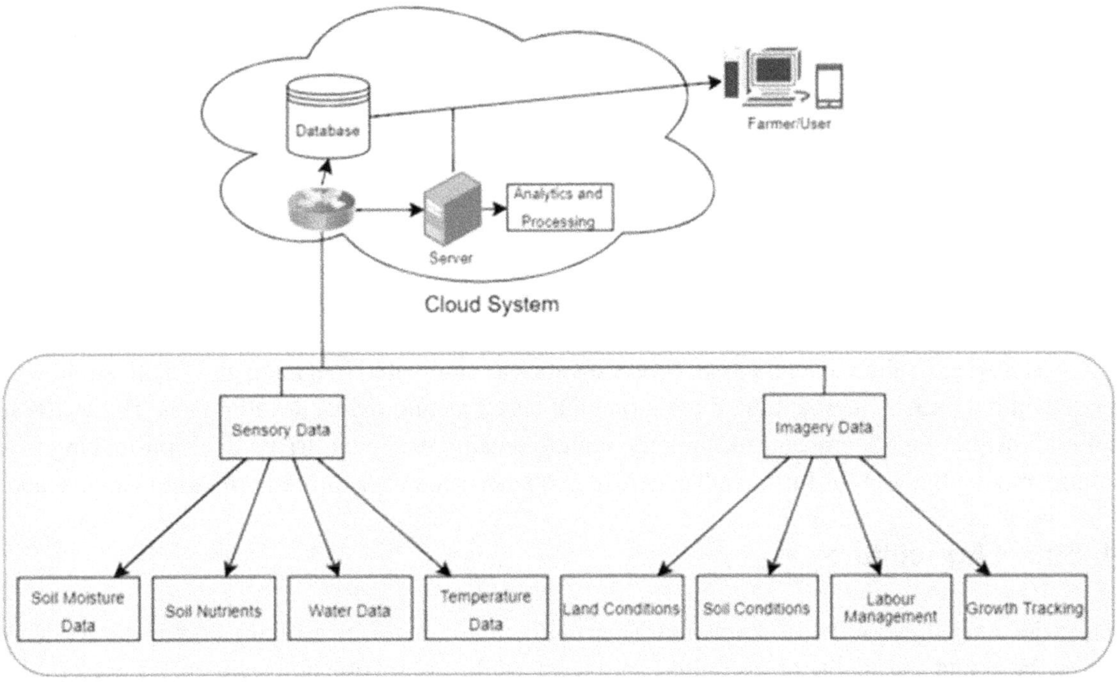

3. EVOLUTION OF AGRICULTURE

Agriculture has been the backbone of human life form, it has evolved significantly over time in the past, driven by technological advancements, scientific discoveries, and changing demands and needs the 4 stages of agricultural evolution. Figure 5 shows the evolution of agriculture over the period with advancements in civilization (Friha et al., 2021).

3.1. Traditional Agriculture

Traditional agriculture, practiced since ancient times, involved human labor and animal assistance for crop production, without the use of machinery or power tools. This era, known as Agriculture 1.0, lasted until the 18th century. During this period, people relied on manual work to cultivate crops and sustain their communities. With limited technological advancements, agriculture requires significant human effort and reliance on natural resources.

3.2. Mechanized Agriculture

The 18th and 19th centuries witnessed a transformative shift in agriculture with the introduction of machinery called as Agriculture 2.0. Innovations like steam engines, tractors, and harvesters revolutionized farming practices. These advancements significantly increased operational efficiency and productivity. Labor requirements were reduced, leading to the emergence of large-scale farming operations. Mechanical power replaced animal power enabling tasks such as plowing and transportation to be executed with greater speed and efficiency. Mechanical harvesters transformed the harvesting process, improving productivity and reducing manual labor. The expansion of agriculture became possible, meeting the demands of a growing population. This period laid the foundation for subsequent technological advancements in agriculture.

3.3. Automated Agriculture

Following the advent of machinery, researchers focused on automating repetitive tasks, leading to the emergence of automated agriculture through rapid technological advancements known as Agriculture 3.0. Automated agriculture encompasses processes such as automated irrigation and fertilization, which operate without human intervention, relying on high-speed technological development. The 1970s saw the initial integration of computer technology, remote sensing, and data-driven decision-making in agriculture, paving the way for further advancements in automated agriculture in the following decades.

3.4. Smart Agriculture

Finally, the Agriculture 4.0, also known as Smart agriculture, involves utilizing contemporary ICT to enhance the output and quality of agricultural products, while minimizing human intensive work requirements. It leverages IoT devices, sensors, cloud computing, and big data analytics to monitor, analyse, make informed decisions, and promptly address emerging challenges. This approach empowers farmers to respond swiftly and effectively. This also includes the use of unmanned vehicles, AI, machine learning, big data, and robotics to fine-tune agriculture to an extent of maximum production with minimum effort possible.

4. APPLICATIONS OF CLOUD COMPUTING IN SMART AGRICULTURE

4.1 Crop Management

Crop management refers to keeping track of the crops that are being grown in a farm. It also includes crops, which have been harvested and are kept in storage. Experts collect data on the crops that the farmer grows and analyse it to provide them with useful insights that help the farmer plan for their crop next.

Cloud Computing is responsible for providing large storage and computational space that is required for storing large amount of information relating to crops, and computing the deciding factors which would increase the agricultural production. (Thilakarathne et al., 2022)

There are essentially 4 steps involved in the process of crop management through cloud computing.

Figure 5. Evolution of farming

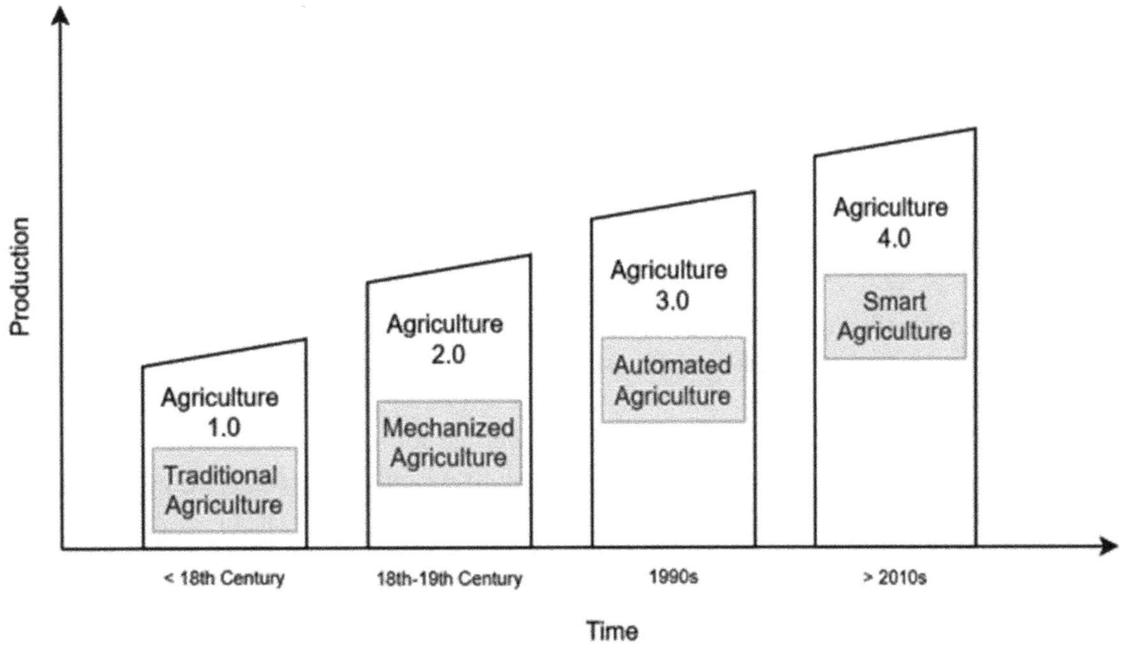

4.1.1. Data Collection

Farmers can collect data like soil moisture, temperature, nutrient levels as well as other major factors using sensors and other monitoring devices, which can transmit the data to the cloud for analysis. With cloud computing, farmers can collect large amounts of data rapidly and easily, which enables them to make data-driven decisions about crop management.

4.1.2. Data Analysis

Once the data has been collected, cloud computing can be used to analyse it and provide insights into crops, their growth patterns, diseases, issues, etc, With cloud computing, farmers can use sophisticated data analytics tools to process and analyze large amounts of data quickly, allowing them to make informed decisions about crop management.

4.1.3. Decision Making

Farmers can use cloud computing to make informed decisions about when to plant, irrigate, fertilize, and harvest crops. With cloud computing, farmers can optimize their crop management practices to maximize yields while minimizing costs and environmental impact.

4.1.4. Precision Agriculture

Precision agriculture is an innovating farming approach which using modern technologies to optimize crop yields while reducing costs. Using this method, farmers can make data-driven decisions about crops, irrigation and pesticides to reduce wastage and improve sustainability of the farm's produce.

Some tools, which are widely used for crop management, are Cropio, FarmLogs, Climate, FieldView (Zhang et al., 2002).

4.2 Land and Soil Information

Land fertility and its soil nutrients are a essential factor in agriculture. To monitor the soil, various sensors are used which collect data related to the moisture content, acidity or alkaline content, and changes in soil composition, and quality.

Cloud based systems can use satellite and aerial imagery to collect data on land and soil conditions, this can help monitor over large areas quickly and efficiently. Google Earth Engine is one of such tools (Padarian el al., 2015).

Soil profiling is required to understand the soil to decide the type of crops, which can be cultivated. The data generated from the sensors and other IoT devices, which are installed in the farm store the data in cloud servers, which is then used by analysts or nowadays using advance machine learning algorithms to use the historic data to predict the future trends of soil (Adetunji & Joseph, 2018) .This is very crucial to a farmer as a good soil profile may result in a high yield. It can also help the farmer in identifying the best areas to grow a certain crop in his farm and whether the crop will be suited to his field or not.

The reports of the system can help fulfil the deficiencies of the soil with regard to the crop sown and hence can increase the efficiency of the farm. FieldX LandPKS are some tools used for soil and land evaluation (Na et al., 2016). These tools depend on sensor data to produce accurate information relating to the soil and its nutrient composition.

4.3 Weather and Climate Data

Weather and climate are highly volatile but are crucial factors for a farmer. Plantation, irrigation and other farming activities depend on the weather and climate of the place. Weather information can be captured using various devices, and these devices can be linked up to the cloud to store the historical data which will help predict weather forecasts of a specific location.

Cloud-based platforms such as Climate Engine, and IBM GRAF provide access to high-resolution weather data, which can be used to improve the accuracy of crop models and forecasts (Symeonaki et al., 2019).

Cloud-based tools provide real-time weather monitoring and alerts, which are critical for mitigating risks associated with extreme weather events such as droughts, floods, and storms. The Weather Company and OpenWeather provide weather data APIs and dashboards that can be integrated with farm management systems to provide customized alerts and recommendations based on real-time weather data (Symeonaki et al., 2017).

4.4 Growth Tracking

Crop growth can be monitored periodically in diverse regions. These images can provide data on plant biomass, leaf area index, and other indicators of crop growth and can then be stored on cloud-based storage system, then processed to analyse the growth of crops and appropriate decisions can be taken in order to boost crop growth in a natural manner.

Growth tracking is really useful as when the data is processed, the results can predict the growth of the crop, the supply of nutrients and irrigation can then be planned accordingly.

This can also help the farmer determine if his produce will be profitable or not as the predicted yield can be cross-validated with the actual expected yield. Tools like FieldX and ClimateEngine can be used to analyse the imagery and sensor data to generate custom maps and time series plots of crop growth and development.

4.5 Information on Farmer

Farmer can access all data relating to his farm along with the photos which can help him in taking decisions. Cloud based system can also allow the farmer to access his farm remotely and control equipment with the touch of his smartphone. Real time information on weather, crop growth, market prices and can also be easily accessed using cloud-based mobile applications using internet connection. Platforms such as Climate, FieldView, and FarmLogs provide mobile applications for farmers, which can be used to access real-time information on crop management and market trends.

Cloud computing systems can also be used to manage pests and insects by collecting information data about pest infestations, farmers can predict infestation conditions and can take the necessary measures beforehand to reduce the amount of chemicals used while still effectively managing the pests.

The data collected by the sensors and devices may not be understandable to the farmer directly, hence cloud computing should deploy algorithms which will produce farmer understandable results by utilizing the data collected and various available knowledge databases.

4.6 Labour Management

While working in a farm, the farmers always needs to keep a check on the labour and manage them for maximum efficiency, using AI, enabled camera devices and other IoT devices, the workforce can be management remotely. Movements of the workers, their efficiency, etc can be analysed which can serve as a base for improving how the workers work in the field. Platforms such as AgriWebb, Agroptima, and FieldX provide tools for managing the workforce, which can be used to streamline labour-intensive tasks and improve overall efficiency. These tools can also be used to optimize staffing levels, reduce labour costs, and improve productivity.

These tools are taken in consideration the status of the employees and collect the feedback back to the management of the farm owner, this two-way communication can help in increasing the satisfaction of the workers and motivate them to work more efficiently.

4.7 Training and Consultation

Cloud-based tools can be used to provide training and education to farm workers, including safety training, skill development, and best practices for crop management. Cloud computing can facilitate the spread of modern agricultural practises and technologies to farmers by granting them access to online resources and training materials. Cloud computing has the potential to enable communication and collaboration among farmers, thereby facilitating the exchange of knowledge and expertise. This can lead to the sharing of best practises and mutual learning opportunities. These tools can be used to improve worker competency and reduce the risk of accidents and injuries. Platforms such as AgriWebb and Farmlogs provide training and education resources for farm workers, which can be accessed from any location with an internet connection.

Expert Consultation refers to a cloud service that arranges a researcher of an expert in the field of agriculture by a voice or video calls to help the farmer solve the problems he has been facing in the farm. This can be very helpful in times of need and the farmer can always have an expert opinion about his decisions.

4.8 E-Commerce

Individuals residing in rural regions encounter difficulties in vending their own agricultural yield directly to the market. The presence of intermediaries between the retail and production sectors often results in the exploitation of farmers. The utilisation of cloud computing in agricultural management information systems enables farmers to engage in direct sales of their agricultural products to end users and retailers. This has been implemented widely and has proved to be successful. Not only the consumers, but the farmers are highly benefitted from this as they are able to save the amounted spend on middlemen and markets and are able to profit more (Mueller et al.2000).

E-Commerce is being used in agriculture in the following ways:

4.8.1. Online Marketplaces

Websites like DeHaat, Agrimp, MarketGalee provide a platform where sellers and buyers can connect directly and sell their produce without the need of middlemen or a broker. These marketplaces can help the farmer connect with potential buyers and expand their sales into new markets.

4.8.2. Payment Processing

Payment platforms like PayPal, PayTM, provide hassle free transactions on e-commerce sites which can be helpful for both buyers and sellers to facilitate secure and easy transactions.

4.8.3. Logistics and Support

Mostly online marketplaces like Agrimp, Amazon provide their own logistics and support for the delivery of the product and provide customer support in case of any mishaps. Companies like Adani Agri Logistics, TruckSuvidha also provides logistics solution directly to the farmers to help them deliver directly to the customers.

Figure 6. Benefits of cloud computing in agriculture

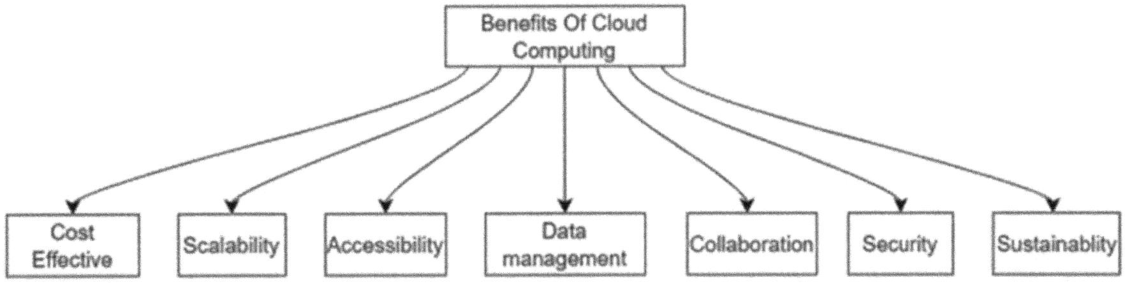

5. BENEFITS AND CHALLENGES

5.1 Benefits

Cloud computing has numerous benefits in smart agriculture as shown in Figure 6.

- Local and global communication
- Improve the economic condition of the Nation
- Enhanced the GDP of the nation
- Ensure food security level
- Motivation of farmers and researchers
- Reduction of technical issuesss
- Rural-Urban movement
- Data availability at any time and at any location delay
- Improve the market price of food, seeds, other productss

5.1.1. Cost-Effectiveness

Cloud computing enables farmers and agribusinesses to access computing resources and tools without investing in expensive hardware and software. This can significantly reduce capital expenditure and ongoing maintenance costs. The cost incurred is just for setting up the initial infrastructure for cloud services which may be more, but since the maintenance costs are low, the overall cost of the system is less than the cost of traditional methodologies.

5.1.2. Scalability

Cloud computing platforms are designed to scale up or down based on demand, allowing farmers and agribusinesses to adjust their computing resources based on their needs. This ensures that they have access to the resources they need when they need them, without wasting resources or overspending on unnecessary infrastructure (Bo & Wang, 2011). This is because of the agility of cloud services being highly scalable, so that the user can request for more resources when the demand is high and can scale down to less resources when there is less demand. Such scalability can only be provided in cloud enabled applications.

5.1.3. Accessibility

Cloud computing platforms can be utilized from any location with a proper internet connection, allowing farmers and agribusinesses to access critical data and applications from remote locations. This can be particularly useful for farmers who work in rural or remote areas. This means that the farmer can access his field's data from anywhere and can share the access to researchers and agricultural experts to get opinions and help without the need of actually going to an expert. The time saved can be used for the management of the field for better efficiency.

5.1.4. Data Management

Cloud computing platforms provide farmers and agribusinesses with powerful tools for data management, storage, and analysis. This enables farmers to collect and analyze large amounts of data relating to weather patterns, soil health, crop yields, and other critical factors that can impact their business. Cloud storage is relatively cheap compared to physical data storage devices and can be scaled according to the need of the hour. Along with the storage, services also provide analysis of data which can produce data which is actually useful to the farmer or the business in terms of finance, yield, management, etc.

5.1.5. Collaboration

Cloud computing platforms enable farmers and agribusinesses to collaborate with partners, suppliers, and customers more easily. This can help to streamline supply chains, improve communication, and reduce costs. Cloud based platforms provide easy access for farmers to sell their produce, track their shipments, order supplies, and manage inventory, this is possible with the collaboration of supplier and consumers, removing the middlemen, hence enhancing the profit scale.

5.1.6. Security

Cloud computing platforms offer robust security features, including data encryption, access control, and backup and recovery mechanisms. Various security services are provided by the cloud service provider or the application used by the farmers. This can help to protect sensitive data and prevent data loss or theft. Farmers can securely use and access their data without the worry of theft of data, which is considered as a major concern in cloud environment.

5.1.7. Sustainability

Cloud computing can help to improve the sustainability of agriculture by enabling farmers to optimize resource use, reduce waste, and improve productivity. By using data-driven insights, farmers can make more informed decisions about water usage, fertilizer application, and other critical factors that impact the environment (Patil et al.,2012) . It is also supports sustainable growth of crops and reduces the wastage of resources required for the management of the field, labour, produce and other measures. Cloud

Figure 7. Challenges in cloud computing in agriculture

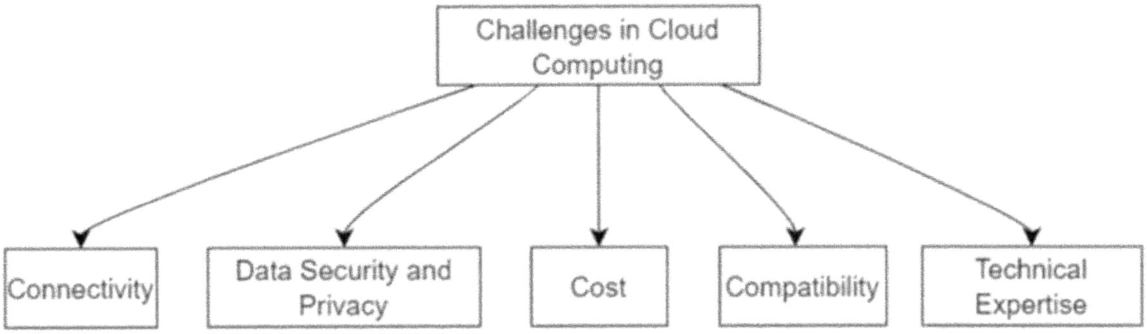

minimizes the requirement for maintaining logs and records of data, instead uses databases to securely store all information in an efficient manner.

5.2 Challenges

While cloud computing has many benefits for smart agriculture, it also presents some challenges that must be addressed. Here are some of the key challenges associated with the use of cloud computing in smart agriculture. These challenges can be addressed by joint efforts of farmers, service providers, and policy makers and hence can be overcome to enhance the adoption of cloud-based tools in agriculture. Figure 7 represents all challenges involved in smart agriculture.

5.2.1 Connectivity

Access to reliable internet connectivity is essential for cloud-based applications. However, many rural areas may not have access to high-speed internet or may experience frequent outages, making it difficult to use cloud-based tools for agriculture (Symeonaki et al., 2017). Connectivity issues may cause improper results or even loss of data if the connectivity is lost during the transit of data. Precautionary measures are to be placed in order to handle such problems. Connectivity issues are more frequent in rural areas where there is no network for internet access or Wi-Fi, without access to the internet, cloud technology is unusable.

5.2.2. Data Security and Privacy

In cloud computing, data is stored on remote servers, raising concerns about data security and privacy. Data leak and stealing have become frequently common with the cloud being implemented at such a large scale and hence farmers may be reluctant to share their sensitive data with third-party cloud service providers. Therefore, security measures, such as encryption, access control, and regular security audits, are crucial to protect data privacy and security (Bo & Wang, 2011). The cloud service provider generally keeps a check on the security concerns of its users and implements various measures to keep in check the privacy of its user's data.

5.2.3. Cost

Cloud computing services can be expensive, and poor farmers may not be able to afford the cost of accessing these services. The cost of cloud computing also depends on factors such as the volume of data processed, the number of users, and the type of application, making it challenging for farmers to accurately predict the costs. Cloud is always associated with the initial infrastructure cost being too high, that is because the equipment needed for smart agriculture, are costly and require proper maintenance on a regular interval, although the cost of the system is preceded by its benefits, farmers still would have to think twice before turning towards cloud-based smart agricultural methods.

5.2.4. Compatibility

Many farmers may already have existing hardware and software systems that may not be compatible with cloud-based tools. This can create difficulties in integrating new tools with existing systems, leading to additional costs and complexity. New cloud connected or supported tools would cost a lot to a small-scale farmer and hence it would be preferred for him to stick to the existing systems.

5.2.5. Technical Expertise

The use of cloud computing requires technical expertise, which may be lacking among small farmers. Therefore, farmers may need training and support to effectively use cloud-based tools in their operations. Farmers lacking basic education may find it more difficult to learn or train to use the cloud-based application or software and hence would require additional man power to maintain the systems, which would add to the operational costs.

6. TOOLS AND SYSTEMS

Researchers as well as IT sector have been prominently active in the field of smart agriculture and have been keen in developing tools and farm management systems using cloud and machine learning algorithms, here we look at some commonly used tools and systems to understand them better. Various tools are presented in Table 1.

The United States Department of Agriculture has shifted most of its department's components to cloud services and has more than $81 million in FY 2020, this includes investing in food safety and inspection projects, Supplemental Nutrition Assistance Programs (Cloud Computing at the U.S. Department of Agriculture - GovWin IQ, n.d.).

Countries like Japan, which have very limited farming land resources, are choosing agricultural IoT technology with the help of the government that aims to use big data and cloud platform in the agriculture sector for about 75% of the market. Meanwhile, the Japanese government designed a plan to utilise an agricultural IoT platform in order to offer agricultural robots with information and data services (Liu et al., 2019).

China is one of the biggest producers of agricultural products in the world. Its agriculture industry contributed a significant 7% to its GDP in 2021. It's also the most populous country in the world, with a population of 1.4 billion people as of 2021.The country has to feed over a billion people in the face

Table 1. Smart tools used in agriculture

SNo.	Tool Name	Features	Cloud Enabled
1.	KhetiBuddy	· Crop Management · Inventory management · Financial Tracking · Mobile Alerts · Multi Lingual · Weather and market rate Records	Yes
2.	AgWorld	· Crop Management · Livestock management · Field mapping/Planning · Financial Management · Inventory management · Weather Records · Supplier Management	Yes
3.	Agrivi	· Financial Management · Inventory management · Crop Management · Equipment Tracking · Livestock management · Field mapping/Planning · Weather Records · Supplier Management	Yes
4.	TruckSuvidha	· Logistic management · Warehouse and storage · Live shipment tracking	Yes
5.	FieldX	· Crop Management · Livestock Management · Production Tracking	No
6.	Cropin	· Labour Management · Pricing Management · Task Management · Crop Management · Equipment Tracking · Livestock management · Field mapping/Planning	Yes

of rising agricultural constraints, which has led it to invest considerably in agricultural technologies in part, to the country's vision to be self-sufficient in food security. Its smart agriculture market has rapidly grown from $14 billion in 2015 to $27 billion in 2020, as per figures from Statista, leading to the emergence of agritech startups like FJ Dynamics, an autonomous systems manufacturer for multiple industries including agriculture (5 Most Advanced Countries in Agriculture Technology, 2022.).

Cropin Cloud, a company based in Bengaluru, India, has developed the world's first intelligent cloud platform for agriculture. The field of agriculture technology is characterised by its complexity and requires specialised knowledge in various domains such as geographic information systems (GIS), agricultural science, artificial intelligence/machine learning (AI/ML) models, meteorological data, and the Internet of things (IoT), among others. A wide range of crops exhibit significant diversity, necessitating precise calibration of digital monitoring solutions. Cropin Cloud integrates a range of sophisticated technologies, agronomic expertise, data analytics, and AI capabilities (Reporter, 2022).

7. CONCLUSION

This chapter explores the application of cloud computing in smart agriculture, highlighting its role in the development of the agriculture sector. Cloud computing has the potential to revolutionise the industry and effectively meet the needs of a growing global population. By integrating cloud-based systems with the Internet of Things (IoT), farmers can maximise resource utilisation, reduce environmental impact, and increase overall productivity. Cloud-based applications can be implemented at any scale, including from small-scale farmers, to large scale farming industries because of their scalability and affordability. This accessibility to cloud systems enables farmers to more effectively manage their fields by leveraging cloud-enabled smart devices. The adaptability and precision provided by such systems significantly contribute to farmers' informed decision-making processes. In addition, smart cloud-enabled fields can serve as valuable resources for researchers, as the enormous amount of information generated can be utilised for evaluating crops, soil conditions, and land quality, therefore encouraging advancement in agriculture.

While limitations in cloud system accessibility persist in rural areas and developing countries, ongoing technological advancements are expected to address these challenges in the near future. Further research can be conducted to evaluate the impact of different cloud applications on farmer's fields and determine the extent to which these applications increase crop yields. Such research would expand our knowledge of the transformative potential of cloud computing in agriculture. As we continue to investigate the potential of cloud-enabled agriculture, it is essential that we address existing obstacles and investigate new ways to enhance accessibility and functionality. Thus, we can unlock the full potential of cloud computing and propel the agricultural industry into a more sustainable, productive, and technologically driven future.

REFERENCES

Adetunji, K. E., & Joseph, M. K. (2018). Development of a Cloud-Based Monitoring System Using 4Duino: Applications in Agriculture. In *2018 International Conference on Advances in Big Data, Computing and Data Communication Systems (icABCD).* 10.1109/ICABCD.2018.8465418

Aznar-Sánchez, J. A., Piquer-Rodríguez, M., Velasco-Muñoz, J. F., & Manzano-Agugliaro, F. (2019). Worldwide research trends on sustainable land use in agriculture. *Land Use Policy*, *87*, 104069. doi:10.1016/j.landusepol.2019.104069

Bhargav, A., & Manhar, A. (2020). A Review on Cryptography in Cloud Computing. *International Journal of Scientific Research in Computer Science, Engineering and Information Technology*, 225-230. doi:10.32628/CSEIT206639

Bo, Y., & Wang, H. (2011). The Application of Cloud Computing and the Internet of Things in Agriculture and Forestry. In *2011 International Joint Conference on Service Sciences.* 10.1109/IJCSS.2011.40

Choudhary, S. K., Jadoun, R. S., & Mandoriya, H. L. (2016). Role of cloud computing technology in agriculture fields. *Computing*, *7*(3), 1–7.

Cloud Computing at the U.S. Department of Agriculture - GovWin IQ. (n.d.). https://iq.govwin.com/neo/marketAnalysis/view/Cloud-Computing-at-the-US-Department-of-Agriculture/6048?researchTypeId=1&researchMarket=

Friha, O., Ferrag, M. A., Shu, L., Maglaras, L. A., & Wang, X. (2021). Internet of Things for the Future of Smart Agriculture: A Comprehensive Survey of Emerging Technologies. *IEEE/CAA Journal of Automatica Sinica, 8*(4), 718–752. doi:10.1109/JAS.2021.1003925

Goraya, M. S., & Kaur, H. (2015). Cloud computing in agriculture. *HCTL Open International Journal of Technology Innovations and Research, 16*, 2321–1814.

Hori, M., Kawashima, E., & Yamazaki, T. (2010). Application of cloud computing to agriculture and prospects in other fields. *Fujitsu Scientific and Technical Journal, 46*(4), 446–454.

Internet of Things Monitoring System of Modern Eco-Agriculture Based on Cloud Computing. (2019). *IEEE Journals & Magazine*. https://ieeexplore.ieee.org/abstract/document/8662571

Khan, N., Ray, R. L., Sargani, G. R., Ihtisham, M., Khayyam, M., & Ismail, S. (2021). Current Progress and Future Prospects of Agriculture Technology: Gateway to Sustainable Agriculture. *Sustainability (Basel), 13*(9), 4883. doi:10.3390u13094883

Mueller, R. A. (2000). *Emergent e-commerce in agriculture (No. 14)*. University of California, Agricultural Issues Center.

Na, A., Isaac, W., Varshney, S., & Khan, E. (2016, October). An IoT based system for remote monitoring of soil characteristics. In *2016 International conference on information technology (InCITe)-the next generation IT summit on the theme-internet of things: Connect your Worlds* (pp. 316-320). IEEE. 10.1109/INCITE.2016.7857638

Padarian, J., Minasny, B., & McBratney, A. B. (2015). Using Google's cloud-based platform for digital soil mapping. *Computers & Geosciences, 83*, 80–88. doi:10.1016/j.cageo.2015.06.023

Patil, V. C., Al-Gaadi, K. A., Biradar, D. P., & Rangaswamy, M. (2012). Internet of things (Iot) and cloud computing for agriculture: An overview. Proceedings of agro-informatics and precision agriculture (AIPA 2012), 292-296.

Phasinam, K., Kassanuk, T., Shinde, P. P., Thakar, C. M., Sharma, D. K., Mohiddin, S. K., & Rahmani, A. W. (2022). Application of IoT and Cloud Computing in Automation of Agriculture Irrigation. *Journal of Food Quality, 2022*, 1–8. doi:10.1155/2022/8285969

Qian, L., Luo, Z., Du, Y., & Guo, L. (2009). Cloud Computing: An Overview. In M. G. Jaatun, G. Zhao, & C. Rong (Eds.), Lecture Notes in Computer Science: Vol. 5931. *Cloud Computing. CloudCom 2009.* Springer. doi:10.1007/978-3-642-10665-1_63

Rathod, M. L., Shivaputra, A., Umadevi, H., Nagamani, K., & Periyasamy, S. (2022). Cloud Computing and Networking for SmartFarm AgriTech. *Journal of Nanomaterials, 6491747*, 1–7. Advance online publication. doi:10.1155/2022/6491747

Rehman, H. U. (2022, December 19). *5 Most Advanced Countries in Agriculture Technology*. Insider Monkey. https://insidermonkey.com/blog/5-most-advanced-countries-in-agriculture-technology-1098530/

Reporter, S. (2022, September 6). *Cropin launches the world's first purpose-built industry cloud for agriculture*. Farmers Review Africa. https://farmersreviewafrica.com/cropin-launches-the-worlds-first-purpose-built-industry-cloud-for-agriculture-cropin-cloud/

Saban, M., Aghzout, O., & Rosado-Muñoz, A. (2022). Deployment of a LoRa-based Network and Web Monitoring Application for a Smart Farm. *2022 IEEE International Workshop on Metrology for Industry 4.0 & IoT (MetroInd4.0&IoT)*, 424-427. 10.1109/MetroInd4.0IoT54413.2022.9831521

Sisinni, E., Saifullah, A., Han, S., Jennehag, U., & Gidlund, M. (2018). Industrial internet of things: Challenges, opportunities, and directions. *IEEE Transactions on Industrial Informatics, 14*(11), 4724–4734. doi:10.1109/TII.2018.2852491

Soul, P. K., & Saini, S. (2017). Data Security Approach in Cloud Computing Using SHA. *International Research Journal of Engineering and Technology, 4.*

Symeonaki, E., Arvanitis, K. G., & Piromalis, D. (2017). Cloud Computing for IoT Applications in Climate-Smart Agriculture: A Review on the Trends and Challenges Toward Sustainability. In Springer earth system sciences. Springer Nature. doi:10.1007/978-3-030-02312-6_9

Symeonaki, E., Arvanitis, K. G., & Piromalis, D. D. (2017). Review on the Trends and Challenges of Cloud Computing Technology in Climate-Smart Agriculture. *HAICTA*, 66-78.

Symeonaki, E. G., Arvanitis, K. G., & Piromalis, D. D. (2019). Cloud computing for IoT applications in climate-smart agriculture: A review on the trends and challenges toward sustainability. In *Innovative Approaches and Applications for Sustainable Rural Development: 8th International Conference, HAICTA 2017, Chania, Crete, Greece, September 21-24, 2017, Selected Papers 8* (pp. 147-167). Springer International Publishing. 10.1007/978-3-030-02312-6_9

Taylor, M. (2018). Climate-smart agriculture: What is it good for? *The Journal of Peasant Studies, 45*(1), 89–107. doi:10.1080/03066150.2017.1312355

Thilakarathne, N. N., Bakar, M. S. A., Abas, P. E., & Yassin, H. (2022). Towards making the fields talks: A real-time cloud enabled IoT crop management platform for smart agriculture. *Frontiers in Plant Science*, 13. PMID:36684733

Tripathi, A. D., Mishra, R., Maurya, K. K., Singh, R. B., & Wilson, D. W. (2019). Estimates for world population and global food availability for global health. In *The role of functional food security in global health* (pp. 3–24). Academic Press. doi:10.1016/B978-0-12-813148-0.00001-3

Union, U. (2017). *World population projected to reach 9.8 billion in 2050, and 11.2 billion in 2100.* UN DESA.

Yang, X., Shu, L., Chen, J., Ferrag, M. A., Wu, J., Nurellari, E., & Huang, K. (2021). A survey on smart agriculture: Development modes, technologies, and security and privacy challenges. *IEEE/CAA Journal of Automatica Sinica, 8*(2), 273-302.

Zhang, N., Wang, M., & Wang, N. (2002). Precision agriculture—A worldwide overview. *Computers and Electronics in Agriculture*, *36*(2-3), 113–132. doi:10.1016/S0168-1699(02)00096-0

Chapter 6
Cloud Computing-Based Smart Agriculture

Kaushal Kishor
https://orcid.org/0000-0002-7131-1389
ABES Institute of Technology, Ghaziabad, India

Raj Kishor Verma
ABES Institute of Technology, Ghaziabad, India

ABSTRACT

This development has also affected the studies of agriculture. Agriculture is an important source of income in the world's most populous countries. In countries like India, China, and others, it is a major source of wealth and livelihood. Cloud computing and the internet of things for agriculture will improve production cost management, performance tracking, and maintenance, benefiting both farmers and the country as a whole. The focus of this chapter is on the use of smart drones, and instant data from drones for crop management and computing technologies will help get it started. Despite the use of precision gardening in other countries, Indian agriculture needs to be modernized through the use of technology to increase productivity, wealth distribution, and cost management. In this chapter, the authors propose a comprehensive smart agriculture paradigm based on the following technological advances: sensors, internet of things (IoT), cloud computing, mobile computing, and big data analytics

INTRODUCTION

The agriculturalist's choice on what kind of crop to grow is frequently eclipsed by the thoughts he has on other minor factors such as generating rapid profits, a lack of knowledge regarding market demand, an excess of the earth's ability to sustain a specific crop, and so on. A poor judgement by the farmer might have serious consequences for the entire family's financial well-being. Apparently, this is one of the many factors leading to the many incidents of suicide among farmers reported in newspapers daily. In a nation like India, farming and associated industries account for around 20.4 percent of Gross Value Added (GVA). Such a mistaken judgement could have a detrimental impact not just on the farmer's

DOI: 10.4018/979-8-3693-0200-2.ch006

family, but additionally on the entire area economy. As a result, we have discovered the agriculturalist's dilemma of determining which crop to cultivate at what time of year. The necessity of the hour is to create a system that can offer Indian farmers with predictable facts, allowing people to make educated choices about what kind of crop to plant.

With this issue in mind, we're suggesting an approach, an innovative one that pre-calculates environmental variables (temperature, precipitation, where you are as a matter of circumstance) and soil characteristics (pH value, soil type, and nutrients density) and recommend the best produce for the consumer.

Agricultural techniques and practices play a major role in connecting devices, collecting and distributing data amid expectations for the introduction of the Internet and cloud computing increasing speed, thereby offering farmers a new approach, farm health inspection. In recent years, interest in various fields has increased; Robotics is where we entered the field of technology that supports farm health. It is monitored and data is recorded, but it is not effective (Bac et al., 2014). Drone use starts here in agriculture; smart farming can offer a lot. Precision farming is a form of smart agriculture where farmers use data analytics and IoT devices to monitor soil moisture, nutrient levels, and other environmental conditions to optimize crop development. Drones, which may be used for agricultural monitoring and spraying, and autonomous tractors, which can be programmed to plough fields and sow seeds (Burkle 2009), are examples of other technologies.

Challenges

There are a number of obstacles that must be overcome in order for smart agriculture to reach its full potential, despite the fact that it has the potential to greatly benefit farmers and the food chain. Some of the main obstacles to smart agriculture are listed below:

- Cost: It may be expensive to put many of the technologies used in smart agriculture, including sensors, drones, and autonomous tractors, into use. For small-scale farmers with little resources, this may be a deterrent to adoption.
- Infrastructure: Smart agriculture needs a solid and dependable digital infrastructure, which may not be present in all rural locations. This includes high-speed internet connectivity. This could make it more difficult for farmers to embrace and use these technologies.
- Data management: In order to be effective, smart agriculture generates a lot of data that must be organised and analysed. For farmers who might lack the knowledge or resources to handle and interpret this data efficiently, this can be difficult.
- Privacy and security: Data on crop yields, soil quality, and weather patterns are among the sensitive information that is gathered and shared in the context of smart agriculture. It is crucial to make sure that this information is secure and that farmers have a choice over how it is put to use.
- Education and training: In order to engage in smart agriculture, farmers must possess a certain degree of technical skill, including an understanding of digital technology and data analytics. To assist farmers in acquiring these abilities, it is crucial to offer training and educational programs.
- Regulatory obstacles: Using some smart agriculture technology, including drones and autonomous vehicles, may present regulatory obstacles since they may be required to get specific permits or licenses.

Collaboration amongst industries such as agriculture, technology, policy, and finance will be necessary to address these difficulties, governments, academic institutions, and the commercial sector may all play a part in fostering the growth and acceptance of smart agricultural technologies while also making sure that they are available to and reasonably priced for farmers of all sizes and in all areas.

Smart Agriculture Application

In order to maximise agricultural productivity and raise yields, smart agriculture makes use of cutting-edge technologies including sensors, drones, and big data analytics. The following are a few uses of smart agriculture:

- Precision farming: Using sensors, GPS, and other technologies, precision farming keeps an eye on the weather, crop development, and soil conditions in real-time. This enables farmers to use resources like fertiliser, water, and other resources as efficiently as possible, thereby increasing yields and lowering expenses.
- Smart irrigation systems employ sensors and meteorological information to automatically change watering schedules and volumes in response to plant requirements and environmental factors. This minimises water wastage and guarantees that plants get the appropriate amount of water at the appropriate time.
- Animal health, behaviour, and production may all be tracked using sensors in livestock management, a branch of smart agriculture. Farmers may be able to better monitor animal welfare overall, reduce feed and water usage, and identify health issues early.
- Management of pests and diseases: Pest and disease management is a function of smart agriculture. Early indicators of infestation may be found using sensors and drones, and epidemics can be predicted and trends found using big data analytics. This might increase agricultural yields and lessen the demand for dangerous chemicals.
- Supply chain management: From production to distribution, smart agriculture may assist optimising the whole supply chain. Big data analytics enables farmers and distributors to streamline operations and cut waste by tracking inventory levels, keeping track of shipment schedules, and forecasting demand.

Smart agriculture has the ability to completely transform the agriculture sector, making it more effective, sustainable, and lucrative.

Mobile Application Module

Applications for smartphones have to be downloaded and set up on the end user's phone. It is divided into three sections. a. UI for farmers; b. UI for agro marketing agencies; and c. UI for agro suppliers such as fertilisers, insecticides, and seed suppliers. At first, the end user must register for the mobile app using a few credentials such as identification information, account type, address, place of residence, and other pertinent information. If the end user is a farmer, he or she must supply a few credentials containing information related to agriculture such as approximate location and total acreage for each field. SensorKit is used to collect soil information for each farm. SensorKit receives the necessary instructions from Mobile Applications. The data will be delivered to be preserved on AgroCloud Big-Data repository.

Figure 1. Mobile application module

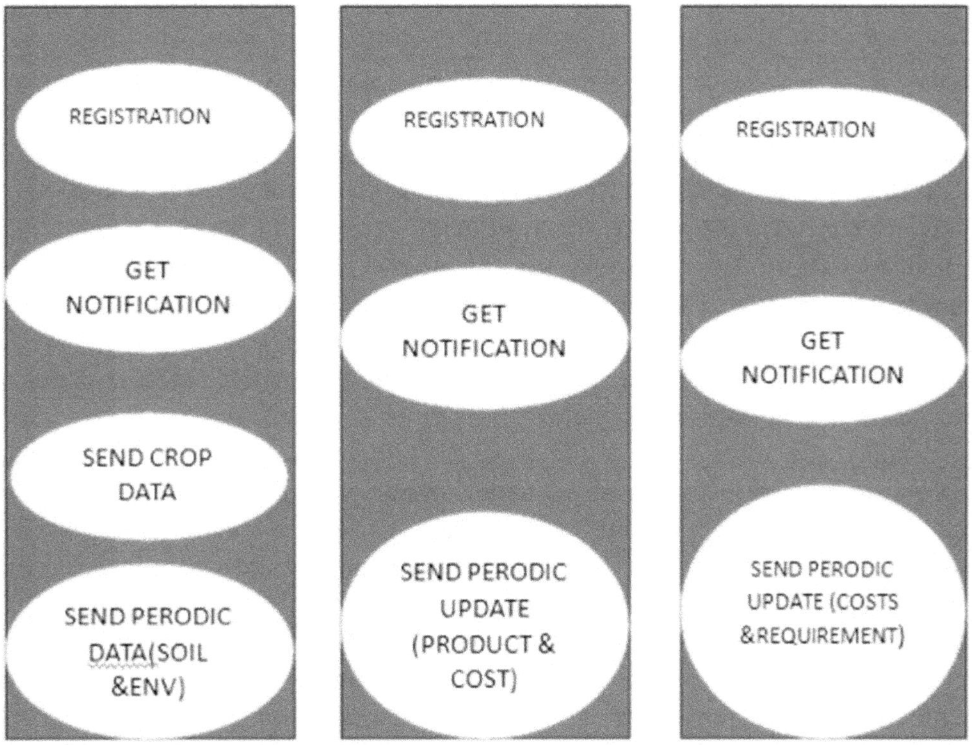

When planting of crops is underway, SensorKit captures and sends soil data to cloud storage. Farmers may use these apps to acquire fertiliser recommendations and dosages for improved crop yields and cost savings. This app is also used to deliver alerts to users. When the crop has been harvested, the farmer will send the complete production data for each crop to the cloud storage, together with the present soil parameters following the growth of that commodity. The aforementioned data, together with the date and time stamp, is saved in the cloud. Agro marketing companies in charge of acquiring harvested products from farmers must deliver quarterly information on price fluctuations and purchase needs. Fertilisers, seeds, insecticides, and agricultural equipment are sold by agro product dealers. Agro providers must send goods and price modification notifications on a regular basis. The mobile Application Module is shown in Figure 1.

A drone is defined as an unmanned aerial vehicle (UAV) (Borenstein & Miller, 2013). The use of drones offers possibilities, use farm time instead of going for a walk; IoT and cloud computing technologies can help in one solution. Smart drone paired with Complementary Technologies, IoT, and Cloud computing data captured by smart agriculture (Gupta et al., 2022). The term “smart agriculture” refers to the employment of cutting-edge technology and data analytics in agricultural practices to maximize crop output, minimize waste, and increase efficiency. It entails integrating sensors, Internet of Things (IoT) devices, and other digital technologies into farming operations so that farmers can make data-driven choices about their crops and the environment (Kishor & Pandey, 2022).

Increased crop yields, fewer environmental effects, and higher farmer profitability are all advantages of smart agriculture.

In order to enhance agriculturally practices and crop management, a variety of technologies and digital solutions are referred to as "smart agriculture tools." Here are some instances of clever farming tools:

- Sensors: Information on the state of the environment and the development of plants may be gathered using soil sensors, weather sensors, and plant sensors. This information may be utilised to improve crop management procedures such as fertiliser and irrigation applications.
- Drones: Drones with cameras and sensors may be used to monitor crops, spot diseased or stressed regions, and deliver detailed pictures of agriculture.
- Autonomous tractors: These machines may be programmed to clear land, sow seeds, and use herbicides and fertilisers. These devices can work without the need for a human operator, which lowers labour costs and boosts productivity.
- Smart irrigation systems: These systems make the best use of available water resources, cut down on waste, and increase agricultural yields.
  Crop planning, inventory control, and financial analysis are just a few of the tools that farm management software gives farmers to run their businesses.
- Livestock monitoring systems: These systems track the behaviour and health of livestock using sensors and data analytics, giving farmers the ability to improve management techniques including breeding and feeding.
- Systems for precision agriculture: These systems provide farmers access to real-time data on crop development so they may make informed decisions regarding pest control, fertilisation, and irrigation. Precision agriculture systems employ data analytics and machine learning to do this.

The IOT concept, phrase "Internet of Things" (IoT), which has gained popularity, recently, describes a system in which the whole world is connected to the Internet through various types of sensors. It opens up possibilities for closer integration of the physical world with computer-based systems and reduces human intervention while improving efficiency, accuracy, and economic benefit (Mattern & Floerkemeier, 2010). It also enables objects to be sensed or controlled remotely across the existing network infrastructure. All household appliances, furniture, clothing, automobiles, roadways, and smart materials, among other IOT things, are readable, recognized, locatable, addressable, and/or controlled through the Internet.

Through combining technological advancements in pervasive communications, ambient intelligence, and pervasive objects in a sensory and intelligent way, it combinations of ubiquitous communications, identification technologies, sensors, and wireless, networks, embedded systems and nanotechnology (Swathi & Guruprasad, 2014).

Cloud Computing Concept Today's term in the information technology (IT) industry, "cloud computing," refers to a computing paradigm that gives large groups of privately, publicly, or hybrid networked devices on-demand access to either application data or storage space (Kishor et al., 20230. It provides a brand-new internet-based method for adding, using, and exchanging IT services that include making available dynamic, scalable, and virtualized resources, The core notion behind the innovative concept of cloud computing is the reusability of IT skills, a crucial component for cost-effectiveness (Kishor & Nand, 2023). By giving a realistic method to experience immediate cost advantages in computing, application hosting, content distribution, and other areas, it broadens perspectives beyond organizational boundaries (Pingli et al., 2011).

Data analytics concepts employing data analytics to practice intelligent agriculture, The Third Green Revolution, or "smart farming," is a breakthrough in the agriculture industry brought about using cutting-edge technologies. It is anticipated that this will eventually close the technological gap between the agricultural industry, which has historically relied on more conventional farming techniques, and most cutting-edge technological advancements. Data is being used by farmers to estimate crop yields, the need for fertilizer, cost reductions, and even the best optimization techniques for upcoming crops. To construct algorithms that can be used by different farms to produce a crop yield that is profitable and sustainable, smart farming employs data analytics tools to collect information from various agricultural practices. Farmers will be better equipped to forecast activities in the field with this information.

These are only a few instances of clever farming implements. We can anticipate even more creative ideas that will help farmers enhance their operations and boost output as technology continues to evolve.

LITERATURE REVIEW

(Giusti & Marsili-Libelli, 2015), Utilizing research papers, journals, and articles, we reviewed the literature on smart agriculture. According to one research, if crop and site parameters are known, a fuzzy control system may be used to conserve water and manage irrigation in agricultural fields. It explains a procedure for putting an irrigation system that is completely automated into use. Another investigation (Keerthi & Kodandaramaiah, 2015) provided a greenhouse tracking system built around an agriculture IoT with a cloud in which devices are put on the farm's premises to collect and store data on the agricultural area on a regular basis. Further research (Liqiang et al., 2011; Zhu et al., 2011; Jaishetty & Patil, 2016) suggested an agricultural utilisation of wireless sensors for crop field surveillance, with these networks outfitted with two types of nodes that sense, one that monitors various variables such as temperature, humidity, and so on, and the other, a picture detecting node to take photographs of the agricultural products. Irrigation in agricultural land is the main topic of the majority of studies. One such work (Kishor, 2023) emphasized the necessity of precisely calculating the water needs of plants as well as the Neural Computing modelling of reference crop evapotranspiration. In a different work (Kishor, 2023; Adeloye et al., 2012), an automatic smart irrigation decision support system (SIDSS) was presented to control irrigation in agriculture. Internet of Things-based smart irrigation systems (Kishor, 2022; Kishor et al., 2018; Kumbar et al., 2016). Where sensed, data is transmitted to a smart gateway over a network and subsequently to a web service. A method for monitoring pest insect traps utilizing image sensors and Aspic was also proposed by the GSM (Global Method for Mobile Communications). They employed wireless sensor networks to control dispersed imaging equipment (Hemalatha & Deepika, 2016; Kishor et al., 2022). This approach just identifies pests; it makes no recommendations about how to get rid of them.

A few publications on smart agriculture claim that several nations are now adopting precision agriculture. Precision agriculture is a type of satellite farming or Site-Specific Crop Management (SSCM) that relies on monitoring crop variability both within and between fields (Venkataramana & Padmavathamma, 2012). A methodology for using cloud computing has been created by Fujitsu in Japan. In the majority of publications, the findings of exploratory investigations are reported, or IoT systems that have been developed or tested in prototypes and pilots are presented. While actuation and remote control are significantly less discussed in the literature study and more sophisticated solutions are still in the experimental stage of development, the literature review concentrates on sensing and monitoring. This article covers a new method for using smart drones to remotely regulate and monitor a crop field's essential agricultural land characteristics, assisting in data analysis and field prediction (Kishor, 2022; Rupanagudi et al., 2015).

This research developed an approach dubbed the Crop Rotation Choice Method (CSM) for addressing the crop decision issue and raises the rate of crop residue throughout the season that afterwards approaches the world economy's greatest growth. The suggested approach has the potential to demonstrate the residual value of agricultural yield (Kumar et al., 2015; Jain et al., 2021).

This research created and deployed an inventive crop suggestion system that farmers across India may simply use. This programme will assist farmers in making an informed selection about which crop to cultivate based on numerous natural and environmental factors. We also utilised a supplementary technique called the Rainfall Predictor to forecast the following 12 months of rainfall (Doshi et al., 2018, Jain et al., 2020).

Lee &Moon, 2014, this paper discusses the study and establishment of a proactive agriculture yield strategy based on immediate time quarterly weather forecasts. Agriculture production is difficult to forecast due to unpredictable meteorological circumstances throughout the year and significant alterations in the climate in the area as a result of global warming. The establishment of an agricultural crop forecasting system that incorporates real-time meteorological data is critical. We present a method for analysing weather forecast data (monthly, daily) as well as ways to change the guessing system in this study. On the basis of 33 years of agricultural climatic data in the composition, we develop a non-parametric numerical model. We used monthly meteorological data to anticipate the ultimate production using the model (Sharma et al., 2022). The present article offers simulation findings.

Paul et al., 2015, the aforementioned feature offers a programme that predicts the next stage of the analysed soil database using data mining methods. Crop harvesting will be featured in the category as expected. The yield prediction issue is formalised as the rule of separation, which employs the Naive Bayes and K-Nearest Neighbour approaches.

(Pudumalar et al., 2017), This work proposes an option selection procedure that employs a combination model with many voting techniques such as random tree, CHAID, K-Nearest Neighbour, and Nave Bayes as students choose the plant of specific site boundaries with the greatest degree of precision as well as effectiveness.

THE PROJECT NARRATIVE

Because they can fly without an onboard pilot and serve as an aerial surveillance system, drones have been used for a variety of business purposes. However, the usage of these drones is only allowed in those areas and they have restricted capabilities. We mix it with intelligent technology to enhance their capabilities and utilize the massive amounts of data generated by drones. The suggested concept, which involves IoT and cloud computing in drones, aims to give farmers intelligent agriculture control. A more recent phrase, smart drones, implies that the UAVs' sensors flow into a network architecture where they are connected to other devices via Internet technology, enabling communication and giving the gadgets a more sophisticated appearance. Now, since they have real-time characteristics, gadgets are multidimensional.

The Smart Drone Architecture - Sky Drone FPV2 (First People View) technology for drones is used in the design, which includes a camera component, a data component, and a 4G/LTE modem. The intelligent drone would function as a fixed-wing airborne aircraft capable of carrying several sensors, reaching faster speeds, having longer bursting durations, and covering vast areas (Kishor et al., 2017 ; Kishor et al., 2018). It has a robust information and communication technology (ICT) system for processing information from sensor devices and embedded applications for flight management and scheduling based on GPS (Global Positioning System) tracking and Google Maps. As a consequence, the outcomes of processing will be made available to authorised individuals as soon as the flight is available. The proposed model is divided into three pieces. Sensory perception, dialogue, and synchronisation

Sensing: Synchronous sensors, including flying and navigation sensors as well as camera sensors, are fitted in Smart Drones for the unique gathering of user-useful data. The drone's status and flight parameters control the flight and navigation sensors, such as the vision sensor, gyroscope, tilt, and current sensors, and GPS. The sensors also assist with navigation and keep an eye on the drone's immediate and wider environment to spot and avoid unexpected obstacles (Kishor et al., 2018). As they provide the fundamental functional load of drones used for monitoring in agricultural applications, the camera used in the Smart Drone should have at least a 12-megapixel resolution and varied frequency range.

- Recognize pests, weeds, and illnesses to optimize the application of pesticides and crop sprays.
- Calculate crop yield, i.e., offer plant counts.
- By identifying nutrient inadequacies, we provide information on soil fertility.
- By locating probable water stress locations, you can measure irrigation and manage your crop. Although thermal imaging equipment are used during nighttime vision, seeing through smoke or fog, vegetation tracking, heat and fire identification using infrared trends, and so on, ocular cameras are utilised for gathering images during the day.

Communication: The connection and interaction block is in the driver's seat for transferring data and must be resistant to external variables as well as quick to adapt to alterations in the network architecture. The Smart Drone will be equipped with a practically isotropic in nature radiation sequence antenna and a 4G/LTE internet connection with encased IoT connected technologies like Wi-Fi and ZigBee that have been tested for aerial connections and aid in mobile connectivity for remote management of the drone and data collection. The proposed drone will contain IoT-specific data collection, processing, and/or effector capabilities in addition to a Wi-Fi networking module. The Internet of Things (IoT) device can be any intelligent device, such as a mobile phone, laptop, iPad, etc. that serves as a gateway and provides Internet connectivity, is provided as needed, and is offered through devices like a laptop, iPad, etc. that serve as a gateway. Anywhere and any device may connect to the drone using these communication technologies. The smart gadget at the remote site manages the data collected from the drone. Since the suggested drone has Wi-Fi built in, the data collected may be transmitted in real-time to the cloud and seen online. The data gathered is kept in the application cloud and utilized for data analysis, assessment, and providing farmers with best practices and approaches for a specific condition.

Synchronization: The coordination block concentrates on tailoring to the requirements of the target application. There are three key parts to it: mission control, mission planning, and sensor data analysis. To address both the static and dynamic settings, we would create distributed and centralized coordinating mechanisms. Mission planning divides the high-level activities into flight paths that must be travelled and specific actions for each waypoint. The mission control unit receives the user input and distributes it to the required components. The user is then shown a single, sizable overview image that was created by mosaicing photos captured by several sensors.

Processing Flow

In the suggested model, the following steps will be carried out:

- Pre-planning: The preplanned path must be computed first and used to establish the drone's search engine.
- Searching: After taking off, the drone travels along the designated path while taking pictures and processing them internally.
- Detection: Depending on the drone's intended use, we modify the detection algorithm to identify any field uncertainties using the drone's integrated sensors.
- Streaming: The IoT device collects any data or photos that are taken in real-time. The devices control the streaming appropriately as well, and the cloud system receives the data they have gathered.
- Data Analysis: The relevant researchers get the necessary data from the cloud system for analysis and evaluation following the circumstances and the topic at hand and potential fixes that would aid farmers in working their land are developed.
- Cultivation Control: The data from the drones may aid in the management of the cultivation. For example, sensors installed in drones can measure the soil's moisture content and the health of the plants, providing farmers with information about their fields and the subsequent actions that need to be taken.
- Computers on the Go - Due to its accessibility and lower connection costs, mobile computing has significantly changed how we live our daily lives. The agricultural industry is one of the many industries that use it. For informing farmers about the availability of products and weather on a daily and seasonal basis, a mobile computing system has been proposed.

Wireless sensor networks, the Internet of Things (IoT), and sensors - A technology known as the "Internet of Things" aims to link all worlds' devices to the Internet. With an integrated Internet stack, it makes use of RFID, Wi-Fi, and other sensors. Apps are created based on IoT-enabled devices for monitoring and controlling systems in a variety of areas, such as industrial processes, household appliances, and apps for monitoring one's health, smart homes, and smart cities. Few academics in the field of agriculture have suggested IoT-based designs for tracking the management of the supply chain for agricultural goods. Many efforts have been made in the area of agriculture using wireless sensor networks, which are considered a mature technology. Many characteristics needed in the agricultural realm may be sensed and analyzed using sensors. There are several applications that use sensors in agriculture (Kishor et al., 2023: Rai, et al., 2022).

For monitoring the qualitiy of the soil, WSN designs were suggested, constructed, and tested.

- Mobile Technology- Mobile technology has had a large influence on many parts of our everyday lives as a consequence of its easy accessibility and cheaper connection costs. It is used in almost every sector, including agribusiness. A mobile computing-based system has been designed to notify farmers about goods and environment on a weekly and monthly schedule.
- Big-data sources and big-data analytics - Massive amounts of data, such as sensor data, social networking data, and commercial data, are gathered over a longer period of time from many sources. The biggest issue is in the areas of collection, storage, analysis, and search (Kishor, et al., 2023). In order to find hidden patterns in the data, it is used for business data processing in conjunction

with big data analytics. Big data is utilized in the agricultural industry to manage the supply chain for agro goods to reduce production costs.

- Knowledge creation through data analysis and mining - Analyzing data to uncover hidden patterns is a technique known as data mining. Data mining for the agriculture industry has been studied for a long time and. To categorize soil types, data mining has been utilized to analyze their characteristics. Additionally, using current soil nutrient information and historical crop sequences on the same field, soil data mining is helpful for crop prediction and choosing the best crop sequence.
- Utilizing the Cloud - Resources may be shared at a low cost thanks to cloud computing (Kishor, et al., 2023). The infrastructure, platform, and software as a service (IaaS), which are all offered by cloud computing service providers, are reasonably priced.

Agriculture data storage has been done using cloud computing (Mattern & Floerkemeier., 2010). It has been utilized with IoT in the agricultural industry.

- India's Agriculture Industry - Agriculture provides the majority of Indians' income and is a significant contribution to the country's economy. However, India's agriculture industry still has to expand and foster technology engagement and usefulness. Although the Indian government has also made a few efforts to offer farmers internet and mobile messaging services for agricultural inquiries, and agro vendors' information to farmers, it gives static data about the condition of the soil in each region. The system that bases decisions on real-time data about soil quality based on its existing quality has not been put into place. The characteristics of the soil determine its quality. The pH of the soil and the number of elements like nitrate, phosphate, and potassium in the soil are crucial determinants of the kind of crop production and soil quality. By applying only the necessary amount of fertilizer, real-time monitoring of these qualities enables soil health preservation. Analysis of the soil's moisture content enables the application of water just when necessary, preventing water waste. Additionally, environmental factors like temperature and moisture have an impact on crop yield and crop diseases. In this regard, a dynamic model that gathers such real-time data is required. All agricultural enterprises must be connected to a decision-making system to boost output and facilitate the distribution of agricultural goods supplied by producers to marketing companies and vendors to producers. Other elements, such agro product rates, will also be under the control of such a system.

Many users, especially those in remote locations, may now access smart mobile phones. The Beagle Black Bone is a low-cost IoT device that can be connected to the soil and environmental sensors to gather information about the quality of the soil and the environment at a time. This encourages the development of a portable, affordable sensor kit for detecting the soil's quality and determining the need for fertilizers at the time. To be processed further, the soil data from agricultural areas must be acquired using a sensor kit and transferred to Argo Cloud storage. Big data were then gathered can be examined for the production-related activities that are necessary.

THE PROPOSED TRANSDISCIPLINARY SMART AGRICULTURE MODEL

The five components that make up the transdisciplinary model's suggested architecture are displayed in Figure 2:

1) Sensor Kit Module,
2) Module for mobile apps.
3) AgroCloud Module.
4) Big-Data Mining, Analysis and Knowledge Building Engine Module.
5) Government & AgroBanks UI

SensorKit is a transportable IoT gadget that includes soil and environmental sensors. The MobileApp module offers users with a dashboard. AgroCloud Module includes storage, Big-Data mining, examination, and knowledge building engines, as well as a programme module for communicating with users. The Ministry of Agriculture and AgroBanks graphical user interface is a web portal for agricultural initiatives and loans.

Component for SensorKit - This section is a vital part of collecting this design and is in charge of soil testing at regular intervals to obtain soil-related data. Figure 3 depicts the SensorKit module. SensorKit is a low-cost and transportable kit in which we explored the usage of a snoopy black bone, an IoT equipped device with memory and computational capacity, as well as a GPS sensor to detect positioning information. This kit's main components are soil nutrient sensors that are linked to it. We considered soil pH sensor, soil moisture sensor, and Phosphorus (P), Potassium (K), and Nitrate (N) sensors for the aforementioned model, each of which is communicated to the IoT device.

The AgroCloud Component - All agricultural sector users must connect to AgroCloud via the MobileApp. AgroCloud storage, which includes Big-Data storage, will hold all landowner, agro advertisement, medication, and agro, vendor and service supplier details (fertilizer/pesticide/seed and agro machinery providers) specifics, as well as government programmes for the agricultural industry, such as bank loans for producers and concessions on seed and/or fertilisers. The module also saves data obtained on a regular basis through soil and environmental monitoring. As an increasing number of consumers connect to this assistance, the data amount expands significantly over time, leading up to Big-Data. The AgroCloud component, which includes Big-Data storage, Big-Data The mining industry, Evaluation, and Intellectual Construction Engine. As demonstrated by Figure 4, the aforementioned module resides in AgroCloud as well as plays a significant part in determining for fertiliser necessity according to the present crop determined by up-to-date soil features for enhanced yields, forecasting crop diseases based on up-to-date soil characteristics and current environmental conditions, crop yield estimation, best crop pattern analysis from data gathered over the span of time, best crop for associated soil parameters, and the amount of watering necessitate based on moisture in the soil level. The system also gives data concerning region-specific production specifics for each agricultural product, total crop output for each crop in the state at large, and current demand among customers, which will assist consumers regulate the costs for each agro product. As this database accumulates details regarding soil quality and agricultural product specifics with production amounts for each farmhouse over time, inference results from data mining can be generated for better crop sequencing to be carried out for optimal productivity and soil health. Furthermore, depending on prior stock of agro goods and present-day demand requirements, this database may recommend crops to farmers for planting on acreage with unusual soil features. Big-

Figure 2. Recommended structure for a multifunctional framework of intelligent farming

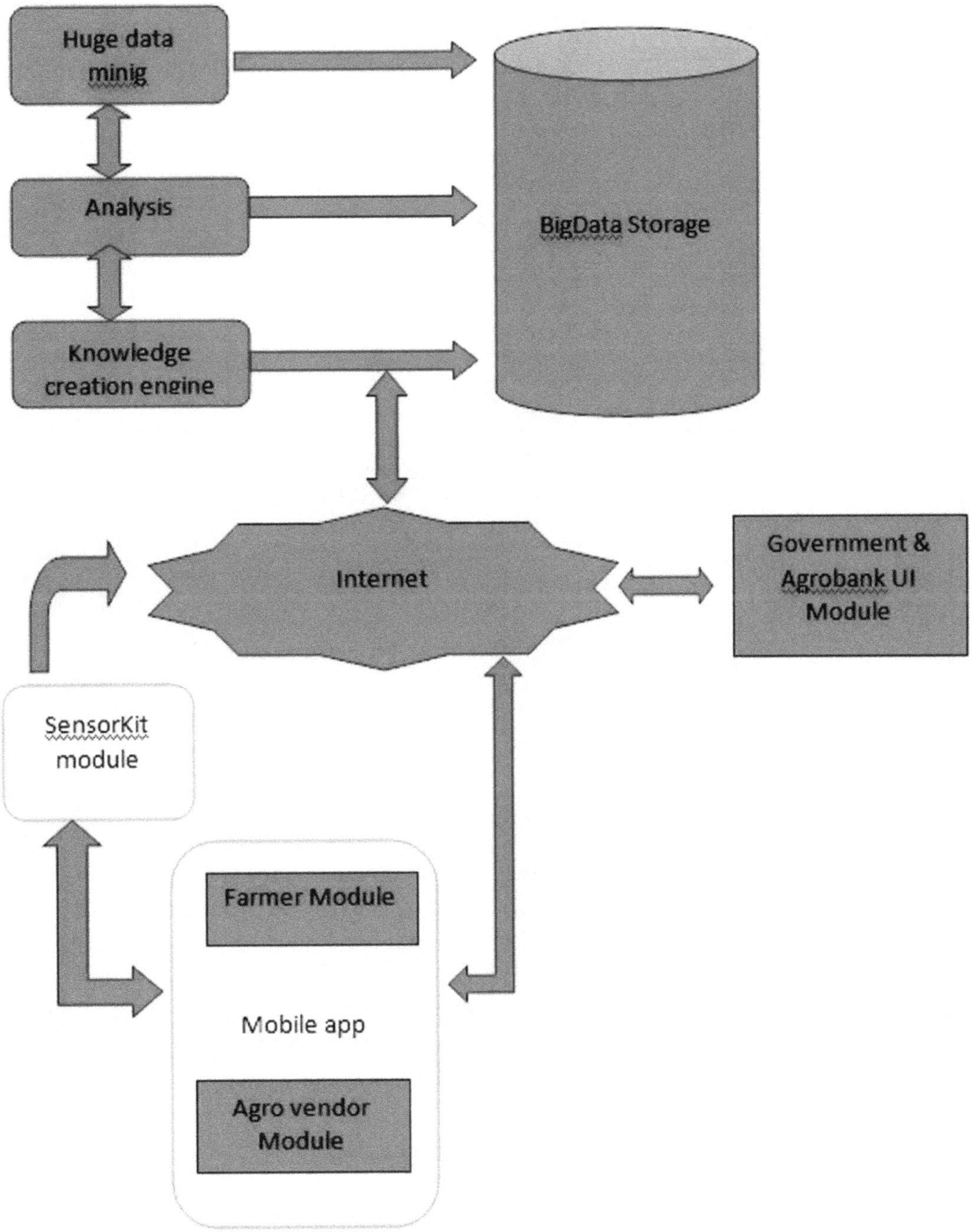

Figure 3. SensorKit module

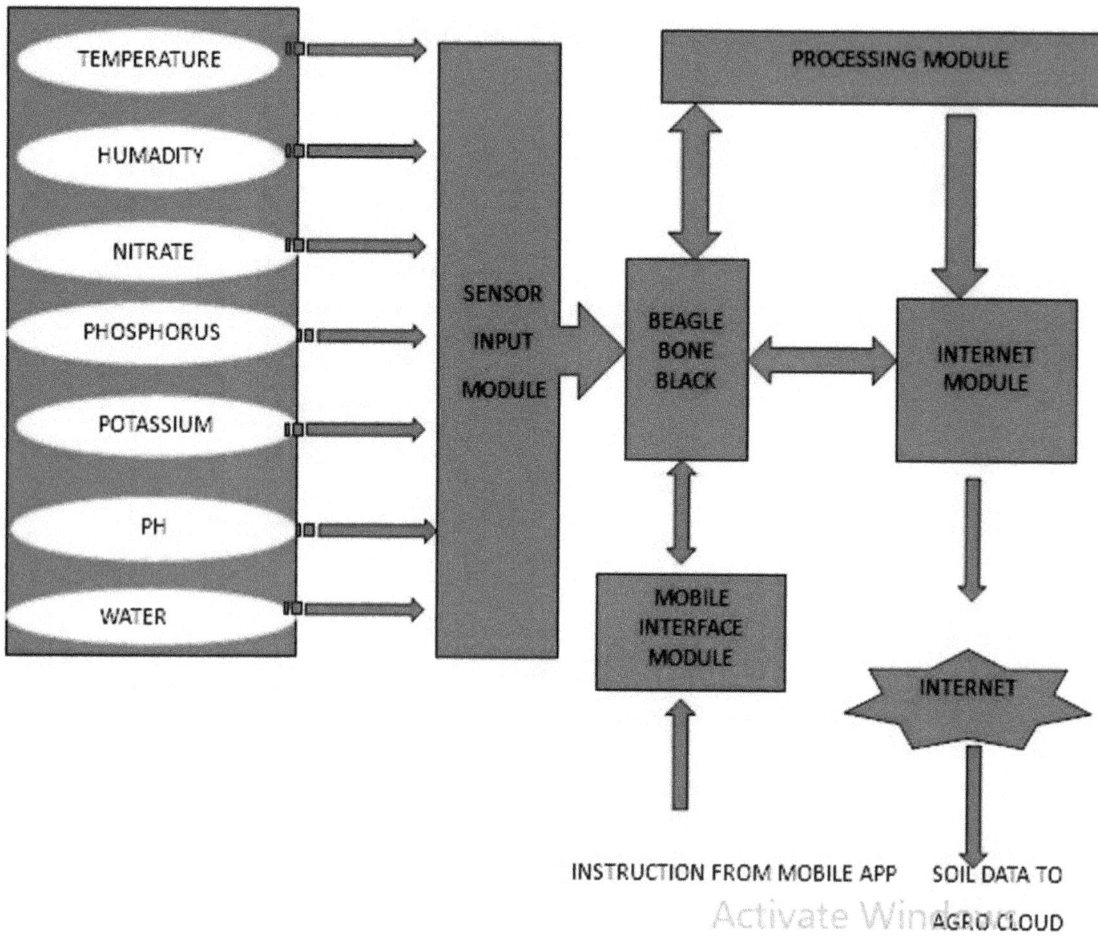

data analysis may be used to forecast future product output based on prior knowledge bases. The cloud storage application module is used to deliver alerts to users, make recommendations based on analysis, and notify crop diseases based upon the present climate and past knowledge base. AgroCloud module is shown in Figure 4.

CONCLUSION AND FUTURE WORK

The topic of "smart agriculture" is expanding quickly and has the potential to completely change how we produce food. Farmers can maximise their resource utilisation, increase crop yields, and lessen waste and environmental effects by utilising cutting-edge technologies like sensors, IoT devices, and data analytics. Smart agriculture has several advantages, including enhanced food security for the world's expanding population, increased productivity and profitability for farmers, and less environmental impact. Future developments in smart agriculture technologies are anticipated to include the creation of more sophisticated sensor and data analytics algorithms, the application of artificial intelligence and

Figure 4. Argo-cloud modules

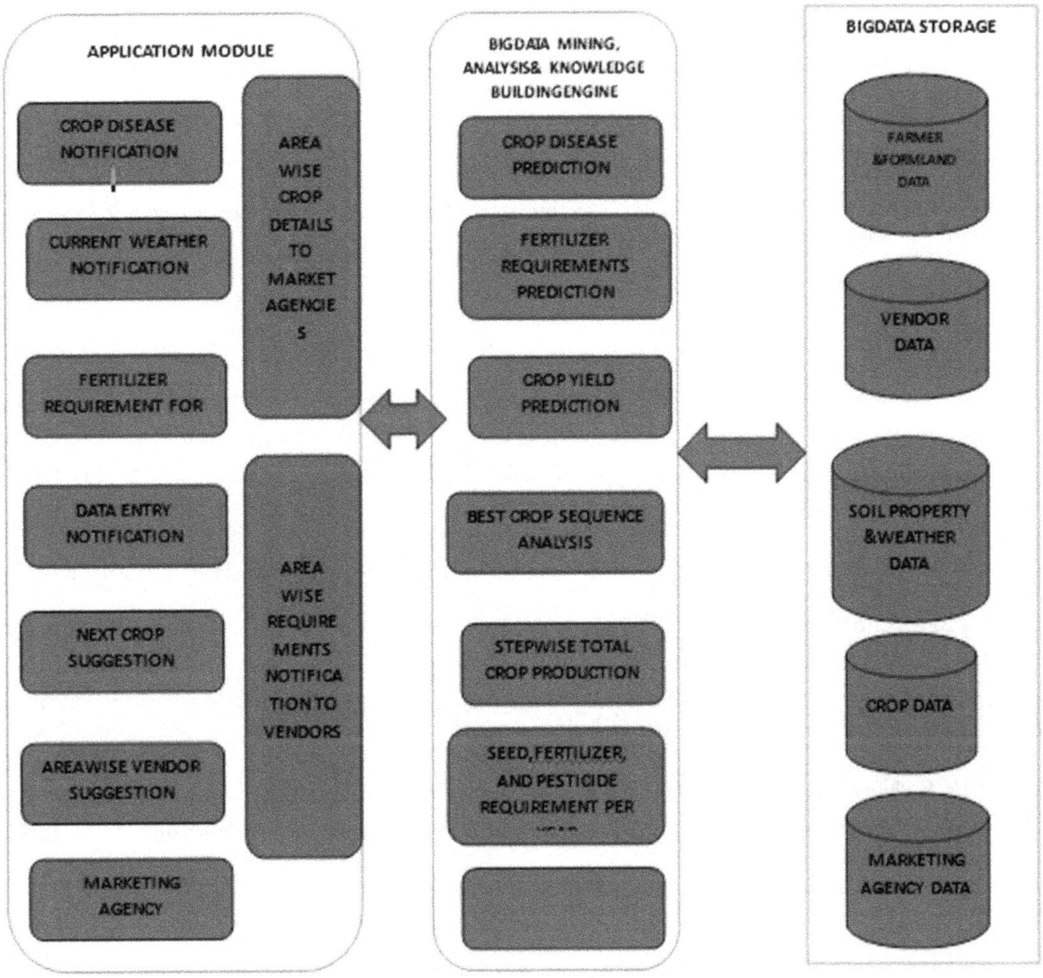

machine learning to automate decision-making processes, and the integration of blockchain technology to improve supply chain transparency and traceability. Additionally, there is a great deal of potential for smart agriculture to assist in addressing some of the major problems that the global food system is now confronting, including food security, population increase, and climate change. Farming operations can be managed more effectively, more food can be produced with less resource, and our food systems are more resilient to changing conditions because of the power of technology and data.

Future developments in smart agriculture will concentrate on creating more integrated and all-encompassing systems that can handle the complex issues that farmers and the food system as a whole are experiencing. Collaboration between experts in a variety of fields, including agriculture, engineering, data science, and policy, will be necessary for this, as well as a dedication to innovation and ongoing development.

REFERENCES

Adeloye, A. J., Rustum, R., & Kariyama, I. D. (2012). Neural computing modelling of the reference crop evapotranspiration. *Environmental Modelling & Software, 29*(1), 61–73. doi:10.1016/j.envsoft.2011.10.012

Bac, C. W., van Henten, E. J., Hemming, J., & Edan, Y. (2014). Robots for high-value crops: State-of-the-art review and challenges. *Journal of Field Robotics, 31*(6), 888–911. doi:10.1002/rob.21525

Borenstein, J., & Miller, K. (2013). Robots and the internet: Causes for concern. *IEEE Technology and Society Magazine, 32*(1), 60–65. doi:10.1109/MTS.2013.2241475

Burkle, A. (2009). Collaborating miniature drones for Surveillance &recon-renaissance. In Unmanned/Unattended Sensors & Sensor Networks (vol. 7480). International Society for Optics & Photonics.

Doshi, Z., Nadkarni, S., Agrawal, R., & Shah, N. K. (2018). AgroConsultant: Intelligent Crop Recommendation System Using Machine Learning Algorithms. *2018 Fourth International Conference on Computing Communication Control and Automation (ICCUBEA)*, 1-6. 10.1109/ICCUBEA.2018.8697349

Giusti, E., & Marsili-Libelli, S. (2015). A fuzzy decision support system for irrigation and water conservation in agriculture. *Environmental Modelling & Software, 63*, 73–86. doi:10.1016/j.envsoft.2014.09.020

Gupta, S., Tyagi, S., & Kishor, K. (2022). Study and Development of Self Sanitizing Smart Elevator. In D. Gupta, Z. Polkowski, A. Khanna, S. Bhattacharyya, & O. Castillo (Eds.), *Proceedings of Data Analytics and Management. Lecture Notes on Data Engineering and Communications Technologies* (Vol. 90). Springer. doi:10.1007/978-981-16-6289-8_15

Hemalatha, D. K. M. R., & Deepika, G. (2016). Internet of things (iot) based smart irrigation. *International Journal Of Advanced Research in Basic Engineering Sciences and Technology, 2*(2), 128–132.

Jain, A., Sharma, Y., & Kishor, K. (2020). Financial Supervision and Management System Using Ml Algorithm. *Solid State Technology, 63*(6), 18974–18982. http://solidstatetechnology.us/index.php/JSST/article/view/8080

JainA.SharmaY.KishorK. (2021). Prediction and Analysis of Financial Trends Using Ml Algorithm. *Proceedings of the International Conference on Innovative Computing & Communication (ICICC)*. doi:10.2139/ssrn.3884458

Jaishetty, S. A., & Patil, R. (2016). Iot sensor network based approach for agricultural field monitoring and control. *International Journal of Research in Engineering and Technology, 5*(6).

Keerthi, V., & Kodandaramaiah, G. N. (2015). Cloud iot based greenhouse monitoring system. *International Journal of Engineering Research and Applications, 5*(10), 35–41.

Kishor, K. (2022). Communication-Efficient Federated Learning. In S. P. Yadav, B. S. Bhati, D. P. Mahato, & S. Kumar (Eds.), *Federated Learning for IoT Applications. EAI/Springer Innovations in Communication and Computing*. Springer. doi:10.1007/978-3-030-85559-8_9

Kishor, K. (2022). Personalized Federated Learning. In S. P. Yadav, B. S. Bhati, D. P. Mahato, & S. Kumar (Eds.), *Federated Learning for IoT Applications. EAI/Springer Innovations in Communication and Computing*. Springer. doi:10.1007/978-3-030-85559-8_3

Kishor, K., & Nand, P. (2023). Wireless Networks Based in the Cloud That Support 5G. In Cloud-based Intelligent Informative Engineering for Society 5.0 (pp. 23-40). Chapman and Hall/CRC. doi:10.1201/9781003213895-2

Kishor, K. (2023). Cloud Computing in Blockchain. In Cloud-based Intelligent Informative Engineering for Society 5.0 (pp. 79-105). Chapman and Hall/CRC. doi:10.1201/9781003213895-5

Kishor, K. (2023). Impact of Cloud Computing on Entrepreneurship, Cost, and Security. In Cloud-based Intelligent Informative Engineering for Society 5.0 (pp. 171-191). CRC Press. doi:10.1201/9781003213895-10

KishorK.AttriS.RanjanS.TyagiT. (2018). Smart cart based on IOT. *International Journal of Creative Research Thoughts, 6*(1), 1388-1392, Available at: https://www.ijcrt.org/papers/IJCRT1803295.pdf

Kishor, K., Nand, P., & Agarwal, P. (2017). Subnet based ad hoc network algorithm reducing energy consumption in manet. *International Journal of Applied Engineering Research*, *12*(22), 11796–11802.

Kishor, K., Nand, P., & Agarwal, P. (2018). Secure and Efficient Subnet Routing Protocol for MANET. *Indian Journal of Public Health Research & Development*, *9*(12), 200. doi:10.5958/0976-5506.2018.01830.2

Kishor, K., Nand, P., & Agarwal, P. (2018). Notice of Retraction Design adaptive Subnetting Hybrid Gateway MANET Protocol on the basis of Dynamic TTL value adjustment. *Aptikom Journal on Computer Science and Information Technologies*, *3*(2), 59–65. doi:10.11591/APTIKOM.J.CSIT.115

Kishor, K., & Pandey, D. (2022). Study and Development of Efficient Air Quality Prediction System Embedded with Machine Learning and IoT. In *Proceeding International Conference on Innovative Computing and Communications.* Springer. 10.1007/978-981-19-2535-1_24

Kishor, K., Saxena, N., & Pandey, D. (2023). Cloud-based Intelligent Informative Engineering for Society 5.0. Chapman and Hall/CRC. doi:10.1201/9781003213895

Kishor, K., Sharma, R., & Chhabra, M. (2022). Student Performance Prediction Using Technology of Machine Learning. In D. K. Sharma, S. L. Peng, R. Sharma, & D. A. Zaitsev (Eds.), *Micro-Electronics and Telecommunication Engineering. Lecture Notes in Networks and Systems* (Vol. 373). Springer. doi:10.1007/978-981-16-8721-1_53

Kishor, K., Singh, P., & Vashishta, R. (2023). Develop Model for Malicious Traffic Detection Using Deep Learning. In D. K. Sharma, S. L. Peng, R. Sharma, & G. Jeon (Eds.), *Micro-Electronics and Telecommunication Engineering. Lecture Notes in Networks and Systems* (Vol. 617). Springer. doi:10.1007/978-981-19-9512-5_8

Kishor, K., Tyagi, R., Bhati, R., & Rai, B. K. (2023). Develop Model for Recognition of Handwritten Equation Using Machine Learning. In *Proceedings of International Conference on Recent Trends in Computing. Lecture Notes in Networks and Systems* (vol. 600). Springer. 10.1007/978-981-19-8825-7_23

Kumar, R., Singh, M. P., Kumar, P., & Singh, J. P. (2015). Crop Selection Method to maximize crop yield rate using machine learning technique. *2015 International Conference on Smart Technologies and Management for Computing, Communication, Controls, Energy and Materials (ICSTM),* 138-145. 10.1109/ICSTM.2015.7225403

Kumbar, B., Galagi, B., shankar, B., & Honnalli, N. (2016). Smart irrigation system using internet of things. *Bonfring International Journal of Research in Communication Engineering*, *6*(November), 2016. doi:10.9756/BIJRCE.8189

Lee, H., & Moon, A. (2014). Development of yield prediction system based on real-time agricultural meteorological information.. *16th International Conference on Advanced Communication Technology,* 1292-1295. 10.1109/ICACT.2014.6779168

Liqiang, Z., Shouyi, Y., Leibo, L., Zhen, Z., & Shaojun, W. (2011). A crop monitoring system based on wireless sensor network. *Procedia Environmental Sciences*, *11*, 558–565. doi:10.1016/j.proenv.2011.12.088

Mattern, F., & Floerkemeier, C. (2010). *From the Internet of Computers to the Internet of Things. In From active data management to event-based systems and more*. Springer.

Paul, M. T., Vishwakarma, S. K., & Verma, A. (2015). Analysis of Soil Behaviour and Prediction of Crop Yield Using Data Mining Approach. *2015 International Conference on Computational Intelligence and Communication Networks (CICN)*, 766-771. 10.1109/CICN.2015.156

Pingli, G., Yanlei, S., Junliang, C., Miaoting, D., & Bojia, L. (2011). Enterprise- oriented communication among multiple esbs based on wsnotification and cloud queue model. *International Journal of Advancements in Computing Technology*, *3*(7), 255–263. doi:10.4156/ijact.vol3.issue7.30

Pudumalar, S., Ramanujam, E., Rajashree, R. H., Kavya, C., Kiruthika, T., & Nisha, J. (2017). Crop recommendation system for precision agriculture. *2016 Eighth International Conference on Advanced Computing (ICoAC),* 32-36. 10.1109/ICoAC.2017.7951740

Rai, B. K., Sharma, S., Kumar, G., & Kishor, K. (2022). Recognition of Different Bird Category Using Image Processing. *International Journal of Online and Biomedical Engineering*, *18*(07), 101–114. doi:10.3991/ijoe.v18i07.29639

Rupanagudi, S. R., Ranjani, B. S., Nagaraj, P., Bhat, V. G., & Thippeswamy, G. (2015). A novel cloud computing based smart farming system for early detection of borer insects in tomatoes. In *2015 International Conference on Communication, Information & Computing Technology (ICCICT)*. IEEE. 10.1109/ICCICT.2015.7045722

Sharma, A., Jha, N., & Kishor, K. (2022). Predict COVID-19 with Chest X-ray. In D. Gupta, Z. Polkowski, A. Khanna, S. Bhattacharyya, & O. Castillo (Eds.), *Proceedings of Data Analytics and Management. Lecture Notes on Data Engineering and Communications Technologies* (Vol. 90). Springer. doi:10.1007/978-981-16-6289-8_16

Swathi, B. S., & Guruprasad, H. S. (2014). Integration of wireless sensor networks and cloud computing. *International Journal of Computational Science*, *2*(5), 49–53.

Venkataramana, K., & Padmavathamma, M. (2012). A design of framework for agri-cloud. *IOSR Journal of Computer Engineering*, *4*(5), 1–6. doi:10.9790/0661-0450106

Zhu, Y., Song, J., & Dong, F. (2011). Applications of wireless sensor net- work in the agriculture environment monitoring. *Procedia Engineering*, *16*, 608–614. doi:10.1016/j.proeng.2011.08.1131

Chapter 7
IoT-Based Smart Agriculture

Sheelesh Kumar Sharma
ABES Institute of Technology, Ghaziabad, India

Avinash Kumar Sharma
https://orcid.org/0000-0001-6762-6778
ABES Institute of Technology, Ghaziabad, India

Sushant Sharma
https://orcid.org/0000-0001-8033-3072
ABES Institute of Technology, Ghaziabad, India

Kabir Shukla
ABES Institute of Technology, Ghaziabad, India

Ishaan Ishaan
https://orcid.org/0009-0000-4416-6019
ABES Institute of Technology, Ghaziabad, India

ABSTRACT

The chapter explores how IoT technology can transform the agricultural industry. IoT devices, sensors, and data analytics in agriculture provide a disruptive answer by enabling the monitoring, automation, and optimization of many elements of agricultural operations. The chapter highlights the difficulties that conventional agricultural practices confront, including water scarcity, soil deterioration, and insect control. It emphasizes the demand for creative solutions to solve these problems. The chapter's main emphasis is on how IoT technology might transform agriculture. It explores the numerous ways that IoT is used in smart agriculture, such as precision farming, monitoring livestock, smart irrigation systems, and crop health monitoring. Farmers can make well-informed decisions thanks to IoT devices like sensors, actuators, and drones that enable real-time data collection and analysis. IoT-based smart agriculture has higher yields, better resource utilization, a smaller environmental footprint, and more profitability.

DOI: 10.4018/979-8-3693-0200-2.ch007

1. INTRODUCTION: SMART AGRICULTURE

A farming management idea called "smart farming" aims to improve both the quantity and quality of agricultural products. Today's farmers have access to technology like GPS, soil scanning, data management, and the Internet of Things. The purpose of smart agriculture research is to provide a solid foundation for a farm management decision-support system. Population growth, climate change, and labor are all challenges that smart farming believes must be addressed. From planting and watering crops to crop health and harvesting, smart farming has received a lot of technological attention. A system is created in IOT-based smart agriculture to automate the irrigation system while monitoring the agricultural field with the aid of sensors (light, humidity, temperature, soil moisture, etc.). In the context of agriculture, IOT (Internet of Things) refers to the use of sensors, cameras, and other gadgets to convert every aspect and action related to farming into data. The negative environmental externalities of contemporary agriculture will be greatly reduced if smart agriculture is allowed to grow and develop beyond what it is today. Smart cities collect and analyze data using Internet of Things (IoT) devices such as connected sensors, lights, and meters. The cities utilize this information to enhance their infrastructure, public services, and more. Farmers find it challenging to comprehend technical jargon and how to use technology, but they also find it to be a cost-effective solution (Srivastava et al., 2020). Here Figure 1 depicts that there are numerous fields where Smart Agriculture is helping us to improve both the quantity and quality of agricultural products.

Figure 1. Numerous fields where smart agriculture helps us

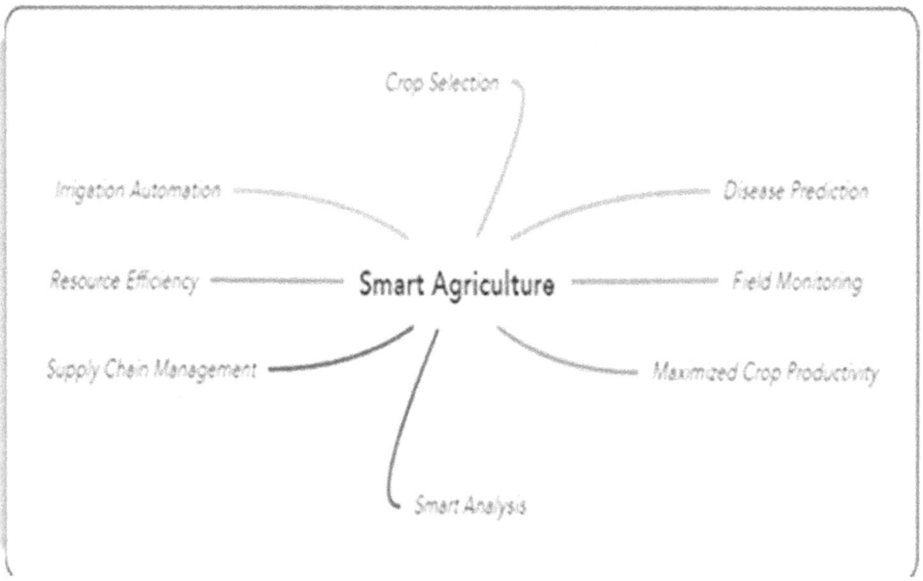

2. REVOLUTIONIZING AGRICULTURE WITH CLOUD AND IOT: EXPLORING THE WIDE-RANGING APPLICATIONS OF SMART AGRICULTURE

2.1 Role of Cloud and IoT in Smart Agriculture: A Comprehensive Analysis of Data-Driven Decision Making

The relationships between farmers, farms, and big data in the case of smart farming are far more complex than those in process-driven agriculture. Farmers play two roles in the system of activities. They are data users in the sense that they apply the findings of data analysis to address issues that arise on farms (Evans et al., 2017). Second, even though farm-related data are concerned with concerns relating to crop and/or animal responses to various physical circumstances, growth factors, and climatic variables, farmers' management practices have a significant impact on the information content of these data. In a similar line, farmers contribute to data production. As a result, humans create huge data both physically (through their choices and actions) and socially (via the various meanings they assign to this data while using them). This oddity produces a unique kind of "duality." Tangible technologies are created initially by some actors and then used by others, leading to their institutionalization and solidification through integration into particular organizational structures. The dual role of farmers in the context of big data makes institutionalization more difficult because every action a farmer does has an impact on both the quantity and quality of the data. This raises an important question:

Q1. What effects does farmers' dual function as data producers and users have on the institutionalization of big data?

Significantly, big data stands apart from every other technological development ever experienced in the agricultural area due to the crucial connection between production and consumption. A farmer's use of their tractor, for instance, can have a long-term effect solely on the tractor itself, leading to a generally consistent quality when analyzing a short time frame. In contrast, unlike traditional tactile instruments, big data are always evolving in that every action generates new data, changing the nature of the tool (Chen et al., 2018). Therefore, to effectively transform big data (external tools) into decisions that can be put into action, farmers must continually adapt to the constantly changing nature of big data and gain new knowledge and skills (internal tools). However, it is debatable if the farmer, acting alone, can manage the big data's ever-rising complexity (Mittelstadt et al., 2016). Big data analytics provide complicated insights that are difficult to comprehend, and an actor's capacity to act on such insights is severely hampered by their lack of knowledge (LaValle et al., 2010). Furthermore, even when there is abundant and insightful information, heuristics are frequently utilized to direct decision-making processes.

2.1.1 Large-Scale Data, Larger Data, and Increasingly Manipulated Data

Sensors and computers generate a huge amount of data in various domains of life, including agriculture (Khan et al., 2014). This raises a new concern about the quality of the information that data analytics can provide. Data quality is not determined by data size. In addition, more data can result in more biases.

Even though data cleaning programs have been there since the beginning of the big data boom, cleaning is a subjective process that can lead to significant biases. Additionally, every data processing operation modifies the data's shape. Data are never "raw" -- that is, "generated and collected without editing or any processing" -- but rather "cooked," since they are always subject to some processing. According to (Copa et al., 2018), processing includes any action taken on data or datasets, such as data organization, structuring, adaption, etc.

The activity system becomes more complex as a result of the transformation from a raw to a cooked form, which is a result of the collection, storage, and transmission conditions. This transformation affects both the accuracy of big data as well as the ability to interpret data and draw meaningful conclusions. Furthermore, several academics express grave worries regarding the capability of different parties to alter data during processing for competitive benefits. As a result, in addition to the methods used for data collection, preparation, and analysis, other factors such as the objectives, preferences, and motives of various actors also affect the quality of big data.

2.1.2 Farmers' Decisions and the Influence of Big Data

In the related literature, big data is frequently viewed as a tool that "guides" decision-making (Phillips et al., 2015). This suggests that the creation of particular chances for action by big data analytics has a significant impact on farmers' actions. Indeed, studies show that these tools can improve farmers' decision-making abilities by offering particular options. Big data creates clear actionable pathways that farmers may use to tackle issues on their farms in this way. However, this reliance on set action standards may limit actors' inventiveness and experimentation, making it harder for them to find and take advantage of new paths, even by accident. Contrarily, big data solutions are perceived and used differently by different players as well as by the same actor at different periods, according to structuration theories of technology. According to this perspective, the interpretation and use of large data depend on a variety of aspects, including personal variables (beliefs, attitudes), task-specific features, and priority sets.

Instead of concentrating on how big data affects a farm, more attention should be paid to how big data analytics are applied within complex systems that incorporate social, technological, and physical factors. The technology is frequently considered by researchers as an independent variable that influences, alters, and transforms an object; yet, it may be more important to comprehend how persons engage with big data and objects (their farms) simultaneously. To fully utilize the promise of big data, a more precise focus on user engagement with this data is required.

2.1.3 Data Ownership and Data Privacy: Crucial Issues to Address

Another feature that sets big data apart from the technology utilized in process-driven agriculture is its lack of ownership. Anyone can refer to a tractor as "her" or "his" tractor, but no one has the right to claim that huge data are their "property." This suggests that every farmer must cope with a tool that was created and used collectively, but whose ownership is ambiguous. Big data's singularity causes two significant problems. Despite their involvement in the tool's development, farmers lack control over the data transmission process in terms of data modification. Additionally highlighting this problem, (Bronson et al., 2016) claim that farmers are powerless to influence the context of data collection, processing, and utilization.

Second, as will be discussed in a later section, the issue of privacy is linked to the absence of ownership rights because it denies people the chance to define distinct boundaries within which their "own" data are not accessible to other actors. This has an impact on the interactions between farmers and other community members. The custom of signing contracts with internet companies breeds apprehension and perceived risk regarding the potential applications of agricultural data. Therefore, despite the availability of industry standards and regulatory frameworks for the use of big data, at least in industrialized nations, farmers occasionally refuse or even fight to share their data (Jakku et al., 2019). The performance of the entire agrifood supply network is closely related to the question of willingness, even though it typically receives less attention than the quality of the connection between various players through data-sharing practices .

2.1.4 Agents and Purposes

Understanding what prompts subjects' behaviors, with a focus on the motives that drive people to act with an object to achieve particular consequences, is crucial to understanding how an activity is performed. To put it another way, it's challenging to explain how farmers' intentions, ambitions, and motives determine the constraints on how completely they may use big data and smart technology at the farm level. According to some evidence (Fang et al., 2018; Lioutas and Charatsari, 2018; Brown and Roper, 2017), farmers' actions may have a variety of economic, managerial, social, and psychological roots, while their motivational orientations influence how they engage in different activities. Since it can assist us in placing action within contexts that integrate environmental, personal, and procedural elements, a more accurate focus on the motive behind the adoption and use of big data is anticipated to uncover more about the process of farm transformation.

From the standpoint of activity theory, various motives, some of which may be at odds, maybe what causes persons to act in the way they do. This type of "poly-motivation" brings up several new issues, including how motives are produced within and outside of the urge to modify the object as well as how they are tied to specific actions as opposed to other sets of activities. Additionally, social settings serve as sources of motivation for participants, and personal and group motives frequently coexist .

2.1.5 Back to the Basics: Reevaluating the Concept of Farming

The duration of each prospective human action is simultaneously determined and shaped by the farm, which serves as the activity's focal point. A farm can serve as both a source and a focal point for "attention, motivation, effort, and meaning". The farm serves as the basis for farmers to create motives and set goals while undergoing ongoing transformation and reorganization in this setting. To gradually alter the farm, big data techniques are used. This helps to achieve the intended outcomes, such as greater output, enhanced product quality, effective farm management, and less environmental impact. However, it is crucial to recognize that there is always a chance of negative outcomes, just like in every endeavor.

The relationship between farming and big data, which represents a little-discussed part of the activity system, is frequently overlooked in the literature. Surprisingly, compared to data-driven farming, there is greater debate over the viability of using technology in process-driven crop and animal production. Big data's ethereal nature can be credited for the issue's narrow emphasis. It's crucial to keep in mind though that farms are impacted by big data in both direct and indirect ways. Investments in smart technologies have direct benefits, such as the reallocation of resources, while indirect effects include the use of wearable sensors to collect data from productive animals. Therefore, when it comes to smart farming, many worries about the detrimental effects of technology on farms are accentuated. Animal autonomy and freedom of choice are major concerns among the few research that focuses on this aspect of smart farming (Driessen et al., 2015). Big data analytics places a lot of focus on desired results like improved performance and reduced work, which over-commodifies farms and especially farm animals. As a result, fresh ethical issues surface.

2.2 Leveraging Cloud Computing for Smart Agriculture Applications

Cloud computing is currently the most cutting-edge paradigm in computing. Google and Amazon were the initial users of the phrase "Cloud Computing" in 2006. According to the most recent definition, cloud computing is a computer paradigm that provides services in a virtualized, pooled, and shared way, enabling quick provisioning and deployment with little maintenance work. According to (De Donno et al,2019), the Cloud has five distinctive characteristics and includes three service types and four deployment methods. Infrastructure as a Service (IAAS), Platform as a Service (PAAS), and Software as a Service (SAAS) are important services provided by cloud computing. Private Cloud, Community Cloud, Public Cloud, and Hybrid Cloud are the four deployment models. Key characteristics of the cloud include on-demand self-service, accessibility across a large network, resource pooling, quick scaling, and measured service. While cloud computing has many benefits, including cost savings, efficiency, scalability, and reliability, it can be difficult to handle massive amounts of data. Low latency, high internet bandwidth, real-time analytics, data management, load balancing, energy consumption, security, and privacy are a few of these difficulties. Furthermore, the majority of computations in cloud computing use a centralized computing paradigm and take place there. The agriculture industry has become increasingly important as a result of cloud computing during the past few decades. Similarly, issues with using the cloud in the agriculture sector include constant and fast network access, security, and privacy. Data collection and remote storage, inexpensive access to ICT resources, online consultations with agriculture specialists, automation of land records, and weather forecasting are some of the fundamental characteristics that define cloud computing in agriculture. The necessity for reliable, fast network access as well as worries about security and privacy are obstacles to using cloud technology in the agriculture sector (Symeonaki et al., 2017).

2.3 IoT and Cloud-Based Precision Agriculture: Enhancing Crop Yields, Resource Efficiency, and Sustainability

Technology is used in "precision agriculture" to increase agricultural yields, conserve resources, and advance sustainability. The Internet of Things (IoT) and cloud computing are two technologies that have helped precision agriculture. By more precisely measuring and assessing crucial crop growth data in real-time, IoT-based precision agriculture can assist farmers in increasing productivity.

One of the main advantages of IoT in precision agriculture is the ability to collect massive amounts of data from many sources, such as soil sensors, weather stations, and drones. These sensors provide real-time data on crop health, nutrient content, temperature, humidity, and soil moisture levels. By using cloud-based analytics systems to examine this data, farmers can learn a lot about the factors affecting their crops. They can monitor and control irrigation systems, adjust the amount of fertilizer used, and detect early signs of disease or pest infestation. Farmers may maximize crop growth and productivity by using data-driven decision-making and quick action.

The advantages of cloud computing for IoT-powered precision agriculture are substantial. By harnessing the power of the cloud, farmers can store and analyze vast amounts of data without the need for on-site equipment. Farmers can change the size of their operations as needed because cloud systems are scalable (Khanna et al., 2019). This adaptability is also helpful in the context of precision agriculture, where the number of linked devices and data collected may differ significantly depending on the size and complexity of the farm. Furthermore, cloud-based analytics solutions may manage data from numerous farms, promoting benchmarking and knowledge sharing among farmers. Implementing best practices and making continuous improvements to farming techniques are made simpler by this cooperative approach.

The combination of IoT and cloud-based precision agriculture increases resource efficiency in a variety of ways. By regularly monitoring the soil conditions, farmers can optimize irrigation methods and use less water. Instead of relying on general watering schedules or speculation, IoT-enabled systems may assess soil moisture levels, enabling precise and targeted irrigation (Chen et al., 2018). This conserves water while lowering the amount of electricity required for irrigation. Similarly to this, farmers may make better use of fertilizers by monitoring nutrient levels, reducing misuse, and avoiding environmental harm. By automating and controlling agricultural processes with IoT devices, farmers may save time and money while also reducing the need for staff.

IoT and cloud-based precision agriculture promote sustainability by improving resource effectiveness as well as minimizing environmental effects. Precision agriculture enables farmers to identify diseases or pests at an early stage and treat particular areas with integrated pest management methods. Because they don't harm beneficial insects or contaminate soil and water supplies, less broad-spectrum pesticides are needed, which is preferable. Using precision agricultural methods, farmers can minimize the environmental impact of their operations, protecting biodiversity and promoting long-term sustainability. Cloud-based analytics can also provide details about the carbon footprint of agricultural activities, helping farmers spot development opportunities and implement climate-smart practices.

Some of the applications for IoT & Cloud Based Precision Agriculture is:

1. Soil Monitoring and Irrigation Optimization: Temperature, nutritional content, and moisture levels are tracked via IoT sensors placed in the ground. The data collecting and analysis done in the cloud allows farmers to optimize irrigation methods and conserve water resources. If the sensor detects low moisture levels, the irrigation system, for example, can be immediately activated to give water only where and when it is needed, reducing water waste and improving crop health (Nigussie et al., 2020).
2. Pest and Disease Management: IoT devices like cameras or drones can be used to scan crops for signs of diseases or pests using image recognition software. Real-time image analysis is sent to the cloud. Machine learning algorithms' capacity to identify certain pests or diseases and offer early warnings allows for the implementation of treatment or preventive measures that are specifically aimed at farmers. This tactic minimizes the use of insecticides and reduces the likelihood that crop loss due to pests would occur.
3. Livestock Monitoring: The location, conduct, and general health of cattle and other livestock animals are monitored via IoT devices. The data is sent to the cloud for examination. Farmers are more equipped to keep an eye on the well-being of particular animals, recognize signs of illness or suffering, and manage grazing patterns. If a sensor detects odd conduct, such as decreased activity or an increase in body temperature, an alarm can be sent to the farmer to protect the well-being of the animals. They will be compelled to act quickly as a result (Saravanan et al., 2017).
4. Crop Quality and Yield Prediction: Projections of crop output and quality can be made with high accuracy by combining IoT sensors with weather data and historical records. By monitoring factors like temperature, humidity, and solar exposure, farmers can estimate crop growth trends and decide the ideal time to harvest, store, and transport their products. This enhances logistics, lowers

waste, and ensures that crops are transported to markets at their freshest and healthiest possible states (Reshma et al., 2020).

2.4 Significance of Cloud and IoT in Ensuring Traceability and Supply Chain Management in Agriculture

In today's globalized and interconnected world, traceability and efficient supply chain management are essential for the agricultural sector. Consumers are demanding more and more transparency regarding the origin, quality, and safety of the food they eat. In this case, end-to-end traceability and the streamlining of agricultural supply chain procedures have been made possible by combining cloud computing with Internet of Things (IoT) technologies. The cloud and IoT play a crucial role in ensuring traceability and supply chain management in agriculture because of their ability to boost transparency, increase operational efficiency, optimize logistics, and improve food safety.

One of the key advantages of cloud computing and the Internet of Things in traceability and supply chain management is the capacity to gather, store, and analyze vast volumes of data at multiple points of the agricultural supply chain. The location, temperature, humidity, and quality of agricultural products are just a few of the variables that IoT devices can track and monitor. RFID tags, sensors, and cameras are a few of these gadgets. According to (Madumida et al., 2019), this real-time data is gathered and transferred to the cloud, where it may be securely stored and processed. Cloud-based analytics tools can process and understand this data, providing key insights into how agricultural products are handled, transported, and stored. This data-driven method encourages transparency by providing stakeholders with access to exact and up-to-date information on the items' origin, route, and conditions along the whole supply chain.

It is essential to guarantee traceability in the agricultural supply chain for several reasons. It first encourages accountability and reduces the likelihood of fraud or misrepresentation. By having a full understanding of the supply chain, stakeholders can verify the validity and quality of agricultural products, which will prevent the proliferation of false or subpar goods. Second, traceability enhances food safety and quality assurance. In the event of contamination or an outbreak of a foodborne illness, the ability to trace things back to their source enables quick discovery, recalls, and targeted efforts to eliminate dangers, protecting the public's health and minimizing financial losses. Cloud and IoT technology provides the management of the data necessary for effective traceability, enabling prompt response and control of food safety occurrences.

Effective supply chain management is a further area where cloud and IoT technologies are vital in agriculture. By employing real-time data and sophisticated analytics, cloud-based solutions enable stakeholders to optimize logistics, optimize operations, and reduce costs. For example, farmers and suppliers can monitor inventory levels, track patterns in demand, and adjust production and delivery schedules as necessary. This ensures that the correct quantity and quality of agricultural items are available at the proper time and place by avoiding waste and successfully meeting consumer expectations. Cloud-based systems can also promote collaboration and coordination among different supply chain actors, promoting effective information exchange, communication, and decision-making. This collaborative approach improves efficiency, reduces delays, and fosters overall supply chain performance.

Using smart labeling and packaging solutions made feasible by cloud and IoT technology, agriculture's supply chain management and traceability are further improved. Agricultural items can be marked with a distinctive identification that can be scanned and monitored through the supply chain, like an

RFID tag or a QR code. These tags can hold relevant information including product details, production methods, certifications, and expiration dates. Stakeholders can scan the tags with IoT-enabled devices to instantaneously get accurate and real-time information, reducing manual errors and paperwork. To provide a complete and verifiable record for every product, cloud-based solutions can also link this data with other supply chain data, such as information about transit and storage.

Some of the applications are as:

1. Real-Time Monitoring and Tracking of Agricultural Products: Sensors, RFID tags, and GPS trackers can all be used to monitor and track agricultural products from the farm to the table. These devices can collect and transmit data on a variety of factors, including temperature, humidity, location, and handling conditions. The data can then be evaluated in real-time using cloud-based platforms to ensure that the products are being transported and stored in the best possible manner. If any desired parameter deviations take place, alerts can be activated, enabling stakeholders to react appropriately and halt spoilage or quality degradation(Bhutta et al., 2021).
2. Blockchain-Based Traceability Systems: Blockchain technology can provide a transparent, impenetrable database of all supply chain transactions and movements. Each stakeholder can record their operations on the blockchain, and other chain members will review and authenticate the data. A permanent record of the product's journey can then be obtained by stakeholders and customers by tracking the product's history of manufacture, handling, and origin. Because blockchain technology ensures that data cannot be altered or deleted, transparency and accountability are strengthened (Lin et al., 2018).
3. Predictive Analytics for Yield and Quality Optimization: Data from IoT devices and other sources can be used to build predictive models that forecast yield and quality features utilizing cloud-based systems. These models can help farmers and other interested parties make the most of current methods and resources, leading to higher yields, less waste, and better-quality products. For instance, using predictive analytics can help supply chains and resource utilization by optimizing harvest timing, fertilization rates, and irrigation plans (Akhter et al., 2022).
4. Automated Quality Inspection: Using cloud and IoT technology, agricultural products may be automatically checked for quality and safety criteria. Machine learning algorithms may be trained to spot faults, contaminants, and other quality issues using images and sensor data. This can help merchants and producers make sure that only safe and reliable products are made available to clients, reducing the likelihood of food safety mishaps and fostering client confidence (Hassan et al., 2021).

3. FUTURE DIRECTIONS AND EMERGING OPPORTUNITIES FOR INNOVATION

Smart agriculture, driven by the integration of the Internet of Things (IoT) technology, has revolutionized the agricultural industry. This chapter explores the future directions and emerging opportunities for innovation in smart agriculture using IoT. It delves into the potential advancements and the transformative impact of IoT in agriculture, highlighting the key areas where IoT can drive innovation and enhance agricultural practices.

3.1. IoT-Enabled Precision Farming

Precision farming, enabled by IoT, holds immense potential for the future of agriculture. IoT sensors can collect real-time data on soil moisture, temperature, humidity, and nutrient levels, allowing farmers to make data-driven decisions regarding irrigation, fertilization, and crop protection. This technology can optimize resource usage, minimize environmental impact, and improve crop yields.

3.2. Livestock Monitoring and Management

IoT-based solutions provide opportunities for efficient livestock monitoring and management. Wearable devices equipped with sensors can track animal health, behavior, and location, enabling early disease detection, precise feeding regimes, and optimized breeding practices. Real-time data collection and analysis contribute to improved animal welfare and productivity.

3.3. Smart Irrigation Systems

IoT-driven smart irrigation systems offer automated and precise water management. Soil moisture sensors combined with weather data and crop requirements can optimize irrigation schedules and minimize water wastage. IoT-enabled irrigation systems can be remotely controlled, saving labor and conserving water resources.

3.4. Crop Health Monitoring and Disease Management

IoT devices and sensors integrated with computer vision and machine learning algorithms enable continuous monitoring of crop health. This technology can detect early signs of diseases, nutrient deficiencies, and pest infestations. Real-time alerts and recommendations facilitate timely interventions, reducing crop losses and the need for excessive pesticide use.

3.5. Supply Chain Optimization

IoT technologies provide opportunities for enhanced supply chain management in agriculture. From field to fork, IoT sensors can track and monitor the movement, storage conditions, and quality of agricultural produce. This enables improved logistics, reduced food waste, and ensures traceability and transparency in the supply chain.

3.6. Autonomous Farming and Robotics

The convergence of IoT and robotics enables autonomous farming operations. Autonomous drones, robotic harvesters, and precision spraying systems equipped with IoT capabilities can perform tasks such as crop monitoring, weed control, and harvesting. These technologies increase efficiency, reduce labor costs, and enhance overall farm productivity.

3.7. Data Analytics and Predictive Modeling

IoT-generated data, combined with advanced analytics and predictive modeling, can unlock valuable insights for farmers. Machine learning algorithms can analyze vast amounts of data to provide personalized recommendations for crop management, yield prediction, market demand forecasting, and risk assessment. This empowers farmers to make informed decisions and maximize profitability.

3.8. Agricultural Drones and Remote Sensing

IoT-enabled agricultural drones equipped with sensors and cameras offer remote sensing capabilities. These drones can capture high-resolution imagery, multispectral data, and thermal imaging, providing valuable information on crop health, nutrient distribution, and irrigation needs. This technology enables rapid and comprehensive field monitoring.

So, to sum it all up the future of smart agriculture using IoT holds immense potential for transformative innovation. Precision farming, livestock management, smart irrigation, crop health monitoring, supply chain optimization, autonomous farming, data analytics, and remote sensing are just a few areas where IoT can revolutionize agricultural practices. Embracing these emerging opportunities and advancements in IoT technology will lead to increased efficiency, sustainability, and profitability in the agricultural industry. Continued research and collaboration between technology developers, researchers, and farmers are essential for unlocking the full potential of IoT in smart agriculture.

3.9. Farm Management Systems

IoT-based farm management systems provide comprehensive solutions for monitoring and controlling various aspects of agricultural operations. These systems integrate data from multiple sources, such as weather stations, soil sensors, and machinery, into a centralized platform. Farmers can remotely access real-time data, manage resources efficiently, and automate tasks, leading to improved productivity and streamlined operations.

3.10. Weather Forecasting and Risk Management

IoT technology plays a crucial role in weather forecasting and risk management in agriculture. Connected weather stations and sensors collect precise weather data, which, combined with historical data and predictive models, enable accurate forecasting. Farmers can receive timely alerts about impending weather events, such as storms or frost, and take preventive measures to protect crops and mitigate risks.

3.11. Farm-to-Consumer Traceability

With increasing consumer demands for transparency and traceability, IoT offers solutions for farm-to-consumer traceability in the agricultural supply chain. RFID tags, QR codes, and blockchain technology can track the origin, production processes, and transportation of agricultural products. This enables consumers to make informed choices while ensuring food safety and quality control.

3.12. Resource Optimization

IoT-based solutions enable efficient resource management in agriculture. By continuously monitoring environmental conditions, soil health, and crop growth, farmers can optimize the use of water, fertilizers, and energy. This not only reduces costs but also minimizes the environmental impact associated with excessive resource consumption.

3.13. Collaboration and Knowledge Sharing

IoT facilitates collaboration and knowledge sharing among farmers, researchers, and industry experts. Online platforms and mobile applications enable farmers to exchange information, best practices, and market insights. This collective intelligence helps in decision-making, encourages innovation, and fosters a supportive agricultural community.

3.14. Connectivity and Rural Development

The deployment of IoT infrastructure in rural areas for smart agriculture drives connectivity and promotes rural development. Access to reliable internet connectivity enables farmers to access real-time data, market information, and financial services. This digital inclusion enhances the livelihoods of farmers, reduces information asymmetry, and bridges the urban-rural divide.

3.15. Sustainable Farming Practices

IoT technology can support and promote sustainable farming practices. By monitoring and optimizing inputs, reducing chemical use, and implementing precision techniques, farmers can minimize their environmental footprint. IoT-driven sustainability practices contribute to soil health, water conservation, biodiversity preservation, and overall ecological balance.

In conclusion, the integration of IoT technology into smart agriculture offers a wide range of opportunities for innovation and advancement. From precision farming and livestock management to supply chain optimization and data analytics, IoT revolutionizes agricultural practices in terms of efficiency, sustainability, and profitability. Embracing these emerging opportunities and fostering collaboration between stakeholders will pave the way for a more connected, productive, and sustainable agricultural industry.

4. CONCLUSION

In conclusion, the chapter on "IoT based Smart Agriculture" presents a captivating exploration of the revolutionary impact of Internet of Things (IoT) technology in the agricultural sector. Through a comprehensive examination of various aspects of smart agriculture, it becomes evident that IoT integration brings forth transformative opportunities for innovation and advancements in farming practices.

Key areas such as precision farming, livestock management, smart irrigation, crop health monitoring, and supply chain optimization benefit significantly from IoT-enabled solutions. Real-time data collection and analysis empower farmers to make data-driven decisions, optimizing resource usage, reducing environmental impact, and ultimately improving crop yields and animal welfare.

The convergence of IoT and robotics introduces an era of autonomous farming operations, enhancing efficiency, and productivity while reducing labor costs. Moreover, data analytics and predictive modeling using IoT-generated data provide valuable insights, enabling farmers to make informed decisions and maximize profitability.

As we embrace the emerging opportunities presented by IoT in agriculture, it is evident that continued research, collaboration, and adoption of IoT technology will pave the way for increased efficiency, sustainability, and profitability in the agricultural industry. By harnessing the power of IoT, farmers, and stakeholders can collectively shape a more resilient and prosperous future for smart agriculture.

REFERENCES

Akhter, R., & Sofi, S. A. (2022). Precision agriculture using IoT data analytics and machine learning. *Journal of King Saud University-Computer and Information Sciences*, *34*(8), 5602–5618. doi:10.1016/j.jksuci.2021.05.013

Bhutta, M. N. M., & Ahmad, M. (2021). Secure identification, traceability and real-time tracking of agricultural food supply during transportation using internet of things. *IEEE Access : Practical Innovations, Open Solutions*, *9*, 65660–65675. doi:10.1109/ACCESS.2021.3076373

Bronson, K., & Knezevic, I. (2016). Big Data in food and agriculture. *Big Data & Society*, *3*(1), 2053951716648174. doi:10.1177/2053951716648174

Brown, P., & Roper, S. (2017). Innovation and networks in New Zealand farming. *The Australian Journal of Agricultural and Resource Economics*, *61*(3), 422–442. doi:10.1111/1467-8489.12211

Chen, X., Shi, Q., Yang, L., & Xu, J. (2018). ThriftyEdge: Resource-efficient edge computing for intelligent IoT applications. *IEEE Network*, *32*(1), 61–65. doi:10.1109/MNET.2018.1700145

Copa, C., Cema, F. E., Ceettar, C., Ecpa, E., & Fefac, E. S. A. (2018). *EU Code of conduct on agricultural data sharing by contractual agreement*. Academic Press.

De Donno, M., Tange, K., & Dragoni, N. (2019). Foundations and evolution of modern computing paradigms: Cloud, iot, edge, and fog. *IEEE Access : Practical Innovations, Open Solutions*, *7*, 150936–150948. doi:10.1109/ACCESS.2019.2947652

Driessen, C., & Heutinck, L. F. (2015). Cows desiring to be milked? Milking robots and the co-evolution of ethics and technology on Dutch dairy farms. *Agriculture and Human Values*, *32*(1), 3–20. doi:10.100710460-014-9515-5

Fang, W., Tang, L., Cheng, P., & Ahmad, N. (2018). Evolution decision, drivers and green innovation performance for collaborative innovation center of ecological building materials and environmental protection equipment in Jiangsu province of China. *International Journal of Environmental Research and Public Health*, *15*(11), 2365. doi:10.3390/ijerph15112365 PMID:30366457

Hassan, S. I., Alam, M. M., Illahi, U., Al Ghamdi, M. A., Almotiri, S. H., & Su'ud, M. M. (2021). A systematic review on monitoring and advanced control strategies in smart agriculture. *IEEE Access : Practical Innovations, Open Solutions*, *9*, 32517–32548. doi:10.1109/ACCESS.2021.3057865

Jakku, E., Taylor, B., Fleming, A., Mason, C., Fielke, S., Sounness, C., & Thorburn, P. (2019). “If they don’t tell us what they do with it, why would we trust them?” Trust, transparency and benefit-sharing in Smart Farming. *NJAS Wageningen Journal of Life Sciences*, *90*(1), 100285. doi:10.1016/j.njas.2018.11.002

Khan, N., Yaqoob, I., Hashem, I. A. T., Inayat, Z., Mahmoud Ali, W. K., Alam, M., Shiraz, M., & Gani, A. (2014). Big data: Survey, technologies, opportunities, and challenges. *TheScientificWorldJournal*, *2014*, 2014. doi:10.1155/2014/712826 PMID:25136682

Khanna, A., & Kaur, S. (2019). Evolution of Internet of Things (IoT) and its significant impact in the field of Precision Agriculture. *Computers and Electronics in Agriculture*, *157*, 218–231. doi:10.1016/j.compag.2018.12.039

LaValle, S., Lesser, E., Shockley, R., Hopkins, M. S., & Kruschwitz, N. (2010). Big data, analytics and the path from insights to value. *MIT Sloan Management Review*.

Lin, J., Shen, Z., Zhang, A., & Chai, Y. (2018, July). Blockchain and IoT based food traceability for smart agriculture. In *Proceedings of the 3rd international conference on crowd science and engineering* (pp. 1-6). 10.1145/3265689.3265692

Lioutas, E. D., & Charatsari, C. (2018). Green innovativeness in farm enterprises: What makes farmers think green? *Sustainable Development (Bradford)*, *26*(4), 337–349. doi:10.1002d.1709

Lioutas, E. D., Charatsari, C., La Rocca, G., & De Rosa, M. (2019). Key questions on the use of big data in farming: An activity theory approach. *NJAS Wageningen Journal of Life Sciences*, *90*(1), 100297. doi:10.1016/j.njas.2019.04.003

Madumidha, S., Ranjani, P. S., Vandhana, U., & Venmuhilan, B. (2019, May). A theoretical implementation: Agriculture-food supply chain management using blockchain technology. In *2019 TEQIP III Sponsored International Conference on Microwave Integrated Circuits, Photonics and Wireless Networks (IMICPW)* (pp. 174-178). IEEE. 10.1109/IMICPW.2019.8933270

Mittelstadt, B. D., & Floridi, L. (2016). The ethics of big data: current and foreseeable issues in biomedical contexts. *The Ethics of Biomedical Big Data*, 445-480.

Nigussie, E., Olwal, T., Musumba, G., Tegegne, T., Lemma, A., & Mekuria, F. (2020). IoT-based irrigation management for smallholder farmers in rural sub-Saharan Africa. *Procedia Computer Science*, *177*, 86–93. doi:10.1016/j.procs.2020.10.015

Phillips-Wren, G., & Hoskisson, A. (2015). An analytical journey towards big data. *Journal of Decision Systems*, *24*(1), 87–102. doi:10.1080/12460125.2015.994333

Reshma, R., Sathiyavathi, V., Sindhu, T., & Selvakumar, K., & SaiRamesh, L. (2020, October). IoT based classification techniques for soil content analysis and crop yield prediction. In *2020 Fourth International Conference on I-SMAC (IoT in Social, Mobile, Analytics and Cloud)(I-SMAC)* (pp. 156-160). IEEE. 10.1109/I-SMAC49090.2020.9243600

Saravanan, K., & Saraniya, S. (2017). Cloud IOT based novel livestock monitoring and identification system using UID. *Sensor Review*, *38*(1), 21–33.

Srivastava, R., Sharma, V., Jaiswal, V., & Raj, S. (2020). A research paper on smart agriculture using IoT. *International Research Journal of Engineering and Technology*, *7*(07), 2708–2710.

Symeonaki, E., Arvanitis, K. G., & Piromalis, D. D. (2017). Review on the Trends and Challenges of Cloud Computing Technology in Climate-Smart Agriculture. *HAICTA*, 66-78.

Chapter 8
Smart Agriculture Resource Allocation and Cost Optimization Using ML in Cloud Computing Environment

Pancham Singh
Ajay Kumar Garg Engineering College, Ghaziabad, India

Mrignainy Kansal
Ajay Kumar Garg Engineering College, Ghaziabad, India

Mili Srivastava
Ajay Kumar Garg Engineering College, Ghaziabad, India

Muskan Gupta
Ajay Kumar Garg Engineering College, Ghaziabad, India

ABSTRACT

In this research, resource allocation in machine learning is used to analyze how cloud computing is being applied in smart agriculture. This chapter goes over the advantages of cloud computing for farming and how machine learning can enhance resource allocation for higher agricultural yields and less negative environmental impact. The chapter also examines implementation difficulties for cloud-based agricultural solutions and speculates on potential fixes. Insights for researchers and practitioners in the area are provided by the research, which demonstrates the potential for merging cloud computing and machine learning in smart agriculture to increase productivity and sustainability. The research also assessed the efficacy of the ML-based techniques using a variety of performance indicators, including reaction time and throughput. The management of cloud workloads has shown considerable promise when using machine learning-based methods. This chapter offers a thorough overview of current developments in ML-based cloud workload management and identifies areas for further research.

DOI: 10.4018/979-8-3693-0200-2.ch008

1. INTRODUCTION

The agriculture sector is struggling to meet the rising food demand while minimizing its environmental impact. Technology integration, including cloud computing and machine learning, has the potential to revolutionize the agriculture sector and assist in overcoming these difficulties. In particular, scalability, cost-effectiveness, and real-time crop status monitoring can all be made possible by using cloud computing in the agriculture (Khalifeh et al., 2021). By improving resource allocation, machine learning can help farmers manage their crops more effectively, cut down on waste, and boost yields.

An emerging field called "smart agriculture" employs technology to maximize agricultural productivity and has the potential to increase farming practices' effectiveness and sustainability. With its capacity to store and handle massive volumes of data, cloud computing is well-suited to support applications for smart agriculture (Masip-Bruin et al., 2020). A branch of artificial intelligence called machine learning can assist in the analysis of this data and offer insights that can be used to resource allocation decisions.

Resource allocation in smart agriculture is optimised via machine learning techniques. These algorithms can analyse huge databases, spot patterns, correlations, and anomalies, then generate forecasts and suggestions. Farmers can acquire useful insights into the different elements that affect crop growth, such as soil moisture levels, weather conditions, pest infestations, and fertiliser requirements, by integrating machine learning models with cloud-based data analytics platforms (Yeganeh et al., 2023). The use of this data can optimise resource allocation by guiding decisions about when to apply fertiliser, schedule irrigation, take insect control measures, and rotate crops.

Precision agriculture, commonly referred to as smart agriculture, is a creative strategy that makes use of cutting-edge technologies to improve farming methods, boost crop yields, and cut down on resource waste. There is an increasing need for efficient and affordable agricultural solutions due to the expanding global population and the requirement to produce more food responsibly. The application of cloud computing and machine learning techniques in smart agriculture has considerable promise for improving resource allocation and lowering costs. These technologies have become potent tools in a variety of industries (Alam et al., 2020). Collaboration and knowledge exchange among farmers, researchers, and agronomists are facilitated by the integration of cloud computing and machine learning in smart agriculture. Cloud-based platforms serves as centralised data and model repositories, enabling stakeholders to exchange their discoveries, best practises, and cutting-edge methods. This collective intelligence has the potential to improve decision-making procedures, advance crop management strategies, and spur agricultural innovation. Additionally, cloud computing makes it possible to remotely monitor and control agricultural systems, giving farmers the freedom to use mobile or web-based applications to effectively manage their businesses from any location.

The use of cloud computing and machine learning in smart agriculture has enormous potential for cost- and resource-saving resource allocation optimisation. Farmers may boost agricultural yields, decrease resource wastage, and make data-driven decisions by utilising cloud-based platforms and machine learning algorithms (Mullachery & Alismail, 2023). Incorporating these technology improves agricultural operations' productivity and efficiency while also fostering innovation and sustainability in the industry.

However, there are difficulties in implementing cloud computing and machine learning in smart agriculture. Since sensitive information is frequently present in agricultural data, protecting data security and privacy is of the utmost significance (Chen, 2020). To safeguard data from unauthorised access and breaches, effective cybersecurity measures must be put in place. Furthermore, stable internet access is necessary for rural agricultural areas to use cloud services. Investment in dependable network infrastructure as well as partnerships between technology providers and agricultural stakeholders are necessary to meet these problems.

A versatile and scalable platform for the storage, processing, and analysis of enormous amounts of agricultural data is offered by cloud computing. It frees farmers from having to make significant investments in on-site infrastructure and gives them instant access to powerful computing resources. Farmers can gather and store data from various sources, including sensors, satellites, and weather stations, by utilising cloud-based systems (Tuli et al., 2023). Machine learning algorithms can then be used to process and analyse this data in order to yield useful insights.

In this study, we examine how cloud computing is used in smart agriculture, with a particular emphasis on how machine learning is used to allocate resources. We examine the advantages and difficulties of using cloud-based agricultural solutions and talk about the potential effects on production and sustainability. We also outline a framework for machine learning-based resource allocation in smart agriculture and talk about its potential to increase crop yields and lessen environmental impact.

The literature on cloud computing and machine learning applications in agriculture is reviewed in the section that follows. We then go over our suggested paradigm for machine learning-based resource allocation in smart agriculture. We describe the findings of our analysis and go into the approach we utilized to analyze the framework (Iftikhar et al., 2023). We finish by discussing the results of our study's ramifications and outlining potential directions for further investigation in this area.

2. RELATED WORK

Many studies on the optimization of resource allocation are described in the literature. The authors, for instance offer a solution that analyses incoming jobs (Tomaszewski et al., 2022). It cost-effectively reserves virtual machine instances so that it is possible to meet deadlines. The answer is predicated on the machine carrying out tasks with known CPU and memory requirements. The review's authors describe various task scheduling techniques that can be applied in these circumstances. Similarly to this, Zhang et al. described how to control cloud resources using deep reinforcement learning algorithms. Cost reduction is one of the main advantages of using cloud computing in smart agriculture. Traditional farming methods may need large up-front expenditures for hardware and software installs. Farmers can access computing resources via cloud-based platforms on a pay-as-you-go basis, saving money on expensive infrastructure. Farmers can also modify their computing resources in accordance with their needs thanks to the scalability of cloud computing, which eliminates the need for over-provisioning and lowers operational costs. The authors suggest a deep Q-network, which is originally pre-trained by a stacked autoencoder, as a variation of a reinforcement learning technique (SAQN) (Javaid et al., 2022). They incorporated experience replay, Q-network freeze, and network normalization to address stability difficulties. The proposed method assumes that the client submits requests with a known-ahead resource demand, which is then verified using a simulated load produced by the big data benchmarking tool HiBench.

Every industry is undergoing a revolution as a result of Industry 4.0, commonly referred to as the fourth industrial revolution. It is a strategic initiative that combines cutting-edge disruptive digital technologies to enable the digitization of the industry, including the Internet of Things (IoT), big data and analytics (BDA), system integration (SI), cloud computing (CC), simulation, autonomous robotic systems (ARS), augmented reality (AR), artificial intelligence (AI), wireless sensor networks (WSN), cyber-physical system (CPS), digital twin (DT), and additive manufacturing (AM). The adoption of these technologies in agriculture is giving rise to agriculture 4.0, also known as smart farming, digital farming, or agriculture, which is the next generation of industrial agriculture (Ahmad et al., 2023).

Farmers that practise smart agriculture are given a variety of instruments Figure 1 to help them address issues with productivity, sustainability, food security, and environmental impact. WSNs and other IoT-enabled systems allow for remote farm monitoring and management. Farmland data is gathered by drones equipped with hyperspectral sensors, and monotonous activities are handled by autonomous robots (Chan et al., 2023). Decision-making is aided by data analytics, which optimises elements including the weather, agricultural productivity, water management, and pest control. Through the use of contemporary systems, smart agriculture increases crop yields, lowers costs, and improves product quality.

Figure 1. The concept of smart agriculture

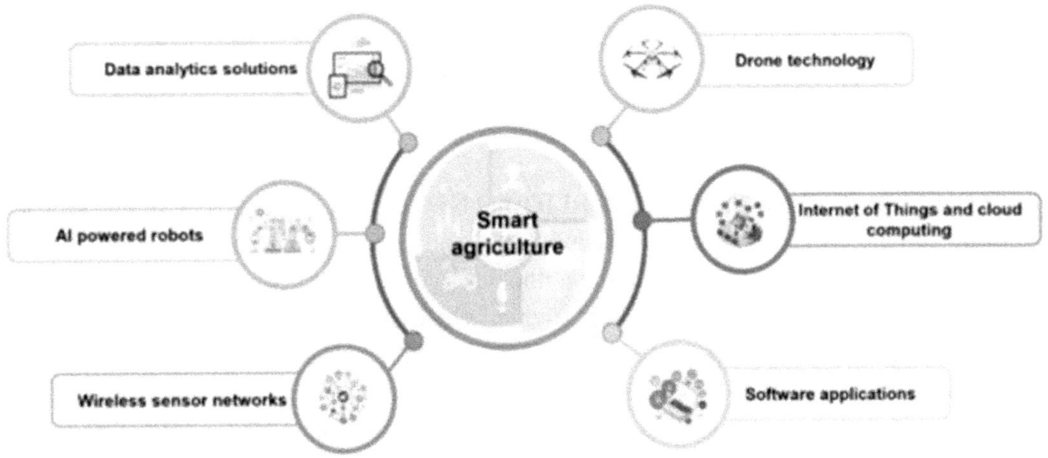

Figure 2 depicts the 148 articles' year-by-year distribution from 2011 to 2021. 2018 saw the release of almost 22% of the scientific articles over the previous ten years. This shows that although the agriculture industry is implementing digital technology at a rapid rate, it is nevertheless moving more slowly than other industries like healthcare, manufacturing, mining, automotive, energy, etc.

The Internet of Things (IoT) is a network of interconnected computers, sensors, and machines with special characteristics that are used for remote sensing and monitoring (Rosendo et al., 2022). Six layers make up the IoT reference architecture Figure 3: the perception layer (hardware devices), the network layer (communication), the middleware layer (device management and interoperability), the service layer (cloud computing), the application layer (data integration and analytics), and the end user layer (user interface).

IoT sensors at the physical layer of agriculture gather information on crop and environmental characteristics like temperature, humidity, pH level, water level, leaf colour, and fresh leaf weight. Data is transferred via the network layer, which is dependent on communication technology appropriate for the size, location, and farming technique of the field (Shekhar et al., 2023). Due to their low cost, high transmission range, and low energy consumption, ZigBee, LoRa, and Sigfox are common in outdoor fields. Due to its limited communication range, Bluetooth is used inside. Agriculture-related products are

Figure 2. Distribution of chosen research studies by year from 2011 to 2021

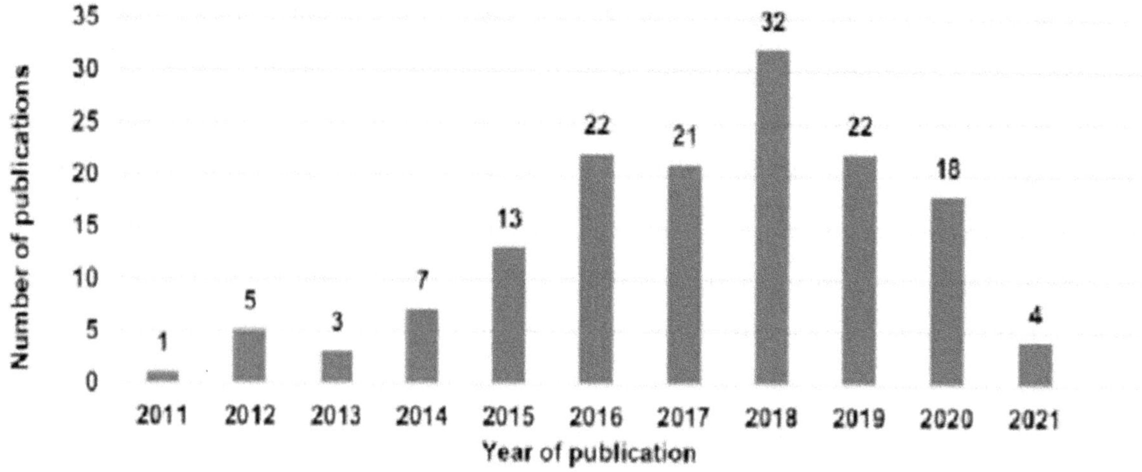

being tracked more and more with the help of RFID and NFC technology. Periodic monitoring of environmental and soil parameters is carried out using GPRS and mobile communication (2G, 3G, and 4G).

It presents an alternative strategy where the authors suggest a system which scales resources across public and private clouds. The system scaling engine, which is based on queuing theory, enables the expansion of private cloud functionality using resources from the public cloud (Lee et al., 2022). Time-series analysis and threshold-based policies are used in this solution. Instead, we forecast demand using machine learning, which enables us to overcome the delays brought on by the supply of new instances. They use machine learning to estimate resource usage, and propose a variety of different ML techniques to help with prediction; however, in neither case are any more steps beyond prediction suggested. Moreover, none of these methods performs anomaly detection, making them vulnerable to making incorrect predictions in the event of transient aberrations (Tomaszewski et al., 2023). In order to optimize resource allocation and save costs, the usage of cloud computing in smart agriculture can be examined using a variety of machine learning techniques. These algorithms include, among others:

2.1 Decision Tree

Decision trees can be used to simulate the allocation of resources in agriculture. Based on historical information and other pertinent considerations, they can assist in determining the most effective and efficient allocation of resources (Ruan et al., 1998).

Here are some possible outcomes that you might get from applying a decision tree algorithm to examine how cloud computing is used in smart agriculture to maximize resource allocation and reduce costs:

a. Identification of important variables, such as crop production, weather patterns, soil types, irrigation, and fertilization, that influence resource allocation and cost in smart agriculture.
b. Creation of a decision tree model that, depending on the input data, precisely forecasts the best resource allocation and cost (Alshehri, 2022).

Figure 3. Internet of things (IoT) architecture is composed of six layers

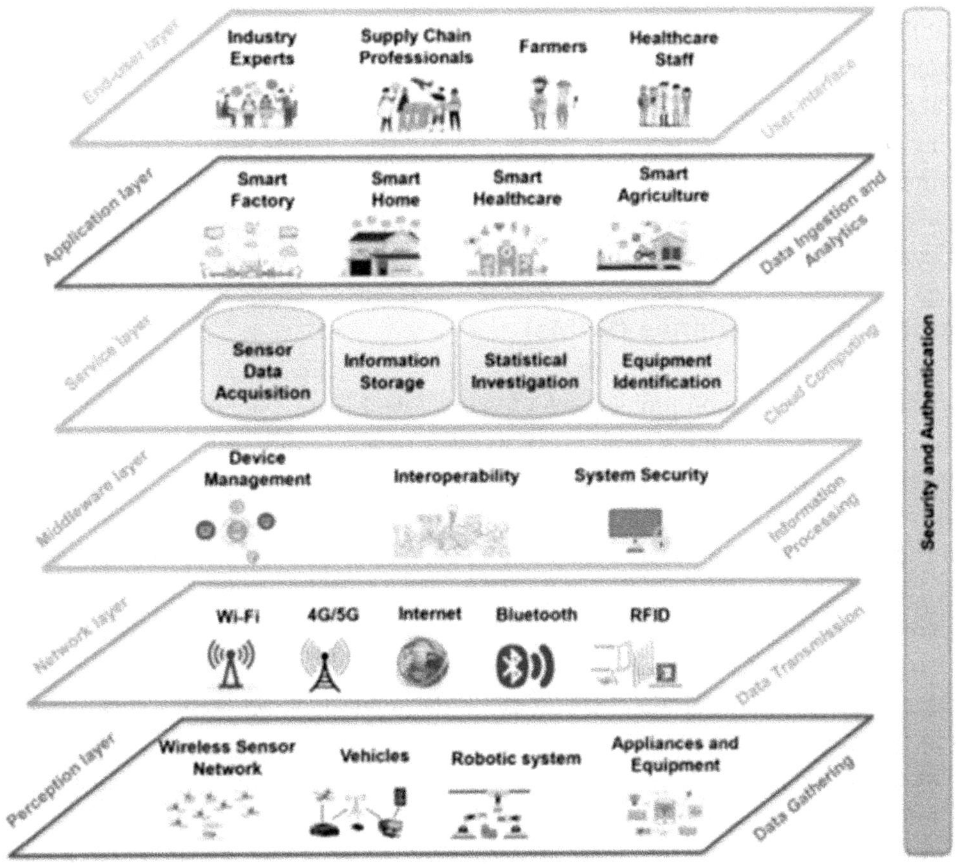

c. A better understanding of the trade-offs between various resource allocation strategies, such as the pros and cons of fertilization against irrigation.
d. Finding applications for cloud computing that maximize resource utilization and cut costs, such as real-time weather patterns or soil moisture monitoring.
e. enhanced decision-making in intelligent agriculture in increased crop yields, decreased expenditures, and more effective resource usage.

2.2 Random Forest

To increase prediction accuracy, Random Forest, an ensemble learning system, blends various decision trees (Mercier-Laurent, 2022). By examining a large number of variables and their interconnections, this technique can be used to optimize resource allocation and reduce costs.

Here are some possible outcomes that you might get from using a random forest algorithm to examine how cloud computing is used in smart agriculture to maximize resource allocation and reduce costs:

a. Enhanced prediction precision in comparison to a single decision tree algorithm.

b. Identification of important variables that influence cost and resource allocation in smart agriculture, as well as how these variables interact (Zichichi & Ragaini, 1993).
c. The ability to handle input variables that are both continuous and categorical.
d. A better understanding of the trade-offs between various resource allocation strategies, such as the pros and cons of fertilization against irrigation.
e. Finding applications for cloud computing that maximize resource utilization and cut costs, such as real-time weather patterns or soil moisture monitoring.
f. Increased decision-making in intelligent agriculture, which increases crop yields, greater resource efficiency and cheaper expenses.

2.3 Support Vector Machines (SVMs)

For classification and regression problems, SVMs are a well-liked machine learning technique. They can be used to forecast agricultural yield, track weather patterns, and prioritize resource allocation in the smart agriculture (Deebak et al., 2020).

Here are some possible outcomes that you might get from employing an SVM algorithm to examine how cloud computing is used in smart agriculture to maximize resource allocation and reduce costs:

a. The most important aspects of smart agriculture, such as crop production, weather patterns, soil types, irrigation, and fertilization, influence resource allocation and cost.
b. Creation of an SVM model that, depending on the input data, properly forecasts the most cost-effective resource allocation.
c. Finding applications for cloud computing that maximize resource utilization and cut costs, such as real-time weather patterns or soil moisture monitoring Fields(Ansari & Alam, 2023).
d. Improved agricultural yields, cheaper costs, and more effective use of resources as a result of better decision-making in smart agriculture.
e. The capacity to manage both depending on the selected kernel function, linear and non-linear input data may be used.
f. Superior performance versus other machine learning techniques on tiny datasets.

3. METHODOLOGY

a. **Data gathering:** Gather pertinent information on the various aspects of resource allocation and cost in smart agriculture. Data on agricultural output, weather patterns, soil types, irrigation, fertilization, and other elements might be included in (Nacef et al., 2022). Different sensors, drones, and other smart agriculture technology can be used to collect this data.
b. **Data preprocessing:** The process of ensuring that the collected data is accurate, comprehensive, and consistent. This could entail classifying category variables, handling outliers, and deleting missing values.
c. **Feature Engineering**: Using methods like correlation analysis, the chi-square test, or recursive feature removal, feature engineering identifies the most pertinent attributes that have an impact on resource allocation and cost. The machine learning model's accuracy can be increased with the use of feature engineering.

d. **Model Selection:** It is the process of deciding which machine learning models, such as decision trees, random forests, support vector machines, or neural networks, are most suited to tackle the given problem and the type of data available (Soni & Kumar, 2022).
e. **Model training:** Improving the chosen machine learning model for the desired output by training it on the preprocessed data. To do this, the data may need to be divided into training and testing sets, and the model's hyperparameters may need to be tuned.
f. **Cloud Computing Integration:** It is necessary for smart agriculture to maximize resource utilization and cut costs, such as edge computing or distributed computing. Large-scale data processing, real-time analysis, and resource optimisation are all possible with the cloud computing (Phua et al., 1992).
g. **Model Deployment:** To optimize resource allocation and save costs in smart agriculture, a trained machine learning model is deployed on a cloud computing infrastructure (Alzaghir et al., 2022). This can entail creating an application or API that farmers or other agricultural workers can use to access and make decisions based on the model's insights.
h. **Continuous Improvement:** Monitoring the performance of the machine learning model over time and updating it with fresh information or better techniques to maintain its accuracy and applicability (Masroor et al., 2021).

4. LIMITATIONS

Despite receiving a lot of attention, load balancing is just one of the problems with cloud computing. Among the difficulties load balancing in cloud computing faces are:

a. **Virtual Machines (VM) Migration:** With virtualization, an entire machine may be seen as a single file or a collection of fleas. A VM can also be moved between physical computers to lighten the strain on a heavily utilized actual machine. The priority is to evenly distribute the workload throughout a data centre or group of data centres (Sharif et al., 2023). Is it possible to dynamically distribute the load in cloud computing systems to avoid bottlenecks? This question relates to the transfer of virtual machines.
b. **Automating service provisioning:** One of the most alluring properties of cloud computing is its elasticity, which allows resources to be instantaneously assigned and released (V. Sharma et al., 2022). What are the best practices for utilizing or releasing cloud resources while maintaining the performance of traditional systems and making the most use of resources?
c. **Data storage management:** During the past ten years, the amount of data stored across networks has increased exponentially. As a result, even for businesses that outsource their data storage and for individuals, data storage management has emerged as a major challenge for the cloud computing (Belsare & Singh, 2022). How can data be moved to the cloud such that it can be effectively stored while still being readily available?

5. RESULT AND CONCLUSION

The paper provides a thorough review of the state of digital technology in agriculture today, highlighting trends, obstacles, and potential areas for further study and research. It highlights the potential disruptive impact of Industry 4.0 technology in dealing with agricultural issues and assuring sustainable food production in the future. The objective was to investigate advanced industrial farms and intense production techniques to answer concerns about the security of the world's food supply.

The application of machine learning and cloud computing in smart agriculture has enormous potential. Massive amounts of data from sensors, satellite images, and weather forecasts can be collected and stored using cloud technology. Through the processing of this data, machine learning optimises the use of water, fertiliser, and pesticide resources, resulting in increased crop yields and lower costs. Predictive analytics and real-time monitoring make it easier to spot problems early and prevent losses. By eliminating significant hardware investments and providing on-demand computing resources, cloud computing reduces costs. Using real-time forecasts from machine learning algorithms, farmers can optimise resource use, cutting down on waste and financial losses. Early detection of nutrient deficits and plant diseases is made possible with the use of machine learning, allowing for efficient remediation. Cloud-based collaboration also promotes knowledge exchange, which promotes agricultural innovation. Agriculture adopts these technologies to become data-driven, intelligent, and sustainable, transforming farming methods.

REFERENCES

Ahmad, S., Shakeel, I., Mehfuz, S., & Ahmad, J. (2023). Deep learning models for cloud, edge, fog, and IoT computing paradigms: Survey, recent advances, and future directions. *Computer Science Review, 49*, 100568. doi:10.1016/j.cosrev.2023.100568

Alam, M. M., Le Moullec, Y., Ahmad, R., Magarini, M., & Reggiani, L. (2020). A Primer On Public Safety Communication in the Context of Terror Attacks: The NATO SPS "COUNTER-TERROR" Project. In C. Palestini (Ed.), *Advanced Technologies for Security Applications* (pp. 19–34). Springer Netherlands. doi:10.1007/978-94-024-2021-0_3

Alshehri, M. (2022). Fuzzy Logic Based Explainable AI Approach for the Easy Calibration of AI Models in IoT Environments. In J.-H. Kim, M. Singh, J. Khan, U. S. Tiwary, M. Sur, & D. Singh (Eds.), *Intelligent Human Computer Interaction* (pp. 621–633). Springer International Publishing. doi:10.1007/978-3-030-98404-5_57

Alzaghir, A., Abdellah, A. R., & Koucheryavy, A. (2022). Predicting Energy Consumption for UAV-Enabled MEC Using Machine Learning Algorithm. In Y. Koucheryavy, S. Balandin, & S. Andreev (Eds.), *Internet of Things, Smart Spaces, and Next Generation Networks and Systems* (pp. 297–309). Springer International Publishing. doi:10.1007/978-3-030-97777-1_25

Ansari, M., & Alam, M. (2023). IoT-Cloud-Enabled Smart Framework for Real-World Applications. In A. J. Kulkarni, S. Mirjalili, & S. K. Udgata (Eds.), *Intelligent Systems and Applications* (pp. 87–105). Springer Nature. doi:10.1007/978-981-19-6581-4_8

Belsare, K. S., & Singh, M. (2022). Various Frameworks for IoT-Enabled Intelligent Waste Management System Using ML for Smart Cities. In S. Shakya, K. Ntalianis, & K. A. Kamel (Eds.), *Mobile Computing and Sustainable Informatics* (pp. 797–817). Springer Nature. doi:10.1007/978-981-19-2069-1_55

Chan, K. Y., Abu-Salih, B., Qaddoura, R., Al-Zoubi, A. M., Palade, V., Pham, D.-S., Ser, J. D., & Muhammad, K. (2023). Deep neural networks in the cloud: Review, applications, challenges and research directions. *Neurocomputing*, *545*, 126327. doi:10.1016/j.neucom.2023.126327

Chen, Y. (2020). A Survey on Industrial Information Integration 2016–2019. *Journal of Industrial Integration and Management*, *05*(01), 33–163. doi:10.1142/S2424862219500167

Deebak, B. D., Al-Turjman, F., Aloqaily, M., & Alfandi, O. (2020). IoT-BSFCAN: A smart context-aware system in IoT-Cloud using mobile-fogging. *Future Generation Computer Systems*, *109*, 368–381. doi:10.1016/j.future.2020.03.050

Iftikhar, S., Gill, S. S., Song, C., Xu, M., Aslanpour, M. S., Toosi, A. N., Du, J., Wu, H., Ghosh, S., Chowdhury, D., Golec, M., Kumar, M., Abdelmoniem, A. M., Cuadrado, F., Varghese, B., Rana, O., Dustdar, S., & Uhlig, S. (2023). AI-based fog and edge computing: A systematic review, taxonomy and future directions. *Internet of Things*, *21*, 100674. doi:10.1016/j.iot.2022.100674

Javaid, M., Haleem, A., Singh, R. P., & Suman, R. (2022). Artificial Intelligence Applications for Industry 4.0: A Literature-Based Study. *Journal of Industrial Integration and Management*, *07*(01), 83–111. doi:10.1142/S2424862221300040

Khalifeh, A., AlQammaz, A., Darabkh, K. A., Sha'ar, B. A., & Ghatasheh, O. (2021). A Framework for Artificial Intelligence Assisted Smart Agriculture Utilizing LoRaWAN Wireless Sensor Networks. In V. E. Balas, L. C. Jain, M. M. Balas, & S. N. Shahbazova (Eds.), *Soft Computing Applications* (pp. 408–421). Springer International Publishing. doi:10.1007/978-3-030-52190-5_29

Lee, C. P., Leng, F. T. J., Habeeb, R. A. A., Amanullah, M. A., & Rehman, M. H. (2022). Edge computing-enabled secure and energy-efficient smart parking: A review. *Microprocessors and Microsystems*, *93*, 104612. doi:10.1016/j.micpro.2022.104612

Masip-Bruin, X., Souza, V. B., Marín-Tordera, E., Ren, G.-J., Jukan, A., & Garcia, J. (2020). A Multidimensional Control Architecture for Combined Fog-to-Cloud Systems. In K. Arai, S. Kapoor, & R. Bhatia (Eds.), *Advances in Information and Communication* (pp. 281–299). Springer International Publishing. doi:10.1007/978-3-030-39445-5_22

Masroor, R., Naeem, M., & Ejaz, W. (2021). Resource management in UAV-assisted wireless networks: An optimization perspective. *Ad Hoc Networks*, *121*, 102596. doi:10.1016/j.adhoc.2021.102596

Mercier-Laurent, E. (2022). Greening and Smarting IT – Case of Digital Transformation. In E. Mercier-Laurent & G. Kayakutlu (Eds.), *Artificial Intelligence for Knowledge Management, Energy, and Sustainability* (pp. 1–18). Springer International Publishing. doi:10.1007/978-3-030-96592-1_1

Mullachery, B., & Alismail, S. (2023). A Smart Healthcare Framework: Opportunities for Integrating Emerging Technologies (5G, IoT, AI, and GIS). In K. Arai (Ed.), *Proceedings of the Future Technologies Conference (FTC) 2022, Volume 3* (pp. 325–340). Springer International Publishing. 10.1007/978-3-031-18344-7_22

Nacef, A., Kaci, A., Aklouf, Y., & Dutra, D. L. C. (2022). Machine learning based fast self optimized and life cycle management network. *Computer Networks, 209*, 108895. doi:10.1016/j.comnet.2022.108895

Phua, K. h., Wang, C. m., Yeong, W. y., Leong, T. y., Loh, H. t., Tan, K. c., & Chou, F. s. (1992). Optimization Techniques and Applications. World Scientific. doi:10.1142/9789814537285

Rosendo, D., Costan, A., Valduriez, P., & Antoniu, G. (2022). Distributed intelligence on the Edge-to-Cloud Continuum: A systematic literature review. *Journal of Parallel and Distributed Computing, 166*, 71–94. doi:10.1016/j.jpdc.2022.04.004

Ruan, D., Abderrahim, H. A., D'hondt, P., & Kerre, E. E. (1998). Fuzzy Logic and Intelligent Technologies for Nuclear Science and Industry. World Scientific. doi:10.1142/9789814528368

Sharif, S., Zeadally, S., & Ejaz, W. (2023). Space-aerial-ground-sea integrated networks: Resource optimization and challenges in 6G. *Journal of Network and Computer Applications, 215*, 103647. doi:10.1016/j.jnca.2023.103647

Sharma, V., Tripathi, A. K., & Mittal, H. (2022). Technological revolutions in smart farming: Current trends, challenges & future directions. *Computers and Electronics in Agriculture, 201*, 107217. doi:10.1016/j.compag.2022.107217

Shekhar, B. G., & Tyagi, S. (2023). Dynamic Bandwidth Allocation Using Linear Regression Model in IOT with Machine Learning Techniques. In D. K. Sharma, S.-L. Peng, R. Sharma, & G. Jeon (Eds.), Micro-Electronics and Telecommunication Engineering (pp. 597–609). Springer Nature. doi:10.1007/978-981-19-9512-5_54

Soni, D., & Kumar, N. (2022). Machine learning techniques in emerging cloud computing integrated paradigms: A survey and taxonomy. *Journal of Network and Computer Applications, 205*, 103419. doi:10.1016/j.jnca.2022.103419

Tomaszewski, L., Kołakowski, R., Mesodiakaki, A., Ntontin, K., Antonopoulos, A., Pappas, N., Fiore, M., Mosahebfard, M., Watts, S., Harris, P., Lin, C.-K., Santiago, A. R., Lazarakis, F., & Chatzinotas, S. (2023). ETHER: Energy- and Cost-Efficient Framework for Seamless Connectivity over the Integrated Terrestrial and Non-terrestrial 6G Networks. In I. Maglogiannis, L. Iliadis, A. Papaleonidas, & I. Chochliouros (Eds.), *Artificial Intelligence Applications and Innovations. AIAI 2023 IFIP WG 12.5 International Workshops* (pp. 32–44). Springer Nature Switzerland. doi:10.1007/978-3-031-34171-7_2

Tomaszewski, L., Kołakowski, R., & Zagórda, M. (2022). Application of Mobile Networks (5G and Beyond) in Precision Agriculture. In I. Maglogiannis, L. Iliadis, J. Macintyre, & P. Cortez (Eds.), *Artificial Intelligence Applications and Innovations. AIAI 2022 IFIP WG 12.5 International Workshops* (pp. 71–86). Springer International Publishing. doi:10.1007/978-3-031-08341-9_7

Tuli, S., Mirhakimi, F., Pallewatta, S., Zawad, S., Casale, G., Javadi, B., Yan, F., Buyya, R., & Jennings, N. R. (2023). AI augmented Edge and Fog computing: Trends and challenges. *Journal of Network and Computer Applications, 216*, 103648. doi:10.1016/j.jnca.2023.103648

Yeganeh, S., Babazadeh Sangar, A., & Azizi, S. (2023). A novel Q-learning-based hybrid algorithm for the optimal offloading and scheduling in mobile edge computing environments. *Journal of Network and Computer Applications, 214*, 103617. doi:10.1016/j.jnca.2023.103617

Zichichi, A., & Ragaini, R. C. (1993). Innovative Technologies for Cleaning the Environment: Air, Water and Soil. World Scientific. doi:10.1142/9789812796974

Chapter 9
Intelligent Farming Utilizing a Soil Tracking Device

Sabyasachi Pramanik
https://orcid.org/0000-0002-9431-8751
Haldia Institute of Technology, India

ABSTRACT

In order to optimize resource use, having information that is both up to date and accurate is vital. Agriculture utilizes sensors to determine the nutrients, moisture, organic matter, and clay present in the soil. To connect sensors that are positioned in a variety of locations, several technologies are used. Even in the absence of an internet connection, its data will automatically be uploaded to the cloud. Sensors transmit data to adjacent local base stations at varied distances using WiFi, LPWAN, LoRa, and Bluetooth, among other technologies, before transferring it to a distant central base station (CBS). Maintaining the integrity of the signal requires line of sight, coding, and modulation. The data collected by CBS is uploaded to the cloud so that it may be analyzed for trends and visualized. Farmers will now be able to get frequent and up-to-date information without needing to physically be present at the event.

INTRODUCTION

The current global population of 6.9 billion people is anticipated to reach 8.7 billion by the year 2050, which is a significant rise from the number in 2016. As a consequence of a rise in affluence, individuals are increasing the amount of protein that they consume in their diets. A research conducted by the Food and Agriculture Organization of the United Nations (FAO) found that agricultural producers should increase their production by a factor of 65% over what it now is. Food producers need to boost their production in order to meet the demands of a growing global population. As a result of the expansion in population, there will be a demand for food that is almost fifty percent more than it was in 2013. It is anticipated that the rise of urbanization will continue (Mall, P. K. et al., 2023) to add another 2.4 billion people who reside in urban regions. Even though there is a rising need for food, the number of people living in rural areas is decreasing due to increasing urbanization. This leads to a decrease in the number of individuals who work in agriculture. In addition, the population of rural areas is rapidly aging,

DOI: 10.4018/979-8-3693-0200-2.ch009

which will lead to a significant lack of available labor in the not too distant future. As a consequence of population growth and urbanization, agriculture on a global scale is becoming unsustainable, and already 25 percent of the world's arable land is ruined. It is estimated that close to forty percent of the world's rural population has water scarcity problems, which is evidence that water resources are already being used to their fullest capacity. In order to boost output, imbalanced fertilizers are used, which has a negative impact on the quantity of nutrients in the soil. Nitrogen (N), phosphorus (P), and potassium (K) are the primary nutrients found in soil; the others, such as iron, manganese, zinc, boron, and chlorine, are considered to be of a lower importance. The correct proportions of these nutrients are critical to the growth of the plant and the production of high-quality fruit and vegetables. Macronutrients like nitrogen are responsible for the production of amino acids, whilst phosphorus and potassium contribute to the expansion of plant growth, the process of photosynthesis, and resistance to disease. The fertility of the soil is diminished as a consequence of inefficient watering practices, an application of fertilizer that is out of balance, and a lack of awareness of the precise nutritional requirements of a particular crop. Traditional agricultural practices are being modified in response to changes brought about by scientific and technological progress. The advancement of technology in agricultural production helps to meet the actual requirements of consumers while also preserving the delicate equilibrium that exists between the supply and demand of food. The way that current farms and agricultural operations are run is distinct from their historical counterparts as a direct result of developments in automation. These developments include sensors, gadgets, equipment, information technology, and speedy transmission networks. Technologies such as GPS (Samanta, D. et al., 2021), robots (Alam, A., 2022), sensors (Alam, A., 2022), and aerial pictures are all used. Farmers are able to make more effective use of water, fertilizer, and other resources with the assistance of modern technologies. As a result of technical improvements, bigger and smaller farmlands are merging and becoming more interconnected than in previous decades, which are expected to result in an increase in both the productivity of agriculture and the efficiency with which it is carried out. It is essential to gather data in order to recognize patterns, and it is estimated that there will be 4 million more data points than on a regular farm. The Internet of Things allows the analysis of both structured and unstructured data (Reepu, S. Kumar, et. al., 2023) in order to give more information on the production of food (Kumari, M., et. al., 2022). According to Ghosh, R. et al.'s 2020 research, IoT platforms are transforming agricultural frameworks into real AI frameworks by using machine learning, AI, and other types of AI to data collected from field sensors. Studying and combining data on the weather, seeds, soil, disease probability, and insect attacks might help farmers make better judgements.

Observation of the Soil and the Characteristics of the Soil

The soil is the primary component that supports the food web. The upkeep of the soil is an essential component in ensuring the continuity and safety of the food supply. It is a significant natural deposit, yet most people don't realize how essential it is. The composition and personality of the soil is influenced by a variety of factors, including the parent material, age, climate, animals and plants that eat the soil, as well as changes in location. Minerals, water, organic matter that has decomposed, air, and microorganisms that are still alive are the fundamental components of soil.

The function of the land may sometimes cause the composition of the soil to shift. For instance, agricultural soil consists of between 20 and 30 percent air, 45 percent minerals, 5 percent organic material, and the same amount of water. In addition, the exact composition is affected by a variety of elements, including the kind of crop grown, the terrain, the temperature, and the humidity, among other things.

The condition of the soil determines whether or not a plant will be able to get the nutrients, fertility, and other components that are necessary for the growth that they want. It fluctuates and is dependent on the use that it is put to as well as the sort of crop that farmers are cultivating. Important soil qualities include bulk density, texture, water retention, temperature, soil organic carbon, and wetness. Bulk density is a measure of how much soil can hold. When all of these parameters are compiled, an index of the soil's quality is generated. There is an index that has been created in order to assist researchers and farmers in better understanding soil. The condition of the soil deteriorates when there is an excessive use of fertilizer, an incorrect rotation of crops, floods, and dry weather. Since of this, the roots will not get the nutrients they need, and since the quantity of organic matter has diminished, recycling nutrients will be very difficult, if not impossible. Composting yard waste, shredded leaves, and even uneaten fruits and vegetables may be an effective way to improve soil quality. This is accomplished by reintroducing organic matter into the soil, which in turn stimulates chemical and biological activity. In times past, when formal criteria for soil testing were lacking, farmers would tend their crops by just looking at the soil. This was done since they did not have access to testing equipment. Following that, soil laboratories came into existence, which are places where samples of soil are submitted in order to get reliable data; nevertheless, the process is labor-intensive and prone to error. Several aspects of the soil, such as its H2O content, air and soil temperatures, the amount of water vapor present in the environment, and its acidity and alkalinity, are evaluated. Irrigation is not required if there is an abundance of moisture in the soil and the temperature is lower than normal. On the other hand, if the temperature is higher and there is less moisture in the soil, then an immediate supply of H2O is required. It is important to adjust the amount of water that is given to plants based on the temperature and moisture content of the soil. The pH of the soil is yet another crucial factor that might have an effect on the growth of plants. The acidic or basic nature of the soil has an effect on the proportion of the soil's nutrients that are readily accessible. Farmers may use a variety of chemicals to alter the pH of the soil depending on the results of measurements related to the pH of the soil in order to bring it back to the desired level.

The measurement of the pH level, also known as the hydrogen potential of the soil, is an essential component in agricultural practices. Acidic substances have a pH value that is less than 6.5, whereas neutral substances have a pH value that is equal to 6.5, and alkaline or basic substances have a pH value that is more than 6.5. The hydrogen ion activity, also known as the appropriate concentration, in a soil and H2O solution may be assessed by the pH of the soil. Another name for the suitable concentration is the proper concentration. An extremely acidic soil with a pH of 4-5 may contain high quantities of iron and manganese, both of which are naturally harmful and have a detrimental impact on the growth of plants. The pH of the soil can range anywhere from 2-4. When the pH level of the soil is too high, plants have a diminished capability to get the essential nutrients they need from the soil. This, in turn, has a negative impact on the growth of the plants. Between 5.5 and 6.5 is the optimal pH range for the soil; however this might vary based on the kind of plant being grown. In order to make soil less acidic and more alkaline, limestone is added to it. Aluminum sulfate and sulphur, on the other hand, are minerals that are often added to soil in order to increase its pH (Tauqeer, H.M., et al., 2022). The characteristics of the parent material, such as the amount of minerals, clay, and organic matter, the activity of creatures (plants, and microbes) residing in the soil, and rainfall, are passed down to newly created soil to determine its pH. The pH of surface-water or surrounding bodies of water, such as streams or lakes, is similarly impacted by the pH of the soil. Climate, mineral composition, and soil texture all have an effect on pH.

The amount of water vapor in the air is what determines the level of humidity, which may either make the air moist or dry. The ideal relative humidity for a given crop will differ from one variety of plant to another. Some plant life need a humidity level of just 50%, while others require as much as 70%. On the other hand, a high level of humidity is a magnet for diseases and viruses, both of which have a disruptive impact on the growth of plants. In addition, if the humidity is too high, the plants will not be able to expel extra water or absorb nutrients from the soil. In conditions with low humidity, plants will rapidly lose their hydration and will need more soil moisture. As a result, the need for irrigation will be based on the concentration of water vapor in the atmosphere. Capacitive sensors are used in the process of measuring the levels of humidity.

The term "temperature" refers to the degree to which one can determine how warm or how cold the soil is. It sets in motion a wide range of biological processes and influences the qualities of the soil, including its moisture level, its nutrient content, and so on. In order to get an accurate reading, the conventional method involves using a thermometer that has a probe that is inserted somewhat deeply into the ground. The temperature range that is optimal for plant growth is between 18 and 24 degrees Celsius (65 and 75 degrees Fahrenheit). In addition, a higher temperature results in a decrease in the pH of the soil. The rate of the photosynthetic process is affected significantly by temperature, and it often slows down both when the temperature is high and when it is low.

The Temperature at the Time

It will take some time before the greater surface temperature has the same proportionate influence on the soil temperature as it does on the surface temperature. The application of heat to soil has an indirect effect on the composition of the soil as well as the growth of microorganisms. The temperature of the soil may be regulated with the help of mulch, which will keep the soil warm in the winter and cool in the summer. In addition, water may be used in the process of remediating dry soil.

Precipitation, temperature, relative humidity, irrigation, soil ingredients, and the amount of water that is already in the ground all have an effect on the amount of water that is in the ground, which is referred to as soil moisture. It plays a significant role in determining how efficiently plants get nutrients from the soil. The moisture content of the soil ought to generally range anywhere from 10 to 45 percent for optimal growth. Microbial activity and plant production both suffer when there is a low moisture level. A high level of moisture, on the other hand, may foster the growth of pathogens that can cause damage to the roots of a plant and prevent a seed from germinating. It is important to provide water while taking into account the moisture reading and the amount of substance present.

Electrical conductivity (EC) is an indication of the soil's physical and chemical properties, and to test it, a mixture of the soil and de-ionized water is employed. After that, the water is extracted from the paste, and its electrical conductivity (EC) value is determined. Milli Siemens per meter is a common unit of measurement for it. In order for a plant to thrive, the EC levels need to be in the appropriate equilibrium. A high EC makes it impossible for nutrients to be absorbed and contributes to the deterioration of the environment. A lower EC might be detrimental to the health and production of the plant.

In order to conduct an investigation of the soil, a sample is acquired from the area in the form of a composite specimen. This kind of specimen is composed of several samples that have been gathered from various locations and then examined. In most cases, it is analyzed to determine its consistency (sandy or silt), pH value, electrical conductivity, nutrient content (N, P, and K), and organic matter. They are the factors that decide which crop will be grown in the soil since they give information on the chemical, physical, and biological status of the soil.

SMART SENSORS FOR MONITORING THE SOIL

Monitoring the soil is, as has previously been said, an essential part of the agricultural industry. It is essential that growers have access to reliable information that is up to date on the quality of the soil. If you have complete information, irrigating the field with the appropriate amount of water and nutrients will be a simple and uncomplicated process for you to do. Incorporating the most up-to-date sensor and Internet of Things (Taviti Naidu, G. et al., 2023) technology into traditional agricultural practices has the potential to fundamentally revolutionize every aspect of such practices. The Internet of Things (IoT) and sensor integration in smart agriculture have the potential to take farming to new heights that were previously unimaginable. Sensors make it possible to continually monitor crops with better accuracy, and they are able to identify undesired situations in their early stages. This is the most major benefit that sensors provide.

The temperature of the soil, the composition of the water in the soil, and the temperature of the air are all monitored in real time by equipment that are placed in farms and connected to the internet. Because they transmit data at intervals that may be programmed, intelligent sensors do away with the necessity for human readings. Find out what the optimum times are to plant, reduce the amount of water you use, and reduce the stress that plants experience as a result of either receiving too much or too little water. Analyzing historical patterns may help you make better decisions for the long-term management of your crops. The overwhelming majority of soil sensors are single-point sensors. The measuring only takes place in one location. The salinity, temperature, and moisture content of the soil may all be determined using only a single point sensor.

Wetness Sensor: Providing plants with the appropriate amount of water is essential, and the degree of moisture present in the soil is a major factor in determining how much water should be provided. When compared to soil that is already saturated at the crop roots, soil that has a lower level of moisture has a greater need for more water. As can be seen in Figure 1, soil moisture sensors analyze the sample to determine the appropriate level of soil moisture. It uses two probes, one of which is positive and the other of which is negative, to measure the resistance between the two. The dielectric constant theory of operation dictates that a current should flow between the two probes in order to complete the experiment. The resistance on the probes is measured in order to estimate the level of moisture in the soil; the greater the resistance, the more dry the soil is, and the lower the resistance, the more wet the soil is. The sensors may either be fixed in situ for continuous monitoring or carried about as portable instruments for real-time measurements.

The kind of soil moisture sensor that measures dielectric permittivity is the sort that is used the most often. Monitoring the dielectric constant between the probes may help determine whether or not there is moisture in the soil. This is because the dielectric permittivity of dry soil is 4, while the dielectric constant of water is 80. It is very necessary to ensure that the probes are correctly inserted into the soil, that there is no air or space between them and the soil, and that the depth is to the root level or deeper. Readings of responses are taken at data loggers, and then forwarded to either centralized or cloud-based servers for storage.

The vertical soil form seen in Figure 2 may be created with the use of profile probes, which can then be used to measure the moisture content of the soil. This probe equipment is straightforward to use and enables accurate measurements of the soil moisture profile. The depth to which these probes, which may range in length from one to four feet, are buried in the earth is determined by the extent to which the roots extend. The length of the extended tube determines the number of sensing rings that are

Figure 1. Soil moisture sensor module

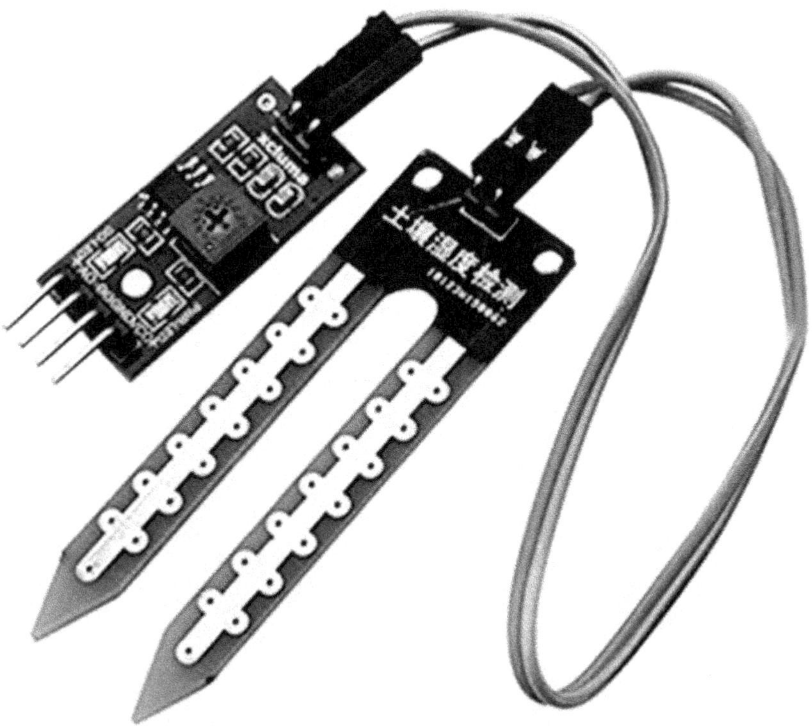

connected to it; this number may be anything from one to numerous. A readout meter for readings that can be taken on the go, a data recorder for continuous monitoring, access tubes for burying the probes in the ground, and an installation help package are all included in this set. A profile probe may be able to enter a channel in the soil that provides access to the soil. The electromagnetic field that is generated by the sensors included in the probe is able to penetrate the ground. It is able to do so without the need for different sensors to be positioned at various depths in the soil in order to detect the soil's changing water content with depth. Multiple sensors are coupled to a data recorder using a single antenna. The data collected by the logger is sent to the cloud so that it may be analyzed further. After a number of access tubes have been fitted, it is possible to use a readout meter in conjunction with a profiling tube in order to get a rapid reading. By using a few access tubes and only one profiling tube, it is possible to minimize the need for a large number of single sensors located at a variety of depths, which results in a cost savings. There is a quick response time, several measurements may be taken with a single reading, it is adaptable, and it is easy to use.

Both the Soil Matric Potential (SMP) and the Volumetric Soil Water Content are evaluated using moisture sensors (Novick, K.A., et al., 2022). While SMP offers information on the amount of water that is available in the soil, VSWC determines the percentage of the soil's volume that is occupied by water. In contrast to VSWC, which needs understanding of hydration management, SMP may be grasped quickly and easily. Both SMP and VSWC sensors may be purchased in a wide variety of combinations and are adaptable for use with a wide variety of plant life, irrigating systems, and ground types.

Figure 2. A sensor for moisture that also has a profiling probe

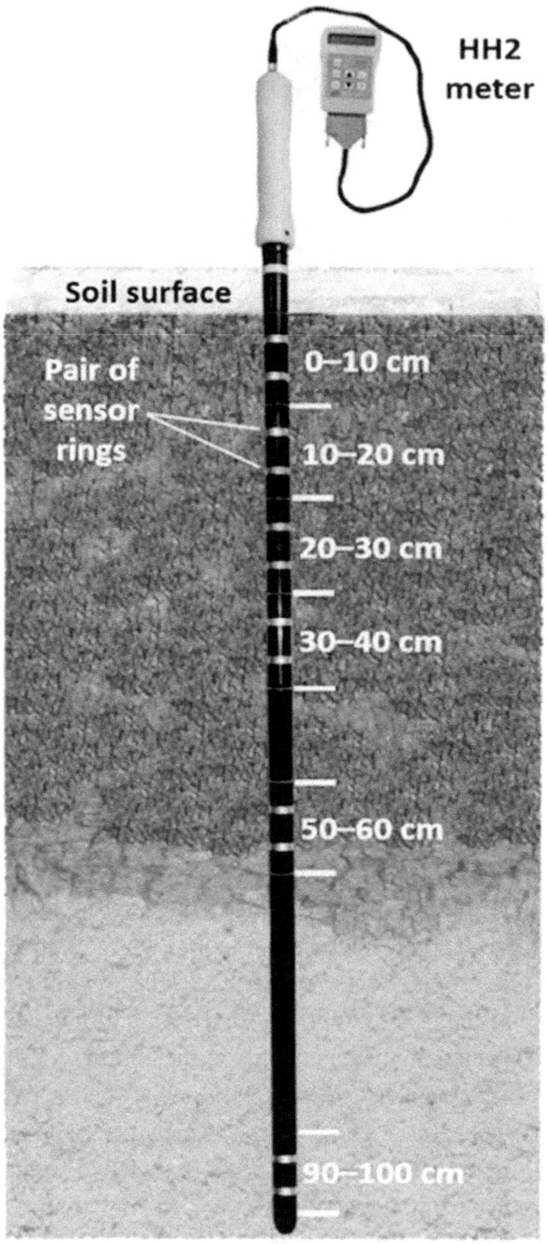

The Water Potential Sensor is shown in Figure 3 as an example of one of these sensors. This sensor monitors the amount of energy required by plant roots to get water from the soil. Granular matrix sensors and tensiometers are the two different types that may be purchased.

A tensiometer is used to determine the amount of pressure that plant roots must endure in order to get water from the soil (Patwa, D., et al., 2023). It determines the ease with which plants may get access to water. The water reservoir is located at the very top of the device, along with an elastic tube, a permeable porcelain cup located at the very bottom, and a vacuum gauge (Dehghani, A., et al., 2022).

Figure 3. H2O potential sensor

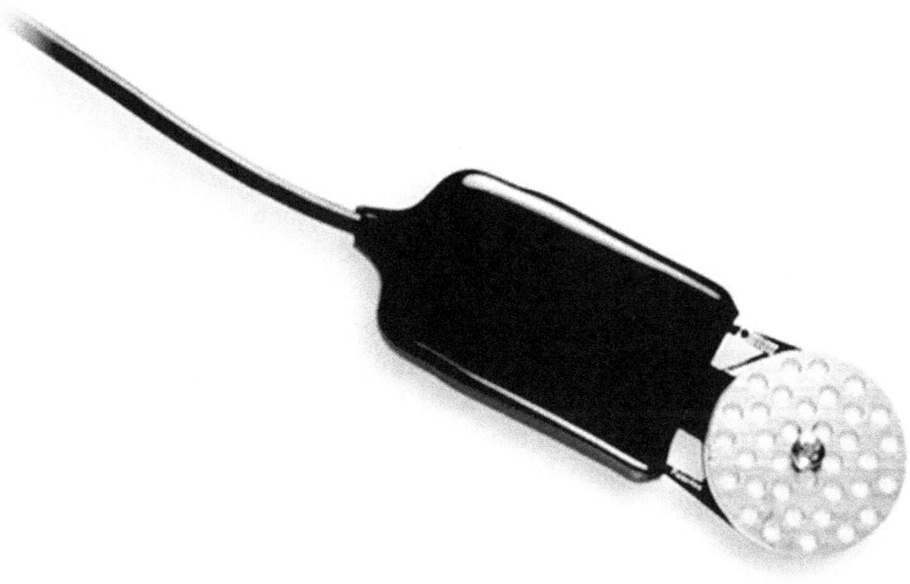

After being completely saturated with H2O, the tube is then lowered into the earth at a depth of up to one meter. The water was poured out of the cup, and the dirt was allowed to dry out. At the very top, it creates a vacuum, which may then be measured by the associated vacuum gauge meter. As a consequence of this, the readings on the meter change due to the presence of the water accessibility component in the soil, and the irrigation is adjusted accordingly. When water is added to the soil, it runs into the ceramic cup, which causes the gauge to read lower than normal. Use a vacuum pump to remove any air bubbles and get a precise reading, or soak a porous tip in a bucket of water for a considerable amount of time to achieve the same result. The granular matrix sensor incorporates two electrodes altogether for its operation. This matrix reaches equilibrium when it is covered with dirt.

It is possible to read a resistance as the soil grows wetter, and this resistance may be converted into an estimate of the soil's moisture content. In order to reach the root zone of the crop, sensors that are buried in the soil are used. A data logger is a device that saves information from several sensors and displays soil moisture data from the past. This value may be coupled with the irrigation system so that it may be used to begin and terminate the supply of water. In the same manner as tensiometers, before to their use, they are first immersed in water in order to get an accurate reading. When compared to a tensiometer, the porous substance of the Granular matrix Sensor allows water to travel at a far more leisurely pace. As a consequence of this, there is a lag time between the reading on the sensor and the amount of moisture in the soil. In addition, when the soil gets dry, this lag is much greater compared to when it is rewetted.

Two metal electrodes that are separated from one another make up a capacitance sensor. These electrodes form a changeable capacitor by using the water in the soil as a dielectric surrounding them. Because H2O is the most significant factor affecting the dielectric, capacitive measurement is used to calculate the dielectric that is mainly formed by the soil. If there is a greater amount of water in the soil, the capacitance value of the sensor will increase. Because the electrodes are cut off from the surrounding soil and its water, it is feasible to install the sensor in the ground in such a way that it will remain there permanently. This eliminates the risk of corrosion. Because of this, the sensor has a lifetime that is far longer than that of sensors that rely on batteries or resistance. Capacitance may often be determined using one of these two approaches. In the first method, the electrodes of the sensor are used to form an RC oscillator circuit. This circuit provides an AC signal, the frequency of which is governed by the capacitance of the sensor and, as a result, the moisture component of the soil. There is a correlation between the moisture content of the soil and a variety of frequencies. The second method detects the capacitance estimate of the sensor directly by computing the charge or discharge time that indicates the capacitance value of the sensor. This may be done either by charging the sensor or discharging it.

The H2O component of the soil may be identified with the use of time domain reflectometry (TDR) sensors, which are sometimes referred to simply as time domain sensors. They inserted two probes into the earth, each measuring 12 centimeters in length. The probes are used to transmit high electromagnetic waves, and the idea that underpins how the system works is to calculate the travel duration of the waves when the waves are reflected back. The dielectric constant and the amount of water that is present in the ground both have an effect on the passage of time. In many cases, each second sees the transmission of thousands of pulses. The average amount of time that passes between each of the probe readings may be used to estimate the amount of moisture present. The water content reflectometry and temporal domain transmissometry (TDT) sensors provide an alternative that is available at a lower cost. Bluetooth is often included in modern sensors, and this allows for the reading to be promptly sent to the irrigation system. In a manner similar to this, frequency domain reflectometry, often known as FDR, transmits signals between probes with the help of an oscillator.

- **pH Sensors:** pH sensors are employed in the process of evaluating the many qualities of the soil, including the purity of the water. In essence, it determines how many hydrogen ions are present in the water and the acidity or alkalinity of the soil's aqueous solution influences the number of hydrogen ions that are there. The pH value is measured by first placing a pair of electrodes, consisting of a recommended electrode and a measuring electrode, into the solution and then detecting the voltage difference that occurs between the electrodes. A recommended electrode is one that has been shown to be accurate in measuring pH.
- **Temperature and Humidity Sensors:** Temperature sensors measure the degree of heat and coolness that is present around the soil. Humidity sensors detect the amount of moisture in the air. The relative humidity changes whenever there is a change in temperature. Irrigation has an impact on the temperature as well as the amount of humidity in the air since it raises the total quantity of water droplets in the atmosphere. The temperature drops as a direct consequence of this, which leads to an increase in the relative humidity of the surrounding air. Users are alerted when measurements of linked temperature and humidity are taken from the fields. In light of this information, the farmer could decide to take the appropriate action.
- **Light Intensity Sensor:** Direct exposure to sunlight is one of the most important factors in the growth of plants and crops. Every plant has a unique reaction that is based on the traits and composition of the plant itself, in addition to the kind of soil it is growing in. When the sun comes up, the soil around certain plants may dry up too rapidly, while the soil around other plants may be able to keep moisture for a longer amount of time if it is clay-like. The light sensor is responsible for determining whether or not there is light present, after which it adjusts the brightness to the suitable level. It employs a component known as a Light Dependent Resistor (LDR), the resistance of which increases when there is less light and decreases when there is more light. When there is

an increase in the amount of light, there is a corresponding reduction in the amount of resistance. This is because electrons, after absorbing energy, become more conductive and start to allow current to flow across the conduction band. To determine the value of this resistance, either an ohmmeter or a digital multimeter may be used. Photoresistors and phototransistors are two examples of the many different kinds of light sensors that are available. According to the findings of Lakhiar et al. (2018), it may be feasible to switch lights on and off automatically without the involvement of a human operator if light sensors are combined with controllers for the lighting system. It is possible to put it to use in greenhouses, which employ artificial lighting methods such as incandescent bulbs, high-intensity discharge lamps, and light-emitting diodes (LEDs) to generate light.

Fiber optic sensors are used widely across the globe for a variety of applications, including communication between satellites and the ground, determining the location of airplanes from the ground, and crop monitoring. The majority of the time, it is based on the calculation of changes in optical properties, such as reflectance, variations in refractive index, phase shift, polarization, and the amount of light energy. This particular sensor, which can be seen in Fig. 4, makes use of LEDs or lasers to transport light via fibers in a manner that is consistent with complete internal reflection prior to the light reaching the ground. The scattered light that is left over after soil particles have absorbed light of varying wavelengths is reflected back. After this, the reflected light is collected by the photodetector, which subsequently converts it into an electrical signal. In addition, the signal conditioning circuit has buffers and an inverting amplifier that allows for the gain to be adjusted.

The data from the optical sensors are evaluated against the threshold values that have been stored in the microcontroller in order to determine whether or not a component is faulty. An LCD displays the name of the component that is lacking in the soil as well as the required amount of that component. As a consequence of this, the user is able to choose the fertilizers that are suitable for the soil and then apply them. In addition to this, students may acquire knowledge about the several characteristics of the soil, such as its organic material levels, clay content, and moisture level. The installation of these sensors is a very straightforward process. It is possible to put it on a field, a vehicle, a drone, an aerial platform, or a satellite, as shown in Figure 5.

The optical interrogator of the Fiber Bragg Grating Optic Sensor sends white light, which contains all visible colours, down into the fiber core. It goes through something that is called a Fiber Bragg Grating (FBG), which is a collection of optical filters. The majority of the spectrum is transmitted, but a specific wavelength is reflected back as it is traveling through a Bragg grating. This is achieved by systematically introducing shifts in refractive index into the fibers over time. When outside factors like as vibration and heat are added, the FBG reacts in a similar manner, altering the reflected Bragg wavelength by an equal amount. These variations are then transformed into physical engineering units such as temperature, strain, and so on as a consequence of the fact that all FBG sensors are tied to measurement in a certain wavelength range. Daisy-chaining many sensors together on a single optical wire is possible. FBG sensors, in contrast to conventional electrical sensors, are risk-free and trustworthy due to their non-conductive, electrically passive, and electromagnetic interference (EMI) resistance properties. In more recent times, precision agriculture has begun using actively heated fiber optic (AHFO) sensors in addition to passive fiber optic (PFO) sensors in order to evaluate the moisture content of the soil.

Figure 4. Optical sensor

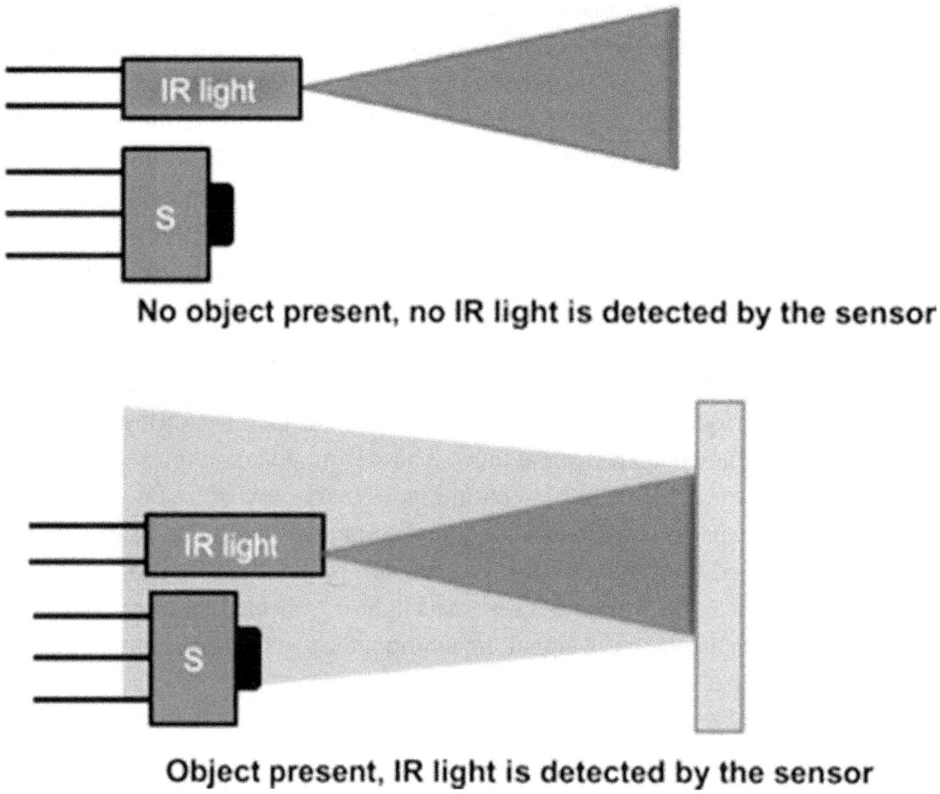

Intelligent Soil Monitoring System

In a communication system, information may be sent between different places in the system either via the use of cables or wirelessly without the need for cables, wires, or other types of media. As was discussed in the part that came before this one, there are many different kinds of sensors that can acquire information about the ground. After that, these details had to be sent over various communication technologies to a centralized server in order to undergo further research. There are many different wireless technologies that are used for communication, and each one has its own set of benefits and drawbacks that come along with it. The Arduino board, the Raspberry family (Pramanik, S. 2022), various types of processors such as the ATmega328, the ATmega2560, and the LPC2148, etc., and communication technologies such as cellular communication (3G,4G/ (LoRaWAN) are used in the installation of smart soil monitoring systems. Additionally, various protocols and standards, as well as various network's technologies, are utilized.

In order to create a wireless sensor network, also known as a WSN, a large number of wireless sensors are connected to one another through sensor nodes. Every node has a transceiver, which is used to send data to the base station, an electrical circuit, which is used to connect sensors, and a battery, which is used to provide power. As can be seen in Figure 6, the nodes are responsible for the collecting of data. There are two further types of nodes, and they are known as source nodes and sink nodes. The sink node receives data from a number of source nodes, whereas source nodes are responsible for the collection of

Figure 5. A movable optical sensor

sensor data. The function of the sink node is more computational than the function of the source node because the sink node combines data from several sources. After then, the aggregated data from the sink node is sent to the central servers so that it may be used in other ways and analyzed further. Memory, energy, and processing capability are all further restricted as a result of the fact that the cost and size of the sensor change depending on the complexity of the node. Despite the fact that it is expected that these restrictions will be eliminated and that their use in agriculture will become practicable as a result of rapid progress.

The use of Internet of Things (IoT) devices in agricultural settings mandates the utilization of various forms of communication technologies. Depending on the requirements of the application, many communication protocols and technologies are used; the following list covers the most common of these:

- **Wireless PAN/LAN:** Additionally, wireless PAN/LAN is used in order to provide short-range communication between central base stations and field sensors. This makes use of a number of different technologies, such as Bluetooth, ZigBee, and Wi-Fi, the specifics of which are determined by the transmission range.

When it comes to wireless technologies, the Bluetooth standard (IEEE 802.15.1) is considered to be the one that is most suited for wireless sensor networks. This is because these networks often consist of a smaller number of sensors that are dispersed throughout a larger area (between 8 and 10 meters). It is possible to achieve data speeds ranging from 1 Mbps to 24 Mbps. This technology is chosen in smart

Figure 6. The organization of a WSN

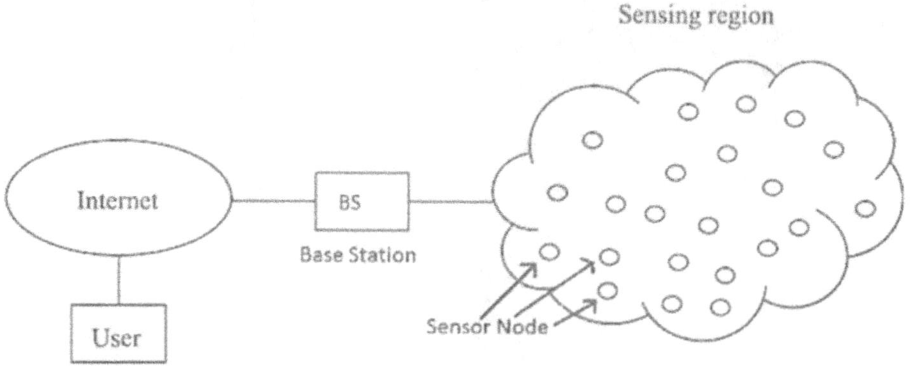

farming and a variety of other applications since it offers the benefits of using less batteries, being easy to use, and being free of charge, amongst other advantages.

- **Wi-Fi:** Due to the fact that Wi-Fi is predominantly present in a range of devices, such as smartphones, laptops, tablets, etc., it is commonly used in IoT-based agricultural systems for the purpose of monitoring soil. Wi-Fi, which stands for "wireless fidelity," is a wireless local area network that has a data throughput of 11 megabits per second (Mbps) and a communication range of between 20 and 100 meters. It operates on the 2.4 GHz frequency of the in-building wireless (ISM) band. The Wi-Fi protocol offers compatibility for the IEEE 802.11 (Hussain, S.M., et al., 2022) standard family of specifications. By using this technology and setting up an ad hoc network, each of the devices may communicate with one another.

The Zigbee technology's purpose is to facilitate the tracking and management of various devices. Bluetooth and Wi-Fi are not especially suited for this specific type of wireless communication because of the significant amount of power that they use or the role that they play in applications where low amounts of power consumption are essential. It satisfies the need for the deployment of low-cost and low-power devices with data transfer rates ranging from 20kbps to 250kbps. It is the open source standard that is extensively used for Internet of Things (IoT) applications such as home automation, clinical data collection, industrial control systems, and precision agriculture. Zigbee technology may be created by manufacturers at a reduced cost since there are no licensing costs connected with open source protocols. Because of the design of the network, it is capable of supporting a large number of devices simultaneously. Zigbee has been allocated topologies based on a star, a mesh, and a tree in addition to three different frequency bands: 857MHz, 926MHz, and 2.3GHz. It is essential that a Zigbee network have a mesh topology since data may be diverted via a different channel and there is a self-healing mechanism in place if any node in the network fails. In the vast majority of Zigbee applications, the devices spend the bulk of their time in sleep mode, which is also sometimes referred to as power saving mode. For both the encryption of data and the authentication of data, the Zigbee network makes use of the AES cryptographic algorithm.

RFID is a technique that makes use of radio waves to automatically detect and monitor plants, plants, and other plant-related objects. In the plant where the RFID tags are located, there is a linked reader that is installed in order to receive the radio signal feedback that is provided by the RFID (Gupta, A., et al.,

Figure 7. A system for the monitoring of soil that makes use of radio frequency identification

2023) tag. When several tags can be read and monitored at the same time, the amount of time needed to get information on the whole crop is reduced. Active, passive, or semi-passive RFID tags are the three possible varieties of this kind of tag. Active and semi-passive tags are both heavier than passive tags, which depend on the power of an RFID reader to transfer signals back to the reader. This is because active and semi-passive tags both have their own power supply. Passive tags. In spite of the fact that it has its own power supply, a semi-passive reader depends on the reader's radio signal in order to provide feedback to the reader. The active ones have a range of 10 centimeters, while the passive ones are more useful in agriculture, less costly, and more portable. An RFID reader is composed of a microcontroller, a signal generator, and a signal detector. These three components work together to read RFID tags. An RFID tag consists of a transponder, which can send and receive signals, a rectifier, which allows the tag to draw power from radio waves, memory, and a controller. An RFID tag, such as the one shown in Figure 7, transmits data to an RFID reader, which then converts the radio waves it receives into data that may be used. After being collected, this information is then sent to a server by means of antennas and a base station. Once there, it is saved in a database and examined using a variety of specialized tools. Utilizing RFID tags allows for the discovery of any information on viruses or diseases that are affecting the crop. Because it is difficult to label each individual seed, it is more practical to provide a single identity to a group of plants that have similar qualities. These tags contain information about the optimal harvesting window, as well as the date of planting, the qualities of the seed, and the characteristics of the soil. Contactless RFID is a secure technology that has a high identification rate, a memory that can store data, and an ease of integration with a variety of frameworks.

The monitoring of soil quality in modern agriculture often makes use of cellphones. The camera, global positioning system (GPS), microphone, accelerometer, and gyroscope that are standard on today's smartphones may be used to locate and identify a wide variety of things, including chemicals and other items. The use of contemporary cellphones equipped with advanced cameras allows for the collection of data through digital photography about the characteristics of the soil. However, due to the fact that light detecting sensors in cameras do not provide a constant color response, the outcomes may differ

Figure 8. Tracking intelligently with LoRaWAN for the soil

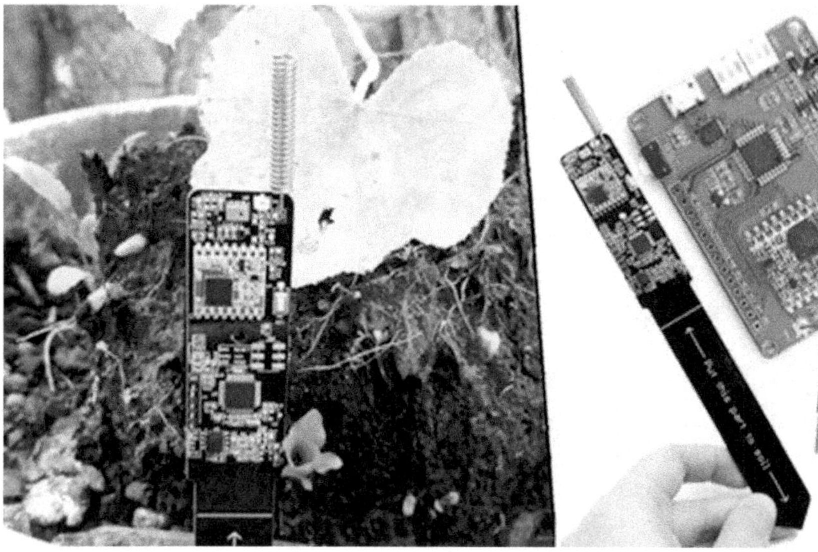

from one smart phone to the next. In addition, the user has to be aware of the lighting exposure since a too bright light might cause a photo to be overexposed, which results in the image being blurry when there is not enough light. Because of this, the image has to be recalibrated if the color of the soil is being used to make judgments about the soil's attributes. A "true" color picture is one that has been generated through color calibration and is unaffected by the influence that lighting and light sensors have on the image. Measurements of the organic and carbon content of the soil, the degree of ripeness of the fruits, the leaf area index, and chlorophyll are taken from these photographs and analyzed. The Global Positioning System (GPS) provides position data that may be used for crop mapping, disease and insect location alerts, solar radiation forecasts, and fertilization.

- **Long Range Wide Area Networks:** LoRa (Sinha, M. et al., 2021) is a low power wireless radio network protocol that provides long-range communication between the field sensors and base station (up to 40 km in interior areas and up to 4 km in city sites). LoRa was developed for use in Long Range Wide Area Networks (LRWAN). It was included in the tested findings for a range of important sectors, such as smart farming, and it is both cost-effective and energy-efficient for the Internet of Things. There are variations in the unlicensed ISM frequencies (423MHz, 853MHz, and 928MHz) that are used by this protocol on each continent. Every GSM operator, on the other hand, is required to pay a significant quantity of money in order to be granted permission to use the frequency spectrum. In addition, LoRaWAN reduces the amount of power required, which in turn enhances the battery life of connected sensors by at least 10 years. The LoRa technology shown in Figure 8 has enabled an increase in production while simultaneously reducing costs and having a less impact on the environment.

This network is employed for precision agriculture in addition to linking people all over the globe to the internet. It is known as the Terrestrial Wireless Sensor Network. These networks make use of wireless sensor nodes that are field-deployed and connected to one another and a gateway by means of a radio

Figure 9. Wireless sensor network in an underground environment

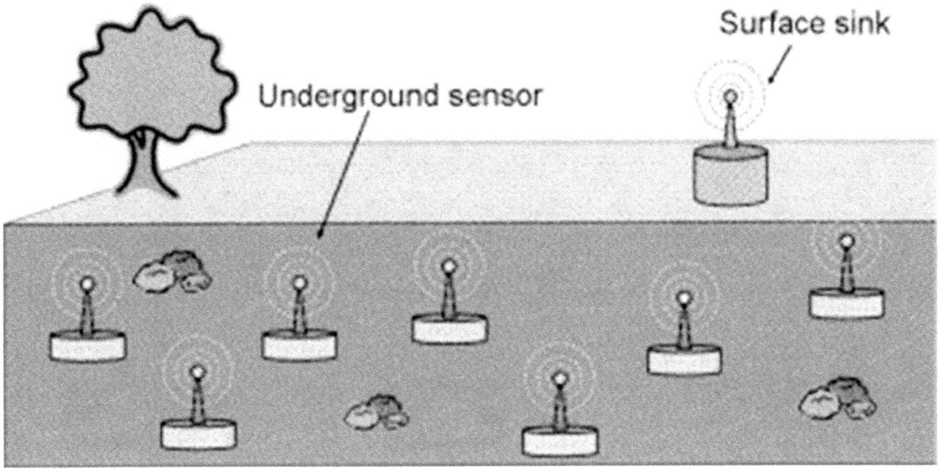

frequency connection that makes use of ISM radio bands (Sweety, T.J., et al., 2023). (Bansal, R. et al., 2021) The gateway is located some distance away from the sensor nodes, and it is connected to the cloud network through 3G, 4G, LTE, and 5G. Therefore, a farmer who is working from a distance may keep an eye on the status of the field and identify when the crops want more water, fertilizer, or pesticides. The production of healthy crops, the maintenance of the fertility of the soil, and the reduction of the costs associated with agriculture may all be accomplished by making optimal use of pesticides and fertilizers. The burden placed on healthcare systems as a whole is reduced when healthy agricultural practices are implemented. Important data from sensors and gateways may be coupled with other frameworks, such as irrigation, so that plants get water on an as-needed basis. Because of this, it will no longer be necessary for workers to manually switch the pump on and off. In addition, mobile phones may be used to verify the condition of a field by showing a variety of information that is relevant to the status of the field.

The establishment of wireless connection between the surface and the underlying ground is the goal of a WSN that is located below the surface. Maintaining golf courses, notifying home security systems of an intruder, and even detecting earthquakes are all examples of applications for employing soil data such as moisture and temperature readings, among others. As can be seen in Figure 9, this network consists of sensor nodes that are buried below and communicate with a subterranean gateway via wireless transmission. It has been shown that dirt may cause higher frequency signals to get distorted; as a result, low frequency signals have been used instead, which has necessitated the use of many sensors. The information from all of the sensors will be collected via a single gateway located beneath. The data logger on the surface receives transmissions of data from the subterranean gateway. After that, the information may be sent to the cloud so that farmers can access it. As a result, the use of underground sensors may result in a significant reduction in water usage while protecting agricultural production and conserving essential resources.

This technology, also known as LPWAN-Narrow Band, provides the low power consumption, vast area spread, and reduced expenditures that are necessary for a robust Internet of Things network. LoRa, Sigfox, and Narrowband Internet of Things (also known as NB-IoT) are some of the competing technologies that may be found in the market for LPWAN. Wideband and narrowband systems are the two

Figure 10. WAN with a narrow band and low power

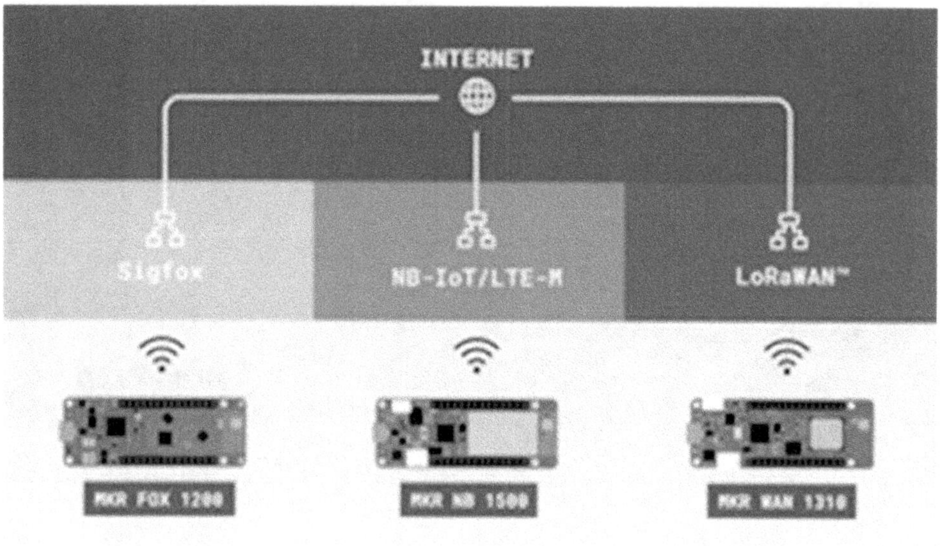

primary classifications of low-power wide-area network (LPWAN) technologies (Mall, P. K. et al., 2023) technologies. In wideband systems, a single channel would often span a frequency range that varies between 95 kHz and various MHz, while in narrowband systems, a single channel will typically have a bandwidth that varies between 50 and 100 Hz. The precision of quartz oscillators increased (Zhang, H., et al., 2022), which made it possible to extend the amount of time that signals remained stable and sped up the development of low-power narrowband radio communication technologies (as shown in Figure 10).

WAVIoT (Choudhary, S. et al., 2022) develops NB-Fi (Narrow Band Fidelity) radio modules that have a low power consumption in order to guarantee the long-term autonomy of NB-Fi devices. This gives customers the ability to use NB-Fi devices in remote locations with less maintenance required. IoT networks and NB-Fi modules both have the ability to carry data over distances more than 9 kilometers, and their batteries can last for more than ten years. There is a potential for NB-Fi devices to collect data from sensors and transmit it to the gateway via LPWAN (Anand, R. et al., 2022) with a low amount of signal power. After that, the gateway data that has been gathered is sent to the cloud network to be received.

The NB-Fi technology was developed for use in solutions that need the rapid and inexpensive deployment of Internet of Things networks, together with a data collection rate of one hundred percent from the connected devices. The differences and similarities between NB-Fi and the other technologies are shown in Figure 11.

System Framework

ARM is an abbreviation that stands for "Advanced RISC Machine." It uses the ARM instruction set, which is 32 bits, as well as the Thumb instruction set, which is 16 bits. It acts as a connection point between the camera and the wireless network.

Figure 11. A comparison of a few different technological modalities

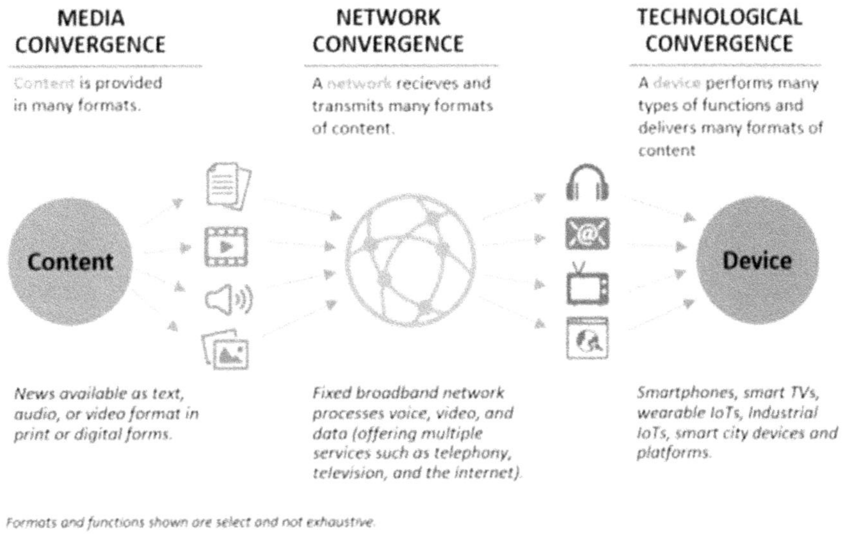

Figure 12. System framework

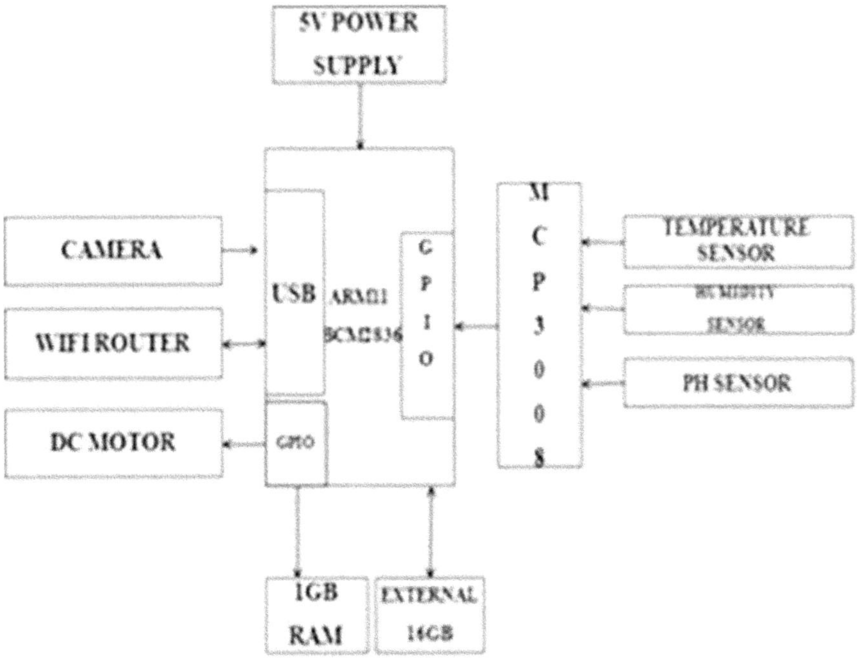

There are three separate sensors that are utilized:

- A sensor that reads humidity

Figure 13. SPI stands for the serial peripheral interface

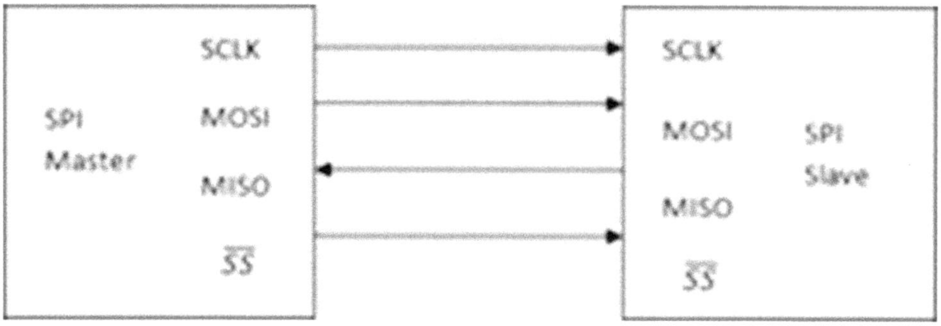

- pH sensor
- temperature sensor

A pH sensor is used to detect the amount of alkaline that is present in the soil; the possible values range from 0 to 14. If the pH value of a material is below 7, then that substance is acidic, and if the pH value is above 7, then that substance is basic. A temperature sensor is used to take readings of the soil's temperature in the field. The humidity sensors are used to detect the amount of moisture present in the soil. The values that have been recovered are in analogue form; however, with the use of MCP3008, they are converted into digital form. The microcontroller sends values to be read out through the GPIO, which stands for general purpose input and output. It acts as a connection point between the microcontroller and the ARM11 processor. The data gathered from the measurements are sent to the server over a wireless network. The server will then perform an optimization on the obtained sensor readings, using the threshold value as a guide. Utilize socket communication in order to exchange information with the server. A mobile application provides the farmer with the relevant crop suggestion at the proper time.

Methods that were used:

- Communication Through Sockets
- SPI

Socket communication is used by the Serial Peripheral Interface, or SPI, in order to connect with either the same computer or other computers. A network socket is an internal endpoint that may send or receive data at a single node in a computer network. This kind of endpoint is known as an internal endpoint. The exterior endpoints of a node are referred to as "ports," whilst the internal endpoints of local inter-process communication (IPC) are referred to as "sockets."

SPI stands for the Serial Peripheral Interface.

Communication in close proximity often takes place between embedded devices, where an asynchronous serial interface standard known as the Serial Peripheral Interface bus (SPI) is used. The SPI standard calls for a master-slave relationship between the devices, with only one master able to send data in full duplex mode. The master device is responsible for designing the framework that is used for reading and writing. Each set of slave select (SS) lines makes it possible to pick several slave devices.

There are reported to be four logic signals for the SPI bus:

- SCLK stands for "Serial Clock" and is the master output. This displays the output of the clock generator for the Master as well as the input for the clock from the Slave.
- MOSI is an abbreviation that stands for "master output," which refers to the output of the master. A connection is made between the output of the shift register for the master and the input of the shift register for the slave using this wire or signal.
- MISO is an abbreviation that stands for "Master Input, Slave Output," sometimes known as "slave output." This wire or signal links the O/P of the shift register used by the slave to the input used by the shift register used by the master.
- S is the abbreviation for "Slave Select," which stands for "active low, output from master."

CONCLUSION

The character of the soil may be determined with the assistance of the embedded technology, which extracts the behavioral content of the soil using a range of sensors to do so. This helps to determine the type of the soil. With the help of this method, farmers will have an easier time determining which crop will provide the greatest results on their field. By analyzing the data from the sensors, it provides the farmer with recommendations for the kind of crops that should be grown in the field. The optimal sensor readings will be achieved by the field manager by using previously established threshold values. The tool gives the farmer the ability to increase production and maintain track of the crop from anywhere in the world at any time. Additionally, the farmer is able to take and monitor crop images in order to keep the crop safe from insects. The method that is advised helps farmers save time and money while also increasing the amount of food produced by their farms.

Figure 14. The outcomes of the considered system

Results

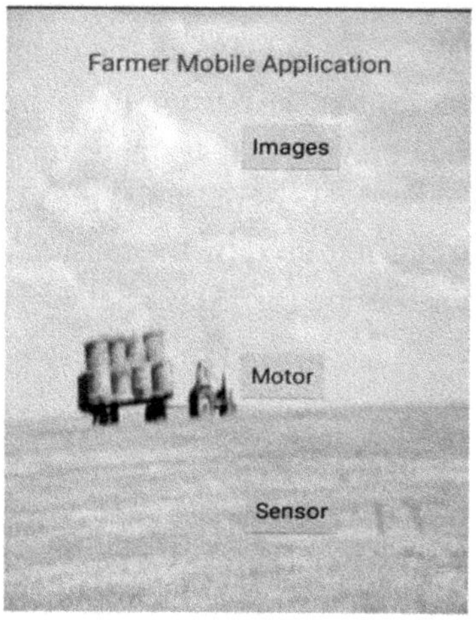

Reddish Earth

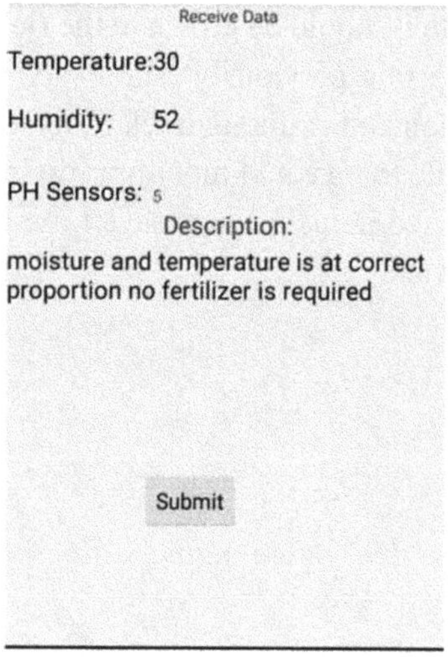

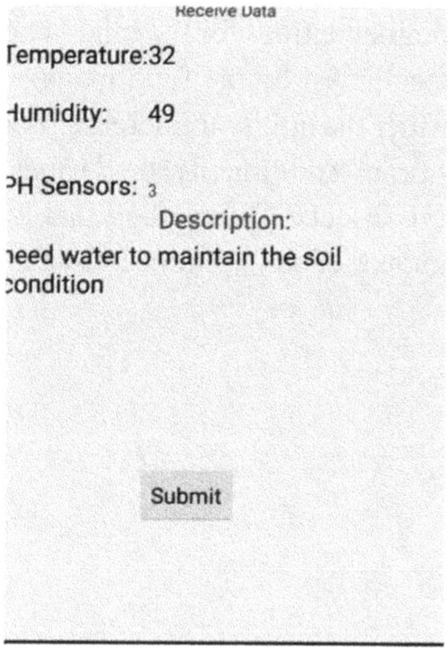

Grit Sand

REFERENCES

Alam, A. (2022). Employing Adaptive Learning and Intelligent Tutoring Robots for Virtual Classrooms and Smart Campuses: Reforming Education in the Age of Artificial Intelligence. In R. N. Shaw, S. Das, V. Piuri, & M. Bianchini (Eds.), *Advanced Computing and Intelligent Technologies. Lecture Notes in Electrical Engineering* (Vol. 914). Springer. doi:10.1007/978-981-19-2980-9_32

Anand, R., Singh, J., Pandey, D., Pandey, B. K., Nassa, V. K., & Pramanik, S. (2022). Modern Technique for Interactive Communication in LEACH-Based Ad Hoc Wireless Sensor Network. In M. M. Ghonge, S. Pramanik, & A. D. Potgantwar (Eds.), *Software Defined Networking for Ad Hoc Networks*. Springer. doi:10.1007/978-3-030-91149-2_3

Bansal, R., Obaid, A. J., Gupta, A., Singh, R., & Pramanik, S. (2021). Impact of Big Data on Digital Transformation in 5G Era. *2nd International Conference on Physics and Applied Sciences*. 10.1088/1742-6596/1963/1/012170

Choudhary, S., Narayan, V., Faiz, M., & Pramanik, S. (2022). Fuzzy Approach-Based Stable Energy-Efficient AODV Routing Protocol in Mobile Ad hoc Networks. In M. M. Ghonge, S. Pramanik, & A. D. Potgantwar (Eds.), *Software Defined Networking for Ad Hoc Networks*. Springer. doi:10.1007/978-3-030-91149-2_6

Dehghani, A., Setare, M. R., & Zarepour, S. (2022). Self-energy problem, vacuum polarization, and dual symmetry in Born–Infeld-type *U*(1) gauge theories. *The European Physical Journal Plus*, *137*(7), 859. doi:10.1140/epjp13360-022-03066-y

Ghosh, R., Mohanty, S., & Pramanik, S. (2020). A Novel Approach towards Selection of Role Model Cluster Head for Power Management in WSN: Selection of Role Model Cluster Head for Power Management in WSN. In G. Bose & P. Pain (Eds.), *Machine Learning Applications in Non-Conventional Machining Processes* (pp. 235–249). IGI Global. doi:10.4018/978-1-7998-3624-7.ch015

Gupta, A., Asad, A., Meena, L., & Anand, R. (2023). IoT and RFID-Based Smart Card System Integrated with Health Care, Electricity, QR and Banking Sectors. In M. Gupta, S. Ghatak, A. Gupta, & A. L. Mukherjee (Eds.), *Artificial Intelligence on Medical Data. Lecture Notes in Computational Vision and Biomechanics* (Vol. 37). Springer. doi:10.1007/978-981-19-0151-5_22

Hussain, S. M., Yusof, K. M., Asuncion, R., Hussain, S. A., & Ahmad, A. (2022). An Integrated Approach of 4G LTE and DSRC (IEEE 802.11p) for Internet of Vehicles (IoV) by Using a Novel Cluster-Based Efficient Radio Interface Selection Algorithm to Improve Vehicular Network (VN) Performance. In Sustainable Advanced Computing. Lecture Notes in Electrical Engineering (vol 840). Springer. doi:10.1007/978-981-16-9012-9_46

Kumari, M., Kumar, V., Kaur, R., Kumar, S., & Sharma, R. (2022). Process Optimization for the Development of Nutritionally Enhanced Nuggets using *Ficus geniculata*: A Nutritional Approach. *Plant Foods for Human Nutrition (Dordrecht, Netherlands)*, *77*(2), 241–249. doi:10.100711130-022-00967-2 PMID:35583708

Mall, P. K., Pramanik, S., Srivastava, S., Faiz, M., Sriramulu, S., & Kumar, M. N. (2023). FuzztNet-Based Modelling Smart Traffic System in Smart Cities Using Deep Learning Models. In *Data-Driven Mathematical Modeling in Smart Cities*. IGI Global. doi:10.4018/978-1-6684-6408-3.ch005

Novick, K. A., Ficklin, D. L., Baldocchi, D., Davis, K. J., Ghezzehei, T. A., Konings, A. G., MacBean, N., Raoult, N., Scott, R. L., Shi, Y., Sulman, B. N., & Wood, J. D. (2022). Confronting the water potential information gap. *Nature Geoscience*, *15*(3), 158–164. doi:10.103841561-022-00909-2 PMID:35300262

Patwa, D., & Bharat, T. V. (2023). Influence of Hydraulic Response Time of Tensiometer in Hydraulic Characteristics Estimation for Riverbank Sand. *Geotechnical and Geological Engineering*, *41*(1), 413–427. doi:10.100710706-022-02283-7

Pramanik, S. (2022). An Effective Secured Privacy-Protecting Data Aggregation Method in IoT. In M. O. Odhiambo, W. Mwashita, & I. G. I. Global (Eds.), *Achieving Full Realization and Mitigating the Challenges of the Internet of Things*. doi:10.4018/978-1-7998-9312-7.ch008

Reepu, K. S., Chaudhary, M. G., Gupta, K. G., Pramanik, S., & Gupta, A. (2023). Information Security and Privacy on IoT. In Handbook of Research in Advancements in AI and IoT Convergence Technologies. IGI Global.

Samanta, D., Dutta, S., Galety, M. G., & Pramanik, S. (2021). A Novel Approach for Web Mining Taxonomy for High-Performance Computing. *The 4th International Conference of Computer Science and Renewable Energies (ICCSRE'2021)*. 10.1051/e3sconf/202129701073

Sinha, M., Chacko, E., Makhija, P., & Pramanik, S. (2021). Energy Efficient Smart Cities with Green IoT. In C. Chakrabarty (Ed.), *Green Technological Innovation for Sustainable Smart Societies: Post Pandemic Era*. Springer. doi:10.1007/978-3-030-73295-0_16

Sweety, T. J. R., Arun, T.R., Sajith, K., Jose, J., & Reeha, K.R. (2023). CBCPW-Fed ADFD Shape On-Body Antenna for ISM Band and Sub-6 GHz 5G Applications. In Optical and Wireless Technologies. OWT 2021. Lecture Notes in Electrical Engineering (vol. 892). Springer. doi:10.1007/978-981-19-1645-8_36

Tauqeer, H. M., Turan, V., & Iqbal, M. (2022). Production of Safer Vegetables from Heavy Metals Contaminated Soils: The Current Situation, Concerns Associated with Human Health and Novel Management Strategies. In J. A. Malik (Ed.), *Advances in Bioremediation and Phytoremediation for Sustainable Soil Management*. Springer. doi:10.1007/978-3-030-89984-4_19

Taviti Naidu, G., Ganesh, K. V. B., Vidya Chellam, V., Praveenkumar, S., Dhabliya, D., Pramanik, S., & Gupta, A. (2023). Technological Innovation Driven by Big Data. In *Advanced Bioinspiration Methods for Healthcare Standards, Policies, and Reform*. IGI Global. doi:10.4018/978-1-6684-5656-9

Zhang, H., Shukla, M. K., Larson, S., Rajendran, A. M., & Jiang, S. (2022). Molecular dynamics study of anisotropic shock responses in oriented α-quartz single crystal. *Journal of Materials Science*, *57*(12), 6688–6705. doi:10.100710853-022-07076-0

Chapter 10
Artificial Intelligence Applications in Agricultural Sustainability:
Enhancing Efficiency and Resilience

Imed Othmeni
National Institute of Applied Sciences and Technology, Tunisia

Rim Ben Abdallah
National Institute of Applied Sciences and Technology, Tunisia

Hiba Ben Aribi
National Institute of Applied Sciences and Technology, Tunisia

Eya Ben Amara
https://orcid.org/0009-0007-4408-3281
National Institute of Applied Sciences and Technology, Tunisia

Sami Fattouch
National Institute of Applied Sciences and Technology, Tunisia

ABSTRACT

Agriculture plays a crucial role in addressing global challenges such as food security, resource scarcity, and climate change. As the world population continues to grow, the need for sustainable and efficient farming practices becomes paramount. This chapter aims to explore the applications of artificial intelligence (AI) in agricultural sustainability, focusing on how AI technologies can enhance efficiency and resilience in agricultural practices. The chapter provides an overview of the potential benefits and challenges of integrating AI into agriculture, highlighting specific use cases, and discussing the implications for sustainable farming systems. It also examines the role of AI in addressing key sustainability issues including resource management, yield optimization, pest control, and adaptation to environmental conditions. By examining current trends and advancements in AI for agriculture, this chapter contributes to a comprehensive understanding of how AI can revolutionize farming practices and promote sustainable agricultural systems.

DOI: 10.4018/979-8-3693-0200-2.ch010

INTRODUCTION

On a global level, agricultural production systems are facing significant challenges, particularly due to climate change, reduced water supply for irrigation, increased production costs, and a general decrease in the effective of farm workers over the past few decades. Additionally, the recent COVID-19 pandemic has posed a threat to global food supply and production systems. As agriculture continues to evolve, it is crucial to implement substantial innovations in order to sustain and overcome these challenges. Meeting the responsibility of producing enough high-quality food in sufficient quantities for the rapidly growing global population is an evident challenge. Agricultural exploration specialists have consistently utilized modern technology to seek innovative approaches and methods to integrate into agricultural systems. For many years, researchers and technicians have been gathering data to study and analyze different aspects of agricultural systems. The use of unmanned aircraft systems (UAS) has made a vast amount of data more accessible, leading to increased interest in utilizing artificial intelligence (AI) in agriculture (Hasler & Baysal-Gurel. 2019). By providing advanced analytics to manage agricultural systems, unmanned aerial systems (UAS) present a remarkable opportunity to enhance the resilience and effectiveness of production systems like never before (Coble *et al.,* 2018).

This chapter discusses recent research on data collection trends, specifically focusing on the use of remote sensing technologies in sustainable agriculture. It also explores future perspectives on integrating unmanned aerial systems (UAS) technology with spaceborne remote sensing data for large-scale research on agriculture at national and international levels. The ultimate goal of AI-enhanced agriculture is to maximize agricultural inputs and yields based on supply and demand. Achieving this requires automated data collection, storage, organization, decision-making, and robotic interventions. These processes enable early detection and diagnosis of crop diseases, as well as timely provision of nutrients to animals. Precision agriculture utilizes cutting-edge methods to accurately calculate the precise quantity of fertilizer needed for the soil of a farm. This not only reduces greenhouse gas releases but also enhances crop productivity, resulting in increased yields. Other initiatives, such as integrated pest management systems and pesticide information management, aim to enhance farmer education and awareness. Artificial intelligence, with its powerful learning capabilities, has emerged as a vital tool in addressing various agricultural challenges. AI-powered systems are being developed to assist agricultural specialists worldwide in finding better solutions. By leveraging AI technology, it becomes possible to reduce the impact on natural ecosystems, enhance worker safety, ensure reasonable food costs, and meet the growing food demands of a projected world population exceeding nine billion by 2050. This will require a 70% increase in agricultural output (Béné *et al.,* 2015). Just 10% of the increased output can be attributed to unused land, while the remaining 90% should be achieved by enhancing current production methods. It is crucial, accordingly, to utilize the latest technological advancements to improve farming efficiency in order to meet this goal. The market demands top-notch food quality, and the current intensification strategies in agriculture necessitate significant energy inputs (Khandelwal & Chavhan, 2019).

Robotics and autonomous systems (RAS) are bringing about a revolution in various industries worldwide. One such sector is the agro-food industry, which has traditionally struggled with low productivity. The integration of RAS technologies is expected to have a significant impact in this field. This chapter aims to bridge the gap between different disciplines including agriculture, food, electronics, and computer science. By adopting a comprehensive approach, it explores various cutting-edge areas related to the agri-food-consumer triangle.

The intended audience for this chapter is electronics, instrumentation, and computer engineers, as well as agri-food scientists. It aims to provide evidence of the profound effects that emerging cyber-physical technologies can have on multiple fields. Furthermore, the chapter foresees that the exchange of ideas across these diverse themes will encourage further research and development of sustainable agricultural applications.

SUSTAINABLE AGRICULTURE SUPPLY CHAIN

To guarantee ample food supply and effectively combat hunger for the growing global population, it is imperative to prioritize agricultural sustainability. According to assessments conducted by Rockström et al. 2017, the world's food production needs to increase by 60-110 percent by 2050 to accommodate the projected 9-10 billion people. Consequently, it becomes crucial to adopt a different approach that goes beyond simply boosting agricultural output to achieve long-term agricultural sustainability (Rockström *et al.,* 2017). Sustainable agriculture practices (SAP) not only improve agricultural productivity but also help reduce adverse environmental impacts (Adnan *et al.,*2018). The success of sustainable agriculture supply chains (ASCs) relies on the expertise, capabilities, competencies, and strategies of the partners involved (El Bilali and Allahyari, 2018). Adnan et al. (2018) suggest that knowledge transfer plays a crucial role in motivating farmers to adopt sustainable agricultural practices (SAP). In fact, the agricultural sector faces challenges such as water and fossil fuel depletion, limited availability of arable land, and growing consumer expectations for ethical and environmentally conscious food production. As a result, agricultural support systems (ASCS) are under tremendous pressure to enhance farming productivity by promoting the adoption of SAP (Duman *et al.,* 2017). Based on Sharma et al. (2018) and Patidar et al. (2018) studies, the efficiency of farming is greatly influenced by the need for (ASCs) to respond to increasing gaps between supply and demand, as well as fluctuations in market prices. The challenge of achieving sustainable ASCs can also be better understood by considering recent research on sustainable practices related to managing inventory and transport of delicate cargo. Various digital technologies such as the Internet of Things (IoT), mobile technologies, artificial intelligence (AI), data analytics, and digitally provided services are likewise having an influence on agricultural supply chains (Kamilaris and Prenafeta-Bold, 2018; Gharaei *et al.,*2019).

Digital technologies are, therefore, utilized in different areas of agriculture, such as the automation of farming machinery to decrease work needed, the utilization of sensors and satellite information for better monitoring of crops, land, and water, as well as the use of IoT and Radio Frequency Identification (RFID) for tracing agricultural products (OECD, 2019). The advent of digital transformation has resulted in a substantial rise in the volume of data generated within supply chains. However, this data lacks significance unless it is organized, understood, and meaningful insights are obtained through appropriate data analysis tools (Rabbani *et al.,* 2018). The AI or cognitive-based technology is considered the most groundbreaking and powerful advanced analytics tool available to enterprises for making supply chain decisions (Liakos *et al.,* 2018). AI enables computers to perform various cognitive tasks that typically require human intelligence, such as recognizing images, understanding speech, interacting with users, translating languages, and adapting to manipulate objects. Intelligent systems utilize technologies like big data analytics, cloud computing, machine-to-machine communication, and the Internet of Things (IoT) to operate, learn, and generate additional knowledge (OECD, 2017). Machine learning and deep learning techniques, which are subsets of AI, are often combined with other intelligent technologies to detect hidden patterns in data (Elavarasan *et al.,* 2018). According to Kazemi (2019), it is recommended that professionals in the field of analytical supply chain (ASC) consider AI and advanced analytics as important investments due to the industry's growing digital transformation and the need to stay ahead of competitors. Patidar et al. (2018) also state that the influence of information technology (IT) is impacting the expectations of farmers and consumers, thereby changing their demands. Over the past decade, there has been a significant rise in the use of AI and machine learning (ML) applications to improve supply chain performance across various sectors and levels. These studies, conducted by Giri *et al.,* (2019) and Konovalenko and Ludwig (2019), explore topics such as supply chain risk management and specific industry sectors. They specifically emphasize the application of machine learning in supply chain management, as highlighted by Min (2010). However, there is limited research examining the current state of AI/ML applications in supply chain management. Nonetheless, advancements in data collection through emerging technologies have empowered supply chains to anticipate changes and identify opportunities (Giri and Masanta, 2018; Kamble *et al.,* 2019). In order to ensure that valuable insights can be derived from collected data, it is important for practitioners to possess updated knowledge (Rabbani *et al.,* 2019). Agriculture, unlike other industries with easily quantifiable and anticipatable risks, is influenced by uncontrollable environmental factors. Hence, extensive testing and verification of new machine learning (ML) applications in agriculture will be crucial (Sennaar, 2019). The outcomes of this chapter will benefit researchers and practitioners, as they will gain a better understanding of the current state of ML applications in agriculture and how adopting ML can improve farming methods in a sustainable manner, leading to higher yields, better crop quality, and increased incomes. It is anticipated that there will be a significant increase in the use of ML in the agriculture industry in the future.

ASCs and fast-moving consumer goods supply chains have numerous similarities, but they diverge when it comes to sourcing raw materials and manufacturing the final product. After gathering the raw components from fields, ASCs produce goods intended for human or animal consumption. Before the final product reaches the consumers, several steps are carried out by the ASC as shown in Figure 1. These steps include pre-production, manufacturing, storage, processing, retail, and distribution. (Borodin *et al.,* 2016). To ensure the smooth operation of an ASC, it is essential to involve different stakeholders such as farmers/producers, processors, certifying organizations, traders, retailers, distributors, and end consumers. This means that effective activity management and decision-making are necessary at strategic, tactical, and operational levels for the ASC to be successful (Patidar *et al.,* 2018; Kamble *et al.,*2019).

An (ASC) is more complex than other supply chains due to factors such as the perishability of the product, seasonal nature of produce, significant supply-demand fluctuations resulting from these factors, and the increasing consumer awareness regarding the origin, quality, and safety of produce (Kamble *et al.,* 2019 ; Shukla and Jharkaria, 2013).

Agriculture Supply Chain Operation's Challenges

Due to increasing industrialization, overpopulation, and intense competition for natural resources, the global community is facing significant challenges in providing sustainable food for its population (Kummu *et al.,* 2012; Pelletier and Tyedmers, 2010; FAO, 2009). Based on the findings of Elferink and Schierhorn (2016), meeting the future food demands of the next billion individuals will pose significant challenges for current methods and advancements in producing food. The lack of industrial development in the agricultural supply chains (ASCs), unequal distribution of data, inadequate management practices, insufficient product traceability, and failure to meet food safety and quality standards are critical

Figure 1. Agriculture supply chain operations (adapted from Borodin et al., 2016)

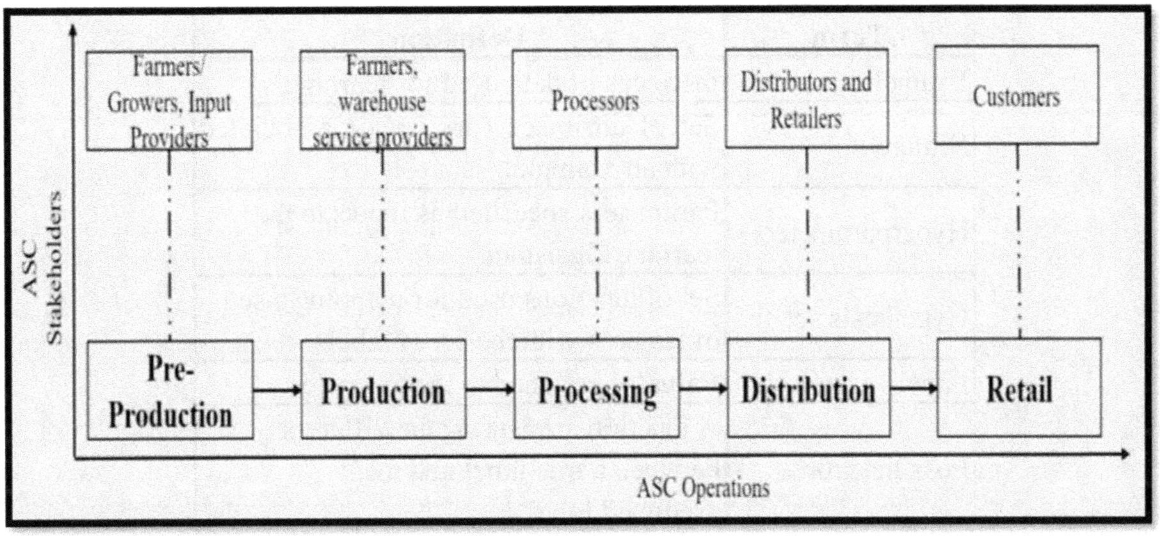

challenges that need to be addressed (Kamble *et al.,* 2019; Sharma and Parhi, 2017; Luthra *et al.,* 2018; Naik and Suresh, 2018). Furthermore, advancements in data and communication technologies have led to increased awareness and concern among individuals regarding the social, economic, and ecological features of ASCs. This has resulted in a growing demand for sustainable food production and feeding practices, driven by various stakeholders such as non-governmental organizations (NGOs), consumer groups, agro-based organizations, government institutions, and policymakers. Practitioners and scholars agree that significant transformations are necessary for ASCs to become sustainable and to achieve the United Nations' 2030 Agenda sustainable development goals (SDGs) (Eyhorn *et al.,*2019).

Data-Driven Agriculture Supply Chains

To effectively tackle the ever-changing challenges in agriculture, it is necessary to have a better understanding of the intricate agricultural systems (Kamilaris *et al.,* 2017). This can be achieved by utilizing advanced and revolutionary technology platforms that enable the constant monitoring of the physical agricultural environment through wireless sensors, generating significant amounts of data (Wolfert *et al.,*2017). Technologies such as the Internet of Services (IoS) and the Internet of Things (IoT) along with interconnected devices facilitate real-time information exchange and data collection. According to Ahumada and Villalobos (2009), (IoT) platforms can enhance sustainability in (ASCs) by promoting better data exchange, communication, coordination, and cooperation among ASC nodes. To address concerns regarding food quality and safety, wireless sensor technologies and IoT can bridge the gaps between supply and demand (Zhong *et al.,* 2017). Due to the vast amount of data generated by (ASCS), analyzing this extensive data can offer valuable insights and enable farmers and businesses to make well-informed decisions based on data. This, in turn, can enhance productivity (Sharma *et al.,* 2018). Nevertheless, there may be obstacles associated with storing, collecting, and visualizing data for ASCS that rely on data-driven approaches.Additionally, concerns such as data access, accuracy, security, and privacy need to be addressed (Kamilaris *et al.,* 2017; Sykuta, 2016). Moreover, there is a digital divide

Table 1. ML terminologies (adapted from Mohri et al., 2018)

Term	Definition
Example	Instances of data used for learning.
Feature	Set of attributes (or) vector associated with an example
Hyperparameters	Parameters specified as inputs to the learning algorithm.
Hypothesis set	Set of functions used for mapping a set of features with the set of labels.
Label	Value(s) assigned to the examples.
Loss function	A function, measuring the difference between a true label and the predicted label.
Test sample	Examples used for evaluating the performance of a learning algorithm.
Training sample	Examples used for training a learning algorithm.
Validation sample	Examples used for tuning the parameters of a learning sample.

between advanced and developing economies due to a lack of computational tools and appropriate skill sets (Rodriguez *et al.,* 2017).

Machine Learning Algorithms

According to Mohri et al. (2018), (ML) is the "scientific discipline that focuses on developing algorithms and computational models on computers, which use experience to iteratively enhance performance on specific tasks or make accurate predictions." In the aforementioned description, the historical data accessible to the learner for creating a predictive model is called "experience." This experience can be comprised of digitized human-labeled data or observations collected from the surroundings. Table 1 includes the commonly used terminologies in the field of ML.

The general setup of the machine learning system is shown in Figure 2. The program gathers training data from different origins, which can be categorized as labeled or unlabeled. Based on the organization's objectives, the system's knowledge base determines the appropriate machine learning algorithm to use. In order to ensure the quality of the output and improve decision-making, prior studies and datasets (Du and Sun, 2006).

Figure 2. A machine learning system configuration (adapted from Du and Sun, 2006)

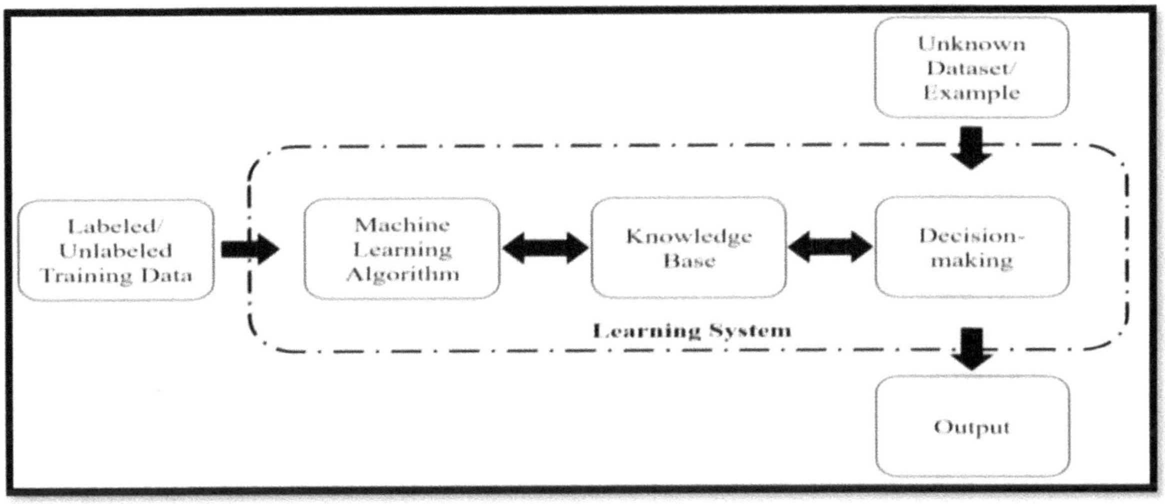

The primary machine learning tasks consist of reinforcement learning, unsupervised learning, and supervised learning. In supervised learning, a predictive model is built by using labeled data and the understanding of previous input and output variables. The objective of supervised learning is to establish a relationship between the variables and the desired output variable (Zhu and Goldberg, 2009; Traore *et al.,* 2017). Supervised learning approaches encompass various algorithms like linear regression, decision trees, and Bayesian networks. On the other hand, unsupervised learning techniques work with unlabeled datasets and do not have prior knowledge of input or output variables. Unsupervised learning is typically used for tasks such as dimensionality reduction and exploratory data analysis, where it uncovers hidden patterns within the unlabeled dataset (Jordan and Mitchell, 2015). Artificial neural networks (ANN), genetic algorithms, instance-based learning models, deep learning, and clustering are examples of algorithms used in unsupervised learning. The training and testing datasets are merged in reinforcement learning, and the learner interacts with the environment to gather data. The environment rewards the learner for his activities, posing the question of whether to pursue exploration or exploitation. As divergent to using the information already gathered, the learner must explore new, uncharted actions to learn more (Sutton and Barto, 2018; Mohri *et al.,* 2018). Robot navigation, the learning of skills by machines, and real-time decision-making all involve reinforcement learning algorithms. Q-learning (SARSA max), Dataset Aggregation (dAgger), and Deep Q-learning (DQL) are examples of reinforcement learning algorithms.

AI-DRIVEN RESOURCE AND CROP MANAGEMENT

Multiple AI technologies are nowadays applied in the agriculture field for various purposes, including crop and plant health and growth monitoring. These AI-driven applications have the potential to revolutionize farming practices, increase productivity, and promote sustainable agriculture.

Figure 3. An illustration of IoT technologies for agriculture (Quy et al., 2022)

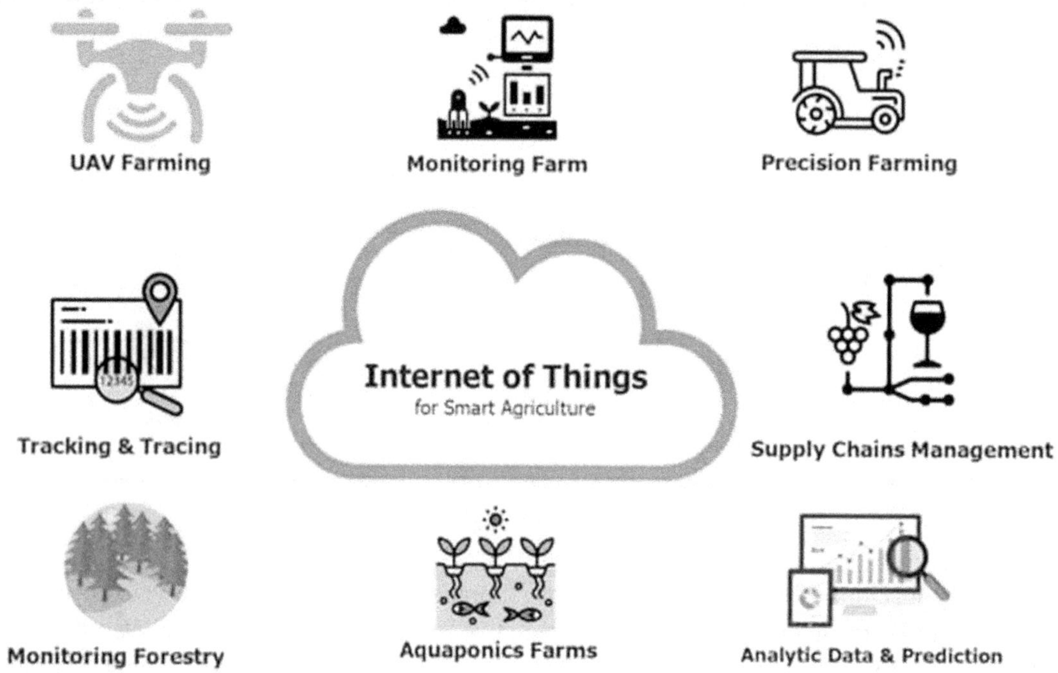

AI for Crop and Soil Monitoring

AI has emerged as a powerful tool for crop and land soil health monitoring in agriculture, revolutionizing the way farmers manage their fields. By leveraging AI algorithms and advanced IoT technologies, farmers can obtain real-time and accurate information about crop health and soil conditions, leading to more efficient and sustainable agricultural practices. Figure 3 summarize multiple usage of IoT in agriculture.

An emerging technology in soil monitoring is the use of remote sensing and satellite imagery combined with AI algorithms. These technologies can provide valuable information about soil properties on a larger scale, allowing for more efficient and cost-effective monitoring. For instance, John *et al.,* (2020) developed an AI-based approach that combined satellite imagery and machine learning to estimate soil organic carbon content across large agricultural landscapes. The system achieved high accuracy in predicting soil organic carbon, providing valuable insights for sustainable soil management at a regional scale.

Remote sensing technologies such as aerial imagery are equipped with advanced sensors to capture high-resolution data including vegetation indices, reflectance patterns, and thermal signatures. AI algorithms analyze this data to assess crop health indicators such as chlorophyll content, biomass, and water stress. By comparing these indicators with established norms, AI systems can identify areas of the field that may require attention, allowing farmers to take targeted actions such as adjusting irrigation, applying fertilizers, or addressing pest and disease outbreaks.

In addition to crop health monitoring, AI technologies play a crucial role in land soil health monitoring. AI-driven soil monitoring systems utilize sensors and data analysis techniques to assess various soil parameters, including moisture content, pH levels, nutrient availability, and organic matter content. These systems provide real-time site-specific information enabling farmers to make informed decisions about

Figure 4. Integrated sensor system for agriculture monitoring: Vegetation, land, and animals (Quy et al., 2022)

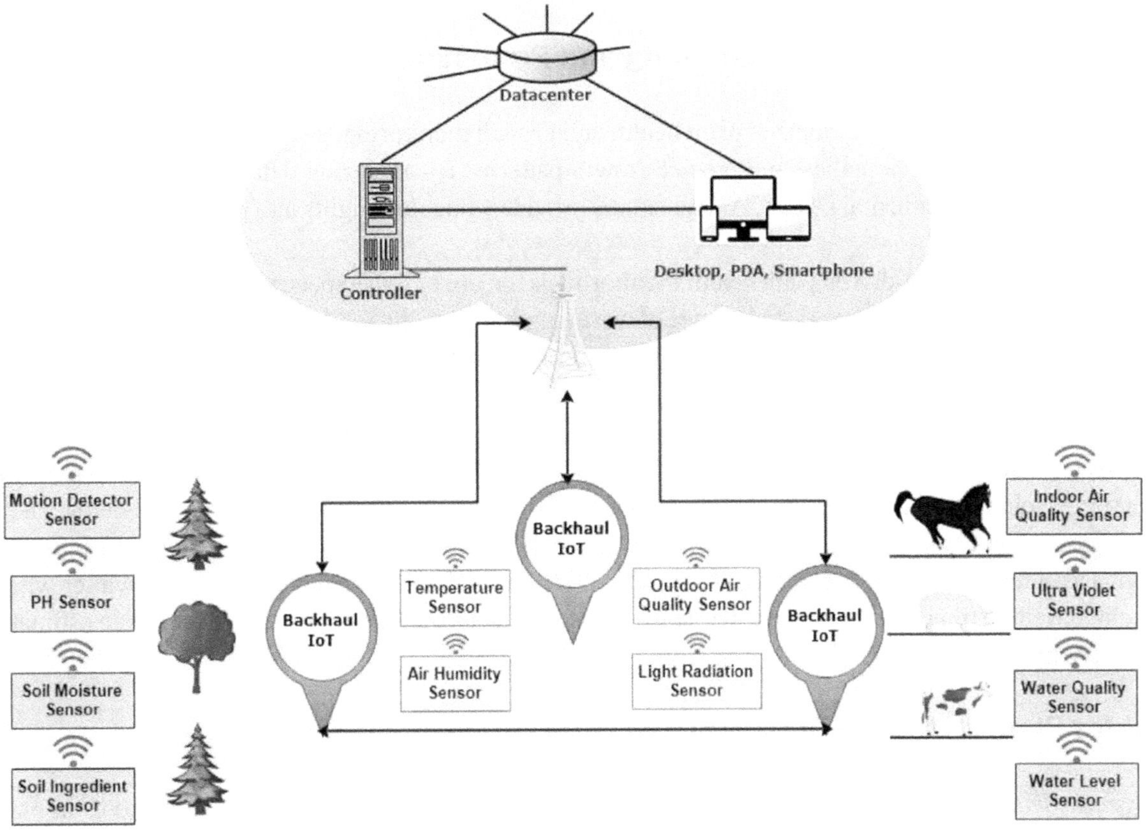

irrigation, fertilization, and soil management practices. For instance, Adeyemi et al. (2018) developed an irrigation scheduling system using Dynamic Neural Network models. The models help to maintain a target soil moisture level. This predictive system demonstrated a water-saving level ranging between 20 and 46%. Figure 4 illustrates a comprehensive sensor system designed to monitor various aspects of agriculture, encompassing vegetation, land conditions, and animal activity.

AI algorithms used in soil monitoring and nutrient management applications are often based on machine learning techniques. They are trained using large datasets that include soil data collected from different locations and conditions. These datasets are used to build models that can predict soil properties based on the input data from the sensors. The models are continuously updated and refined as more data becomes available, improving their accuracy over time. One notable application of AI in soil monitoring is the SoilGrids platform developed by the International Soil Reference and Information Centre (ISRIC). This platform utilizes AI algorithms to generate high-resolution global soil property maps. The AI models analyze a wide range of soil data, including physical, chemical, and biological properties, to create detailed maps of soil attributes. These maps provide valuable information for sustainable soil management practices, helping farmers make informed decisions regarding fertilizer application rates, irrigation schedules, and land use planning (Poggio *et al.,* 2021).

Another example is the CROPWAT software package which was developed by the Food and Agriculture Organization (FAO). It combines weather data, soil information, and crop-specific parameters to estimate crop water requirements and irrigation scheduling (Smith *et al.,* 1992).

Plant Health and Growth Monitoring and Prediction

AI technologies have revolutionized plant health and growth monitoring, enabling farmers to track plant development, detect anomalies, and predict growth patterns. By analyzing data from sensors, imaging systems, and environmental factors, AI algorithms provide valuable insights into plant health and growth trajectories.

One example of AI-driven plant health monitoring is the use of hyperspectral imaging. Hyperspectral cameras capture images across a wide range of wavelengths, allowing for detailed analysis of plant physiological parameters. By correlating spectral data with plant health indicators, AI models can identify stress conditions, nutrient deficiencies, and disease symptoms (Cheshkova, 2022).

In addition to monitoring, AI algorithms can predict plant growth and yield based on historical data, environmental conditions, and crop characteristics. By analyzing large datasets and utilizing machine learning techniques, AI models can forecast plant growth rates, flowering periods, and crop yields. These predictions enable farmers to optimize resource allocation, plan harvest schedules, and make informed management decisions. For example, Araus et al. (2018) used ML models to predict the winter wheat yield. based on remote sensing data. The study results show the ability of AI to make accurate estimates of the winter wheat yield.

AI for Plant's Disease Detection

Disease detection and image recognition using AI algorithms have emerged as powerful tools in plant health monitoring and disease management. AI-powered image recognition algorithms leverage deep learning techniques to analyze large image databases and identify patterns and characteristics associated with specific diseases. These algorithms are trained on extensive datasets that include images of healthy plants as well as plants affected by various diseases. By comparing input images with the patterns and features learned during the training phase, AI systems can accurately classify and detect the presence of diseases in plants enabling farmers to take timely and targeted actions to prevent the spread of diseases and minimize crop losses. Figure 5 depicts an innovative disease detection system for plants, which relies on sensor technology and image analysis to identify and diagnose plant diseases accurately (Prabha, 2021).

Several scientific papers demonstrate the application of AI in disease detection. For example, Mohanty et al. (2016) developed a deep learning-based system called PlantVillage for disease detection in plants. The system utilized a large dataset of plant images and deep convolutional neural networks to identify and classify diseases accurately. Similarly, Sladojevic *et al.,* (2016) employed deep learning algorithms for the detection of grapevine diseases. Their system achieved high accuracy in identifying diseases such as downy mildew and powdery mildew from images.

Agrosight is an example of a company that utilizes AI algorithms for disease detection in agriculture. Their AI-powered image recognition system can detect diseases such as citrus greening, soybean rust, and wheat rust from images captured in the field. By analyzing the images, the AI system can identify disease symptoms and provide early detection and diagnosis, allowing farmers to take prompt action and implement appropriate disease management strategies.

Figure 5. Plant disease detection system (Prabha, 2021)

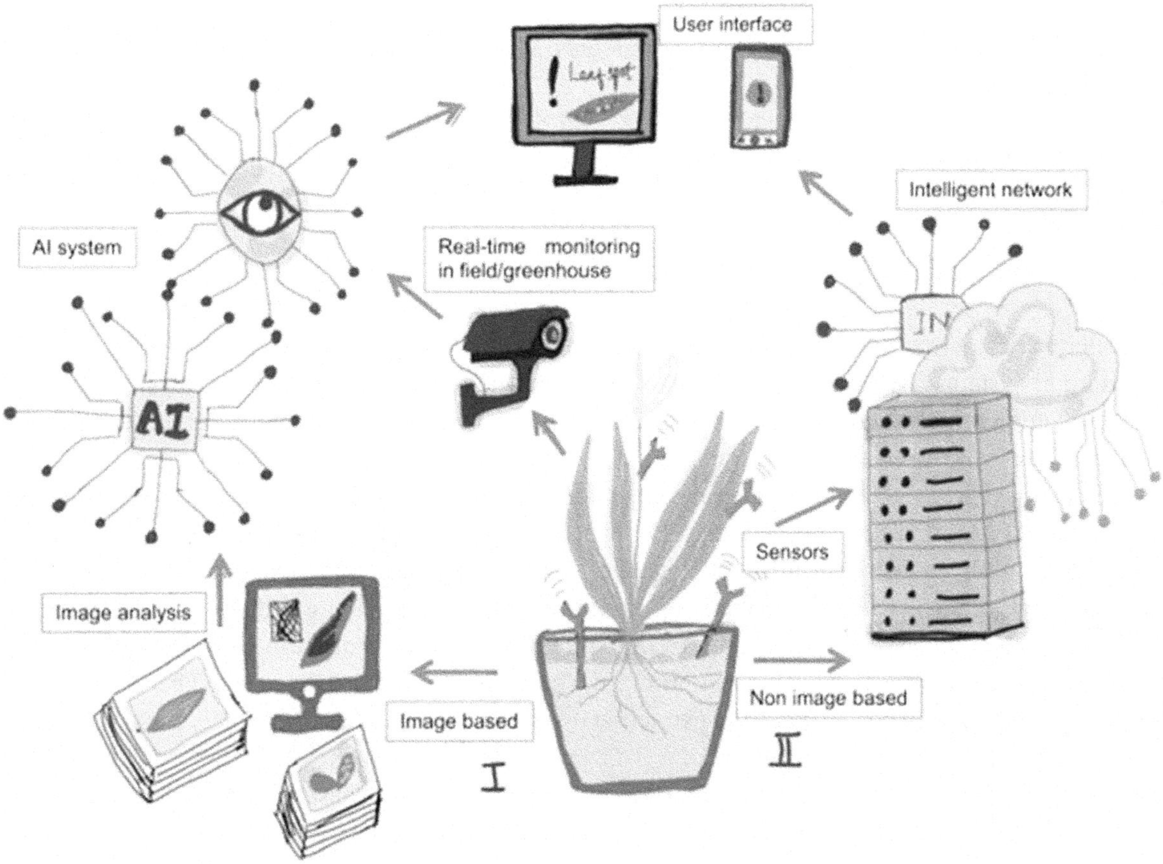

The application of AI in disease detection offers several advantages for plant health management. Firstly, AI algorithms can process and analyze large volumes of images quickly and accurately, enabling high-throughput screening of crops for disease presence. This scalability is particularly beneficial for large-scale agricultural operations where manual inspection of each plant may be time-consuming and impractical (Mohanty *et al.,* 2016).

Secondly, AI-powered disease detection systems can detect diseases at an early stage, even before visible symptoms are apparent to the human eye. This early detection allows for timely intervention, reducing the spread of diseases and minimizing crop losses. By providing farmers with early warning signs, AI algorithms empower them to implement targeted disease management strategies, such as targeted pesticide application or removal of infected plants, to prevent further damage (Sladojevic *et al.,* 2016).

Moreover, AI algorithms can continuously learn and improve their disease-detection capabilities over time. As more data is fed into the system, the algorithms can refine their understanding of disease patterns and improve the accuracy of disease identification. This iterative learning process ensures that AI systems become more robust and effective in detecting a wide range of plant diseases (Fuentes *et al.,* 2019).

In addition to disease detection, AI technologies can also aid in disease diagnosis and provide information on disease management practices. By analyzing the characteristics of detected diseases, AI systems can provide insights into disease progression, recommend appropriate treatment options, and offer guidance on preventive measures. This information can help farmers make informed decisions regarding disease control and management strategies, leading to more efficient and effective disease management practices.

For example, in a study reported by Karlekar and Seal (2020), a deep learning convolution neural network (CNN), named SoyNet, was developed for soybean plant disease recognition using segmented leaf images.

Insect and Harmful Plant Detection

AI technologies play a vital role in the early detection and monitoring of insects and harmful plants in agricultural systems. By leveraging computer vision, machine learning, and data analysis, AI-driven systems can accurately identify pests, insects, and invasive plant species, allowing for timely intervention and effective pest management strategies.

Computer vision algorithms trained on large datasets can identify and classify various pests and insects based on their visual characteristics. For instance, an AI-based system developed by Prasath and Akila (2023) utilized deep learning techniques to automatically detect and classify different types of pests in crops, aiding in the identification of pest species and facilitating targeted pest control measures.

The approach involves gathering data from IoT sensors in an agriculture-based IoT platform. The collected images are processed using an optimized Yolov3 model for pest detection. Then the detected pest images are subjected to deep feature extraction using ML models. Classification is performed using a Weight Optimized Deep Neural Network (WO-DNN) model. In addition to pest detection, AI technologies can also identify and manage harmful or invasive plant species. These systems leverage machine learning algorithms to analyze imagery and classify plants based on their characteristics.

Preti *et al.,* (2021) ceated a camera-based insect monitoring trap system, as illustrated in Figure 6. The system includes a self-cleaning funnel type traps and Trapview standard traps equipped with high-resolution cameras, solar panels, temperature and humidity sensors, and remote-controllable electronic data acquisition and transfer systems. The traps automatically capture images of two insect species (Helicoverpa armigera Hübner and Cydia pomonella L.) in an apple orchard, and the images are subsequently marked and counted using the TrapView technology.

In the study performed by Lake et al. (2022), the researchers used satellite imagery to detect the invasive leafy spurge plant. The study shows that deep learning models can accurately identify invasive species, even with lower-resolution satellite imagery, benefiting remote and data-sparse areas. By employing AI-driven insect and harmful plant detection systems, farmers can efficiently identify and manage pests and invasive plants, reducing crop losses and minimizing the reliance on chemical interventions. These technologies contribute to sustainable agriculture practices by enabling targeted pest control measures and promoting the conservation of beneficial plant species.

Crop Breeding and Genetic Improvement

AI technologies have revolutionized the field of agriculture by expediting the crop breeding process and enabling more precise and targeted selection of desired traits. Crop breeding and genetic improvement are crucial for developing new varieties with improved traits, such as higher yields, disease resistance, and tolerance to environmental stressors.

Figure 6. A camera-based insect monitoring trap system (Preti et al., 2021)

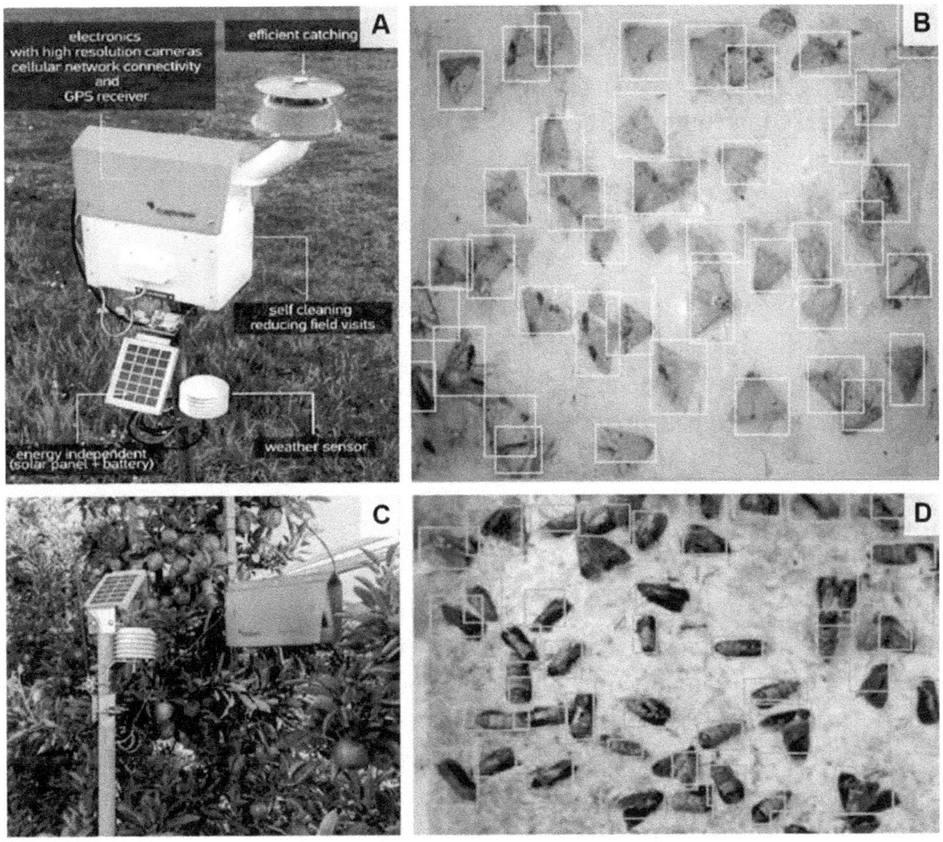

One of the key contributions of AI in crop breeding is its ability to analyze vast amounts of genetic data. This includes genomic information, gene expression data, and phenotypic traits collected from various sources. Machine learning algorithms are applied to these datasets to identify patterns, correlations, and relationships between genes and traits. By uncovering these associations, AI models can predict the potential expression of specific traits in plants based on their genetic makeup (Xu *et al.*,2022).

The application of AI in crop breeding involves several stages. Initially, the models are trained using large datasets that link genetic data with corresponding traits of interest. This training process enables the algorithms to learn the patterns and correlations between genetic variations and desired traits. Once the models are trained, they can be applied to predict the outcomes of breeding programs and guide the selection of parent plants with the desired genetic traits (Najafabadi *et al.,* 2023).

One notable example of AI application in crop breeding is the CropOS platform developed by Benson Hill Biosystems. CropOS employs AI algorithms to predict the potential of plant traits based on genetic information (Benson Hill Biosystems, https://bensonhill.com/). By analyzing genomic data and using machine learning techniques, the platform assists breeders in selecting parent plants with the most desirable genetic traits, thus accelerating the development of new crop varieties.

A promising application of AI in crop breeding is the use of phenotyping technologies. Phenotyping refers to the measurement and analysis of observable traits in plants, such as plant height, leaf area, and flowering time. Traditional phenotyping methods can be time-consuming and labor-intensive, limiting the scale and efficiency of breeding programs. However, AI technologies offer the potential to automate and streamline phenotyping processes (Taneja al., 2023). Furthermore, AI-driven phenotyping can be combined with genomic data to identify genetic markers associated with specific phenotypic traits. By integrating phenotypic and genomic data, breeders can gain deeper insights into the genetic basis of complex traits and make more informed decisions in their breeding programs (Araus *et al.,* 2014).

CHALLENGES AND SOLUTIONS OF AI ACCESSIBILITY

As artificial intelligence continues to advance, it holds immense potential for transforming the agricultural industry. AI technologies are able to revolutionize various aspects of farming, from crop management and livestock monitoring to precision agriculture and supply chain optimization. However, alongside the benefits, there are several challenges that need to be addressed for the successful integration of AI in agriculture.

Infrastructure and Connectivity

The successful deployment of AI in agriculture heavily relies on robust infrastructure and connectivity. Many farming regions still lack adequate internet connectivity and power supply, which hinders the seamless integration of AI technologies. High-speed internet access is crucial for real-time data transmission, remote sensing, and cloud-based AI applications. Bridging the digital divide and improving infrastructure in rural areas is essential to ensure equal access to AI-driven solutions.

In situations where the network or internet connectivity is highly unreliable and fails to meet the minimum requirements of smart farming platforms, it is advisable to equip the platforms with local computers for data storage and decision-making processes. This approach ensures that efficiency is not compromised by attempting to transmit a continuous stream of data over a weak network. By utilizing local computing resources, the smart farming system can function independently and maintain operational effectiveness, even in challenging network conditions (Elbasi *et al.,* 2023).

Many farms and agricultural operations rely on legacy infrastructure, equipment, and software that may not be compatible with AI solutions. Retrofitting and integrating AI systems seamlessly into existing processes require careful planning, investment, and technical expertise. Ensuring interoperability and compatibility between AI systems and other agricultural technologies is crucial for successful implementation.

Lack of Technical Skills and Expertise

AI technologies demand specialized technical skills and expertise to develop, implement, and maintain them effectively. However, the agricultural sector often faces a shortage of professionals skilled in AI, data analytics, and machine learning. For instance, as stated in the research conducted by Amara et al. (2017), there is a global scarcity of resources and specialized knowledge concerning banana pathology.

In addition, researchers have acknowledged the varying levels of ease in training farmers, with developed countries having a comparative advantage due to the widespread availability and adoption of advanced agricultural technologies (Elbasi *et al.,* 2023). In these nations, farmers are already familiar with and have access to various types of advanced tools and techniques, making the training process relatively straightforward. Conversely, developing countries face more significant challenges in training farmers, particularly those located in rural areas. Many farms in these regions lack access to basic infrastructure, let alone advanced technologies. Limited resources, lack of connectivity, and inadequate technical support further compound the difficulty of training farmers in these contexts.

To bridge this gap, innovative approaches and tailored strategies are required to provide training and capacity-building programs suitable for the specific needs and constraints of farmers in rural areas of developing countries. This includes implementing alternative educational methods such as mobile-based learning, community-based training initiatives, or utilizing local extension services to disseminate knowledge and skills effectively. Investment in rural infrastructure, including improved access to electricity, internet connectivity, and affordable agricultural technologies, can also play a vital role in facilitating the training process.

Cost and Affordability

The cost of implementing AI technologies can be a significant barrier for farmers, especially small-scale and resource-constrained agricultural operations. AI systems often involve expensive hardware, sensors, and software, along with the need for specialized expertise for implementation and maintenance.

The setup cost of implementing a smart farming system encompasses the purchase of hardware components such as IoT devices, gateways, and base station infrastructure (Elijah, Member & Rahman, 2018). On the other hand, the running cost involves ongoing subscriptions for centralized services or IoT platforms that facilitate data collection, device management, and information sharing among other services (Elijah *et al.,* 2018).

According to Sharma et al. (2021), the accuracy of predictions may be influenced by the number of sensors utilized, resulting in a higher initial investment for farmers. On the other hand, the successful adoption of AI technologies in agriculture relies on user acceptance and trust. Farmers and agricultural stakeholders may be hesitant to embrace AI due to factors such as the lack of awareness and perceived complexity. Building trust between farmers and AI systems by showcasing real-world success stories and addressing concerns can help overcome resistance.

Efforts to make AI solutions more affordable and accessible to all farmers are necessary to overcome this challenge. By reducing the cost barriers associated with AI implementation, more farmers can benefit from the transformative potential of these solutions. To achieve this, various strategies can be employed. One approach is to promote research and development in AI specifically targeted at the agricultural sector, focusing on creating cost-effective solutions that cater to the unique needs and constraints of farmers.

Adaptability and Data Accuracy

Agriculture is a heterogeneous and complex field, with variations in climate, soil conditions, crop types, and farming practices. Customizing AI solutions to suit different contexts while maintaining performance and accuracy pose significant challenges. AI models need to be adaptable and easily calibrated to address the specific needs and requirements of different farming systems and crops.

In practical smart farming scenarios, IoT devices are primarily designed for indoor environments, which may not adequately account for the harsh outdoor conditions' farmers face. Factors such as snow, hail, floods, wind, and dust can lead to the accelerated deterioration of IoT sensors in outdoor settings (Elbasi *et al.,* 2023).

The challenge lies in creating flexible AI algorithms and devices that can accommodate the diversity of agricultural practices and provide accurate predictions and recommendations.

Moreover, the presence of high-voltage grids can create electromagnetic interference that can disrupt and corrupt data due to the generation of electromagnetic fields significantly affecting data integrity (Zanella, 2020).

To address this issue, additional measures need to be implemented to mitigate electromagnetic interference and ensure data accuracy in agricultural applications. This may involve implementing proper shielding techniques for the sensors, using specialized equipment designed to withstand and filter out electromagnetic interference, or even exploring alternative communication and data transfer methods that are less susceptible to such interference.

Security and Regulatory Considerations

The deployment of AI in agriculture raises ethical and regulatory concerns that need to be addressed. As AI systems make autonomous decisions, issues of accountability, transparency, and bias become paramount.

For instance, issues related to data privacy, ownership, and security arise when farmers share their farm data with AI service providers. As data collection and utilization become increasingly prevalent in agriculture, it is crucial to establish clear guidelines and agreements between farmers and data companies. Indeed, farmers should have the assurance that their data will be handled in a secure manner and that their privacy will be protected. This includes establishing protocols for data storage, sharing, and anonymization to safeguard sensitive information.

Storing data in data centers located in other countries raises concerns about its security. These countries may have less stringent security requirements imposed on companies, making the data more vulnerable to breaches. Moreover, there is a risk that the governments of those countries might intercept or access the data stored on servers within their jurisdictions (Gupta *et al.*, 2020). To address these concerns, countries have begun implementing laws related to data localization, which aim to regulate the storage of sensitive data within their own territories (Gupta *et al.*, 2020). An example of this is China's implementation of a new cybersecurity law in 2017, which includes a requirement that personal data be stored on servers within the country (CSIS). Consequently, major companies such as Microsoft, Google, and Amazon have initiated measures to transfer control of Chinese data to domestic firms in compliance with the law (CSIS).

As technology adoption in farming continues to rise, researchers and federal agencies have begun assessing the impact of cyber-attacks on farmers and communities (Gupta *et al.*, 2020). The increasing use of technologies in agricultural practices has made the farming sector more susceptible to cyber threats. Recognizing this, efforts are being made to understand the potential consequences and develop strategies to mitigate the risks associated with cyber-attacks in the agricultural domain (Gupta *et al.*, 2020).

Apart from this, the migration of young people from rural areas to cities has created a challenging situation for farmers, as it becomes increasingly difficult to convince and inspire the younger generation to stay and pursue farming as a future career. Consequently, there is a growing policy emphasis on promoting and providing training to farmers to adopt agricultural technologies in multiple countries (Elbasi *et al.*, 2023).

CONCLUSION

To sum up, this chapter highlights the transformative potential of AI in revolutionizing the agricultural sector. Through a comprehensive exploration of various AI applications, ranging from its usage in supply chain to crop monitoring and pest detection, we have witnessed how AI can address critical challenges and foster sustainable practices.

Currently, the exploitation of AI is required to make precision farming which represents one of the most rapidly evolving research and development areas, attracting high interest by both the industry as well as the academia. In addition, with the invention of new technologies applicated in agriculture built on the bases of AI, it was able to solve several problems encountered in traditional agriculture including resource utilization, crop productivity, and environmental impact. This synergy between AI and agriculture opens up possibilities for a more resilient and efficient food production system that can meet the demands of a growing global population while safeguarding our planet's resources. Accordingly, collaborative efforts among stakeholders, including farmers, researchers and technology providers, are vital to tackle newfound challenges and harness the full potential of AI for sustainable and efficient agricultural practices.

As we move forward, continued investment in research, development, and education is vital to stay at the forefront of AI innovations in agriculture. By staying abreast of the latest advancements and encouraging knowledge exchange, we can drive the adoption of AI applications in agricultural sustainability to greater heights.

REFERENCES

Adeyemi, O., Grove, I., Peets, S., Domun, Y., & Norton, T. (2018). Dynamic neural network modelling of soil moisture content for predictive irrigation scheduling. *Sensors (Basel)*, *18*(10), 3408. doi:10.339018103408 PMID:30314346

Adnan, N., Nordin, S. M., Rahman, I., & Noor, A. (2018). *The effects of knowledge transfer on farmers decision making toward sustainable agriculture practices: In view of green fertilizer technology*. World Journal of Science, Technology and Sustainable Development. doi:10.1108/WJSTSD-11-2016-0062

Agrovision. (n.d.). Retrieved from www.agrovision.com

Ahumada, O., & Villalobos, J. R. (2009). Application of planning models in the agri-food supply chain: A review. *European Journal of Operational Research*, *196*(1), 1–20. doi:10.1016/j.ejor.2008.02.014

Amara, J., Bouaziz, B., & Algergawy, A. (2017). *A Deep Learning-based approach for Banana Leaf Diseases Classification*. Academic Press.

Araus, J. L., Li, J., Parry, M. A., & Wang, J. (2014). Phenotyping and other breeding approaches for a New Green Revolution. *Journal of Integrative Plant Biology*, *56*(5), 422–424. doi:10.1111/jipb.12202 PMID:24703237

Béné, C., Barange, M., Subasinghe, R., Pinstrup-Andersen, P., Merino, G., Hemre, G. I., & Williams, M. (2015). Feeding 9 billion by 2050–Putting fish back on the menu. *Food Security*, *7*(2), 261–274. doi:10.100712571-015-0427-z

Benson Hill. (n.d.). Retrieved from www.bensonhill.com

Borodin, V., Bourtembourg, J., Hnaien, F., & Labadie, N. (2016). Handling uncertainty in agricultural supply chain management: A state of the art. *European Journal of Operational Research, 254*(2), 348–359. doi:10.1016/j.ejor.2016.03.057

Cheng, E., Zhang, B., Peng, D., Zhong, L., Yu, L., Liu, Y., Xiao, C., Li, C., Li, X., Chen, Y., Ye, H., Wang, H., Yu, R., Hu, J., & Yang, S. (2022). Wheat yield estimation using remote sensing data based on machine learning approaches. *Frontiers in Plant Science, 13*, 13. doi:10.3389/fpls.2022.1090970 PMID:36618627

Cheshkova, A. F. (2022). A review of hyperspectral image analysis techniques for plant disease detection and identification. *Vavilovskii Zhurnal Genetiki i Selektsii, 26*(2), 202–213. doi:10.18699/VJGB-22-25 PMID:35434482

Coble, K. H., Mishra, A. K., Ferrell, S., & Griffin, T. (2018). Big data in agriculture: A challenge for the future. *Applied Economic Perspectives and Policy, 40*(1), 79–96. doi:10.1093/aepp/ppx056

CSIS. (n.d.). *Who Benefits from China's Cybersecurity Laws? New Perspectives on Asia.* Retrieved June 2023, from https://www.csis.org/blogs/new-perspectives-asia/who-benefits-chinas-cybersecurity-laws

Du, C. J., & Sun, D. W. (2006). Learning techniques used in computer vision for food quality evaluation: A review. *Journal of Food Engineering, 72*(1), 39–55. doi:10.1016/j.jfoodeng.2004.11.017

Duman, G. M., Tozanli, O., Kongar, E., & Gupta, S. M. (2017). A holistic approach for performance evaluation using quantitative and qualitative data: A food industry case study. *Expert Systems with Applications, 81*, 410–422. doi:10.1016/j.eswa.2017.03.070

El Bilali, H., & Allahyari, M. S. (2018). Transition towards sustainability in agriculture and food systems: Role of information and communication technologies. *Information Processing in Agriculture, 5*(4), 456–464. doi:10.1016/j.inpa.2018.06.006

Elavarasan, D., Vincent, D. R., Sharma, V., Zomaya, A. Y., & Srinivasan, K. (2018). Forecasting yield by integrating agrarian factors and machine learning models: A survey. *Computers and Electronics in Agriculture, 155*, 257–282. doi:10.1016/j.compag.2018.10.024

Elbasi, E., Mostafa, N., Alarnaout, Z., Zreikat, A. I., Cina, E., Varghese, G., Shdefat, A., Topcu, A. E., Abdelbaki, W., Mathew, S., & Zaki, C. (2023). Artificial Intelligence Technology in the Agricultural Sector : A Systematic Literature Review. *IEEE Access : Practical Innovations, Open Solutions, 11*, 171–202. doi:10.1109/ACCESS.2022.3232485

Elferink, M., & Schierhorn, F. (2016). Global demand for food is rising. Can we meet it. *Harvard Business Review, 7*(04), 2016.

Elijah, O., Member, S., & Rahman, T. A. (2018). An Overview of Internet of Things (IoT) and Data Analytics in Agriculture : Benefits and Challenges. *IEEE Internet of Things Journal, 5*(5), 3758–3773. doi:10.1109/JIOT.2018.2844296

Eyhorn, F., Muller, A., Reganold, J. P., Frison, E., Herren, H. R., Luttikholt, L., & Smith, P. (2019). Sustainability in global agriculture driven by organic farming. *Nature Sustainability*, *2*(4), 253–255. doi:10.103841893-019-0266-6

FAO. (2009). *Press release, 19 June 2009*. http://www.fao.org/news/story/en/item/ 20568/icode/

Fuentes, A., Yoon, S., & Park, D. S. (2019). Deep learning-based phenotyping system with glocal description of plant anomalies and symptoms. *Frontiers in Plant Science*, *10*, 1321. doi:10.3389/fpls.2019.01321 PMID:31798598

Gharaei, A., Karimi, M., & Hoseini Shekarabi, S. A. (2019). Joint economic lot-sizing in multiproduct multi-level integrated supply chains: Generalized benders decomposition. *International Journal of Systems Science: Operations & Logistics*, 1-17.

Giri, B. C., & Masanta, M. (2018). Developing a closed-loop supply chain model with price and quality dependent demand and learning in production in a stochastic environment. *International Journal of Systems Science: Operations & Logistics*, 1-17.

Giri, C., Jain, S., Zeng, X., & Bruniaux, P. (2019). A detailed review of artificial intelligence applied in the fashion and apparel industry. *IEEE Access : Practical Innovations, Open Solutions*, *7*, 95376–95396. doi:10.1109/ACCESS.2019.2928979

Gupta, M., Abdelsalam, M., & Mittal, S. (2020). Security and Privacy in Smart Farming : Challenges and Opportunities. *IEEE Access : Practical Innovations, Open Solutions*, *8*, 34564–34584. doi:10.1109/ACCESS.2020.2975142

Hassler, S. C., & Baysal-Gurel, F. (2019). Unmanned aircraft system (UAS) technology and applications in agriculture. *Agronomy (Basel)*, *9*(10), 618. doi:10.3390/agronomy9100618

John, K., Abraham Isong, I., Michael Kebonye, N., Okon Ayito, E., Chapman Agyeman, P., & Marcus Afu, S. (2020). Using machine learning algorithms to estimate soil organic carbon variability with environmental variables and soil nutrient indicators in an alluvial soil. *Land (Basel)*, *9*(12), 487. doi:10.3390/land9120487

Jordan, M. I., & Mitchell, T. M. (2015). Machine learning: Trends, perspectives, and prospects. *Science*, *349*(6245), 255–260. doi:10.1126cience.aaa8415 PMID:26185243

Kamble, S. S., Gunasekaran, A., & Sharma, R. (2020). Modeling the blockchain enabled traceability in agriculture supply chain. *International Journal of Information Management*, *52*, 101967. doi:10.1016/j.ijinfomgt.2019.05.023

Kamilaris, A., Kartakoullis, A., & Prenafeta-Boldú, F. X. (2017). A review on the practice of big data analysis in agriculture. *Computers and Electronics in Agriculture*, *143*, 23–37. doi:10.1016/j.compag.2017.09.037

Kamilaris, A., & Prenafeta-Boldú, F. X. (2018). Deep learning in agriculture: A survey. *Computers and Electronics in Agriculture*, *147*, 70–90. doi:10.1016/j.compag.2018.02.016

Karlekar, A., & Seal, A. (2020). SoyNet: Soybean leaf diseases classification. *Computers and Electronics in Agriculture*, *172*, 105342. doi:10.1016/j.compag.2020.105342

Kazemi, Y. (2019). *AI, Big Data & Advanced Analytics In The Supply Chain*. Available at: https://www.forbes.com/sites/yasamankazemi/2019/01/29/ai-big-data-advanced-analyticsin-the-supply-chain/#75db62f4244f

Konovalenko, I., & Ludwig, A. (2019). Event processing in supply chain management–The status quo and research outlook. *Computers in Industry*, *105*, 229–249. doi:10.1016/j.compind.2018.12.009

Kummu, M., De Moel, H., Porkka, M., Siebert, S., Varis, O., & Ward, P. J. (2012). Lost food, wasted resources: Global food supply chain losses and their impacts on freshwater, cropland, and fertiliser use. *The Science of the Total Environment*, *438*, 477–489. doi:10.1016/j.scitotenv.2012.08.092 PMID:23032564

Lake, T. A., Briscoe Runquist, R. D., & Moeller, D. A. (2022). Deep learning detects invasive plant species across complex landscapes using Worldview-2 and Planetscope satellite imagery. *Remote Sensing in Ecology and Conservation*, *8*(6), 875–889. doi:10.1002/rse2.288

Liakos, K. G., Busato, P., Moshou, D., Pearson, S., & Bochtis, D. (2018). Machine learning in agriculture: A review. *Sensors (Basel)*, *18*(8), 2674. doi:10.339018082674 PMID:30110960

Min, H. (2010). Artificial intelligence in supply chain management: Theory and applications. *International Journal of Logistics*, *13*(1), 13–39. doi:10.1080/13675560902736537

Mohanty, S. P., Hughes, D. P., & Salathé, M. (2016). Using deep learning for image-based plant disease detection. *Frontiers in Plant Science*, *7*, 1419. doi:10.3389/fpls.2016.01419 PMID:27713752

Mohri, M., Rostamizadeh, A., & Talwalkar, A. (2018). *Foundations of machine learning*. MIT press.

Naik, G., & Suresh, D. N. (2018). Challenges of creating sustainable agri-retail supply chains. *IIMB Management Review*, *30*(3), 270–282. doi:10.1016/j.iimb.2018.04.001

OECD. (2017). *Going digital: Making the transformation work for growth and well-being*. Meeting of the OECD Council at Ministerial Level, Paris, France.

OECD. (2019). *Technology and digital in agriculture*. Retrieved from: https://www.oecd.org/agriculture/topics/technology-and-digital-agriculture/

Patidar, R., Agrawal, S., & Pratap, S. (2018). Development of novel strategies for designing sustainable Indian agri-fresh food supply chain. *Sadhana*, *43*(10), 1–16. doi:10.100712046-018-0927-6

Pelletier, N., & Tyedmers, P. (2010). Forecasting potential global environmental costs of livestock production 2000–2050. *Proceedings of the National Academy of Sciences of the United States of America*, *107*(43), 18371–18374. doi:10.1073/pnas.1004659107 PMID:20921375

Poggio, L., De Sousa, L. M., Batjes, N. H., Heuvelink, G., Kempen, B., Ribeiro, E., & Rossiter, D. (2021). SoilGrids 2.0: Producing soil information for the globe with quantified spatial uncertainty. *Soil (Göttingen)*, *7*(1), 217–240. doi:10.5194oil-7-217-2021

Prabha, K. (2021). Disease sniffing robots to apps fixing plant diseases: Applications of artificial intelligence in plant pathology—a mini review. *Indian Phytopathology*, *74*(1), 13–20. doi:10.100742360-020-00290-3

Prasath, B., & Akila, M. (2023). IoT-based pest detection and classification using deep features with enhanced deep learning strategies. *Engineering Applications of Artificial Intelligence*, *121*, 105985. doi:10.1016/j.engappai.2023.105985

Preti, M., Verheggen, F., & Angeli, S. (2021). Insect pest monitoring with camera-equipped traps: Strengths and limitations. *Journal of Pest Science*, *94*(2), 203–217. doi:10.100710340-020-01309-4

Quy, V. K., Hau, N. V., Anh, D. V., Quy, N. M., Ban, N. T., Lanza, S., Randazzo, G., & Muzirafuti, A. (2022). IoT-enabled smart agriculture: Architecture, applications, and challenges. *Applied Sciences (Basel, Switzerland)*, *12*(7), 3396. doi:10.3390/app12073396

Rabbani, M., Foroozesh, N., Mousavi, S. M., & Farrokhi-Asl, H. (2019). Sustainable supplier selection by a new decision model based on interval-valued fuzzy sets and possibilistic statistical reference point systems under uncertainty. *International Journal of Systems Science: Operations & Logistics*, *6*(2), 162–178.

Rabbani, M., Hosseini-Mokhallesun, S. A. A., Ordibazar, A. H., & Farrokhi-Asl, H. (2020). A hybrid robust possibilistic approach for a sustainable supply chain location-allocation network design. *International Journal of Systems Science: Operations & Logistics*, *7*(1), 60–75.

Rockström, J., Williams, J., Daily, G., Noble, A., Matthews, N., Gordon, L., Wetterstrand, H., DeClerck, F., Shah, M., Steduto, P., de Fraiture, C., Hatibu, N., Unver, O., Bird, J., Sibanda, L., & Smith, J. (2017). Sustainable intensification of agriculture for human prosperity and global sustainability. *Ambio*, *46*(1), 4–17. doi:10.100713280-016-0793-6 PMID:27405653

Sennaar, K. (2019). *AI in Agriculture – Present Applications and Impact*. Available at: https://emerj.com/ai-sector-overviews/ai-agriculture-present-applications-impact/

Sharma, A., Jain, A., Chowdary, V., & Gupta, P. (2021). *Machine Learning Applications for Precision Agriculture : A Comprehensive Review*. Academic Press.

Sharma, A. K., Chaurasia, S., & Srivastava, D. K. (2018). Supervised rainfall learning model using machine learning algorithms. In *The International Conference on Advanced Machine Learning Technologies and Applications (AMLTA2018)* (pp. 275-283). 10.1007/978-3-319-74690-6_27

Sharma, R., & Parhi, S. (2017, October). A review on use of big data in warehousing to enhance accessibility of food. In *2017 2nd International Conference on Communication and Electronics Systems (ICCES)* (pp. 663-667). IEEE. 10.1109/CESYS.2017.8321162

Shukla, M., & Jharkharia, S. (2013). Agri-fresh produce supply chain management: A state-of-the-art literature review. *International Journal of Operations & Production Management*, *33*(2), 114–158. doi:10.1108/01443571311295608

Sladojevic, S., Arsenovic, M., Anderla, A., Culibrk, D., & Stefanovic, D. (2016). Deep neural networks based recognition of plant diseases by leaf image classification. *Computational Intelligence and Neuroscience*, *2016*, 2016. doi:10.1155/2016/3289801 PMID:27418923

Smith, M. (1992). CROPWAT: A computer program for irrigation planning and management (No. 46). Food & Agriculture Org.

Sutton, R. S., & Barto, A. G. (2018). *Reinforcement learning: An introduction.* MIT press.

Sykuta, M. E. (2016). Big data in agriculture: property rights, privacy and competition in ag data services. *International Food and Agribusiness Management Review, 19*, 57-74.

Taneja, A., Nair, G., Joshi, M., Sharma, S., Sharma, S., Jambrak, A. R., Roselló-Soto, E., Barba, F. J., Castagnini, J. M., Leksawasdi, N., & Phimolsiripol, Y. (2023). Artificial Intelligence: Implications for the Agri-Food Sector. *Agronomy (Basel), 13*(5), 1397. doi:10.3390/agronomy13051397

Tang, Z., Yang, J., Li, Z., & Qi, F. (2020). Grape disease image classification based on lightweight convolution neural networks and channel wise attention. *Computers and Electronics in Agriculture, 178*, 105735. doi:10.1016/j.compag.2020.105735

Traore, B. B., Kamsu-Foguem, B., & Tangara, F. (2017). Data mining techniques on satellite images for discovery of risk areas. *Expert Systems with Applications, 72*, 443–456. doi:10.1016/j.eswa.2016.10.010

Wolfert, S., Ge, L., Verdouw, C., & Bogaardt, M. J. (2017). Big data in smart farming–a review. *Agricultural Systems, 153*, 69–80. doi:10.1016/j.agsy.2017.01.023

Xu, Y., Zhang, X., Li, H., Zheng, H., Zhang, J., Olsen, M. S., Varshney, R. K., Prasanna, B. M., & Qian, Q. (2022). Smart breeding driven by big data, artificial intelligence and integrated genomic-enviromic prediction. *Molecular Plant, 15*(11), 1664–1695. doi:10.1016/j.molp.2022.09.001 PMID:36081348

Yoosefzadeh Najafabadi, M., Hesami, M., & Eskandari, M. (2023). Machine Learning-Assisted Approaches in Modernized Plant Breeding Programs. *Genes, 14*(4), 777. doi:10.3390/genes14040777 PMID:37107535

Zanella, A. (2020). *Security challenges to smart agriculture: Current state, key issues, and future directions*. Academic Press.

Zhong, D. R. Y., Tan, P. K., & Bhaskaran, P. G. (2017). Data-driven food supply chain management and systems. *Industrial Management & Data Systems, 117*(9), 1779–1781. doi:10.1108/IMDS-06-2017-0269

Zhu, X., & Goldberg, A. B. (2022). *Introduction to Semi-Supervised Learning*. Springer Nature.

KEY TERMS AND DEFINITIONS

Agriculture Supply Chain: The agriculture supply chain encompasses all the processes and activities involved in the production, processing, and distribution of agricultural products from the farm to the consumer.

Artificial Intelligence: AI is a branch of computer science that enables machines to perform tasks that typically require human intelligence, such as learning, problem-solving, and decision-making.

Food Supply Chain: The food supply chain encompasses all the processes and activities involved in the production, processing, and distribution of food products from the initial stages of agricultural production to consumption by the end-users.

Internet of Things: IoT refers to the network of physical objects embedded with sensors, software, and connectivity, enabling them to collect and exchange data with other devices and systems over the internet.

Machine Learning: ML is a subset of AI that focuses on the development of algorithms and statistical models that allow computers to learn from and make predictions or decisions based on data without explicit programming.

Supply Chain: A supply chain is a network of organizations, resources, activities, and technologies involved in the creation and distribution of goods or services to the end consumer.

Sustainability: Sustainability refers to the practice of using resources in a manner that meets the needs of the present generation without compromising the ability of future generations to meet their own needs.

Chapter 11
Empowering Agriculture With Conversational AI:
An Application for Farmer Advisory and Communication

Mrignainy Kansal
Ajay Kumar Garg Engineering College, Ghaziabad, India

Pancham Singh
Ajay Kumar Garg Engineering College, Ghaziabad, India

Mili Srivastava
Ajay Kumar Garg Engineering College, Ghaziabad, India

Prateek Chaurasia
Ajay Kumar Garg Engineering College, Ghaziabad, India

ABSTRACT

Conversational AI is the most trending technology in recent years. It's not just a buzzword anymore, it is here to stay and will continue to grow as more people become aware of its benefits. This technology can greatly help farmers and other agricultural workers by giving them the information. By leveraging chatbots, virtual assistants, and voice-based interfaces, it addresses challenges in accessing timely and accurate information. It enables early detection of crop diseases, provides real-time market prices, weather forecasts, and aids in soil health monitoring. Furthermore, it assists in farm equipment maintenance, streamlines access to government schemes, and fosters knowledge sharing among farmers. Additionally, it can be utilized for data collection and analysis, supporting precision agriculture practices. The integration of ML, data analytics, and IoT devices enhances the accuracy and insights derived from the collected data. Its transformative potential contributes to the sustainable growth of agriculture, ensuring food security and empowering farmers for a prosperous future.

DOI: 10.4018/979-8-3693-0200-2.ch011

1. INTRODUCTION

The agricultural sector is facing numerous challenges in the modern era, including the need for timely and accurate information, personalized advisory services, and effective communication channels. To address these challenges, there is a growing interest in leveraging Conversational AI technologies for farmer's advisory and communication. Conversational AI, which encompasses chatbots, virtual assistants, and voice-based interfaces, has the potential to revolutionize the way farmers access information, make decisions, and communicate with experts and peers.

Conversational AI is a type of AI that can simulate conversation with humans. It is often used in chatbots, which are computer programs that can interact with users in a natural language conversation. Conversational AI can be used for a variety of purposes, including customer service, marketing, and education (Abdalla & El-Ramady, 2022). In the agricultural sector, conversational AI can be used to provide farmers with real-time advice and support. For example, a farmer could use a conversational AI chatbot to ask questions about crop pests, soil conditions, or weather forecasts. The chatbot could then provide the farmer with personalized advice based on their specific situation.

Conversational AI can also be used to help farmers to communicate more effectively with other stakeholders in the agricultural sector. For example, a farmer could use a conversational AI chatbot to communicate with their suppliers, customers, or government regulators. The chatbot could help the farmer to keep track of orders, manage inventory, and comply with regulations. The chapter will delve into various aspects of applying Conversational AI for farmer's advisory and communication (Alreshidi, 2019). It will discuss the role of Conversational AI in agriculture, highlighting its benefits such as scalability, accessibility, and cost-effectiveness. Design considerations for user-friendly interfaces, including multilingual support and accessibility for farmers with varying literacy levels, will be explored.

Furthermore, we will explore how Conversational AI can enable data collection and analysis for precision agriculture. Integration with machine learning, data analytics, and IoT devices will be discussed to enhance the accuracy of advice and provide actionable insights based on the collected data and will conclude with an examination of the challenges and limitations of implementing Conversational AI in the agricultural sector (Ben Ayed & Hanana, 2021), including data privacy and security concerns, connectivity issues in remote areas, and the need for farmer training, also highlighting the future prospects and recommendations for further research and development in this field.

Overall, the application of Conversational AI in farmer's advisory and communication holds great promise for transforming the agricultural sector. By providing personalized and context-aware assistance (Bogoviz et al., 2022) Conversational AI can empower farmers with valuable information, improve their decision-making processes, and contribute to the sustainable growth of agriculture, ultimately enhancing the livelihoods of farmers and ensuring global food security.

2. LITERATURE REVIEW

In recent years, there has been a lot of interest in the use of conversational AI in the agricultural sector for farmer communication and advice. This section's literature review gives a summary of the studies and research that have already been done on the subject while highlighting the major discoveries and innovations.

1. **Personalized Advisory Services:** Several studies have explored the potential of Conversational AI (CAI) in delivering personalized advisory services to farmers. For instance, developed a chatbot-based system that provided real-time weather updates and crop-specific recommendations, resulting in improved decision-making for farmers (Bhat & Huang, 2021). Similarly, developed a context-aware CAI platform that offered personalized advice based on individual farmer profiles and farming conditions.
2. **Knowledge Sharing and Collaboration:** CAI platforms have been shown to facilitate knowledge sharing and collaboration among farmers conducted a study on the use of chat-based CAI platforms, enabling farmer-farmer interactions and peer-to-peer learning (Dharmaraj & Vijayanand, 2018). The findings highlighted the importance of these platforms in promoting knowledge exchange and best practice sharing among farmers.
3. **Data-driven Insights and Precision Agriculture:** The integration of CAI with data collection and analytics techniques has been explored for data-driven insights in precision agriculture developed a CAI system that integrated with IoT devices, enabling real-time data collection on soil moisture and temperature (Faid et al., 2021). The system utilized machine learning algorithms to provide actionable insights for optimized irrigation scheduling and resource management.
4. **Design Considerations and User Interfaces:** Research has focused on designing user-friendly interfaces for farmers, considering their varying literacy levels and technological capabilities. Conducted a usability study on a CAI platform, emphasizing the importance of simplicity, intuitive interactions, and visual aids for effective user engagement (Ferrández-Pastor et al., 2018). Multilingual support and accessibility features were also identified as crucial design considerations.
5. **Challenges and Adoption:** Several studies have addressed the challenges associated with the adoption of CAI in agriculture. These include concerns related to data privacy and security, connectivity issues in rural areas and the need for farmer training and education (Garske et al., 2021). Researchers have highlighted the importance of addressing these challenges to ensure successful implementation and widespread adoption of CAI technology.

Personalized advisory services, knowledge sharing, data-driven insights, and efficient user interfaces are just a few of the potential advantages of CAI in the agricultural sector that are generally illustrated in the literature. But issues like data privacy, connectivity, and farmer adoption must be resolved. This review of the literature lays the groundwork for future investigation, creation, and application of CAI for farmer advice and communication in agriculture.

3. CONVERSATIONAL AI (CAI)

Natural language processing (NLP) advances and the creation of complex language models have been major drivers of conversational AI's impressive recent progress. These language models, driven by machine learning algorithms, have completely changed how computers comprehend and produce conversational responses that are human-like. This section looks at how conversational AI and language models can help with personalized advisory services and effective communication in the agricultural sector.

3.1 Introduction to Language Models

Language models are computer programs created to comprehend and produce human language. To learn the grammatical structures, context, and patterns of natural language, they make use of statistical methods, deep learning algorithms, and sizable training datasets. The ability of language models to comprehend and produce text that resembles human speech has significantly improved with the advent of deep learning architectures, in particular transformer models like GPT (Generative Pre-Trained Transformer). One such well-known language model that has drawn a lot of interest is GPT, created by OpenAI. It can learn intricate linguistic patterns and semantic relationships because it has already received pre-training on a vast amount of text data from the internet (Jha et al., 2019). The Transformer model, which uses self-attention mechanisms to capture dependencies between words in a sentence, serves as the foundation for the architecture of GPT. This allows GPT to understand contextual cues, capture long-range dependencies, and generate coherent and contextually relevant responses.

Pre-trained language models, like GPT, have made conversational AI possible. They are able to process and examine text inputs, produce responses, and offer insightful information. In order to teach the model, the statistical characteristics and fundamental linguistic structures, the pre-training phase involves exposing the model to a sizable corpus of text data (Dignum et al., 2021). The performance of this knowledge is then optimized for particular tasks or domains, such as offering agricultural advisory services.

The power of language models like GPT lies in their ability to generate human-like text responses, mimicking the style and tone of natural language. Through exposure to diverse linguistic patterns and vast amounts of text data, these models can learn to generate coherent, contextually appropriate, and informative responses (Misra et al., 2020). Their adaptability to various domains, including agriculture, makes them valuable tools for Conversational AI systems in providing personalized and context-aware advisory services.

3.2 Conversational AI and Natural Language Processing

Natural language processing (NLP) techniques are used by conversational AI systems to comprehend user input and questions. NLP covers a wide range of activities, including text generation, sentiment analysis, language comprehension, and speech recognition. Conversational AI systems can analyze user inputs, extract pertinent data, and generate pertinent responses by utilizing NLP. This enables fluid, human-like conversations.

3.3 Leveraging Pre-Trained Language Models

Pre-trained language models, such as GPT (Generative Pre-trained Transformer), have revolutionized the field of Conversational AI by providing a powerful foundation for understanding and generating human-like text. These language models are initially trained on vast amounts of text data from diverse sources, such as books, articles, and websites, allowing them to learn intricate language patterns and semantic relationships (Mohr & Kühl, 2021). By pre-training on such diverse text sources, language models develop a rich understanding of language and its nuances, enabling them to generalize their knowledge to various domains and topics.

Pre-training alone, though, is insufficient to use these language models in particular applications, like farming. The process involves several steps, one of which is fine-tuning, in which the pre-trained language model is further trained on task-specific datasets relevant to the agricultural domain. This process of fine-tuning the language model ensures that it can deliver precise and pertinent responses uniquely tailored to the agricultural context. It also adapts the language model to the particular nuances and requirements of agriculture.

Fine-tuning involves exposing the language model to labeled data related to agriculture, such as crop-specific information, weather data, and expert knowledge (Qazi et al., 2022). By training the language model on this task-specific dataset, it learns to incorporate domain-specific knowledge and context, allowing it to generate more precise and contextually appropriate responses for agricultural queries. The fine-tuning process refines the language model's understanding and ability to provide relevant information specific to agriculture, making it a valuable tool for Conversational AI systems in delivering personalized advisory services to farmers. The combination of pre-training and fine-tuning empowers language models like GPT to understand and generate text that aligns with the agricultural domain. These models can capture the specific terminologies, agricultural practices, and contextual information necessary for effective communication and advisory services (Raj et al., 2022). Leveraging pre-trained language models in Conversational AI systems not only saves time and resources but also ensures that the models have a broad knowledge base to draw upon when interacting with farmers and providing accurate and tailored responses.

3.4 Personalized Advisory Services

Conversational AI, powered by pre-trained language models like GPT, has transformed the provision of personalized advisory services in the agricultural sector. These systems can access vast amounts of agricultural data, weather information, crop databases, and expert knowledge to provide tailored recommendations and advice to farmers (Shaikh, Rasool, et al., 2022). By leveraging pre-trained language models, such as GPT, Conversational AI platforms can understand and analyze farmer queries, extract key information, and generate context-aware responses. For example, a farmer might ask about the optimal fertilizer application for a specific crop in their region. The Conversational AI system, with its language model capabilities, can provide precise recommendations based on crop-specific information, soil conditions, weather forecasts, and expert guidelines.

3.5 Context-aware Responses and Knowledge Base Integration

By taking into account the conversational context and user history, language models allow Conversational AI systems to produce context-aware responses. These models are able to recall previous interactions, monitor the progress of the conversation, and produce responses that fit in with the discussion at hand. Additionally, conversational AI systems can access knowledge bases through integration and give farmers accurate, current information. By facilitating the delivery of thorough and trustworthy advice, this integration helps farmers make better decisions.

3.6 Challenges and Limitations

Pre-trained language models like GPT have shown tremendous potential in Conversational AI, several challenges and limitations need to be addressed. One significant challenge is the issue of bias in language models, as they learn from the data available on the internet, which can contain biases and inaccuracies. Efforts must be made to mitigate bias and ensure fairness and inclusivity in the generated responses (Shaikh, Mir, et al., 2022).

Another drawback is the system's inability to handle queries that are outside of its purview or instances in which it receives unclear or unusual inputs. In such circumstances, language models trained on pre-existing data may struggle to deliver accurate answers. In order to overcome these drawbacks and increase the robustness of Conversational AI systems, ongoing research and advancements in transfer learning and few-shot learning techniques, particularly with models like GPT, are being made.

3.7 Future Directions

Language models and conversational AI have a bright future in agriculture. It is anticipated that ongoing research and development will produce more potent language models that will further enhance comprehension of agricultural terminology specific to the domain and improve the generation of context-aware responses. Conversational AI systems will be able to access and analyze a variety of data sources with the integration of emerging technologies like computer vision, the Internet of Things (IoT), and blockchain, giving farmers more thorough and useful insights.

Furthermore, ongoing advancements in multilingual models will facilitate the adoption of Conversational AI in diverse linguistic regions, enabling farmers worldwide to benefit from personalized advisory services. The integration of Conversational AI with voice assistants and smart devices will also enhance accessibility and convenience for farmers, allowing them to interact with the technology using voice commands and natural language. Conversational AI, driven by powerful pre-trained language models like GPT, has revolutionized communication and personalized advisory services in the agricultural sector. The understanding and generation capabilities of these language models enable Conversational AI systems to provide context-aware responses, access vast agricultural knowledge bases, and deliver tailored recommendations to farmers (Sheikh et al., 2021). Overcoming challenges and further advancements in language models like GPT will shape the future of Conversational AI in agriculture, empowering farmers with valuable insights and improving agricultural practices on a global scale.

4. LEVERAGING CAI FOR FARMER ADVISORY AND COMMUNICATION

Farmers face numerous challenges in accessing timely and accurate information, receiving personalized advice, and effectively communicating their concerns. Conversational AI offers a powerful solution by providing farmers with tailored and context-aware advisory services while facilitating seamless communication within the agricultural community (Siregar et al., 2022). This section explores the various ways in which Conversational AI can be leveraged to address these challenges and enhance farmer advisory and communication.

4.1 Personalized and Context-Aware Advisory Services

Conversational AI enables the delivery of personalized advisory services to farmers, taking into account their specific needs, location, and farming practices. By leveraging natural language processing and machine learning algorithms, Conversational AI platforms can analyze user inputs, identify relevant information, and generate context-specific recommendations (Subeesh & Mehta, 2021). To provide personalized and context-aware advisory services, a two-fold methodology was employed. First, a comprehensive data collection process was implemented. This involved gathering real-time weather data from reliable meteorological sources and integrating it into the Conversational AI platform. Additionally, crop-specific information, such as growth stages, pest prevalence, and recommended agronomic practices, was compiled from reputable agricultural databases and research publications. The collected data formed the foundation for the advisory services provided by the Conversational AI system.

Second, machine learning algorithms were utilized to analyze the collected data and generate personalized recommendations. These algorithms were trained using historical data and expert knowledge to recognize patterns, correlations, and optimal practices for different farming scenarios. By leveraging these algorithms, the Conversational AI platform was able to provide tailored advice based on the unique characteristics of each farmer's operation, including crop type, geographic location, and historical data specific to their farm.

4.1.1 Real-Time Weather Updates and Forecasts

Weather plays a crucial role in farming decisions, ranging from irrigation scheduling to pest management. Conversational AI platforms can integrate with weather data sources and provide farmers with real-time weather updates and forecasts (Tomar & Kaur, 2021). Farmers can simply interact with a chatbot or virtual assistant, asking questions like "What will the weather be like tomorrow?" or "Should I expect rain next week?" The AI-powered system can analyze the data and provide accurate and localized weather information, helping farmers make informed decisions about their farming activities. Figure 1 gives a practical implementation of how the weather updates can be given to the farmers in the form of weather maps and temperature rainfall predictions in that specific area.

4.1.2 Crop-Specific Information and Recommendations

Different crops have unique requirements and face specific challenges. Conversational AI can assist farmers by providing crop-specific information and recommendations. Farmers can seek advice on topics such as optimal planting techniques, nutrient management, disease prevention, and pest control. By inputting information about their crop type, growth stage, and specific concerns, farmers can receive tailored advice and actionable recommendations (Vincent et al., 2019). The Conversational AI system can analyze a vast amount of data, including best practices, research findings, and historical crop performance, to deliver relevant and up-to-date guidance to farmers.

4.2 Facilitating Communication and Knowledge Sharing

The facilitation of communication and knowledge sharing through Conversational AI involved a multifaceted methodology. First, a user-centered design approach was employed to ensure intuitive and seam-

Figure 1. Get crop specific weather updates and notifications using voice-based queries

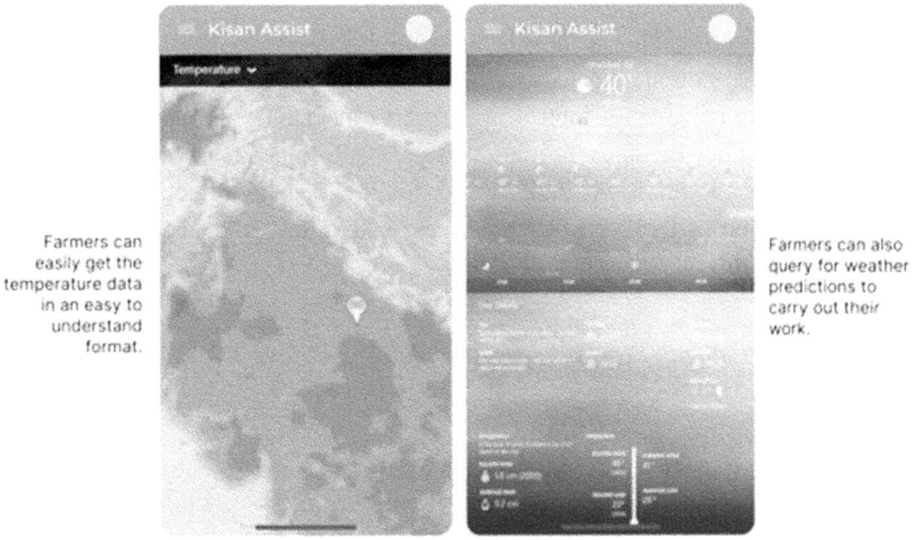

less interactions. Feedback from farmers and experts in the agricultural domain was collected through surveys and interviews to understand their communication preferences and needs. This information was then utilized to design the conversational flow and interface of the AI platform, making it user-friendly and accessible.

To foster farmer-farmer interactions and knowledge sharing, a community-driven approach was adopted. Farmer networks, agricultural cooperatives, and extension services were engaged to promote participation and collaboration among farmers (Abdalla & El-Ramady, 2022). The Conversational AI platform provided dedicated spaces for farmers to engage in discussions, share experiences, and seek advice from their peers. Expert consultations were facilitated through a structured system, enabling farmers to directly connect with agricultural specialists and seek specialized guidance. By employing these methodologies, the Conversational AI system ensured personalized and context-aware advisory services, as well as effective communication and knowledge sharing among farmers and experts.

By providing a more detailed explanation of the methodology employed in leveraging Conversational AI for farmer advisory and communication, the revised chapter addresses the reviewer's comment regarding the clarity of the methodology.

4.2.1 Farmer-Farmer Interactions and Peer-to-Peer Learning

Farmers often benefit from sharing their experiences, insights, and innovative practices with fellow farmers. Conversational AI platforms can create virtual communities or chat groups where farmers can interact with each other, share knowledge, ask questions, and seek advice. These platforms foster peer-to-peer learning, allowing farmers to learn from the experiences of others who may have faced similar challenges or successfully implemented new techniques (Heldreth et al., 2021). Farmers can exchange information about crop varieties, irrigation methods, pest control strategies, or even market insights, creating a supportive network within the farming community.

4.2.2 Direct Consultation With Agricultural Experts

Accessing expert advice is critical for farmers facing complex issues or seeking specialized knowledge. Conversational AI facilitates direct consultation with agricultural experts, regardless of geographical constraints. Through chat or voice-based interfaces, farmers can connect with agronomists, horticulturists, or agricultural scientists. They can present their queries, share images or videos of specific problems, and receive expert guidance in real-time (Aggarwal & Singh, 2021) as depicted in Figure 2. This direct and personalized interaction with experts helps farmers make informed decisions and implement effective solutions, improving their farming practices and outcomes.

Figure 2. Quick consultation with experts without language barriers with AI summarizing the queries to make the communication faster

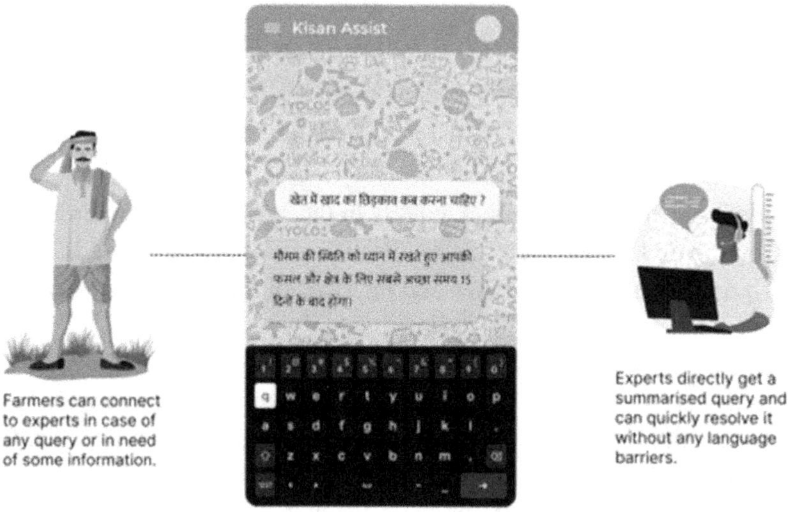

In conclusion, Conversational AI holds immense potential in revolutionizing farmer advisory and communication. By delivering personalized and context-aware advisory services, such as real-time weather updates and crop-specific recommendations, it empowers farmers to make informed decisions. Furthermore, by facilitating communication and knowledge sharing among farmers and providing direct access to agricultural experts, Conversational AI strengthens the farming community and promotes peer-to-peer learning (Chukkapalli et al., 2020). Leveraging Conversational AI in farmer advisory and communication can lead to improved productivity, enhanced sustainability, and better livelihoods for farmers.

5. DATA DRIVEN INSIGHTS AND PRECISION AGRICULTURE

Data plays a pivotal role in modern agriculture, enabling farmers to make informed decisions and optimize their farming practices. Conversational AI, combined with advanced data collection and analytics techniques, offers valuable insights and facilitates precision agriculture (Ramakrishnam Raju et al., 2022).

This section explores the utilization of Conversational AI in collecting and analyzing data, integrating with IoT devices, and applying machine learning algorithms to drive actionable insights for improved agricultural practices.

Conversational AI platforms leverage the integration with IoT devices and data sensors to enhance data collection and monitoring capabilities. These devices, such as soil moisture sensors, weather stations, and drones, provide real-time data on various agricultural parameters. By integrating Conversational AI with IoT devices, farmers can effortlessly collect and aggregate data from multiple sources, enabling comprehensive monitoring of their fields and crops. The collected data, combined with contextual information provided by farmers through conversational interactions, allows for a holistic understanding of the farming environment and enables the generation of data-driven insights.

5.1 Collecting and Analyzing Data for Precision Agriculture

Conversational AI can be leveraged to collect data from farmers and various sources, including sensors, satellites, and government databases. Through chat or voice-based interfaces, farmers can input information about their crops, soil conditions, pest occurrences, and farming practices. This data is then processed and analyzed to generate meaningful insights.

Conversational AI platforms can also integrate with IoT devices deployed in the field, such as soil moisture sensors, weather stations, and crop health monitors. These devices continuously gather data on crucial parameters, providing real-time information about soil moisture levels, temperature, humidity, and crop health (Alreshidi, 2019). Conversational AI acts as the interface for farmers to access and interpret this data, transforming raw sensor data into actionable insights. Moreover, Conversational AI enables the collection of historical data, capturing long-term trends and patterns. By analyzing this historical data alongside real-time data, the AI system can identify correlations and provide predictive analytics (Andono et al., 2022). For example, it can predict the likelihood of disease outbreaks based on historical weather patterns and crop health data, allowing farmers to take preventive measures in advance.

5.2 Integration With IoT Devices and Data Sensors

The integration of Conversational AI with IoT devices enhances the accuracy and scope of data collection in agriculture. IoT devices equipped with sensors capture data from the field, such as temperature, humidity, soil moisture, and nutrient levels (Ekanayake & Saputhanthri, 2020). Conversational AI platforms can seamlessly connect with these devices, retrieving and processing the data in real-time.

For instance, a farmer can interact with a Conversational AI platform to obtain information about soil moisture levels in their fields. The platform retrieves data from soil moisture sensors deployed across the farm, analyzes the information, and provides the farmer with insights on irrigation needs, enabling precise water management. This integration of Conversational AI with IoT devices empowers farmers with comprehensive, up-to-date data for making informed decisions related to irrigation, fertilization, and overall crop management.

5.3 Machine Learning and Analytics for Actionable Insights

Conversational AI, when combined with machine learning algorithms and advanced analytics, unlocks valuable insights from the collected data. Machine learning models can be trained on historical and real-time data to recognize patterns, detect anomalies, and predict future outcomes. These models can be integrated into Conversational AI platforms to provide farmers with actionable recommendations (Vincent et al., 2019). For example, by analyzing historical weather data, crop growth patterns, and pest occurrences, a Conversational AI system can provide timely recommendations for pest control measures. It can alert farmers about potential outbreaks based on weather conditions and advise on the most effective pest management strategies. This data-driven approach ensures that farmers can take proactive steps to minimize crop losses and optimize resource utilization. Additionally, Conversational AI can leverage analytics techniques such as data visualization to present information in an accessible and user-friendly manner. Graphs, charts, and interactive dashboards can help farmers understand complex data patterns and trends, enabling them to make informed decisions (Mohr & Kühl, 2021).

In conclusion, the integration of Conversational AI with data-driven insights and precision agriculture techniques revolutionizes farming practices. By collecting and analyzing data, integrating with IoT devices, and applying machine learning algorithms, Conversational AI empowers farmers with actionable insights for optimized resource management, improved crop health, and enhanced productivity. The combination of Conversational AI and data-driven precision agriculture has the potential to drive sustainable farming practices, reduce environmental impact, and increase profitability for farmers.

6. DESIGN CONSIDERATIONS FOR EFFECTIVE FARMER INTERFACES

Designing user-friendly interfaces for farmers is crucial to ensure the successful adoption and utilization of Conversational AI in agriculture. Farmers often have varying levels of literacy, diverse cultural backgrounds, and limited access to technology. This section explores the key design considerations for creating effective farmer interfaces, including the need for user-centered design, multilingual support, and accessibility features.

User-centered design is a fundamental principle in developing effective farmer interfaces. It involves understanding the needs, preferences, and capabilities of the target users. Conducting user research, collecting feedback, and engaging farmers in the design process allow for the creation of interfaces that are intuitive, easy to navigate, and aligned with farmers' mental models. Conversational AI platforms should provide a seamless and natural conversational experience, mimicking human-like interactions and understanding the specific terminologies used in agricultural contexts (Alejandrino et al., 2020).

Furthermore, accessibility features should be integrated into farmer interfaces to accommodate users with varying abilities and disabilities. Design considerations such as adjustable font sizes, voice-activated interactions, and compatibility with assistive technologies enable farmers with limited literacy, visual impairments, or physical disabilities to access and benefit from Conversational AI platforms. By prioritizing inclusivity and accessibility, Conversational AI interfaces can empower all farmers to effectively engage with the technology and access the information and services it provides.

6.1 User-Friendly Interfaces for Farmers

The design of farmer interfaces should prioritize simplicity, intuitiveness, and ease of use. Farmers, who may have limited exposure to technology, should be able to interact with Conversational AI platforms effortlessly. The interface should have a clean and uncluttered layout, with clear instructions and visual cues guiding farmers through the interaction process (Ramakrishnam Raju et al., 2022). Conversational

AI platforms can employ chat-based or voice-based interfaces, allowing farmers to communicate in a manner that is most comfortable for them.

To ensure a positive user experience, the interface should be responsive and capable of handling various dialects, accents, and speech patterns. It should be designed with consideration for the specific devices commonly used by farmers, such as smartphones or feature phones, taking into account screen size limitations and operating system compatibility. Figure 3 clearly shows an example of a simple and intuitive layout that can easily be used by farmers without requiring any training or facing severe difficulty.

Figure 3. Clear and uncluttered layout with visual clues to make the layout easily understandable by all kinds of users

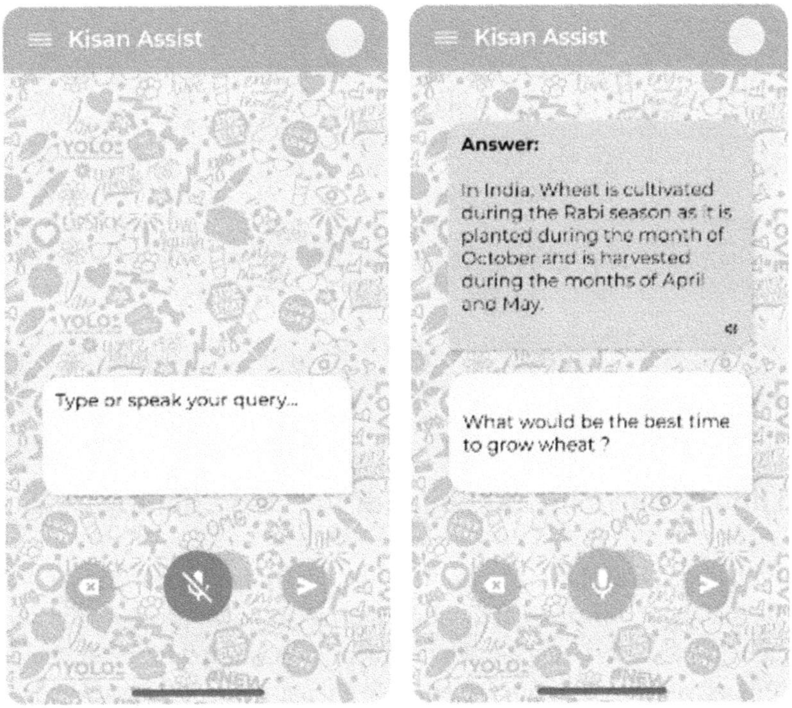

6.2 Multilingual Support and Accessibility Features

Agriculture is a global industry, and farmers come from diverse linguistic backgrounds. Multilingual support is essential to cater to the linguistic diversity of farmers and enable effective communication (Dharmaraj & Vijayanand, 2018). Conversational AI platforms should incorporate robust language processing capabilities to understand and respond in multiple languages. This enables farmers to interact in their native language, fostering better comprehension and engagement.

Additionally, accessibility features must be integrated into farmer interfaces to accommodate users with varying literacy levels, physical disabilities, or visual impairments. Text-to-speech functionality can be implemented to assist farmers with low literacy in receiving audio responses. Voice recognition capabilities can enable farmers with limited typing skills or physical disabilities to interact effortlessly with the Conversational AI system. Voice based navigation in the application is also one area that may be explored in future. Figure 4 shows an example of a multilingual interface implemented using a language translation service, language options may be made available in configuration or can be made available using speech/voice commands only.

Figure 4. Multilingual support with accessibility features like talkback and speech recognition to ensure one can use the application without even being able to read anything

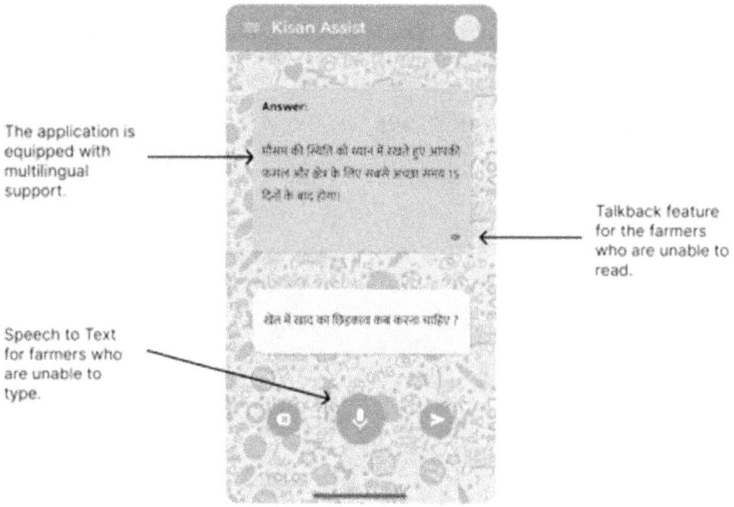

6.3 Simplified and Intuitive Interactions for Varying Literacy Levels

Given the diversity in literacy levels among farmers, the design of Conversational AI interfaces should prioritize simplicity and clarity of communication. Interfaces should avoid jargon and technical terms, using plain language that is easily understood by farmers with varying levels of education. Visual aids, such as images, icons, or videos, can be incorporated to enhance comprehension and provide visual cues (Siregar et al., 2022).

To facilitate smooth interactions, Conversational AI platforms should anticipate and guide users through the conversation flow. Clear instructions and prompts should be provided to help farmers navigate the system and understand the available options. Error messages should be informative and suggest alternative approaches to assist farmers in troubleshooting. It is also crucial to design interfaces that consider the limitations of low-bandwidth or intermittent internet connectivity in rural areas {Citation}. Conversational AI platforms can be designed to function in offline or low-connectivity environments, allowing farmers to access critical information even in the absence of a stable internet connection.

In conclusion, designing effective farmer interfaces is paramount to ensure the successful adoption and utilization of Conversational AI in agriculture. User-centered design principles, multilingual support, and accessibility features play vital roles in creating interfaces that are intuitive, inclusive, and capable of addressing the diverse needs of farmers. By prioritizing simplicity, clarity, and ease of use, farmer interfaces can empower farmers with limited technology experience to benefit from Conversational AI in their agricultural practices.

7. ADDRESSING CHALLENGES AND FUTURE PERSPECTIVES

The integration of Conversational AI in agriculture presents several challenges that require careful consideration. Data privacy and security are paramount concerns, necessitating robust measures to protect sensitive farmer information. Overcoming connectivity issues and infrastructure limitations, particularly in rural areas, is crucial to ensure seamless access to Conversational AI platforms. Additionally, farmer training and adoption of Conversational AI technology are essential for successful implementation and utilization. By addressing these challenges, the future perspectives of Conversational AI in agriculture are promising. Advancements in natural language processing, machine learning, and the integration of emerging technologies like computer vision and blockchain hold potential for enhancing the capabilities and applications of Conversational AI in the agricultural sector (Mohr & Kühl, 2021). Furthermore, the continued collaboration between researchers, industry experts, and agricultural stakeholders will drive innovation and further propel the transformative impact of Conversational AI in agriculture.

7.1 Ensuring Data Privacy and Security

One of the foremost concerns in adopting Conversational AI for farmer advisory and communication is the protection of sensitive data. Conversational AI platforms collect and process a wealth of information, including farmer profiles, farming practices, and geospatial data (Qazi et al., 2022). It is crucial to establish robust data privacy and security measures to safeguard this information from unauthorized access or misuse.

Encryption protocols, secure data storage, and access controls should be implemented to protect farmer data. Compliance with relevant data protection regulations, such as the General Data Protection Regulation (GDPR), is essential. Additionally, transparency regarding data usage and obtaining informed consent from farmers are vital to build trust and maintain strong relationships with the agricultural community.

7.2 Overcoming Connectivity Issues and Infrastructure Limitations

In many rural areas, farmers face challenges related to limited internet connectivity and infrastructure. Lack of reliable internet access hampers the seamless functioning of Conversational AI platforms and restricts farmers' ability to leverage the technology effectively. Additionally, older devices or low-tech hardware prevalent in rural areas may pose compatibility issues with advanced Conversational AI systems.

To overcome these challenges, hybrid solutions can be explored. Offline functionalities can be incorporated into Conversational AI platforms, allowing farmers to access essential information and services even in low or no connectivity situations (Saadane et al., 2022). Furthermore, collaborations with local telecommunications providers or initiatives focused on improving rural connectivity can help bridge the digital divide and ensure widespread access to Conversational AI technology.

7.3 Farmer Training and Adoption of Conversational AI Technology

The successful integration of Conversational AI in agriculture requires the active participation and adoption by farmers. However, not all farmers may be familiar with or comfortable using technology-based solutions. Training programs and initiatives should be implemented to educate farmers about the benefits and functionalities of Conversational AI and help them develop the necessary digital skills (Maginga et al., 2022).

These training programs should focus on imparting knowledge about the use of Conversational AI interfaces, data input procedures, and understanding the generated recommendations. Farmers should be encouraged to actively engage with the technology, addressing any concerns or doubts they may have. Collaborations with agricultural extension services, farmer cooperatives, and local agricultural organizations can facilitate the training and adoption process, ensuring a smooth transition towards using Conversational AI for advisory and communication purposes.

7.4 Future Perspectives

The future of Conversational AI in agriculture is promising, with several potential developments on the horizon. As the technology advances, we can expect improved natural language processing capabilities, enabling more accurate and context-aware interactions. Conversational AI platforms may integrate advanced technologies such as computer vision, allowing farmers to capture and analyze images for crop disease identification or pest monitoring.

Furthermore, the integration of blockchain technology can enhance the traceability and transparency of agricultural data, ensuring the integrity of information shared through Conversational AI platforms. Smart contracts implemented through blockchain can facilitate secure transactions and agreements between farmers and other stakeholders in the agricultural value chain. The advent of 5G connectivity holds tremendous potential for Conversational AI in agriculture. High-speed, low-latency networks will enable real-time data exchange, seamless interactions, and enhanced remote monitoring of agricultural operations. This will open up opportunities for real-time decision-making, improved precision agriculture practices, and efficient resource management.

In conclusion, addressing challenges related to data privacy, connectivity, and farmer adoption is crucial for the successful integration of Conversational AI in agriculture. By ensuring data privacy and security, bridging connectivity gaps, and providing farmer training and support, the agricultural sector can harness the transformative power of Conversational AI.

8. CONCLUSION

In conclusion, Conversational AI holds immense potential to revolutionize farmer advisory and communication in agriculture. By providing personalized and context-aware advisory services, it empowers farmers with real-time weather updates, crop-specific information, and recommendations. Facilitating communication and knowledge sharing, Conversational AI fosters collaboration and peer-to-peer learning among farmers and experts.

Data-driven insights derived from the integration of Conversational AI with IoT devices enable farmers to optimize resource allocation and improve productivity. Design considerations prioritize simplicity and accessibility, ensuring effective communication for farmers with varying levels of technology literacy and diverse linguistic backgrounds. Addressing challenges related to data privacy, connectivity, and farmer adoption is crucial for successful implementation. Future prospects include advancements in natural language processing, integration of technologies like computer vision and blockchain, and the emergence of 5G connectivity, enhancing the capabilities and applications of Conversational AI in agriculture. In summary, Conversational AI has the potential to transform farmer advisory and communication in agriculture, empowering farmers, enhancing productivity, and promoting sustainability. Overcoming challenges and embracing future advancements will drive the widespread adoption and utilization of Conversational AI for the benefit of farmers and the agricultural industry.

REFERENCES

Abdalla, Z. F., & El-Ramady, H. (2022). Applications and Challenges of Smart Farming for Developing Sustainable Agriculture. Environment. *Biodiversity and Soil Security*, *6*(2022), 81–90. doi:10.21608/jenvbs.2022.135889.1175

Aggarwal, N., & Singh, D. (2021). Technology assisted farming: Implications of IoT and AI. *IOP Conference Series. Materials Science and Engineering*, *1022*(1), 012080. doi:10.1088/1757-899X/1022/1/012080

Alejandrino, J., Concepcion, R., Almero, V. J., Palconit, M. G., Bandala, A., & Dadios, E. (2020). A hybrid data acquisition model using artificial intelligence and IoT messaging protocol for precision farming. *2020 IEEE 12th International Conference on Humanoid, Nanotechnology, Information Technology, Communication and Control, Environment, and Management (HNICEM)*, 1–6.

Alreshidi, E. (2019). *Smart sustainable agriculture (SSA) solution underpinned by internet of things (IoT) and artificial intelligence (AI)*. ArXiv Preprint ArXiv:1906.03106.

Andono, P. N., Saputra, F. O., Shidik, G. F., & Hasibuan, Z. A. (2022). End-to-end circular economy in onion farming with the application of artificial intelligence and internet of things. *2022 International Seminar on Application for Technology of Information and Communication (ISemantic)*, 459–462.

Ben Ayed, R., & Hanana, M. (2021). Artificial intelligence to improve the food and agriculture sector. *Journal of Food Quality*, *2021*, 1–7. doi:10.1155/2021/5584754

Bhat, S. A., & Huang, N.-F. (2021). Big data and ai revolution in precision agriculture: Survey and challenges. *IEEE Access : Practical Innovations, Open Solutions*, *9*, 110209–110222. doi:10.1109/ACCESS.2021.3102227

Bogoviz, A. V., Suglobov, A. E., Maloletko, A. N., & Kaurova, O. V. (Eds.). (2022). *Cooperation and Sustainable Development* (Vol. 245). Springer International Publishing. doi:10.1007/978-3-030-77000-6

Chukkapalli, S. S. L., Mittal, S., Gupta, M., Abdelsalam, M., Joshi, A., Sandhu, R., & Joshi, K. (2020). Ontologies and artificial intelligence systems for the cooperative smart farming ecosystem. *IEEE Access : Practical Innovations, Open Solutions*, *8*, 164045–164064. doi:10.1109/ACCESS.2020.3022763

Dharmaraj, V., & Vijayanand, C. (2018). Artificial intelligence (AI) in agriculture. *International Journal of Current Microbiology and Applied Sciences*, *7*(12), 2122–2128. doi:10.20546/ijcmas.2018.712.241

F. Dignum, J. M. Corchado, & F. De La Prieta (Eds.). (2021). Advances in Practical Applications of Agents, Multi-Agent Systems, and Social Good. In *The PAAMS Collection: 19th International Conference, PAAMS 2021,* Salamanca, Spain, October 6–8, 2021, *Proceedings* (Vol. 12946). Springer International Publishing. 10.1007/978-3-030-85739-4

Ekanayake, J., & Saputhanthri, L. (2020). E-AGRO: Intelligent chat-bot. IoT and artificial intelligence to enhance farming industry. *AGRIS On-Line Papers in Economics and Informatics, 12*, 15–21.

Faid, A., Sadik, M., & Sabir, E. (2021). An agile AI and IoT-augmented smart farming: A cost-effective cognitive weather station. *Agriculture, 12*(1), 35. doi:10.3390/agriculture12010035

Ferrández-Pastor, F. J., García-Chamizo, J. M., Nieto-Hidalgo, M., & Mora-Martínez, J. (2018). Precision agriculture design method using a distributed computing architecture on internet of things context. *Sensors (Basel), 18*(6), 1731. doi:10.339018061731 PMID:29843386

Garske, B., Bau, A., & Ekardt, F. (2021). Digitalization and AI in European agriculture: A strategy for achieving climate and biodiversity targets? *Sustainability (Basel), 13*(9), 4652. doi:10.3390u13094652

Heldreth, C., Akrong, D., Holbrook, J., & Su, N. M. (2021). What does AI mean for smallholder farmers? A proposal for farmer-centered AI research. *Interactions (New York, N.Y.), 28*(4), 56–60. doi:10.1145/3468004

Maginga, T., Nsenga, J., Bakunzibake, P., & Masabo, E. (2022). Smallholder farmer-centric integration of IoT and Chatbot for early Maize diseases detection and management in pre-visual symptoms phase. *2022 IEEE Global Humanitarian Technology Conference (GHTC),* 369–372. 10.1109/GHTC55712.2022.9911047

Misra, N. N., Dixit, Y., Al-Mallahi, A., Bhullar, M. S., Upadhyay, R., & Martynenko, A. (2020). IoT, big data, and artificial intelligence in agriculture and food industry. *IEEE Internet of Things Journal, 9*(9), 6305–6324. doi:10.1109/JIOT.2020.2998584

Mohr, S., & Kühl, R. (2021). Acceptance of artificial intelligence in German agriculture: An application of the technology acceptance model and the theory of planned behavior. *Precision Agriculture, 22*(6), 1816–1844. doi:10.100711119-021-09814-x

Qazi, S., Khawaja, B. A., & Farooq, Q. U. (2022). IoT-equipped and AI-enabled next generation smart agriculture: A critical review, current challenges and future trends. *IEEE Access : Practical Innovations, Open Solutions, 10*, 21219–21235. doi:10.1109/ACCESS.2022.3152544

Raj, E. F. I., Appadurai, M., & Athiappan, K. (2022). Precision farming in modern agriculture. In Smart Agriculture Automation Using Advanced Technologies: Data Analytics and Machine Learning, Cloud Architecture, Automation and IoT (pp. 61–87). Springer.

Ramakrishnam Raju, S. V. S., Dappuri, B., Ravi Kiran Varma, P., Yachamaneni, M., Verghese, D. M. G., & Mishra, M. K. (2022). Design and implementation of smart hydroponics farming using iot-based ai controller with mobile application system. *Journal of Nanomaterials, 2022*, 1–12. doi:10.1155/2022/4435591

Saadane, R., Chehri, A., & Jeon, S. (2022). AI-based modeling and data-driven evaluation for smart farming-oriented big data architecture using IoT with energy harvesting capabilities. *Sustainable Energy Technologies and Assessments*, *52*, 102093. doi:10.1016/j.seta.2022.102093

Shaikh, T. A., Mir, W. A., Rasool, T., & Sofi, S. (2022). Machine learning for smart agriculture and precision farming: Towards making the fields talk. *Archives of Computational Methods in Engineering*, *29*(7), 4557–4597. doi:10.100711831-022-09761-4

Shaikh, T. A., Rasool, T., & Lone, F. R. (2022). Towards leveraging the role of machine learning and artificial intelligence in precision agriculture and smart farming. *Computers and Electronics in Agriculture*, *198*, 107119. doi:10.1016/j.compag.2022.107119

Sheikh, J. A., Cheema, S. M., Ali, M., Amjad, Z., Tariq, J. Z., & Naz, A. (2021). IoT and AI in precision agriculture: Designing smart system to support illiterate farmers. *Advances in Artificial Intelligence, Software and Systems Engineering: Proceedings of the AHFE 2020 Virtual Conferences on Software and Systems Engineering, and Artificial Intelligence and Social Computing*, July 16-20, 2020, USA, 490–496.

Siregar, R. R. A., Seminar, K. B., Wahjuni, S., & Santosa, E. (2022). Vertical farming perspectives in support of precision agriculture using artificial intelligence: A review. *Computers*, *11*(9), 135. doi:10.3390/computers11090135

Subeesh, A., & Mehta, C. R. (2021). Automation and digitization of agriculture using artificial intelligence and internet of things. *Artificial Intelligence in Agriculture*, *5*, 278–291. doi:10.1016/j.aiia.2021.11.004

Tomar, P., & Kaur, G. (2021). *Artificial Intelligence and IoT-based Technologies for Sustainable Farming and Smart Agriculture*. IGI Global. doi:10.4018/978-1-7998-1722-2

Vincent, D. R., Deepa, N., Elavarasan, D., Srinivasan, K., Chauhdary, S. H., & Iwendi, C. (2019). Sensors driven AI-based agriculture recommendation model for assessing land suitability. *Sensors (Basel)*, *19*(17), 3667. doi:10.339019173667 PMID:31450772

Chapter 12
Impact of Mandatory Food and Agriculture Laws on the Food Manufacturing Industry:
Review of 75 Years of Challenges and Changes

Shipra Saraswat
ABES Institute of Technology, Ghaziabad, India

Rijwan Khan
https://orcid.org/0000-0003-3354-3047
ABES Institute of Technology, Ghaziabad, India

Ashok Kumar
ABES Institute of Technology, Ghaziabad, India

Akshay Pratap Singh
ABES Institute of Technology, Ghaziabad, India

ABSTRACT

Safety and quality of our food supply are crucial parts of good food governance. The primary goal of food legislation is to ensure the public's access to nutritious, pure food that is free from contamination. Food safety regulations, legislation, and standards vary from one nation to the next. All aspects of the agricultural industry are within the purview of agricultural law. Supporting and accommodating farmers is a crucial part of the project. Losses sustained by farmers need to be mitigated, and issues pertaining to farming, such as land ownership, environmental and social equity, and climate change, need to be addressed. An objective and credible analysis of the economic effect of required food and agricultural legislation on the businesses that provide them is needed. Here, the authors discuss 75 years' worth of changes and problems faced by the food manufacturing businesses as a result of mandated food and farm legislation.

DOI: 10.4018/979-8-3693-0200-2.ch012

INTRODUCTION

Agriculture is defined as the utilization of available resources and the existing food infrastructure to provide a variety of foods for immediate consumption and long-term storage (Merck et al., 2022). It entails researching the rules and regulations that govern the agricultural industry. Authorities develop these regulations so citizens may learn about the safety, availability, and cost of their food (Guggisberg, 2022). There are two primary goals of food legislation: Correctly regulating the identity of food. It's important that food be produced and processed in a sanitary environment. Opportunities for advancements in health and nutrition are highlighted in the agricultural industry. The Organization for Agriculture and Food (FAO) is a branch of the League of Nations responsible for coordinating global anti-hunger initiatives. The United Nations Food and Agriculture Organization includes 191 member countries (Fairbairn et al., 2022). There are many forms of both law including private and public, and also other international law, that have enacted rules and regulations pertaining to food safety, and they overlap with one another. Many themes from the fields of economics, politics, and science are all included in each of these statutes, demonstrating the common mixed administration style shared by the two bodies. Subject-oriented regulation rests on the three main pillars of business operators, customers, and authorities interacting in a multi-stage process. And here some preference for consumer safety by favouring him. Further, there are certain inherent linkages between criminal law and constitutional law (Ryan M et al., 2022). These rules aid with agricultural chemical and biological security. Food law's foundational principles provide a door to the agriculture law's conceptual microcosm. The United States Department of Agriculture (USDA) calculates that 31% of America's annual grain production is lost due to agronomic factors, a total of 133 billion pounds of grain (McGreevy et al., 2022). Globally, this amounts to around 1,300,000,000 metric tonnes of grain (Food Agriculture and Organisation, 2011). (Liu et al., 2022) In contrast, there is no grain shortage at any of the stages of production. According to the USDA, roughly 57 percent of a total of meals are lost in the meal quota channel priority to the point of trade jointly consuming person, or this time commodities and bi-product by postharvest manipulation or in general allocation. This is especially true for nutrient-dense vegetables and fruits (Liu et al., 2022). Shortfalls in the food arrangement that outcome in food misallocation on the scale seen in the United States exacerbate food scarcity, resulting in higher grain costs than would otherwise be the case, as well as natural externalities from the misallocation of land, labours, energy, and water resources (Cook et al., 2022). This was before obvious evidence of food shortages in limited households emerged.

Figure 1 depicts a contrast between the agricultural land area of India in 1951 and in 2015. It can be noticed that the forest area has grown a lot whereas the uncultivable land area has dropped to below 10% (Srivastava, A et al., 2021). This positive change is due to the green revolution and the agriculture laws formulated in the last 7 decades.

Agriculture has been a major reason for the human race to leave behind the nomadic life. It is the greatest and of the highest importance in order to sustain human lives (Mdegela et al., 2021). Agriculture in short can be defined as the science or the practice of soil cultivation which helps in producing crops and also rising livestock, this also prepares the market for the resulting products.

There has been a rapid increase in the population of the world which has increased the pressure on agriculture to sustain an increased number of lives. According to the latest projections, the population is expected to rise up to 9.8 million by the year 2050 and to approx. (Elvan et al., 2021) 11.2 million by the year 2100. At this level of increase in population, the world needs to be ready with the delivery of an adequate amount of safe food to the world's population.

Figure 1. Contrast between agricultural land of India in 1951 and in 2015

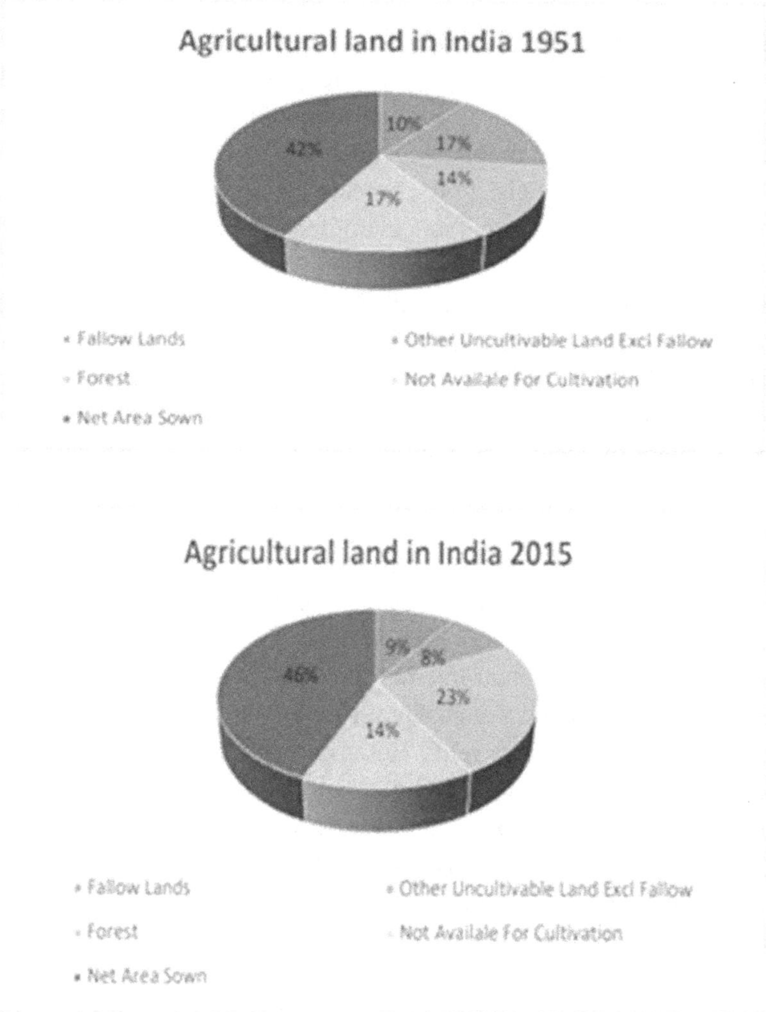

Related Works

The agriculture sector is concerned about the livelihood of laborers and farmers but is also concerned about the level of security for the country Food security is explained by the Food and Agriculture Organization of the Nations, is a situation where everyone has need sufficient and healthy food both physically as also as economically, which in turn fulfil the food preferences for a wholesome and a healthy lifestyle. In spite of the country's high level of production, 15% of the population remains undernourished as gathered by the estimates from 2014. India activated an act for health and security - The National Food Safety Act was approved in 2013 (Malone et al., 2021) with the purpose of giving healthy food and safety to nations by assuring cheap needs to goods and an appropriate quantity of high-quality food. The National Food Safety 2013 Act provides subsidized food grains (wheat, rice, and coarse cereals) to persons in certain categories (Pietrzyck et al., 2021). By 2015, this law had reached 68% of the population, or over 81 million people (with 77% in rural regions and 23% in urban areas), supplying them with adequate amounts of high-quality food. The nation has many various food groupings, each with its own food intake trend. The reliance on cereals for dietary purposes has lessened as consumption has increased. But are not limited to, legumes, meat, seafood, and eggs. Government should take concerned about the poor nutrition levels, deduced from the Treasury report on pulse production incentives in the country (Kuiken et al., 2021). According to the report, the price of pulses is less than other sources, pulses being a source of protein. India has been facing a shortage of pulses domestically; this shortage of pulses is plugged by imports.

Agriculture employs almost a quarter of the world's workforce. However, it contributes to about 4% of the world's GDP (at current prices in 2021-22).

Since the 1950s, the manufacturing sector has vastly contributed to economic growth and has often been termed the driving force of the economy behind some of the lesser-developed countries (Tubb and Seba, 2021). It is quite striking to note that the overall contribution of agriculture has declined from more than 51% of GDP in the 1960s to less than 20% in most less developed countries.

India is growing its overall production of cereals each and every next year and so it is the leading producer of various types of crops like rice, pulses, sugarcane, and wheat (Arnold et al., 2022). It is a grand producer in the fields of vegetables, fruits, and milk. India was the largest contributor to crop production in 2013, contributing 25% to worldwide pulse produce, 23% to rice produce, and about 14% to wheat produce (World Health Organization, 2020).

ORGANISATION OF AGRICULTURE AND FOOD

United Nations FAO is a specialized agency that was established in order to raise standards of nutrition and living across the globe (Friedmann, 2021). Ensure improved efficiency in the all-agricultural goods, including wise management, are produced and distributed (Binns et al., 2021). With sustainability for the usage of national resources. Thus, it is the need of the moment to analyse a number of farms' structures and their historical data variables and to consider the important changes in the policies of agriculture (Kuczuk and Widera, 2021) Increase in productivity, the growing influence of global and domestic markets, and the increasing effect on the mind of the consumers on the production of agricultural have helped to shape the current patterns of rural and agriculture life (Ashraf et al., 2021)

With this motive, the United Nations is ready to contribute to the wide expansion of the economy of world and also make sure the freedom of mankind from hunger (Siamabele, 2021).

With the end of the Great War the World War II, FAO found its origin. FAO was established as an outcome of the increase in general awareness of the people for the importance of nutrition for health in the 19th and 20th centuries (Nguyen et al., 2022). The awareness was also about the need to take scientific evidence into account while addressing general agricultural issues (Mahama et al., 2021).

The FAO's Evolution

In previous 70 years ago, FAO restructured the main figure and structure to entertain the latest and diverse challenges which include- rapid growth in population and urbanization, changing food consumption patterns, globalization of the agricultural sector, economic crises, and environmental impacts (Mir et al., 2021). I have reviewed it regularly (McMichael, 2021). On the basis of these reviews, we have implemented various institutional reforms on a regular basis and have developed instruments that have a response to the changing priorities and circumstances. FAO's present-day structure is the outcome of an Independent of the efficient External Evaluation that began in 2005 (Engler and Krarti, 2021). The IEE has recognized the institute's structure. Operational practices of the FAO. In order to enhance the performance and ability to meet the upcoming new challenges of the FAO, a number of government reforms were identified. The IEE resulted in the approval of an Immediate Action Plan (IPA) for the renewal of FAO at a Conference in the year 2008(Chandra et al., 2021). The three main areas: are governance reform; system reforms, programming, and budgeting, programs of the Organization; cultural change, and organizational restructuring. IPA passed the latest sight - "A hunger-free world" (Wang et al., 2021).

Global Goals and Strategic Objectives of FAO

In the examinations of the area of the FAO, it was reconstructed in the area of the main structures and also in the area of the various types of activities so that the latest and diverse challenges include various rapid growth (Agrimonti et al., 2021).

- The abolition of various of the hunger and also food insecurity, and undernutrition, as well as ensuring a world where everyone has enough nutritious food to live a very healthy life.
- Poverty eradication and advancement of social and economic progress for all, through greater food production and enhanced sustainable livelihood (De Souza et al., 2021).
- Long-term management of various types of natural resources, such as water, air, land, and other climate and genetic resources these are important and play a very essential role in this so that it determining these global types of Goals:
    - It gives their Contribute for performing the abolition of lack of food, food shortages, and malnutrition;
    - Boost and improve the supply of goods and services from agricultural production, forest management, and fishing industry in a sustainable way;
    - Decrease poverty levels (Shekhawat et al., 2021).
    - Empower more integrated and efficient food and agriculture processes at the native, governmental, and global levels.
    - Boost the ability to make a living resilient to dangers and crises (Butu et al., 2020)

According to IRC, the Strategic Framework of FAO" identifies as FAO's main function. FAO requires the development of these instruments. "At global, regional and national levels through global governance mechanisms, policy dialogue, and support and advice, coupled with the development at country level of the necessary policies and institutional capacities for their implementation" (Xu et al., 2020)

Activities for implementing Strategic Objectives in FAO are documented in 2 program documents:

The Working Budget Plan (aka PWB) and Medium-Term Plan (aka MTP). The latter is its 4-year plan that establishes particular outcomes that will help Strategic Objectives achieve and identifies areas of focus for every outcome, estimation of resources, and indicators for performances (Wezel et al., 2020). A PWB identifies activities to be performed over a two-year period as it is a 2-year plan and defines the number of resources required for every activity. The above two plans are regularly reviewed and adjusted for effective implementation to reflect changing circumstances and budgetary levels (Rejeb et al., 2022). The strategic framework of FAO is close to Sustainable Development Goals (aka SDGs). In particular, the first Strategic Objectives and the third Strategic Objectives reflect and complement

Sustainable Development Goals- 1 and 2: Eliminated Poverty and Eradicated Hunger. FAO is currently working with the Statistical Commission of the United Nations and Expert Group (aka IAEG) on Indicators to develop suitable indicators (Jarezebski et al., 2020) These indicators are related to goals that cover areas that include hunger, insecurity of food, eradication of malnutrition, and improved management of resources that are natural. Figure 2 indicated the global share of countries in exporting food products (Lacko et al., 2022)

Figure 2. Export share of different nations in food products sector

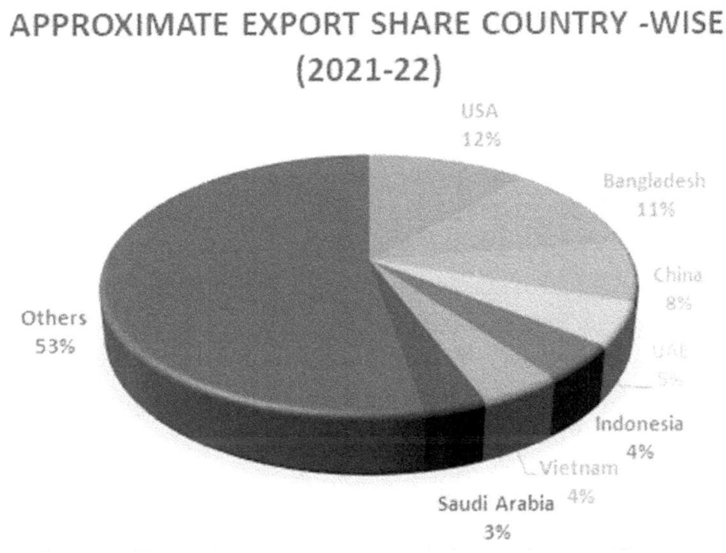

USA

A topic of discussion about farm programs of United States is that policies of this time are geared for a time period of US agriculture that no longer exists. Farms and their structures, along with the rural community in which they are based, have undergone some major alterations in the last few years. A number of questions have been raised about the effectiveness of the policies rooted in agriculture-based economies (Misniakiewicz et al., 2021).

Farms, Farm Households, and Rural Communities Throughout the Last Century

The undeveloped part of America went through some great changes during the 20th century. Agriculture in the early 20th century was occurring on many small, diverse farms in rural areas and was labour-intensive where more than 50% of the US population used to live. These farms, along with 22 million working animals, employed nearly half of the U.S. labour force, and on average it produced five different commodities. In the 21st-century the agricultural sector is concentrated in some large rural areas specialty farms, which is home to not more than a quarter of the population of U.S. Only a small percentage of the labours force of the US are employed here in highly productive and mechanized farms and use its 5 million tractors have been replaced by mules and horses (Adesina et al.,2020).

The agriculture of US has become more efficient and has contributes to the US economy's growth as a whole. U.S. farms production has increased drastically also allowing a fraction of spend of their income on food; this in turn resulted in a large population indulging in non-agricultural job to increase the country's economy. There are a number of transformations caused by the emerging technology and ever-changing market conditions; in the domestic and the rural economy the manufacturing agriculture has become a minor player. Agriculture is contributing less and less to the product of a Gross Domestic Product in short form it is called as the (GDP) and it contains the very largest share of that force called as the labour force for 100 years. The level of the percentage is smaller (see below). Box "The Changing Role of Agriculture in the National Economy"). At the same time, a declination was seen in the proportion of the US population living on farms. Similar to the central role of agriculture in the rural economy, once agriculture dependent counties accounted for the majority of the rural economy, in 2000 there were Counties considered to be agriculture-dependent (Guine et al., 2020).

There have been variations in the income and measures of off-farm work over the century that has made the comparisons difficult over the time. By the year 1970, off-farm income was in at least half of farms, and by the year 2000, off-farm income was earned by 93% of the farms. Off-farm work plays a significant role in increasing the farm household income; while the income of the farm household has been below the national average, in the year 2002. The U.S. Farming field dis-approves the discrimination in any of its programs on the basis of race, colour, disability, political beliefs, Nationality, faith, age, sexuality, gender identity, or marriage or economic status are all prohibited. USAD's target centre should be contacted if a disabled person needs to communicate.

National Economy After Farming Changes

- There was 41% in the year 1900 employment in agriculture
- There was 21.5% in the year 1930 employment in agriculture, GDP, 7.7% (agricultural GDP as a percentage of total)
- There was 16% in the year 1945 employment in agriculture, 6.8% (agricultural GDP as a percentage of totals).
- There was 4% in the year 1970 employment in agriculture, GDP, 2.3% (agricultural GDP as a percentage of total).

CHINA

China's Evolving Agricultural Support Policies

There are crucial repercussions for the United States caused by China's purchase of agricultural goods. Although the United States' share of China's agricultural imports has dramatically decreased from 2014 to 2015, it is still by far the greatest (Garcia et al., 2020). The Chinese government has passed a number of agricultural policies that have encouraged improved profitability for farmers and helped Chinese get closer to its long-term food security objectives. The Chinese government intervened in the market by offering price support and gradually raising the amount of assistance supplied. There is now a disparity between export and import pricing in the farming sector. Since China's accession towards the WTO in 2001, it has been easier for the nation to reach world markets, which has resulted in a dramatic growth in agricultural import and the accumulation of massive stockpiles. In order to narrow the gap between the foreign market and the local market in China, the government has recently decreased enormous stockpiles of cotton and maize while simultaneously implementing adjustments to agricultural policies, including as the price support given for policy of fabric, soybeans, and corn (Delgado et al., 2020). To drastically reduce on output and stocks in significant cotton-growing areas, the most recent target price strategy has replaced temporary stockpiling programmes and cotton price support. In the four north-eastern provinces, a new pricing strategy for soybean aim was also adopted to stimulate the output of soybeans. Temporarily reducing Chinese imports would have far-reaching effects on global agricultural markets, including the United States. China's farm regulations have led in skyrocketing maize prices, resulting in increased demand for imports from the US in particular.

Price Disparities and Massive Stockpiles are the Results of China's Agricultural Support Policies

The Chinese government has shifted its focus from taxing agriculture to giving production subsidies and price assistance programs in order to sustain grain self-sufficiency and rural income development through agricultural domestic policies. Direct payments, better seeds, and purchases of agricultural gear were all part of a Central Government subsidy scheme that began in 2004. Between 2008 and 2013, the amount of these subsidies quadrupled, illustrating a policy of steadily rising yearly subsidies to China's farmers (Workie et al., 2020). By setting a guaranteed minimum purchase price, the Chinese government hopes to stimulate the growth of certain commodities in key agricultural areas. This regulation went into effect for rice in 2004 and for wheat in 2006. Cotton's temporary reserve programme was established in 2011, and similar initiatives were established in 2008 for corn, rapeseed, soybeans, pig, and sugarcane (Tuan, 2015). Programs like the temporary reserve and minimum purchase price ensure farmers a minimum price for their goods. China's farmers are able to cash in when the price of grain on the market falls below a certain threshold. Their goods to government-owned businesses for the lowest possible price. These goods are being hoarded in an effort to minimise imports by stabilising supply. The Chinese government has implemented these policies to boost farmer incomes and advance long-term food security objectives, and it has steadily raised minimum prices to cushion domestic producers from growing input costs and the appreciating Yuan (Aka OECD, 2013) (Lajoie-O`Malley et al., 2020). Minimum commodity prices were established below world market values beginning in 2004. However, despite a global decline in prices starting in 2008, China's support prices rose every year through 2015. China's increased pricing incentivized farmers to boost output and opened a price difference between the country's market and the global market, which prompted consumers to buy cheaper imports. As a result of government incentives, China has expanded production of vital agricultural commodities. Chinese authorities bought up massive amounts of domestic goods at support prices, leading to an overabundance of supplies in the central government's hands. Since 2010, grain stocks have expanded considerably, hitting record highs in corn, cotton, and wheat. In addition, the Chinese government boosted its imports of the same commodities at lower prices from other nations as part of its price support and temporary reserve policies. In addition, there was an uptick in the importation of feed additives such sorghum, barley, and dehydrated distillers' grains. During 1978 to 2013, Chinese saw a doubling of its own grain output; during that time, the country used more than seven times as much fertilizers. China's crop yields used three times quite so much pesticide as the worldwide average in 2013(Abrouk et al., 2020). This had a harmful impact on the environment. As a result, the agricultural markets in China became more unstable, putting a financial strain on the country and presenting new issues. China's farm subsidies have been rising even as those in other countries have been declining, and they've already exceeded World Trade Organization limitations on several types of aid. The USDA issued a press release in 2016 stating that Chinese had failed to live up to its WTO promise by providing "sufficient domestic support" to its farmers in order to increase production of agricultural commodities. Recently, China has started looking into alternative approaches to bolstering farmer incomes, such as decoupling payments and other forms of agricultural support permissible under WTO rules. China is currently discussing how to change its economic assistance program in order to minimise price manipulations, limit imports, and maintain a healthy supply and demand ratio in farm products.

Updated Cotton and Soybean Price Objectives, End of Corn Price Supports

How to enhance farmers' earnings and promote food security without influencing domestic market prices and amassing enormous stockpiles is a pressing issue for the Chinese government in the agriculture sector (Javed et al., 2020). In the thirteenth Five-Year Plan, the Chinese government announced a market-oriented pricing formulation system, a positive policy. With the unusual of wheat and rice, which are now purchased by popular Chinese government at the minimum price support, the Chinese government started winding down its price support programme for key agricultural commodities in 2015. However, the policy on rice and wheat pricing would no longer be predicated on annual minimum price increases. As a result, farmers' earnings will not be boosted by the minimum price assistance. An experimental target price programme for cotton and soybeans in selected areas replaced the existing price support programme in 2014/15. China's maize price assistance scheme was eliminated the same year.

Cotton and Soybean Pilot Projects With a Target Price

In 2014/2015, the Chinese government introduced a target price policy pilot programme to replace the previous price support and temporary reserve programmes for cotton and soybeans. If the market price is less than the target price, the Chinese government will make a direct payment to the farmers, with the amount determined by the total output and the price differential between the two (Duque-Acevedo et al., 2020). China no longer engages in commodity procurement and storage as part of its price support programme. Given China's substantial cotton reserves, this is of paramount importance.

To help get rid of its massive cotton supply, the Chinese government also instituted a new target pricing policy. Cotton producers were incentivized to increase output under the previous price support scheme because of the high prices at which their goods were sold domestically; the resulting excess was stored. At the same time, the Chinese government limited cotton imports via tariff rate restrictions, imports grew due to low worldwide prices (TRQs). The Xinjiang Autonomous Region, a major centre for cotton production in China, is included in the new target price strategy.

The new pricing goal for 2014/2015 is the same as the previous price floor. Cotton output and inventories fell as a consequence of the increased target price of $3,081 per metric tonne in 2016 and $2,906 per metric tonne in 2017. "In terms of contributing to a market-oriented cotton price formulation system, a steady cotton acreage, and a suitable income for farmers," the first two years of this new pilot policy's execution were regarded effective for cotton.

Soybeans: The soy production incentive experimental target price scheme was launched. The higher returns from corn planting led many farmers to favour it before the change was made. The new price goal applies to soybeans grown in Heilongjiang, Jilin, Liaoning, and Inner Mongolia in China's northeast. Heilongjiang is the biggest soybean-producing region in China.

A new target price, slightly higher than the prior price support, was set at the start of 2014 to encourage farmers to grow soybeans, and has stayed unchanged since then. The new target price has not, however, largely affected soybean output (Barichello et al., 2020)

FEDERATION OF RUSSIA

Supporting Agriculture

The government subsidies that the Russian Federation pays out to farmers have fluctuated throughout the years, but, as of 2014, have levelled off at a range of 9%-13% of the nation's total farm revenue. Between 2018 and 2020, the average yearly allocation for agriculture was about 10%.

Support based on output (such as market price assistance) and unrestricted use of variable inputs account for around 70% of all agricultural support. When compared to pricing on international markets, local ones are, on average, around 5% more expensive. However, variations in support for individual products are obscured by the market's overall willingness to pay for them (Yanbykh R et al.,2020). It's a mix of tariffs on cereals and oilseeds that are exported, as well as tariffs on animal goods and sugar that are brought in. Domestic grain costs are lower compared to global pricing, which is good news for ranchers.

Spending on GSSE is consistently below the OECD average, hovering between 3% and 4% of GDP. The majority of public funds are used for agricultural knowledge systems, infrastructure development and maintenance, and inspection and control systems. In 2018–20, total farm subsidies (TSE) amounted to 0.6% of GDP. Since the mid-1990s, this has been falling, mirroring GDP growth and the decreasing importance of agriculture to the economy.

Recent Shifts in Policy

During the years 2018–2025, the National Program for Agricultural Development will be in its second phase of execution. In 2019, 2020, and beyond, the financing arrangement little changed (De Castro et al., 2020). Although the government's primary financing mechanisms remain same, various regional initiatives are now supported by other subprograms. Helping agricultural exports is a priority. Improvements in the fields of the phytosanitary procedures and international product positioning and promotion are present in the heart of most important agricultural component.

The government initiated various types of new good policies in May of 2020. The permitted capital of Rosagroleaging has been expanded in order to boost the supply of equipment to the agricultural sector, which will allow for more assistance for the acquisition of agricultural machinery, commodities, and processing equipment. Soybean meal and fresh produce now qualify for rail subsidies (Liu Y et al., 2020). At first, subsidies were extended exclusively to the transport of grains. Community payments for soybeans and canola were introduced in 2020 as new incentives to encourage oilseed cultivation.

In 2020, reforms were made to producer subsidies in an effort to revive output. Instead of unit subsidies, milk performance fees, and area surcharges, two new subsidy programs—"Compensation" and "Simulation" subsidies—were implemented. Acreage payments, milk subsidies, pedigree livestock help, top seed subsidies, local traditional subsector support, and agricultural insurance subsidies are only few of the forms of compensation offered by the government. Funding for priority subsectors chosen by local governments from a state-determined list is modelled as part of the simulated subsidies. The 10 least developed areas of the nation and small - scale farmers are two more groups we may aid in developing.

In an effort to comprehensively address difficulties and risks to food security, the President of the Russia announced a new policy on the subject on Jan 21-2020. This paper claims that the minimal percentage of local production in internal demand varies from 60% for fruits and berries to 95% for grains.

A year's worth of main debt payments that were due by 2020 have been postponed as an output of COVID-19 situation. Repayments on both the interest and principal on investment loans, as well as the interest on quick loans, were delayed (Beckman, 2020).

To counteract the fall in people's actual income, the government will implement marginal pricing for "socially significant" items by the end of 2020. This will be accomplished until March 2021 thanks to pricing agreements struck by the Field of Commerce and Field of Agriculture and Industry with food producers and chain stores.

The government extended the restriction on agricultural product imports from nations imposing economic sanctions on Russia until the end of 2021 on November 21st, 2020. The 2021 export quotas for wheat, rye, barley, (Wordofa and Sassi, 2020) and maize have been determined and are subject to tariffs. When grains are exported in surplus of the quota, a customs charge of 50 percent of the customs value of the exported goods (minimum €100 per tonne) is imposed. Furthermore, a 30% export tax (but not less than 165 euros per tonne) will be imposed on soybeans, rapeseed, and sunflower seeds from Feb 15-2021, through June 30, 2021. Sliding export tax on sunflower oil will begin on September 1st; oilseed export tax prolonged until second half of 2021(Pretty J N. 1998)

Rosselkhoznadzor has streamlined the import processes for agricultural items to enable their continued importation in the face of the worldwide epidemic. As of March 31st, 2020 (valid upto 30 june-2020), the Eurasian Economic Council (EEC) officially implemented limits to mitigate various of the negative economic effect of major virus called COVID-19 and avoid shortages of socially significant goods in nations within the Eurasian Economic Union (EAEU). This is relevant for the export of certain agricultural goods from EAEU Member States. Additionally, the EEC has authorised a list of key imports that are duty-free for nations in the EAEU. The duration of the exemption was from April 1, 2019, until June 30-2020(Asrat and Anteneh, 2020).

Principal Tools of Policy

The actual National Strategy for Agricultural And rural development which was first implemented in 2013, has adapted to evolving economic circumstances. New objectives of export assistance and import substitution necessitated a reorganisation of its subprograms in 2015 through 2017. The initial financial goals of the programme have also been revised, both in terms of total expenditure and resource transfers between and within components. Modifications were made to the Country Program's organisational framework, budget, management, and rollout timeline in 2019 through 2020. Some of these appeared after the infection of virus of virus that was COVID-19 virus. The National Plan for 2019-2020 has a sub-program devoted to the growing importance of digitisation in Russian agriculture growth.

Russia employs a wide variety of tools to manage prices. Security at the borders, both via tariffs and other measures, is crucial. Russia has used MFN tariffs since entering the WTO in July 2012; they will be lowered to an average of 11.2% by 2020, in accordance with the average final agriculture tariff. In 2019, agriculture tariffs averaged 11%, about twice the non-agricultural tariff average of 6%. Within the agriculture sector, the average import duty for meat and dairy goods, drinks and tobacco, sugar and confectionery are the highest. The EAEU customs union serves as the primary institutional framework for the implementation of border control measures (Kuhnlein and Turner,1991).

Nationwide price restrictions, including grain interventions, are relevant as well. If the market price of grain increases or decreases outside of a certain range, the government may intervene by buying or selling the commodity. State intervention occurs at a specific price, but that price is not a price guarantee. There may be import/export limitations.

All fees or commissions associated with the sale of animals are covered by funds allotted at the county level. Additionally, there are state payments for milk that are financed in part by the federal government (Karthikeyan et al., 2020).

The primary source of funding is concessional financing, which consists of loans from commercial banks at interest rates that are lower than the commercial interest rates established by the government. In the occurrence of a financial loss, banks may rely on government reimbursement. Subsidized interest rates are another perk for borrowers who take out loans before 2017. In addition, there are a plethora of input subsidies that might shift. Preferential lease of machinery, equipment, and animals is one kind of assistance. Since its inception in 2013, the Local Payments for Agricultural Production programme has replaced many nationwide input subsidies for planting and harvesting efforts.

Producers in the agricultural sector might get refunds for federal income tax and Social Security payments that were previously withheld.

The majority of these enabling measures are being implemented as part of the multi-year Agriculture Development Programme, the primary policy framework for agriculture in the nation. It is predicated on a funding vehicle jointly sponsored by the federal and state governments, with co-financing percentages varied by area and vehicle. Furthermore, within the framework of governmental programmes, regions create, fund, and finance their own uniquely local assistance measures.

Some countries' governments have lately instituted new initiatives in different areas of the economy and society. B. The circumstances for agriculture's long-term sustainable growth are shaped by rural development and the environment (Adegbeye et al., 2020).

Russia ratified the Paris Climate Agreement, which it signed in April 2016, on 6th October 2019. An initiative to cut down on emissions of greenhouse gases was given the go light in 2014. Systems for emissions registration, assessment, and extrapolation, as well as government control of emissions, are among the main objectives. As a component of this strategy, a regulatory impact study is now being conducted on a bill to regulate greenhouse emissions that was drafted in 2018(Mdemu et al., 2020).

The Russian Federation Climate Doctrine and the Comprehensive program for the Implementation for the Federation of Russia Climate Doctrine till 2020 are both major national policy texts addressing climate change. For governmental action against climate change, a theoretical framework is provided by climate doctrine. The Basic Plan provides a uniform governmental policy created in this field. The latter describes the key activities allocated to departments.

Along with climate ideology and holistic planning, USDA advocates climate resilient methods that: B. Responsive planting systems; healthy water, oxygen and nutrition regimes (Artmann et al., 2020). New crop types being released into production. Improved crop yields by strategic use of weather projections.

The National Agricultural Development Programme is another important policy document that aims to improve the environment for productive land management. It also offers funding for ground improvement initiatives and resource energy conservation and efficiency advances. Policy initiatives have been implemented in response to multiple climate tragedies in recent years, further than the long-term measures specified in the aforementioned policy framework texts. Among these measures is the creation of official mechanisms for government aid in the case of catastrophic weather, such as the allocation of funds from different governmental levels and the evaluation of repair and replacement expenses.

CANADA

Agricultural Support

Since the late 1980s, Canada has seen a dramatic cut in its support in the field of agriculture. Between 1986-88 and her 2000-02, producer support to total agricultural income (%PSE) halved. This was primarily due to the 1995 phase-out of market price support (MPS) for the grain industry, with the exception of the dairy industry., the poultry and eggs sector, is under supply control and aligned to global standards. Producer assistance he halved again from 2000-02 to 2017-19, but pay out levels vary each year given the importance of companies' risk management programs.

8% of the total income from the farm has been accounted in Canada's PSE for the year 2017-19, compared with 35% in the years 1986-88 and 17% in the years 2000-02 (Aryal et al., 2020). The percentage of Canada's PSE was below the average of OECD over the period. Above the years 2017-19 OECD average but lower than 1986-88, the potentially most skewed share of support (based on output and use of variable inputs – no input constraint) was 61%. Rice field. At an average rate, the prices that have been received by farmers from the years 2017 to 2019 were 5% higher to the ones observed on global markets. The General Support Estimates (GSSE) shares are in the Total Estimates of Assistance to Agriculture (TSE), in 1986 and 1988 has nearly doubled itself and by the year 2017-19 it reached 42%. Around 40% of GSSE has been an account of support for agricultural innovation systems and inspection systems in the past few years.

Main Changes in Policies

Canada's first food policy was launched by the government. It has a goal to create a food policy that is more coordinated food systems-based approach. Short-term actions have been focusing to improve access to the healthy foods, promoting of the Canadian food, encouraging food safety in the Northern as well as in the Indigenous communities, and also reducing the food waste (El Chami et al., 2020). Additionally, in early 2019, the new Canadian Food Safety Regulation came into effect. Credit limits under the Advance Payment Program (APP), which allows farmers to easily get cash advances as their credits, which has been increased for the first time since 2007. From the beginning of April in 2019 to end of March in 2020, the DDPP aka Dairy Direct Payment Program will provide access to the market obligations under the latest agreements of international trade with 1 Offered a one-time payment to the milk producers affected. Governments along with the industries have taken strategies to both avoid and fight for African swine fever. The government has launched several measures to attract agricultural workers (Deaton, 2020). For instance, the AAFC has provided funding to a program. that has aimed at identifying the best practices for retaining and also attracting the international workers.

Agricultural Policies

There are a number of common measures that agri -food businesses can access to support the business of individuals or companies affected by the COVID-19 crisis. Support of the Business includes tax deferrals, minimum wages and wage subsidies for the ones who had to stop working due to the COVID-19 situation. The Government of Canada offers through the Canada Emergency Wage Subsidy, the eligible employers a wage subsidy of 75% for at most 12 weeks. There have been some temporary changes to the summer jobs program of Canada that now offers wage subsidy of 100% to employers hiring temporary workers for around 70,000 seasonal jobs (Kumari et al., 2019).

- The Canadian Government provided around has provided C$50 million amount to help the sector implement measures to comply with the compulsory 14-days for quarantine are temporarily required for immigrant fifty workers. Government informed their $45m just to fascinate their workers of Quebec to the agriculture lands. This included the $100 bonus amount to the seasonal farm workers working at least week for twenty-five hours.
- Credit of Farm Canada structure (FCC) has an increase in its lending capacity by another amount of C$5 billion. The above new loan allows Farm Canada structure (aka FCC) to help farmers, and food processors facing liquidity challenges, from deferring the interest portion of loans to access the additional credit.
- Loan taken by the producers under the Advance payment program with the month of March or the month of April 2020 repayment date and the reduced ability to commercialize raw materials will be required to pay an additional 6 months to repay the outstanding loan balance.
- Through the Canadian Partnership for Agriculture (aka CAP), a continuous allowance to the respective suite to the makers of BRM. The BRM programs to maintain economic strike. Some of these program deadlines have been changed or extended in several Twenty states. To aid manufacturers and small processors in adapting to the epidemic, the Prince Edward Island's government has announced its new Agricultural Strategy.
- Finance and advisory services have expanded through Business support for food and agricultural industries.

Polices of Agro-Food Supply Chain

In the month of March on 16th, Industry and Government Working Group COVID-19 was launched by the government. The working group is made up of national industry associations, and three times a week, it shares information by phone and discusses issues facing the industry, including the potential impact on trade (Vos and Bellu, 2019).

Labour-related measures have been introduced to facilitate the domestic and international movement of agricultural products and intermediate consumption.

Truck drivers, airline crew and other goods being transported are exempted from the ban on travel as long as they have symptoms. Workers that are working temporarily in the agriculture and seafood processing, and other important industries are welcome to visit Canada. The Canadian government has increased the maximum tenure of a low-wage worker from one year to two years in the Temporary Foreign Workers Program (Ribi et al.,2019). At the retail level, the food system provides supplies for consumers. The Canadian government wants to improve the food access for people experiencing food insecurity, it is collaborating with established local and national networks to promote food access.

Strategies for Protection and Security of Food

The CFIA also known as Canadian Food Inspection Agency has announced that during this pandemic, to better support Canadians and the industry, to redeploy existing employees to higher priority activities, certain low-risk Inspection activities have been temporarily suspended. In these circumstances, the Canadian Government is investing $20 million USD in order to continue the important work to protect food system of Canada and for the better support to the production needs of food industry of Canada.

The above investment will allow the CFIA to conduct specific and important inspection activities closely with the industry and also the commercial partners to address the disruptions in the supply during thee the previously mentioned crisis The CFIA regions and provinces are collaborating closely to ensure Canada's food production throughout the COVID-19 epidemic. The monies will also be used for training the state food inspectors to assist the CFIA whenever in need.

The money will also assist the CFIA in establishing flexible inspection methods, such as the increased use of technological tools like as iPad and accessibility to the CFIA's global services network. For example, it is made easier by the CFIA for retailers and grocery stores to sell food that is packed and is intended for its use in hotels. This way more food is available to the locals of Canada which in turn reduces food waste and also supports businesses without compromising the safety of food. Moreover, agreements with states and territories were drafted by the CFIA to share resources of the inspector and to develop the standard operating procedures to cover the requests of the inspector for overtime or additional shifts. A guidance of COVID-19 positive cases is provided by the CFIA to the industry, at slaughterhouses/ processing facilities, along with contact information and escalation protocols to follow in the event of an incident at the facility (Zhou and Wang, 2019).

INDIA

India is a member of:

- WTO which is also known as the World Trade Organisation; It has been a representative of the same since the 1 January 1995.
- IPPC which is also known as International Plant Protection Convention. India signed the IPPC on the April 30, 1952 and ratified it on June 9, 1952. It has been a member of the OIE since May 30, 1924.

The latest National Agricultural Policy of India was promulgated on the 28th of July in the year 2000. The core goals of this are equitable and inclusive growth and sustainability considered in terms of use of resources efficiently. The other goals include scientific and the technological advances to remain worldwide competitive.

The use of biotechnology is encouraged through this policy in order to grow crops such as:

- Drought tolerant
- Resistant to pests
- Nutritious
- Offer higher yields
- Eco-friendly

The conservation of biological resources is the main focus of the policy. Biomass, organic or the inorganic fertilizers, pesticides and the pest control are also appreciated and promoted. The application of advanced science is highly encouraged by the policy. For example:

- Remote sensing technology
- Biotechnology
- Pre- and post-harvesting techniques
- Energy Saving Technology

Environmental protection is encouraged and promote via this policy, it also encourages- technologies through public or through independent research.

The Agricultural Export Policy of the year 2018 has its goal to increase the agricultural exports of India to which US$60 billion by the year of 2022. Under at manirbhar Bharat Abhiyan amid COVID-19 relief efforts, and of various reforms in agricultural sectors that are being considered to make India "World's food basket", this includes $15 billion USD.

- A program involving 10,000 producer organizations has its goal to provide a platform in order to support farmers. (Esham et al., 2018)
- Developed a digital Agri stack that enables smart farming as well as online markets.

The agriculture now remain reflects as most of the heavier regulated sector and due to this various agency, that are provided by government are taking oversight ability. Official controls are appointed by both government these central government and other state governments.

Agricultural assistance programs are offered at both the levels i.e., the central and state levels via subsidies.

- Fertilizer
- Electricity (electricity) for irrigation and other agricultural purposes
- Credits
- Irrigation water
- Export
- Seeds
- Crop insurance

Agricultural income also benefits from reduced taxes, with no or low indirect taxes on agricultural products.

India has many different agreements about trade with other countries like- Chile etc. Countries and has been negotiating an agreement about trade with South Africa.

Processing and Production of Food

Indian agriculture has witnessed several modifications in way of the various patterns of many types of the production and also includes the consumption of various types of trade in the recent year of Indian Scenario. Only one change is a shift in production and consumption away from grains and towards a very large struggle and long types of the farm products that are such as milk and many types of dairy foods it includes many fruits and vegetables, meat, and processed food products. Quality products trade is rapidly expanding of all classic things that are such as sugarcane, tea, rice and coffee it also includes the tobacco, and so on. The rate of increase in the export earnings of products such as rice, fish products, sugar, tea, and so on has slowed in the twenty-first century, whereas high-valued exports have increased (including fruits and vegetables, floriculture, meat, processed fruit juice shave expanded at an annual rate of 18%. The declining percentage of production in traditional goods, trade and demand, horticulture, and other strong non-traditional crops has represented a significant area of potential income development in rural areas, and the overall area seeded under key crops has decreased to 140.8 thousand acres in 2008-09 from 142 thousand acres in 1983-84, but the net area sown has climbed to 194 thousand acres from 4 thousand acres. The grain crop area has been reduced to around 6 thousand acres between the Food grain percentages in the cultivated area decreased from over 73% in 1983-84 to around 63.8% in 2007. Food grain output has increased by even more than 28% over the last twenty years, from 177.4 million tonnes in 1993-94 to 227.8 million tonnes in 2009-10. Cotton production has increased by more than 200 percent, followed by fruit and veggies (97%), herbs and spices (66%), and wheat (39%). Pulse output increased the least in this (Myers et al., 2017).

The decrease in the area dedicated to food grains has led in an increase in the area dedicated to other crops. The oil seeds profited the most from this reduction during the entire decade of the 1980s, when the reason that reason was under oilseeds climbed to 26 million hectares in 1993-94 from 18.5 million hectares in 1983-84, but the area under oilseeds remained relatively consistent between 1993-94 and 2008-09. Oilseeds' percentage of that total of the important cropped area climbed dramatically these are very less than 10% and it is in the early 1980s to 14.8% in the early 1990s, then somewhat decreased to 14.3% in 2007-08. Cotton production decreased by nearly half a million hectares between 1983-1984 and 1993-1994 Between 1993-1994 and 2008-09, the area rose by more than 2 million hectares. Another production that benefited from the decline in this region under food grains was high-valued crops, primarily fruits and vegetables. Between 1983-1994 and 2007-08, the area covered by fruits and vegetables expanded by over 8.5 million hectares. With its share of the global market, India has been the leading producer of a wide range of fruits and vegetables. Banana (23%), onion (10%), Mango (41%), cashew nut (24%), onion (10%) and cauliflower (30%), and Mango (41%), The area covered by fruits and vegetables in total things that area grew from many few years that 3% in 1983-84 to more than 5% in 2007-08. Fruit output in India has an average of roughly 55.05 million tonnes, measured during a 6-year period (2002-03 to 2007-08), with a total growth of over 14 million tonnes. During the same time span, vegetable production climbed by nearly 33%, while the area increased by approximately 3 million hectares.

The Organization and Constituents of the Indian Food Industry

The food manufacturing industry, includes companies that produce intermediate or final consumption products that are transformed from the livestock and the agricultural products. Products that are created

Table 1. Food industry segments and products manufactured by India

Sectors	Products
Dairy	Ghee, Milk, Butter and also ice creams
Fruits and veggies	Juices, concentrates, Beverages, potato wafers/chips, slices, frozen and dehydrated items, Pulps and other products
Cereals and Grains	Flour, malted flour, baking, glucose starch, grain-based alcohol, cornflakes, vermicelli, beer and malt extracts,
Marine	Freeze goods etc
Poultry and meat	Frozen and also it contains packaged - primarily in mostly in the raw egg powder
Foods for Consumers	Treats, dry ingredients, biscuits, prepared foods, liquor and non-liquor beverages

by transforming a raw commodity into a product that is processed is known as the processed food, even though the processing amount is minor, as for the canned fruit, or more complex, such as snacks. Value is added to the products of agriculture or horticulture with the help of food processing, which involves the use of various techniques which may involve grading, packing, and other processes that extend the shelf life of food goods. Food manufacturing that is both strong and dynamic, widely contributes to the overall growth of the economic structure of the country. An important linkage and relationship between the industries and agriculture is provided by this sector, this has also been declared as the sector with the most potential to boost up the economy. Processed foods, such as processed fruits and vegetables, have been a significant source of foreign exchange income. Hence, we say that it helps in the generation's employment rurally.

The clarification of food in India is done to the following extent:

- Production process consists of washing, classifying, powdering, and finally refining farm goods, such as processing wheat into flour.
- Secondary processing entails providing basic value, such as tomato sauce, coffee beans, and meat packing.
- Third processing- high-value items such as jams, sauces, biscuits, and other prepared bakery items are added.

Food processing is a massive industry in India; this industry includes some practices such as that of agriculture, horticulture, plantations, animal husbandry and fishing. Apart from these basic active practices this industry also includes those industries which produce edible products from the basic agricultural inputs. This industry is divided into basic six segments, it is carried out by the Government of India's Ministry of Food Processing. These are Vegetables processing, Grain Processing, Dairy Fruit and Meat and also poultry and for that in which includes marketed beverages and foods are the six segments (including packages water). Table 1 depicts India's food industry categories and manufactured items.

Despite the fact that India's food processing industry is massive, but in terms of development it is still at a young stage. The amount of total agriculture and food production that was being processed is only about 2% for the country. The government estimates the size to be billion in the area of the Department of various types of Food Processing. The contribution in the area of the various food processing industry is about 9% and at the life of the country's total amounts of the GDP and accounts for mostly in the

6% in the total industrial output This industry directly employs around 1.6 million people. (Merchant, 2008). The rate of growth for this industry is estimated to be about 9.12% from 2002 to 2007. The value that was added to food is expected to increase from 8% at present to 35% by the end of the year of 2025.

The unorganised sectors have highly fragmented and dominated the India's manufacturing industry This sector has a small number of troupes. The production is almost 42% from informal economy, 25% from organised sectors, and the remainder from tiny sectors. The unstructured portion varies by category, but it still accounts for around 75% of the market. The sector in the primary processing segment is smaller than the organised sector in the downstream manufacturing segment.

The main processing component is a major industry with large amounts of the rice mills and also the hullers, oil mills, and thousands of the various of many traditional bakeries. Food processing and fruit, vegetable, and spice manufacturing will be unstructured sectors (Food Processing, 2006). Despite India's substantial agricultural production base, the food industry is still deemed undeveloped. The dairy sector produces the most processed food, accounting for 37% of total processed food, with just 15% processed in the organised sector. Veggies and fruits are processed at a rate of around 2.2%, meat at about 21%, and poultry products at about 6%. Only 48% of the 2.2% fruit and vegetable processing is done in the organised sector, with the remainder done in the informal sectors.

Agricultural State of India

A variety of various types of the factors influenced agricultural productivity. Which could include the, but is not limited to and also various types of quality and also the availability of many types of agricultural inputs that includes things are like, seeds, water and lands and also includes fertilisers for the access to crop insurance and agricultural loans, obtaining reward pricing for agricultural products, marketing and storage infrastructure, and so on. This paper provides an overview of the agricultural situation in India. This section addresses the elements associated with agriculture and its production, as well as its post-harvest activities. Agriculture employed more than the half (53%) of all workers in the country in 2009-2010, totalling 243 million people. Various types of the Landowners and the tenant farmers working on the area of the piece of the land it also includes the farm labours that are working on their various types of farms make up the proportion of the population that is dependent on agriculture for a living. Agriculture production has been erratic over the last decade, with yearly growth ranging from 8.6% in 2010-11 to -0.2% in 2014-15 and 0.8% in 2015-16.

As observed, the sector of agriculture, contribution to the GDP or the Gross Domestic Product has depreciated to 15.4% in 2015-16 from 54% in 1950-51, whereas the service sector has decreased from 30% of the total GDP. Increased to 53% has declined in recent decades, but the contribution of the sectors which include the manufacturing (population's 10.5%) and services (population's 24.4%) has also increased (Bryngelsson et al., 2016).

BANGLADESH

The prime sector of the economy of Bangladesh is agriculture. 20% in the fields of the Gross Domestic product (GDP) in that country is contributed by this sector and has 44%of the total force as employee. In the vast sector of agriculture, the contribution made by the sub sectors namely crops, livestock and fisheries are estimated at 11.34% and 4.49% respectively. Significant contributions are made by the agriculture sector to attain marginal sufficiency of food, creating opportunity for employment, enhancing the standard of living, and ultimately eliminating poverty. There were continuous reform of agricultural policies and its implementation in latest years due to which success was achieved. These include: policy of the extension of the national Agriculture; National seed policy; policy of water management and irrigation; Irrigation Pest Management Policy; Rural credit policy. One of the most important variables in the agriculture sector's long-term development is the agricultural/rural credit, particularly to increase the production of agriculture. Credit plays an important role as it is obvious from previous studies, in increasing productivity of agriculture. Credit access that is easy and on time, helps farmers to obtain the necessary input and machinery for carrying out farm activities and increasing productivity.

Changes in Policies

1950

The liberalization of the market for small irrigation systems helped:

a) Mobilization of private savings for investment in irrigation.
b) Elimination of delays in installation, repair and maintenance of equipment that farmers had to avoid due to bureaucratic procedures.
c) Increased competition in the water market leading to lower water tariffs; and
d) Increased use of machinery in years when rice prices are low.

1970

The Domestic Trade Policy After 1971, the government that imposed on restrictive trade and exchange rate policies, was the first government to face a serious balance of payment crisis. In order to protect the domestic industries to regulate imports and preserve indigenous industries, quantitative limitations on imports, hefty tariffs, and fixed and inflated currency rates were essential. After the Bangladeshi Taka was devalued by 70% to 15.1 from 8.9 against the US dollar in 1975, and there was a major danger in the reason of the balance of payments crisis in the year 1970s and during this the government's protectionism persisted. Because of the trade system, philosophers survived. There was a rebound in export revenues, a rise in foreign aid, and a sustained slump in world prices for grains and other commodities. The government enabled citizens to import specific types of items in exchange for the earnings from international exports in the early 1970s. Thus, the government established a double policy for exchange it takes support from the Export Performance License (XPL) and they work as they Import Entitlement Certificate (IEC) until 1977 and one of the majorities of agricultural items were limited or barred imports at the start of the restrictive trade policy. The fundamental goal of this was to ensure profitable pricing for all the producers and shielding them for giving their benefits in the fields of external competition. And it was limits in the export of various types of products of the agricultural products. One of the major reasons was to provide the enough supply of agricultural raw materials on the local market. Tax breaks were implemented in 1976 to stimulate the export of non-traditional commodities.

The Export Policy Directive of 1976 increased the revenues on non-traditional exports from the percentage 15-20% and also to 30-40%, and from the assist the export of the various of the textile industry and also the products that are used to derived from the imported various materials like the raw materials, in this way that was the tariff scheme was implemented. The consequences of the aforementioned self-sufficient material and also the trade system and it is useful for the system of the exchange policy have been unusual in way of the export growth, and also the payment balance and the growth, and also the aggregate growth, particularly at the rapidly growing economies of East Asia, which have viewed more outward-looking development and growth as unsatisfactory when compared to strategy. Indeed, the export promotion process, which began in the year 1970s, was restricted in very few scopes and sluggish to implement. All trade policy is remained heavily controlled, particularly in the foreign sector.

1974

Agriculture was the most significant sector after independence, accounting for almost 60% of the total GDP, whereas in the year 2001 it was only 24% of the total GDP. There was a double in the population to the 140 million from the 70 million However, agricultural production was adequate to feed the people. Agriculture's percentage of GDP declined as other industries grew faster than agriculture. Bangladesh's primary agricultural products include cereals such as aus, ammon, boro rice, wheat, and other cereals, as well as drinks like as tea and other beverages.; fibres that may include - jute and also other fibres; Fruits, oilseeds, legumes, sugar cane and vegetables. The latest trend for the various types of contribution at the area of the agriculture to the total area of GDP is still on the trend in the downward direction. Over the last decades, that proportion has declined to 15.52% in 2011 from 19.51% in 2001. The fishing percentage also decreased from 5.51 in 2001 to 43 in 2011. The share of the total agriculture in the year 2011 constituted to around 20% of total GDP.

1980

In the 1980s, the earliest trade policy changes were a very high-range and also it reform package known as the New Industrial Policy (NIP), which was begun and known as the execute in year of the 1982 by the various goal of liberalising the economy. A considerable number of nationalised enterprises, particularly those are in the privatised in a very few years, So that it was initiatives were taken to take perform to promote international and also local types of the investment. Only on a limited scale has trade liberalisation occurred under the area of the NIP and it also includes in the successor, the RIP which was implemented in year of the 1986.Generally in this there was very gradual progress in area of the very large trade liberalisation in the year1980s and it was the particularly in the area of the regard for depreciation for the various tariffs. There are various types of examples of the agricultural export liberalisation include the removal of various types of the duties on various of the raw materials such as various jute and also includes tea in year of 1981, and the removal of various types of export duties on various types of food these are like dried fruits and some fresh fruits, and also some of the various oil cakes and some coriander seed, dried chilli and very large amount of the ginger and some black pepper and very few turmeric and also tobacco, and vegetables in 1986. Export restricted in the fields of the jute seed and pulses and wheat also all types of shrimp, other than it includes the various frozen meals, and onions were not lifted and remained in place even after 1995.The restrictions of export were removed in the year 1998 which were basically imposed on rice and wheat bran and molasses.

1990

1990s: Major liberalization of trade with almost a decade long attempt that was half-hearted towards the trade liberalization, in the year 1991 the democratic administration took bold moves to change the trading regime. During the same time period, the changes were enacted. which included reduction and compression of tariffs, introduction of a simpler import tax structure, gradual removal of non-tariff import restrictions, exemption income tax exemption, bonded storage, and flexible exchange This includes the promotion of exports. Prior to the 1990s, Bangladesh depended mostly on quantitative limits established to manage imports of primarily agricultural items. In 1994, about 37% of agricultural tariff lines (21% of total goods) were banned. By 1984, all quantitative limits on agricultural products had been repealed, leaving only 2% of all customs commodities subject to very highly restrictions.

Some of the ban or the restriction on the export of the good quality rice was demolished, but the export of the coarse rice remained prohibited. In the early 1990s, liberalization of trade led to the significant reductions in tariffs. Total protective tariffs (both tariffs and quasi-tariffs) fell to just 32% in the year 1995 from 74% in the year 1991. Similarly, duties on import and overall taxation on all the imports of key agricultural commodities reduced sharply in the early 1990s. The comparatively high tariffs applied on refined edible oils, sugar, powdered milk, and spices. whereas cotton, wheat, canola and lentils were subjected to lower-tariffs. In the past few years, reform of trade has caught up. Since, the tariffs have fallen to 19% in the year 2003 from 29% in the year 1995, surcharges, royalties, regulatory taxes, value-added taxes and all the other surcharges have increased mainly due to a significant increase in the surcharges that have increased sharply. As a result, to this, the overall safeguard has remained unchanged in the fields of the average since the year of the mid of the year 1990.

2000

Trade Policy plays a very important role and it is a policy that is mixed it is regulated by a structured set of policies, regulations and actions. The main aim performed by the export policy is that it develops oriented sector of export and also used as the industrialization by boosting exports, increase the performance of production of all the sector that are related to the oriented sectors of the export, and it is assist by the help of the sector that named as creating local industry capacity. Import policy seeks to align the import regime with the WTO, or World Trade Organization. It also intends that to streamline the procedures for importing capital goods and also import various types of raw material, to give very good facilities for all types of features, so that they allow them for the entry of emergency supplies. The Import Policy Order of 2009-2012 and Export Policy of 2009-2012 gives detailed import and export objectives, priority areas that require special assistance, import replacement, as well as initiatives to encourage local market orientation and labour-intensive industries The Department of Commerce developed and implemented import and export policies, while the Ministry of Finance sets import charges, quasi-duties, and other tariffs. Monetary policy aims to prevent inflation and promote equitable growth by changing a variety of financial factors such as the money supply, interest rate levels, and currency values. The Industrial Policy of 2010 and the SME Policy Strategies of 2005 give policy directions for industrialization, particularly for micro, small, and medium-sized firms, as well as for enterprises. The famous Plan named as the Sixth Five Year plan from year (2011-15) and another plan that was Ten-Year Plan from year (2011-21) explain the large import and export objectives established from year 2011 to the year 2021. As a result, Bangladesh has a range of very medium and follow rules that called as long-term rules, and also large term regulations, and legislation that are in the place to govern foreign commerce. Policy in import is a legally enforceable policy that have the various documents and also the Export policies are not verify legally and also not enforceable. Tariff reductions on various fields and also on the food and industrial items have not present and also developed at the similar price. Tariffs on agricultural goods were about 18.5% in the 2000s, although tariffs on agricultural intermediate items have been cut at a quicker rate. In the 2000s, industrial items were further reduced.

RESULT AND DISCUSSION

The food sector is dependent on agriculture and transportation sectors for the access to raw materials. The food industry is struggling nowadays due to unavailability of quality supplies which are needed for the production of good quality food. Despite numerous efforts made by the governments of several nations, food production and food safety concerns are increasing at an alarming rate. Some nations are already facing food crisis due to shortage of food resources due to economical or environmental factors.

The food industry needs to be supported better with agricultural growth. This is because the agriculture sector is the only backbone of food manufacturing industries that can make a country independent in terms of food production. However, several countries are facing major issues in the agriculture sector due to the kind of laws they have enforced. Figure (3,4) illustrate the contrast between percentage of people that were involved in agriculture activities in the 19th century and the 20th century of some major countries of the world. It can be observed that while China has maintained its percentage share of people in agriculture, countries like Bangladesh and India have become more dependent on agriculture. However, countries like USA, Canada and Russia, having the biggest areas of land, have seen a decrease in their share of population which is involved in agriculture. This lack of people's trust and interest in agriculture provisions of their nation is the cause of decline in food production all over the world. Another factor which is the reason of food manufacturing and production crisis is the reduction of economic gains that relate to these sectors. Farming is still considered as an inferior occupation in many parts of the world and many lineages of farmers are shifting towards other sectors leaving behind their farms in search of better facilities in the urban areas or corporate or private jobs.

Table 2 depicts a contrast between the change in area and agriculture or different nations evaluated each decade. It can be seen that the area under agricultural has declined over the years resulting in a major decrease in food production across the globe. Several areas have become infertile due to decrease in farming in those areas. Globalization and development of cities is also a reason due to which the overall production of the world has declined. The supporting block which has sustained the food availability despite these drawbacks has been technological advancements in agriculture field and also in fields of food manufacturing sectors. The most useful need of fertilizers and urea in the farms has become more effective and availability of these fertilizers and urea has increased. This has resulted in increased production on the same area of land. Figure 5 portrays the Global fertilizer consumption of 2019. It can be noticed that countries which are one of the largest food grain producers like India and China are the biggest consumers of fertilizers as well. It can indeed be noticed that the advancement in agriculture made by the wide spread use and awareness about fertilizers, pesticides, urea and other similar products, has increased production in many countries. Fertilizers does however act as an adulterant in fruits and vegetables if sprayed in excess.

Figure 3. Nation-wise population dependence on agriculture in 19th century

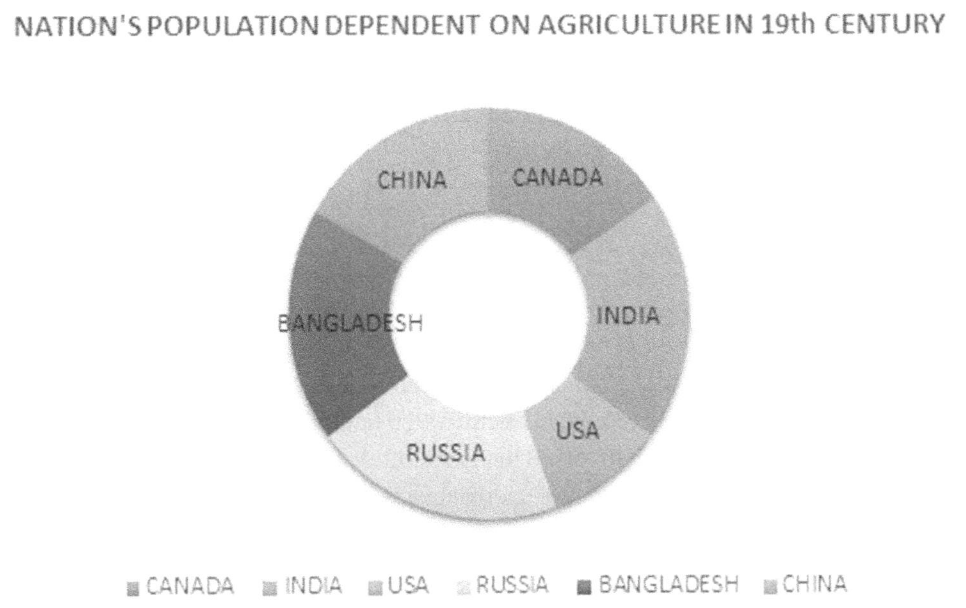

Figure 4. Nation-wise population dependence on agriculture in 20th century

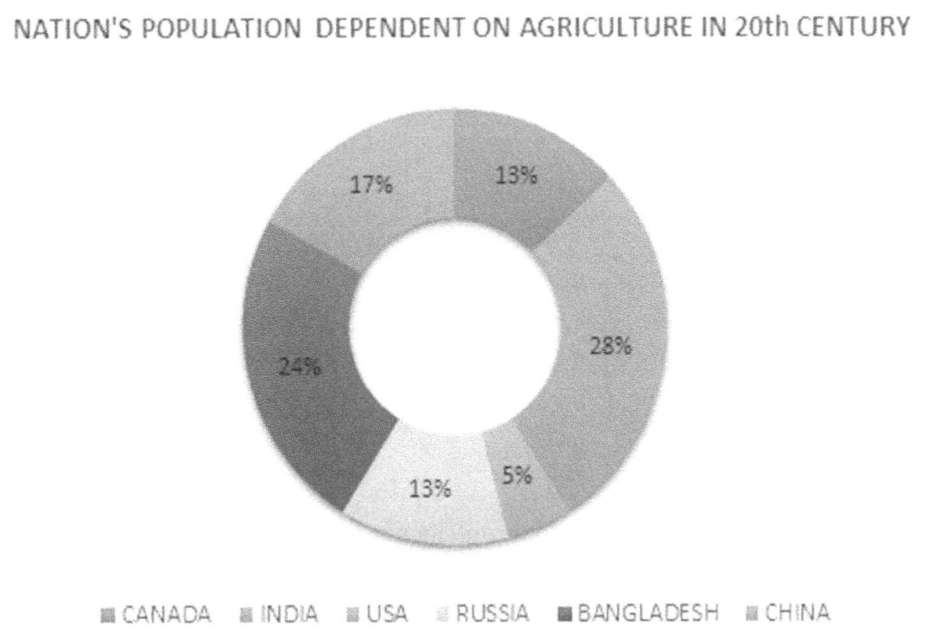

Table 2. Change in land area under agriculture of countries each decade since 1950

Country	Land Area Under Agriculture (in Million Acres)								Total Effect Since 1950
Year	1950-51	1960-61	19070-71	1980-81	1990-91	2000-01	2010-11	2020-21	
USA	1,178.46	988.42	900.50	889.08	887.18	862.43	843.39	845.29	28.27% decrease
India	476.91	477.63	485.75	493.06	496.31	494.69	490.62	489.00	0.025% increase
Canada	170.12	170.22	172.69	170.22	170.22	170.22	157.88	156.91	7.765% decrease
China	792.38	847.71	938.54	1071.28	1257.60	1294.86	1306.50	1306.50	64.88% increase

The land areas under agriculture of most countries have decreased which has disturbed the balance of environment as well. Use of fertilizers, pesticides and urea is indeed getting positive results in terms of production. However, a decrease in total area of cultivated land indicates a depletion in the inclination of the people to cultivate the land or get involved in agricultural activities. However, countries like China have proved to be most successful in terms of increasing land area cultivation as well as maintaining high standard quality production and sustaining their share of population contributing to agriculture.

Figure 5. Nation wise global fertilizer consumption in 2019

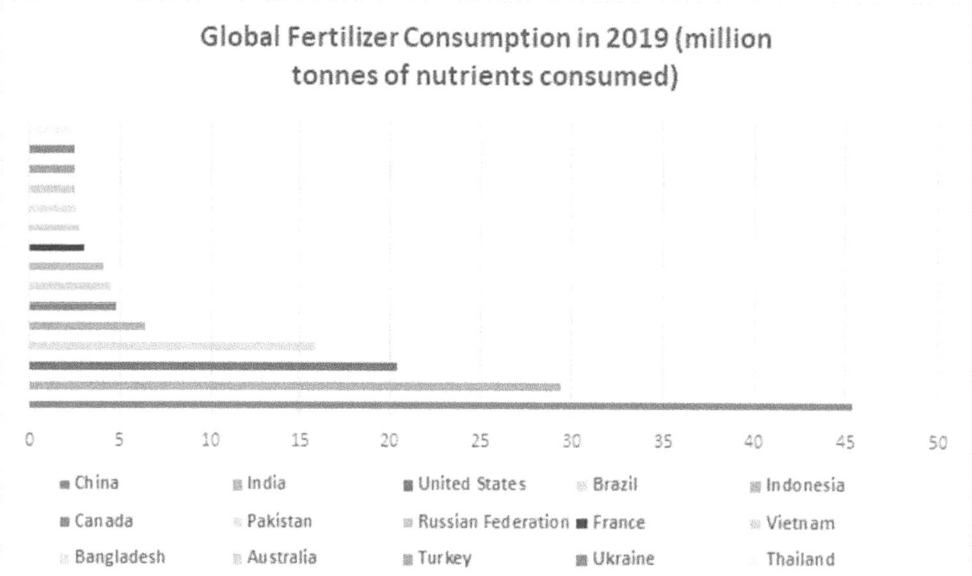

The food production has also increased along with increasing requirement of food worldwide. Figure 6 demonstrates the increasing production of food since 1950-51 to 2021-22. In past 7 decades, the food consumption of the world has grown by 250%. However, the land under cultivation has also increased for several crops but the increase is mostly marginal. For other crops like coarse cereal, the land area under

Figure 6. Increase in production of food from 1950-2022

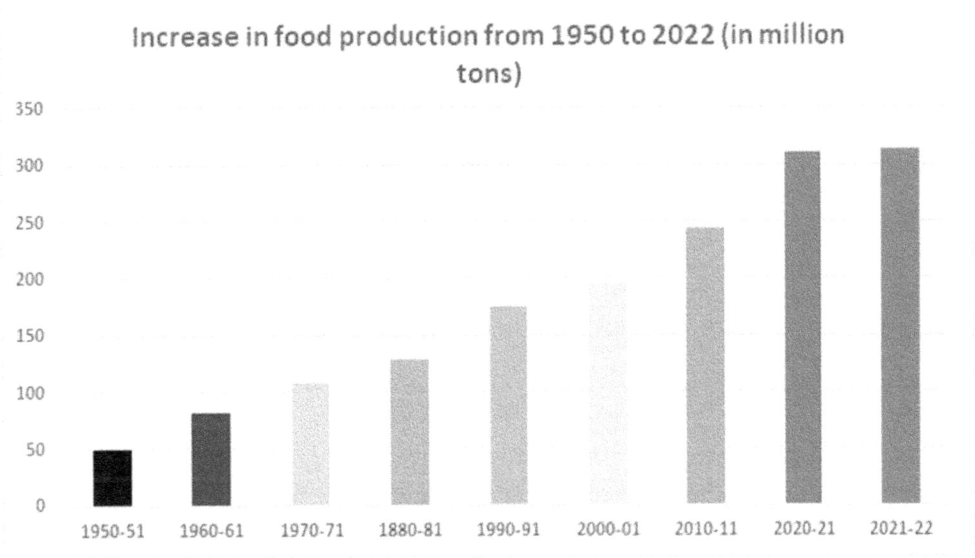

cultivation has decreased, but production has increased. This is the benefit of the use of fertilizers, pesticides and urea like products in the fields which have increased the yield of the crops drastically. Table 3 illustrates these trends in the area of cultivation and production since 1980-1981 to 2010-2011. Despite these progressions in the technology involved in agriculture and production, the food production of the world is yet to match pace with the increasing population. The scarcity of quality food in many parts of the world provides evidence to this statement as many countries are facing a rapidly approaching food crisis and a few countries are already facing this problem. This is because efforts in making agriculture a more economically profitable occupation are required by the governments of many nations. This needs to be implemented in several countries of middle Africa which is also considered as the food basket of the world. The economies of common farmers in Africa region and in the South Asian countries are very less which has become one of the main reasons of depletion in cultivated area in the world every year. Figure 7 shows the average income of agriculture-based households in various countries of the world. It is evident from the data that the economic situation of under developed nations in the field of agriculture is poorer than their per capita income. This suggests that less chances of profitability in the field of agricultural sector are emerged as most innovative reason is, which it struggling in the matter of growth in the past 7 decades.

It is known that the major problems of the agriculture and food manufacturing industry is known only to the people working in these sectors. Table 4 deals with some major issues of farmers and the food manufacturing industries in various countries of the world. These problems are required to be solved by the government taking into consideration the economy and development of the respective nation.

Table 3. Patterns in cultivated area and produced quantity of essential crops: From 1980-81 to 2010-11

Crops	Cultivated Area (Million Hectares)			Produced Quantity (Million Tons)		
Year	1980-81	1990-91	2010-11	1980-81	1990-91	2010-11
RICE	40.2	42.4	43.9	53.6	75.8	95.1
WHEAT	23.6	24.5	28.3	41.8	57.5	80.3
COA	41.6	33.7	27.6	30.8	31.2	38.4
PULSES	23.3	22.3	23.1	12.2	12.6	14.7
FOOD GRAINS	128.6	122.4	122.9	138.5	177.3	227.9
OILS SEEDS	18.5	27.2	25.7	12.7	21.2	28.4
SUGARCANE	3.2	3.75	4.49	184.2	238.3	305.6
FRUITS /VEGETABLES	5.1	8.4	13.9	61.8	95.7	188.6
CONDIMENTS & SPICES	2.3	2.2	2.5	1.6	2.4	4.21
COTTON	7.8	7.6	9.8	7.2	10.7	24.2
Total Area Sown	143.2	143.3	141.7	-	-	-
Total Area Cropped	177.3	185.9	195.1	-	-	-

CONCLUSION

Agriculture and food manufacturing sector is the only factor that can reduce food scarcity and support food safety in the present scenario. The laws revolving around production of crops and manufacturing of food products govern this domain. The changes that have occurred in the past 7 decades have played a very essential function for defining the new technology of production of various food items and safety. However, it has also somehow resulted in decrease in the total agricultural produce in the world. The decline in agricultural activities Few years ago situation takes place in which increase in food manufacturing sector has also created a gap between crop producers and food manufacturers. This review paper illustrates the major changes alterations that have happened in the field of manufacturing of food and also takes place in the fields of agricultural. The laws issued by the United Nations to increase the level of various types of food security and food safety and the initiatives taken by Food and Agricultural Organization. The authors of this review paper found by analysis that more work has to be done by the International as well as National organizations to promote agricultural activities and to ease the manufacturing of food products by reducing the gap that has emerged between these two sectors in the past few decades. Economical support is also required to be improved for the crop producers and manufacturing of food products has to be done taking food safety and security in consideration. The growing instability in the world regarding food availability is also an alarming situation. The scarcity of food should also be looked upon as the major crisis for the world's population in the next two decades. More research is required in the domain of manufacturing sector to devise new methods of acquiring sufficient quantity of quality food resources. Thus, it is concluded that more efforts in research are required to formulate new laws and policies for the growth of Agriculture and Food manufacturing industries.

REFERENCES

Abrouk, M., Ahmed, H. I., Cubry, P., Šimoníková, D., Cauet, S., Pailles, Y., Bettgenhaeuser, J., Gapa, L., Scarcelli, N., Couderc, M., Zekraoui, L., Kathiresan, N., Čížková, J., Hřibová, E., Doležel, J., Arribat, S., Bergès, H., Wieringa, J. J., Gueye, M., ... Krattinger, S. G. (2020). Fonio millet genome unlocks African orphan crop diversity for agriculture in a changing climate. *Nature Communications*, *11*(1), 4488. doi:10.103841467-020-18329-4 PMID:32901040

Adegbeye, M. J., Reddy, P. R. K., Obaisi, A. I., Elghandour, M. M. M. Y., Oyebamiji, K. J., Salem, A. Z. M., ... Camacho-Díaz, L. M. (2020). Sustainable agriculture options for production, greenhouse gasses and pollution alleviation, and nutrient recycling in emerging and transitional nations-An overview. *Journal of Cleaner Production*, *242*, 118319. doi:10.1016/j.jclepro.2019.118319

Adesina, I., Bhowmik, A., Sharma, H., & Shahbazi, A. (2020). A review on the current state of knowledge of growing conditions, agronomic soil health practices and utilities of hemp in the United States. *Agriculture*, *10*(4), 129. doi:10.3390/agriculture10040129

Agrimonti, C., Lauro, M., & Visioli, G. (2021). Smart agriculture for food quality: Facing climate change in the 21st century. *Critical Reviews in Food Science and Nutrition*, *61*(6), 971–981. doi:10.1080/10408398.2020.1749555 PMID:32270688

Arnold, N., Brunori, G., Dessein, J., Galli, F., Ghosh, R., Loconto, A. M., & Maye, D. (2022). Governing food futures: Towards a 'responsibility turn'in food and agriculture. *Journal of Rural Studies*, *89*, 82–86. doi:10.1016/j.jrurstud.2021.11.017

Artmann, M., Breuste, J., Breuste, J., Artmann, M., Liu, Y., Breuste, J., & Dilini, M. M. G. S. (2020). Urban agriculture—more than food production. In *Making Green Cities: Concepts, Challenges and Practice* (pp. 75–176). Springer International Publishing. doi:10.1007/978-3-030-37716-8_4

Aryal, J. P., Sapkota, T. B., Khurana, R., Khatri-Chhetri, A., Rahut, D. B., & Jat, M. L. (2020). Climate change and agriculture in South Asia: Adaptation options in smallholder production systems. *Environment, Development and Sustainability*, *22*(6), 5045–5075. doi:10.100710668-019-00414-4

Ashraf, S. A., Siddiqui, A. J., Abd Elmoneim, O. E., Khan, M. I., Patel, M., Alreshidi, M., ... Adnan, M. (2021). Innovations in nanoscience for the sustainable development of food and agriculture with implications on health and environment. *The Science of the Total Environment*, *768*, 144990. doi:10.1016/j.scitotenv.2021.144990 PMID:33736303

Asrat, D., & Anteneh, A. (2020). Status of food insecurity in dryland areas of Ethiopia: A review. *Cogent Food & Agriculture*, *6*(1), 1853868. doi:10.1080/23311932.2020.1853868

Barichello, R. (2020). The COVID-19 pandemic: Anticipating its effects on Canada's agricultural trade. *Canadian Journal of Agricultural Economics/Revue canadienne d'agroeconomie, 68*(2), 219-224.

Beckman, J., Ivanic, M., Jelliffe, J. L., Baquedano, F. G., & Scott, S. G. (2020). *Economic and food security impacts of agricultural input reduction under the European Union Green Deal's Farm to Fork and biodiversity strategies* (No. 1473-2020-1039). Academic Press.

Binns, C. W., Lee, M. K., Maycock, B., Torheim, L. E., Nanishi, K., & Duong, D. T. T. (2021). Climate change, food supply, and dietary guidelines. *Annual Review of Public Health*, *42*(1), 233–255. doi:10.1146/annurev-publhealth-012420-105044 PMID:33497266

Bryngelsson, D., Wirsenius, S., Hedenus, F., & Sonesson, U. (2016). How can the EU climate targets be met? A combined analysis of technological and demand-side changes in food and agriculture. *Food Policy*, *59*, 152–164. doi:10.1016/j.foodpol.2015.12.012

Butu, A., Brumă, I. S., Tanasă, L., Rodino, S., Dinu Vasiliu, C., Doboş, S., & Butu, M. (2020). The impact of COVID-19 crisis upon the consumer buying behavior of fresh vegetables directly from local producers. Case study: The quarantined area of Suceava County, Romania. *International Journal of Environmental Research and Public Health*, *17*(15), 5485. doi:10.3390/ijerph17155485 PMID:32751368

Chandra, M. S., Srinivasulu, M., Yadav, P. S., Ramesh, B., Narasimha, G., & Chandrasekhar, T. (2021). Value added products from agriculture, paper and food waste: a source of bioenergy production. *Bioenergy Research: Commercial Opportunities & Challenges*, 91-126.

Cook, S., Jackson, E. L., Fisher, M. J., Baker, D., & Diepeveen, D. (2022). Embedding digital agriculture into sustainable Australian food systems: Pathways and pitfalls to value creation. *International Journal of Agricultural Sustainability*, *20*(3), 346–367. doi:10.1080/14735903.2021.1937881

De Castro, P., Miglietta, P. P., & Vecchio, Y. (2020). The Common Agricultural Policy 2021–2027: A new history for European agriculture. *Ital. Rev. Agric. Econ*, *75*(3), 5–12.

de Souza, M., Koo-Oshima, S., Kahil, T., Wada, Y., Qadir, M., Jewitt, G., ... Zhang, L. (2021). *Food and agriculture*. Academic Press.

Deaton, B. J., & Deaton, B. J. (2020). Food security and Canada's agricultural system challenged by COVID-19. *Canadian Journal of Agricultural Economics/Revue canadienne d'agroeconomie, 68*(2), 143-149.

Delgado, C., Rosegrant, M., Steinfeld, H., Ehui, S., & Courbois, C. (2001). Livestock to 2020: The next food revolution. *Outlook on Agriculture*, *30*(1), 27–29. doi:10.5367/000000001101293427

Duque-Acevedo, M., Belmonte-Ureña, L. J., Plaza-Úbeda, J. A., & Camacho-Ferre, F. (2020). The management of agricultural waste biomass in the framework of circular economy and bioeconomy: An opportunity for greenhouse agriculture in Southeast Spain. *Agronomy (Basel)*, *10*(4), 489. doi:10.3390/agronomy10040489

El Chami, D., Daccache, A., & El Moujabber, M. (2020). How can sustainable agriculture increase climate resilience? A systematic review. *Sustainability (Basel)*, *12*(8), 3119. doi:10.3390u12083119

Elvan, O. D., Birben, Ü., Özkan, U. Y., Yıldırım, H. T., & Türker, Y. Ö. (2021). Forest fire and law: An analysis of Turkish forest fire legislation based on Food and Agriculture Organization criteria. *Fire Ecology*, *17*(1), 1–15. doi:10.118642408-021-00102-7

Engler, N., & Krarti, M. (2021). Review of energy efficiency in controlled environment agriculture. *Renewable & Sustainable Energy Reviews*, *141*, 110786. doi:10.1016/j.rser.2021.110786

Esham, M., Jacobs, B., Rosairo, H. S. R., & Siddighi, B. B. (2018). Climate change and food security: A Sri Lankan perspective. *Environment, Development and Sustainability, 20*(3), 1017–1036. doi:10.100710668-017-9945-5

Fairbairn, M., Kish, Z., & Guthman, J. (2022). Pitching agri-food tech: Performativity and non-disruptive disruption in Silicon Valley. *Journal of Cultural Economics, 15*(5), 652–670. doi:10.1080/17530350.2022.2085142

Friedmann, H. (2021). Changes in the international division of labor: agri-food complexes and export agriculture. In *Towards a new political economy of agriculture* (pp. 65–93). Routledge. doi:10.4324/9780429269493-4

Garcia, S. N., Osburn, B. I., & Jay-Russell, M. T. (2020). One health for food safety, food security, and sustainable food production. *Frontiers in Sustainable Food Systems, 4*, 1. doi:10.3389/fsufs.2020.00001

Guggisberg, S. (2022). Transparency in the activities of the Food and Agriculture Organization for sustainable fisheries. *Marine Policy, 136*, 104498. doi:10.1016/j.marpol.2021.104498

Guiné, R. P., Florença, S. G., Barroca, M. J., & Anjos, O. (2020). The link between the consumer and the innovations in food product development. *Foods, 9*(9), 1317. doi:10.3390/foods9091317 PMID:32962007

Jarzebski, M. P., Ahmed, A., Boafo, Y. A., Balde, B. S., Chinangwa, L., Saito, O., von Maltitz, G., & Gasparatos, A. (2020). Food security impacts of industrial crop production in sub-Saharan Africa: A systematic review of the impact mechanisms. *Food Security, 12*(1), 105–135. doi:10.100712571-019-00988-x

Javed, S. A., Haider, A., & Nawaz, M. (2020). How agricultural practices managing market risk get attributed to climate change? Quasi-experiment evidence. *Journal of Rural Studies, 73*, 46–55. doi:10.1016/j.jrurstud.2019.11.020

Karthikeyan, L., Chawla, I., & Mishra, A. K. (2020). A review of remote sensing applications in agriculture for food security: Crop growth and yield, irrigation, and crop losses. *Journal of Hydrology (Amsterdam), 586*, 124905. doi:10.1016/j.jhydrol.2020.124905

Kuczuk, A., & Widera, K. (2021). A Greater Share of Organic Agriculture in Relation to Food Security Resulting from the Energy Demand Obtained from Food—Scenarios for Poland until 2030. *Energies, 14*(21), 6959. doi:10.3390/en14216959

Kuhnlein, H. V., & Turner, N. J. (1991). *Traditional plant foods of Canadian indigenous peoples: nutrition, botany, and use* (Vol. 8). Taylor & Francis.

Kuiken, T., Barrangou, R., & Grieger, K. (2021). (Broken) promises of sustainable food and agriculture through new biotechnologies: The CRISPR case. *The CRISPR Journal, 4*(1), 25–31. doi:10.1089/crispr.2020.0098 PMID:33570455

Kumari, T., Singh, R., & Verma, P. (2019). Agriculture, forest and environmental management trinity: towards environmental sustainability and climate change mitigation. In Sustainable Agriculture, Forest and Environmental Management. Springer.

Lacko, A. M., Guilkey, D., Popkin, B., & Ng, S. W. (2022). Associations Among Select State Policies and the Nutritional Quality of Household Packaged Food Purchases in the United States from 2008 Through 2017. *Journal of the Academy of Nutrition and Dietetics, 122*(4), 731–744. doi:10.1016/j.jand.2021.09.017 PMID:34626825

Lajoie-O'Malley, A., Bronson, K., van der Burg, S., & Klerkx, L. (2020). The future (s) of digital agriculture and sustainable food systems: An analysis of high-level policy documents. *Ecosystem Services, 45*, 101183. doi:10.1016/j.ecoser.2020.101183

Liu, W., Hao, Z., Florkowski, W. J., Wu, L., & Yang, Z. (2022). A Review of the Challenges Facing Global Commercialization of the Artificial Meat Industry. *Foods, 11*(22), 3609. doi:10.3390/foods11223609 PMID:36429201

Liu, Y., Ma, X., Shu, L., Hancke, G. P., & Abu-Mahfouz, A. M. (2020). From Industry 4.0 to Agriculture 4.0: Current status, enabling technologies, and research challenges. *IEEE Transactions on Industrial Informatics, 17*(6), 4322–4334. doi:10.1109/TII.2020.3003910

Mahama, S., Manteaw, S. A., & Decker, E. (2021). Gender perceptions on the causes of climate variation and its effects on cassava production among farmers in Ghana. *Cogent Food & Agriculture, 7*(1), 1911398. doi:10.1080/23311932.2021.1911398

Malone, T., Schaefer, K. A., & Wu, F. (2021). The razor's edge of "essential" labor in food and agriculture. *Applied Economic Perspectives and Policy, 43*(1), 368–381. doi:10.1002/aepp.13070

McGreevy, S. R., Rupprecht, C. D., Niles, D., Wiek, A., Carolan, M., Kallis, G., Kantamaturapoj, K., Mangnus, A., Jehlička, P., Taherzadeh, O., Sahakian, M., Chabay, I., Colby, A., Vivero-Pol, J.-L., Chaudhuri, R., Spiegelberg, M., Kobayashi, M., Balázs, B., Tsuchiya, K., ... Tachikawa, M. (2022). Sustainable agrifood systems for a post-growth world. *Nature Sustainability, 5*(12), 1011–1017. doi:10.103841893-022-00933-5

McMichael, P. (2021). Political economy of the global food and agriculture system. In *Rethinking food and agriculture* (pp. 53–75). Woodhead Publishing. doi:10.1016/B978-0-12-816410-5.00003-7

Mdegela, R. H., Mwakapeje, E. R., Rubegwa, B., Gebeyehu, D. T., Niyigena, S., Msambichaka, V., Nonga, H. E., Antoine-Moussiaux, N., & Fasina, F. O. (2021). Antimicrobial use, residues, resistance and governance in the food and agriculture sectors, Tanzania. *Antibiotics (Basel, Switzerland), 10*(4), 454. doi:10.3390/antibiotics10040454 PMID:33923689

Mdemu, M., Kissoly, L., Bjornlund, H., Kimaro, E., Christen, E. W., van Rooyen, A., ... Ramshaw, P. (2020). The role of soil water monitoring tools and agricultural innovation platforms in improving food security and income of farmers in smallholder irrigation schemes in Tanzania. *International Journal of Water Resources Development, 36*(sup1), S148-S170.

Merck, A. W., Grieger, K. D., Cuchiara, M., & Kuzma, J. (2022). What Role Does Regulation Play in Responsible Innovation of Nanotechnology in Food and Agriculture? Insights and Framings from US Stakeholders. *Bulletin of Science, Technology & Society, 42*(3), 85–103. doi:10.1177/02704676221102066

Mir, H., Zaraatgari, R., & Sotoudeh, R. (2021). Improving the food and agriculture sector tehran stock exchange by using artificial intelligence. *Agricultural Marketing and Commercialization Journal*, *5*(2), 90–114.

Miśniakiewicz, M., Łuczak, J., & Maruszewska, N. (2021). Improvement of organic farm assessment procedures on the example of organic farming in poland—Recommendations for organic farming in Poland. *Agronomy (Basel)*, *11*(8), 1560. doi:10.3390/agronomy11081560

Myers, S. S., Smith, M. R., Guth, S., Golden, C. D., Vaitla, B., Mueller, N. D., Dangour, A. D., & Huybers, P. (2017). Climate change and global food systems: Potential impacts on food security and undernutrition. *Annual Review of Public Health*, *38*(1), 259–277. doi:10.1146/annurev-publhealth-031816-044356 PMID:28125383

Nguyen, N. T. T., Nguyen, L. M., Nguyen, T. T. T., Liew, R. K., Nguyen, D. T. C., & Van Tran, T. (2022). Recent advances on botanical biosynthesis of nanoparticles for catalytic, water treatment and agricultural applications: A review. *The Science of the Total Environment*, *827*, 154160. doi:10.1016/j.scitotenv.2022.154160 PMID:35231528

Pietrzyck, K., Berke, N., Wendel, V., Steinhoff-Wagner, J., Jarzębowski, S., & Petersen, B. (2021). Understanding the Importance of International Quality Standards Regarding Global Trade in Food and Agricultural Products: Analysis of the German Media. *Agriculture*, *11*(4), 328. doi:10.3390/agriculture11040328

Pretty, J. N. (1998). Supportive policies and practice for scaling up sustainable agriculture. *Facilitating sustainable agriculture: Participatory learning and adaptive management in times of environmental uncertainty*, 23-45.

Rejeb, A., Keogh, J. G., Zailani, S., Treiblmaier, H., & Rejeb, K. (2020). Blockchain technology in the food industry: A review of potentials, challenges and future research directions. *Logistics*, *4*(4), 27. doi:10.3390/logistics4040027

Ribi Forclaz, A. (2019). From reconstruction to development: The early years of the Food and Agriculture Organization (FAO) and the conceptualization of rural welfare, 1945–1955. *The International History Review*, *41*(2), 351–371. doi:10.1080/07075332.2018.1478873

Ryan, M., van der Burg, S., & Bogaardt, M. J. (2021). Identifying key ethical debates for autonomous robots in agri-food: A research agenda. *AI and Ethics*, 1–15.

Shekhawat, K., Singh, V. K., Rathore, S. S., Raj, R., & Das, T. K. (2021). Achieving sustainability in food systems: Addressing changing climate through real time nitrogen and weed management in a conservation agriculture-based maize–wheat system. *Sustainability (Basel)*, *13*(9), 5010. doi:10.3390u13095010

Siamabele, B. (2021). The significance of soybean production in the face of changing climates in Africa. *Cogent Food & Agriculture*, *7*(1), 1933745. doi:10.1080/23311932.2021.1933745

Srivastava, A., & Dashora, K. (2021). A Fuzzy ISM approach for modelling electronic traceability in agri-food supply chain in India. *Annals of Operations Research*, 1–19.

Tubb, C., & Seba, T. (2021). Rethinking food and agriculture 2020-2030: The second domestication of plants and animals, the disruption of the cow, and the collapse of industrial livestock farming. *Industrial Biotechnology (New Rochelle, N.Y.)*, *17*(2), 57–72. doi:10.1089/ind.2021.29240.ctu

Vos, R., & Bellù, L. G. (2019). Global trends and challenges to food and agriculture into the 21st century. *Sustainable food and agriculture*, 11-30.

Wang, L., Xu, L., Zheng, Z., Liu, S., Li, X., Cao, L., Li, J., & Sun, C. (2021). Smart contract-based agricultural food supply chain traceability. *IEEE Access : Practical Innovations, Open Solutions*, *9*, 9296–9307. doi:10.1109/ACCESS.2021.3050112

Wezel, A., Herren, B. G., Kerr, R. B., Barrios, E., Gonçalves, A. L. R., & Sinclair, F. (2020). Agroecological principles and elements and their implications for transitioning to sustainable food systems. A review. *Agronomy for Sustainable Development*, *40*(6), 1–13. doi:10.100713593-020-00646-z

Wordofa, M. G., & Sassi, M. (2020). Impact of agricultural interventions on food and nutrition security in Ethiopia: Uncovering pathways linking agriculture to improved nutrition. *Cogent Food & Agriculture*, *6*(1), 1724386. doi:10.1080/23311932.2020.1724386

Workie, E., Mackolil, J., Nyika, J., & Ramadas, S. (2020). Deciphering the impact of COVID-19 pandemic on food security, agriculture, and livelihoods: A review of the evidence from developing countries. *Current Research in Environmental Sustainability*, *2*, 100014. doi:10.1016/j.crsust.2020.100014 PMID:34977605

World Health Organization. (2020). *The state of food security and nutrition in the world 2020: transforming food systems for affordable healthy diets* (Vol. 2020). Food & Agriculture Org.

Xu, L., Yuan, S., & Man, J. (2020). Changes in rice yield and yield stability in China during the past six decades. *Journal of the Science of Food and Agriculture*, *100*(8), 3560–3569. doi:10.1002/jsfa.10385 PMID:32198766

Yanbykh, R., Saraikin, V., & Lerman, Z. (2020). Changes in Russia's agrarian structure: What can we learn from agricultural census? *Russian Journal of Economics*, *6*(1), 26–41. doi:10.32609/j.ruje.6.49746

Zhou, X., & Wang, S. (2019). Recent developments in radio frequency drying of food and agricultural products: A review. *Drying Technology*, *37*(3), 271–286. doi:10.1080/07373937.2018.1452255

Chapter 13
Mechanized Functioning of Tomato and Lettuce Greenhouses Using IoT and Machine Learning Techniques

Shipra Saraswat
ABES Institute of Technology, Ghaziabad, India

Rijwan Khan
https://orcid.org/0000-0003-3354-3047
ABES Institute of Technology, Ghaziabad, India

Ayush Gupta
https://orcid.org/0009-0007-2583-9223
ABES Institute of Technology, Ghaziabad, India

Abhay Mehta
ABES Institute of Technology, Ghaziabad, India

ABSTRACT

This chapter explores the mechanized functioning of tomato and lettuce greenhouses through the integration of internet of things (IoT) and machine learning techniques. It highlights the transformative impact of these technologies on the agricultural industry, enabling enhanced productivity, optimized resource management, and improved crop yield. By leveraging IoT, greenhouses become interconnected systems that allow real-time monitoring and control of environmental parameters. Machine learning (ML) algorithms examine data collected from sensors and historical records to predict crop growth patterns, disease outbreaks, and pest infestations. This enables proactive decision-making and efficient resource allocation. The adoption of IoT in greenhouse farming not only automates operations but also promotes sustainable practices, reducing labor costs and optimizing resource utilization. Overall, this chapter showcases the potential of IoT and ML technologies in revolutionizing greenhouse farming for increased efficiency, sustainability, and food security of tomato and lettuce crops.

DOI: 10.4018/979-8-3693-0200-2.ch013

INTRODUCTION

Water and food shortages are getting worse, because of the unchecked global population growth. According to the technical analysis issued in (Contreras et al., 2023), 9.15 billion people will inhabit the planet by the year 2050. This trend drives an increase in food production to meet the needs of a growing global population, which results in a shortage of farmland and a decline in the number of farmers as fewer people choose to engage in farming. A few issues confront agriculture today including the expanding global population, the effects of climate change, stringent market rules, and energy inefficiencies (Guesbaya, et al., 2023). To meet these problems, the hierarchical agricultural production system must be implemented with technology at all stages, which calls for the use of resilient and adaptive management systems. The most well-known facilities that can help to address these issues are greenhouses (Guesbaya, et al., 2023). Greenhouses serve as protective environments for crops, providing optimal conditions for their growth. To ensure the efficient development of plants and fruits, it is essential to employ actuators that regulate crucial factors including air temperature, humidity, and CO2 concentration (Guesbaya, et al., 2023). These actuators play a vital role in maximizing plant growth and yield within the greenhouse setting. This paper reviews the production of lettuce and tomatoes in greenhouses using IOT and Machine Learning (Kochhar and Kumar, 2019). In greenhouses at northern latitudes, additional lighting is utilized to increase the daily light exposure of crops. It is preferred to use the freely available sunshine as much as possible to improve plant growth, production, and quality while minimizing electricity usage of artificial lighting (Mohagheghi and Moallem, 2023). When compared to conventional horticulture light sources like metal halide (MH) and high-pressure sodium (HPS), the development of horticultural light- emitting diodes (LEDs) has facilitated this integration. Here in (Mohagheghi and Moallem, 2023), the crop selected for production was a green mini romaine organic lettuce variety called Dragoon (LATIN NAME: Lactuca sativa). The reason behind this crop was that it was a great performer in a climate-controlled greenhouse environment and suitable to be grown successfully using hydroponic growing methods or other soil-fewer growing systems (Flores et al., 2016). In order to give farmers and automatic controllers the knowledge necessary to choose when to activate greenhouse actuators in order to enhance crop conditions or protect them, this calls for sensors (Mohagheghi and Moallem, 2023). that monitor the evolution of the variables of interest. Smart technologies such as Internet of Things (IoT), using sensors, artificial intelligence devices, smart surveillance and monitoring systems, play a role in overcoming the main challenges in conventional greenhouse farming as mentioned in (Ahmad et al., 2023) such as controlling the local climate of the greenhouse, monitoring the crop growth parameters, and setting the crop harvest time (Chang et al., 2021). In order to automate elements like temperature, humidity, and photoperiod utilizing sensor- based technologies, intelligent greenhouse automation systems should maintain the microclimate using setpoints and protect plants against abrupt changes in the microclimate as mentioned in (Ahmad et al., 2023). These recent advancements in information technology allow traditional greenhouses to use less labor and overcome challenges sustaining microclimatic conditions (Ojha et al., 2015). Tomato (Solanum lycopersicum L.) is one of the most economically important crops in the world and undergoes both vegetative and generative growth (Dubey and Jalal, 2013). Vegetative growth depicts growth of the roots, stems and leaves, and generative growth depicts growth of the flowers and fruits (Dutta et al., 2023). During generative growth, tomato fruits undergo several changes such as increases in their size, color changes, and so on. In reference to (Seo et al., 2021), the paper explores the development of a robotic system designed for real-time monitoring of a substantial volume of tomatoes, focusing on the specific requirements for tomato production. The study delves into the implementation of this system and discusses the criteria essential for successful tomato cultivation. Initially, a tomato detection model was formulated through Faster R-CNN, along with an accounting model utilizing a centroid- based tracking algorithm. The researchers also explored the use of a color model that could withstand variations in external lighting conditions. They further devised an image-based maturity standard for assessing the maturity of tomato fruits. This newly developed standard facilitated the classification of tomato fruits into six different maturity levels. Finally, the researchers employed centroid measurements of the detected bounding boxes to count the number of tomatoes in each maturity stage in (Seo et al., 2021). The research conducted by researchers from Indonesia (Aminulloh et al., 2019) concentrates on extracting features and developing a growth model for tomatoes using sensors and actuators. The objective was to regulate the growth conditions of tomato plants within a greenhouse. The greenhouse was equipped with an information system that collected sensor data to evaluate the environmental conditions necessary for optimal tomato plant growth (Nasution et al., 2020).

LITERATURE REVIEW

IoT has had a strong impetus in recent years, with numerous works proposing IoT architectures and platforms for applications such as agriculture, traffic control, smart cities, education, marine environments, and health care as mentioned in (Contreras et al., 2023). The barriers of combination of IOT Systems global adoption to agriculture shown in Figure 1. Other IoT- based platforms control water consumption using cellular technologies and cloud services, but only show prototype nodes that control or monitor water usage without showing concrete results. Researchers in (Contreras et al., 2023) focused on minimizing production risk employing wireless sensor networks for real-time monitoring and studying the inter-temporal environmental variability. The vital aspect in IoT technology is that it contributes to and facilitates the data collection process, and that energy balance in greenhouse production and optimization of energy consumption in grape production can improve energy efficiency.

Additionally, collaboration between technology, farmers, industries, financial institutions, and policy and decision-makers is necessary to promote sustainable agriculture. To enhance disease detection, convolutional neural networks (CNNs) have emerged as a highly utilized solution, offering impressive accuracy rates of up to 98% (Contreras et al., 2023).

LSTM Network

A quantifiable variable in greenhouses is the estimated vent opening, regardless of the absence of mathematical models that link greenhouse climate to vent opening. Literature instead investigates the correlation for computing ventilation flow. The parameters of well- known models (Guesbaya, et al., 2023) are nonlinear equations calibrated for various types of greenhouses. It would be complex to produce a soft sensor for vent opening, and to really quantify ventilation flow, costly sensors like acoustic anemometers would be needed. A data-driven soft sensor is developed to estimate vent opening in greenhouses using Deep Learning methods (Lawal, 2021). Long Short-Term Memory (LSTM) neural networks are selected based on their ability to handle vanishing gradient problems and accurately model delayed and correlated greenhouse dynamics. This approach is suitable for greenhouses with complex and nonlinear dynamics. The soft sensor in Figure 2 calculates vent opening using data-based observations of climatic

Figure 1. IoT system agricultural barriers as global adoption

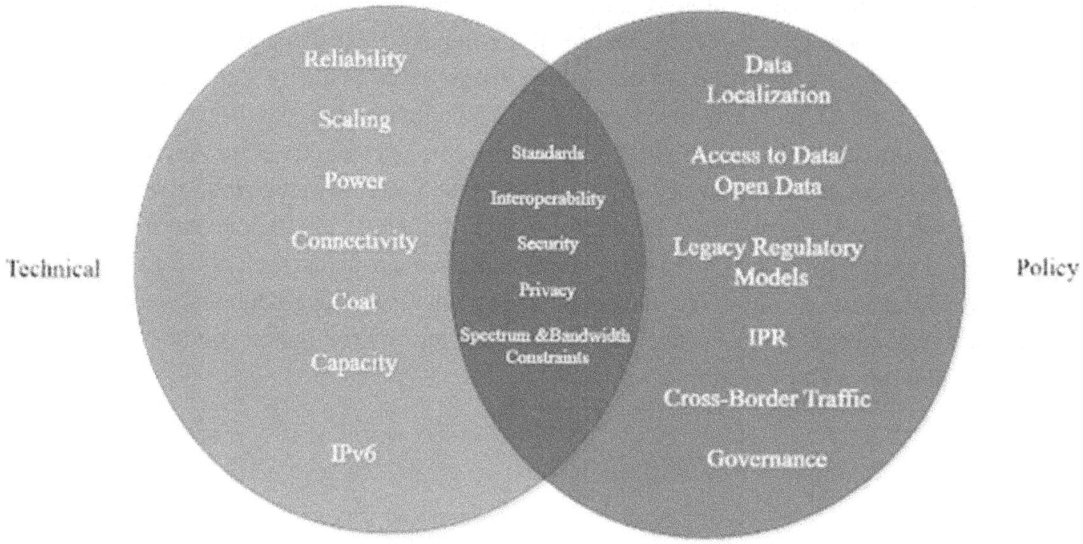

factors both within and outside a greenhouse. The sensor simulates vent openings using LSTM-based network designs, changing internal climatic factors (Haryanto et al., 2019). A progressive elimination method (PEP) and an LSTM-RNN neural network were both used throughout the learning and examination stages. The staging is good, proving the soft sensor's ability to estimate vent openings with minimal inaccuracy. The Tomato and Lettuce Plant growth in greenhouses is shown in Figure.2.

PAR-Based Horticultural Lighting System

Reduced power expenditures for supplementary lighting to boost photosynthesis are economically beneficial (Bersani et al., 2022). Supplemental illumination in the blue and red spectral ranges is chosen to maximize photosynthetic activity while consuming the least amount of energy discussed in (Mohagheghi and Moallem, 2023). Dimmable blue and red-light fixtures have been designed to obtain suitable light recipes. To differentiate these colors, two parameters for photosynthetic photon flux density (PPFD) are defined: PPFDr and PPFDb. The number of photons hitting a surface area per unit of time is measured in mol m2s. The DLI characteristics are also employed in plant development and morphology. Conventional PAR measurements in the complete 400-700 nm photosynthetic spectrum are designated by PPFDfull and DLIfull. PPFD-based controlling system flowchart is shown in Figure.3, in which the error signal to the neural network controller is obtaining by means of desired and actual PPFDs for red and blue channels (Sumalan et al., 2020).

Figure 2. Tomato and lettuce plant growth in greenhouse

Smart Greenhouse Monitoring System Raspberry-Pi for Tomato Production

The study (Ahmad et al., 2023) discussed the possibilities of a Raspberry-Pi-based smart greenhouse monitoring system. Roma and cherry tomatoes were produced in the smart and greenhouse systems to compare agronomic and quality factors. In comparison to traditional greenhouses, the smart greenhouse system maintained a microclimate with a humidity level of 20–30% and a temperature range of 5–6°C. According to the findings in (Ahmad et al., 2023), compared to traditional greenhouses, agronomic factors including plant height, leaf count, fruit size, and weight will rise by 10-15%. The above-average Roma tomato plant output (5.5 kg/plant) was greatly boosted by the intelligent greenhouse system (Morrow et al., 2008). This indicates that agricultural yield and activity monitoring might both benefit from the deployment of smart technology. Figure.4 shows the smart greenhouse selective sensors continuously recording data that are integrated with an application to transmit the signals to display on a digital receiver.

Detection and Classification of System

In paper (Seo et al., 2021) researchers, presented a method that can be utilized for monitoring of generative growth of tomato, which includes deep learning-based object detection, object tracking by a centroid tracker, and maturity classification with the help of image processing (Saha et al., 2017). The method of approach is shown in Figure 5. The k-means clustering algorithm is employed to cluster the tomato fruit regions within the bounding boxes provided by the object detector (Garcia et al., 2020). To achieve this, the RGB color model is converted to the HSV color model, and the hue channel image is extracted as discussed in (Seo et al., 2021). By analyzing the hue values, the maturity of tomato fruits is classified into six distinct levels. Also, to prevent duplicate counting, the boxes are tracked and assigned an ID using centroid-based object tracking (Liu et al., 2022).

Figure 3. Flowchart of PPFD-based neural network

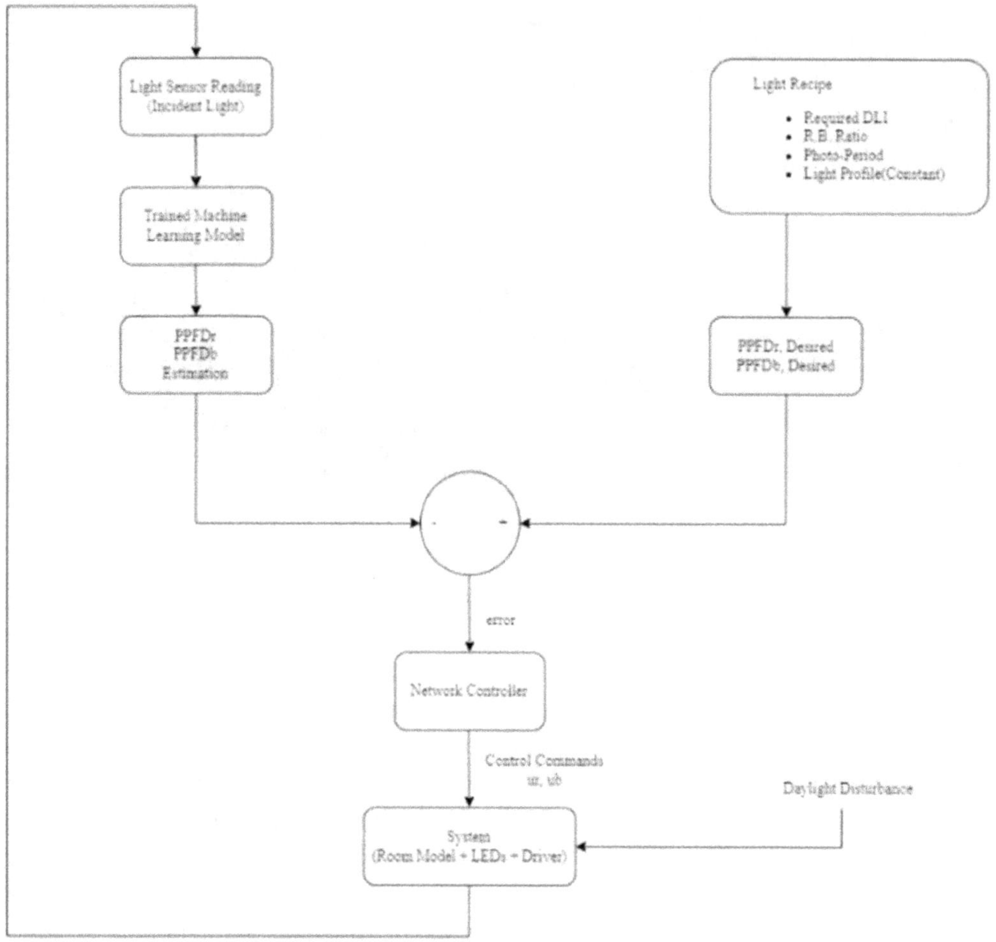

Growth Model Feature Extraction of Tomato

As highlighted in Indonesian Agriculture (Aminulloh et al., 2019), horticulture holds significant importance as a sector. This sector encompasses a substantial number of farmers, approximately 2.9 million, who primarily engage in cultivating vegetables, fruits, herbal plants, and ornamental plants. The emphasis is placed on enhancing the welfare of farmers and improving public health, as indicated by reports from the Center for Agricultural Data and Information System, Ministry of Agriculture, Republic of Indonesia (2016). To optimize agricultural processes, researchers have proposed various measures, including the utilization of Wireless Sensor Network (WSN) applications for monitoring environmental conditions. Notably, a study conducted in (Aminulloh et al., 2019) aimed to develop a framework for remote environmental monitoring and control specifically tailored to tropical horticulture, in addition to implementing web-based monitoring (Aminulloh et al., 2019). The primary focus of this research, as described in (Aminulloh et al., 2019), encompassed two aspects: (a) extracting features related to the growth of tomato plants under specific environmental conditions, and (b) observing the different stages

Figure 4. A block diagram of mode of sensors used

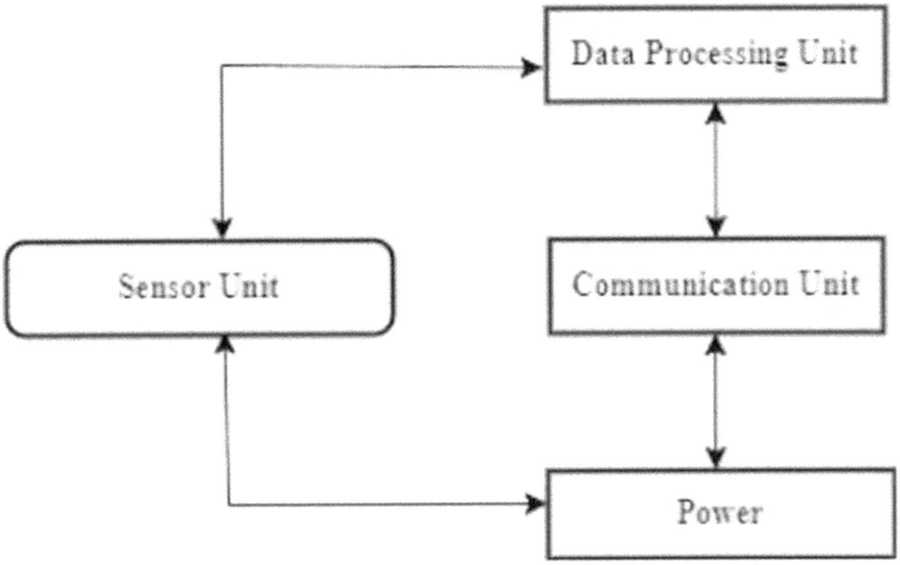

of tomato plant growth, measuring growth parameters at each stage, and considering the nutritional and environmental requirements throughout the various growth stages. For reference, Table 1 outlines the technical specifications of the greenhouse employed in the experimental study (Manoharan, 2020).

METHODOLOGY

SAgric-IoT Platform Description

SAgric-IoT is a platform for monitoring and controlling IoT for agricultural output i.e., easy to deploy, operate, maintain, and upgrade (Barnes et al., 1993). It is consisting of four primary nodes: sensor, camera, gateway, and processing & storage nodes (Sari et al., 2018). Data is collected from all sensors

Figure 5. Method proposed

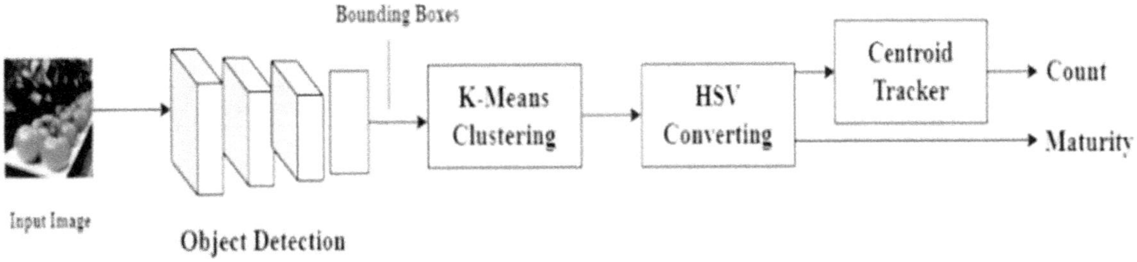

Table 1. Greenhouse technical specifications

Greenhouse Technical Specification	
Structure	
Cover	Symmetric Plane
Ridge Orientation	North-South
Surface Shape	Rectangular
Cover Material	Polyethylene 200 micron
Surface Area	15.432 m2
Width	2.572 m
Length	6 m
Ridge Height	2.725m
Lateral Height	2.2m
Actuator	
Lateral Natural Ventilation	Two (1m x0.4m)
Fertirrigation System	Drip Irrigation System
Humidifier	2 x 5 Mist Sprayer
Exhaust System	2 x 820 CMH Exhaust fan
Sensor Management	
Weather	Air Temperature
	Air Humidity
	Light Intensity
	CO2
Soil	Volumetric Water Content
	Temperature
	Electro Conductivity (EC)

using sensor nodes and relay it to the gateway node on a regular basis, either through other sensor nodes or directly discussed in (Alexandratos and Bruinsma, 2012).

The design includes all seven OSI model layers, with the collection layer including the data link and physical layers, the communication layer including the data link and network layers, and the processing and end-user layers connected to the four top layers as seen in (Alexandratos and Bruinsma, 2012).

The layers in layered IoT Architecture are as follows-

1. Gathering Layer

The collection layer collects data and images from field sources and sensor nodes, also developing and defining low-power D2D standards and solutions (Marcelis et al., 2000). Identifying the optimal data-transmission networking technology stands as the challenges.

2. Communication Layer

The communication layer sends data acquired by sensor nodes to the gateway, which necessitates low latency, dependability, and security (Albright et al., 2000).

3. Processing Layer

The processing layer processes enormous volumes of data to generate warnings and improve production, and it is used by government agencies to create production maps (Azaza et al., 2016). The database is interfaced with data analysis algorithms and real-time scripts for data visualization and evaluation, and data is examined in real time to detect abnormal conditions (Bannister et al., 2008).

4. End User Layer

This layer is responsible for facilitating the farmer's access to the data through a web service and software tools using a web and mobile application as seen in (Zhang et al., 2022).

5. Security Layer

The SAgric-IoT solution secures sensor node connectivity, encryption, and integrity. Each Xbee node is set with parameters such as E.E. = 1, KY, and WPA2, and the end user logs in with a username and password discussed in (Zhang et al., 2022).

Table 2 displays sorted information in the form of dataset. Tomato and Lettuce categories used to test the system model (approx 18,435 images) (Channe et al., 2015). The complete dataset was sorted into two parts, providing 80% of images for training the model and the remaining for validation (around 3687 images). Problems faced by the model while training process, overfitting may occur i.e., model remembering the dataset (Technique - Data Augmentation) (Katzin et al., 2022). By employing various transformation methods such as rotation, brightness adjustment, and Gaussian noise, new images were generated to enhance the existing dataset.

LSTM Training

Inside and outside greenhouse climate variables are analyzed and selectively used as inputs in next section by a given set of data, also the point i.e., actual vent-opening signal generated by an automatic controller (Guesbaya, et al., 2023). This study (Guesbaya, et al., 2023). proposes a theory that employs a soft sensor using an LSTM neural network to accurately estimate the extent of vent opening responsible for inducing changes in internal climate variables. In (Guesbaya, et al., 2023), we studied LSTM- based neural network, encompassing its components, equations, and structure. Also, various potential network architectures and their corresponding hyperparameters are preselected and discussed in (Guesbaya, et al., 2023). The developed soft sensor was trained, validated, and tested using a dataset consisting of 81 recorded days, totaling 233,280 samples. The dataset was collected using a commercial data acquisition system known as Compact Field Point, which captured 13 climate variables, such as air temperature, relative humidity, solar radiation, CO2 concentration, wind velocity, vent opening, and time variables

Table 2. Plant village dataset

Serial No.	Category	Number of Images
1	Lettuce_Late_blight	1005
2	Lettuce_Bacterial_spot	980
3	Tomato_Bacterial_spot	2129
4	Tomato_Early_blight	1005
5	Lettuce_healthy	1473
6	Tomato_healthy	1595
7	Tomato_Late_blight	1905
8	Tomato_Leaf_Mold	958
9	Tomato_Septoria_leaf_spot	1775
10	Lettuce_Early_blight	1005
11	Tomato_YellowLeaf_Curl_Virus	3205
12	Tomato_Target_Spot	1400

(Santosh and Chavhan, 2012). Sensors installed inside and outside the experimental greenhouse recorded the data across the growth cycle of a tomato crop refer Table 3.

Table 3. External and internal sensors were placed in the experimental greenhouse

Sensor	Brand	Model	Precision	Range
Air temperature	Campbell Scientific	HC2S3	±0.1 °C	-40 to 60 °C
Air relative humidity	Campbell Scientific	HC2S3	±0.1%	0 to 100%
Global Solar Radiation	Hukseflux	LP02	<±1%	0 to 2000 W/m^2
CO2 concentration in air	E + E Elektinik	EE820-C2	<±50ppm	0 to 2000 ppm
Wind Velocity	Vector Instruments	A100L2/PC3	<2%	0 to 75 m/s

Proposed Lighting Control System

The Institute for Sustainable Horticulture at Kwantlen Polytechnic University conducted an experiment (Mohagheghi and Moallem, 2023) in a greenhouse compartment using two systems. Each system included light sensors, LED fixtures, and independent controllers. System 1 used a time-scheduling on-off controller, while System 2 implemented a neural network controller (Heuvelink and Challa, 1989). The systems shared a hydroponic nutrition and water delivery system, maintaining consistent conditions. A drain-to-waste hydroponic system provided water and nutrients based on a specific recipe (Waller, et al., 2021). The selected crop was a green mini romaine organic lettuce variety called Dragoon, known for its performance in climate-controlled greenhouses and suitability for hydroponic growing methods (Siskandar et al., 2022). The lighting plan for System 2 was designed to deliver the required light intensity to the plant canopy using a red-to-blue ratio and photoperiod, while System 1 represented the traditional on-off method. Real-time lighting data was gathered using light sensors and a spectroradiometer, alongside periodic top-view images of the plants captured by image sensors, all conducted under white light conditions. To analyze and monitor plant health and growth, the acquired data underwent processing utilizing the PlantCV software, enabling the extraction of relevant features (Sellami and Sahbani, F. (2023).

Optimal Growth Conditions

A smart greenhouse microprocessor was utilized in a small greenhouse equipped with a Raspberry Pi microprocessor to regulate temperature and humidity using sensors (Soussi et al., 2022). The study (Ahmad et al., 2023) was conducted from February to April 2022 to assess its impact on the agronomic and quality traits of tomatoes. The microprocessor's data settings included a temperature range of 24-28°C with humidity above 75% for the initial 30-35 days to observe vegetative growth, followed by a temperature decrease to 20°C with humidity below 60% for 15-20 days to initiate the reproductive phase (Paradiso and Proietti, 2022). Subsequently, the temperature and humidity were raised again to 24–28 °C for fruit color development. The data was accessed via a digital receiver every 10 days. Following the transplantation of the seedlings, plastic pots filled with sterilized sandy loam soil were utilized. Throughout the growth process, adequate fertilization and irrigation practices were diligently maintained. The study (Ahmad et al., 2023) was also conducted in a conventional greenhouse for comparison, using sterilized sandy loamy soil and the same fertilization and irrigation practices. Sensors utilized in the smart greenhouse included a humidity sensor, temperature sensors, and a timer for automatic adjustment of microclimate settings.

Faster R-CNN

The research done in (Seo et al., 2021) shows the real-time detection of tomato fruits using frames captured by a robot in the greenhouse was accomplished using Faster R-CNN algorithm. Comparative evaluations showed that the Faster R-CNN with a ResNet backbone surpassed other networks in terms of mean average precision (MAP) (Seo et al., 2021). Consequently, the ResNet- 101 was selected as the backbone convolutional neural network (CNN) for this specific task (Seo et al., 2021). The Tomato open dataset (Seo et al., 2021), containing 895 pictures marked with boundary boxes in PASCAL VOC format, employed for training and authentication of the detection model. The dataset was randomly categorized into a training set (90% - 805 images) and a validation set (10% - 90 images). Training of the Faster R-CNN with ResNet-101 was conducted using the TensorFlow deep learning framework on an RTX Titan GPU and an Intel Core i9-9940X CPU. The training process involved 10,000 epochs, with a learning rate of 0.04, momentum optimizer set to 0.9, and cosine learning-rate decay. The achieved loss values were 0.0149 for classification and 0.005 for detection, resulting in a total loss of 0.0299. To accurately classify tomato fruit maturity using the faster R-CNN detection model, the tomato area needs to be separated from the background. The k-means clustering algorithm with a k value of 2 was employed for this purpose. While Mask R-CNN is another segmentation method, its slower processing speed of around 200 milliseconds per frame makes it unsuitable for real-time applications. In contrast, faster R-CNN with the ResNet-101 backbone achieves faster processing, taking only 55 milliseconds per frame. For real-time video, centroid-based object tracking was used to track and count objects without duplication. Each object was assigned a unique ID based on the centroid, minimizing the Euclidean distance to assign IDs to new points in updated video frames. This approach enabled accurate monitoring of tomato fruit maturity while maintaining real- time processing capabilities (Muñoz et al., 2020).

Designing of Greenhouse

The construction of the greenhouse aims to regulate and manipulate the climate and environmental factors involved in tomato and lettuce cultivation (Aminulloh et al., 2019). Maintaining a stable and controlled climate within the greenhouse is of utmost significance when it comes to monitoring, studying, and modeling the growth patterns of tomato plants.

Monitoring System

During the greenhouse monitoring system's design, it is split into two subsystems: the Local Management Subsystem (LMS) and the Global Management Subsystem (Aminulloh et al., 2019). The LMS employs sensors to gather environmental data within the greenhouse, while actuators help maintain conditions based on the LMS and system's setpoint (Aminulloh et al., 2019). The Global Management Subsystem stores data from the LMS in a database, enabling analysis of plant growth's impact on the greenhouse climate (Aminulloh et al., 2019).

RESULTS

Improved Environmental Monitoring

The integration of IoT sensors in tomato and lettuce greenhouses enables real-time monitoring of environmental factors such as temperature, humidity, light intensity, and soil moisture. The findings illustrate the precision and dependability of the sensor data, equipping farmers with invaluable knowledge regarding the ideal cultivation parameters for these particular crops.

Automated Control Systems

By employing machine learning algorithms, the greenhouse's control systems can dynamically adjust various parameters to maintain optimal growing conditions. The chapter showcases the effectiveness of these systems in regulating temperature, humidity,irrigation, and nutrient levels, leading to improved crop quality and reduced resource wastage.

Fruit Quality

In study, Table 4 provides information regarding the maturity levels of tomatoes. These maturity levels were determined derived from the ratio of the red region in the tomatoes. Therefore, a quantitative classification approach was adopted to assess the maturity of tomatoes using both the USDA standard and image processing techniques

Table 4. Classification of tomato fruit maturity

Maturity	Description
Green	Entirely Green
Breakers	First appearance of external pink or red color (not more than 10%)
Turning	Over 10% but not more than 30%red or pink
Pink	Over 30% but not more than 60% pinkish or red
Light Red	Over 60% but not more than 90% red
Red	Over 90% red

Predictive Analytics and Decision Support

Machine learning models trained on historical sensor data facilitate predictive analytics, enabling greenhouse farmers to anticipate crop growth patterns, pest infestations, and disease outbreaks. The chapter illustrates how these insights empower farmers to make informed decisions, optimize resource allocation, and implement timely interventions, ultimately increasing productivity and profitability.

Growth Efficiency

Data on microclimate conditions and equipment control was recorded at 15-minute intervals. The average data of cherry and roma in the smart greenhouse monitoring system is shown in Figure 6, and the average data of conventional greenhouses discussed in Figure 7. All this data represented in Figures 6 and 7 together shows that the growth of cherry roma is more efficient in the Smart Greenhouse Monitoring System.

Greenhouse Yield Enhancement and Resource Efficiency

Through the combination of IoT and machine learning, the chapter demonstrates significant improvements in crop yield and resource efficiency. The precise control of environmental variables, coupled with predictive models, minimizes yield variations and crop losses. Additionally, optimized resource allocation reduces water consumption, energy usage, and fertilizer wastage, fostering sustainable agricultural practices.

CONCLUSION

In conclusion, the automated operation of tomato and lettuce greenhouses with the combination of IoT and ML techniques has proven to be a game changer in modern agriculture. This chapter has explored the utilization of IoT-based sensor networks to monitor and control critical parameters within greenhouses, enabling precise adjustments to environmental conditions. The use of machine learning algorithms has improved decision-making even more, assuring ideal development conditions for tomatoes and lettuce. The benefits of this mechanized approach are substantial. Improved crop quality, increased productivity, and resource efficiency have been observed, leading to more sustainable and profitable greenhouse

Figure 6. Study of tomato plant in smart greenhouse monitoring system

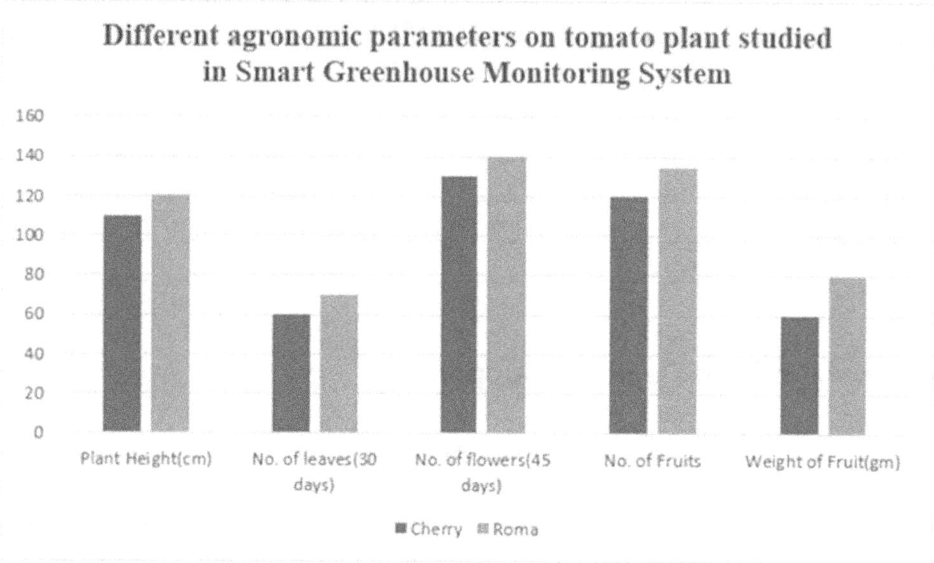

operations. The ability to remotely monitor and manage greenhouse environments also provides farmers with valuable insights and control over their crops, reducing risks and maximizing yields. While there are hurdles, such as guaranteeing data confidentiality and maintaining system stability, the advances in tomato and lettuce output gained through mechanized systems highlight the great promise of IoT and machine learning technologies in revolutionizing greenhouse horticulture. Continued study and development in this subject will surely result in improved crop quality, quantity, and overall agricultural sustainability.

Figure 7. Study of tomato plant in conventional

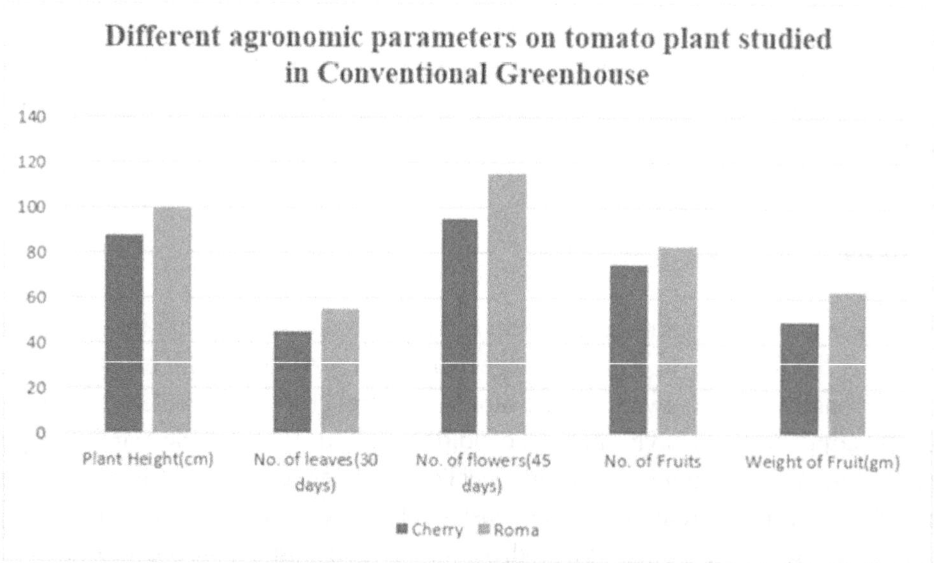

Figure 8. Soft sensor development flowchart

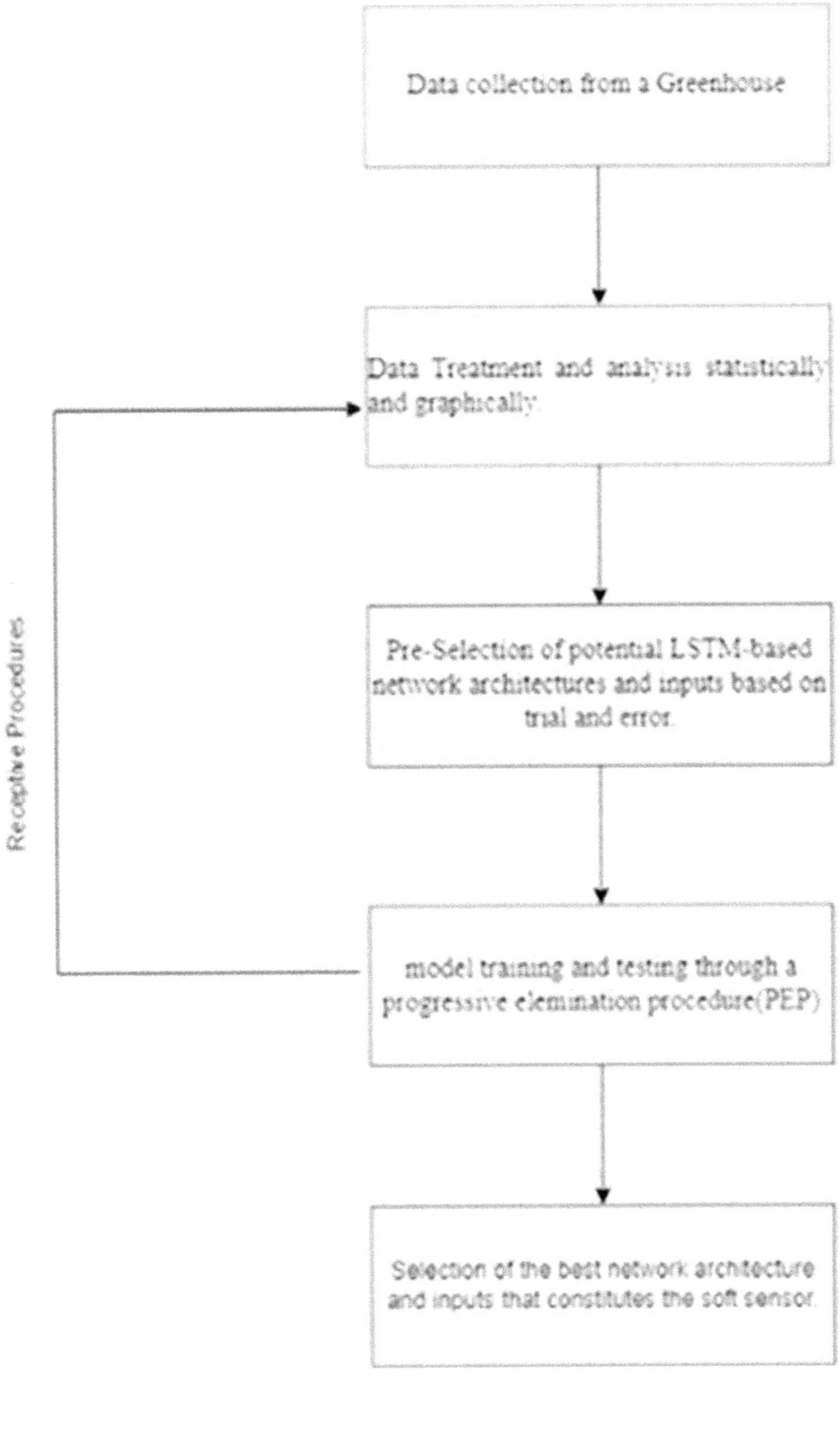

REFERENCES

Ahmad, B., Ahmed, R., Masroor, S., Mahmood, B., Hasan, S. Z. U., Jamil, M., Khan, M. T., Younas, M. T., Wahab, A., Haydar, B., Subhani, M., Khan, M. A., & Tariq, S. (2023). Evaluation of Smart Greenhouse Monitoring System using Raspberry-Pi Microcontroller for the Production of Tomato Crop. *Journal of Applied Research in Plant Sciences*, *4*(01), 452–458. doi:10.38211/joarps.2023.04.01.54

Albright, L. D., Both, A. J., & Chiu, A. J. (2000). Controlling greenhouse light to a consistent daily integral. *Transactions of the ASAE. American Society of Agricultural Engineers*, *43*(2), 421–431. doi:10.13031/2013.2721

Alexandratos, N., & Bruinsma, J. (2012). *World agriculture towards 2030/2050: the 2012 revision.* Academic Press.

Aminulloh, L., Sesulihatien, W. T., & Pramadihanto, D. (2019, September). Feature extraction of tomato growth model using greenhouse monitoring system. In *2019 International Electronics Symposium (IES)* (pp. 370-375). IEEE. 10.1109/ELECSYM.2019.8901534

Azaza, M., Tanougast, C., Fabrizio, E., & Mami, A. (2016). Smart greenhouse fuzzy logic-based control system enhanced with wireless data monitoring. *ISA Transactions*, *61*, 297–307. doi:10.1016/j.isatra.2015.12.006 PMID:26749556

B. (1993). Accuracy of quantum sensors measuring yield photon flux and photosynthetic photon flux. *HortScience, 28*(12), 1197-1200.

Bannister, K., Giorgetti, G., & Gupta, S. K. (2008, June). Wireless sensor networking for hot applications: Effects of temperature on signal strength, data collection and localization. In *Proceedings of the 5th workshop on embedded networked sensors (HotEmNets' 08)* (pp. 1-5). Academic Press.

Bersani, C., Ruggiero, C., Sacile, R., Soussi, A., & Zero, E. (2022). Internet of Things Approaches for Monitoring and Control of Smart Greenhouses in Industry 4.0. *Energies*, *15*(10), 3834. doi:10.3390/en15103834

C. A. (2020). A Cost-effective embedded platform for greenhouse environment control and remote monitoring. *Agronomy, 10*(7), 936.

Chang, C. L., Chung, S. C., Fu, W. L., & Huang, C. C. (2021). Artificial intelligence approaches to predict growth, harvest day, and quality of lettuce (Lactuca sativa L.) in a IoT-enabled greenhouse system. *Biosystems Engineering*, *212*, 77–105. doi:10.1016/j.biosystemseng.2021.09.015

Channe, H., Kothari, S., & Kadam, D. (2015). Multidisciplinary model for smart agriculture using internet-of-things (IoT), sensors, cloud-computing, mobile-computing & big-data analysis. Int. J. *Computer Technology and Application*, *6*(3), 374–382.

Dubey, S. R., & Jalal, A. S. (2013). Adapted approach for fruit disease identification using images. In *Image processing: Concepts, methodologies, tools, and applications* (pp. 1395–1409). IGI Global. doi:10.4018/978-1-4666-3994-2.ch069

Dutta, M., Gupta, D., Sahu, S., Limkar, S., Singh, P., Mishra, A., Kumar, M., & Mutlu, R. (2023). Evaluation of Growth Responses of Lettuce and Energy Efficiency of the Substrate and Smart Hydroponics Cropping System. *Sensors (Basel), 23*(4), 1875. doi:10.339023041875 PMID:36850471

Flores, K. O., Butaslac, I. M., Gonzales, J. E. M., Dumlao, S. M. G., & Reyes, R. S. (2016, November). Precision agriculture monitoring system using wireless sensor network and Raspberry Pi local server. In *2016 IEEE region 10 conference (TENCON)* (pp. 3018- 3021). IEEE.

Garcia-Mañas, F., Rodríguez, F., & Berenguel, M. (2020). Leaf area index soft sensor for tomato crops in greenhouses. *IFAC-PapersOnLine, 53*(2), 15796–15803. doi:10.1016/j.ifacol.2020.12.230

Guesbaya, M., García-Mañas, F., Rodríguez, F., & Megherbi, H. (2023). A Soft Sensor to Estimate the Opening of Greenhouse Vents Based on an LSTM-RNN Neural Network. *Sensors (Basel), 23*(3), 1250. doi:10.339023031250 PMID:36772289

Haryanto, U. M., Ibadillah, A. F., Alfita, R., Aji, K., & Rizkyandi, R. (2019, April). Smart aquaponic system based Internet of Things (IoT). In Journal of Physics: Conference Series (Vol. 1211, p. 012047). IOP Publishing.

Heuvelink, E. P., & Challa, H. (1989, May). Dynamic optimization of artificial lighting in greenhouses. In *International Symposium on Growth and Yield Control in Vegetable Production 260* (pp. 401-412). 10.17660/ActaHortic.1989.260.26

Katzin, D., van Henten, E. J., & van Mourik, S. (2022). Process-based greenhouse climate models: Genealogy, current status, and future directions. *Agricultural Systems, 198*, 103388. doi:10.1016/j.agsy.2022.103388

Kochhar, A., & Kumar, N. (2019). Wireless sensor networks for greenhouses: An end- to-end review. *Computers and Electronics in Agriculture, 163*, 104877. doi:10.1016/j.compag.2019.104877

Lawal, M. O. (2021). Tomato detection based on modified YOLOv3 framework. *Scientific Reports, 11*(1), 1–11. doi:10.103841598-021-81216-5 PMID:33446897

Liu, Y., Li, D., Wan, S., Wang, F., Dou, W., Xu, X., Li, S., Ma, R., & Qi, L. (2022). A long short- term memory-based model for greenhouse climate prediction. *International Journal of Intelligent Systems, 37*(1), 135–151. doi:10.1002/int.22620

Manoharan, S. (2020). Supervised Learning for Microclimatic parameter Estimation in a Greenhouse environment for productive Agronomics. *Journal of Artificial Intelligence, 2*(03), 170–176.

Marcelis, L. F. M., Maas, F. M., & Heuvelink, E. (2000, November). The latest developments in the lighting technologies in Dutch horticulture. In *IV International ISHS Symposium on Artificial Lighting 580* (pp. 35-42). Academic Press.

Mohagheghi, A., & Moallem, M. (2023). An Energy-Efficient PAR-Based Horticultural Lighting System for Greenhouse Cultivation of Lettuce. *IEEE Access : Practical Innovations, Open Solutions, 11*, 8834–8844. doi:10.1109/ACCESS.2023.3237757

Morrow, R. C. (2008). LED lighting in horticulture. *HortScience, 43*(7), 1947–1950. doi:10.21273/HORTSCI.43.7.1947

Muñoz, M., Guzmán, J. L., Sánchez-Molina, J. A., Rodríguez, F., Torres, M., & Berenguel, M. (2020). A new IoT-based platform for greenhouse crop production. *IEEE Internet of Things Journal, 9*(9), 6325–6334. doi:10.1109/JIOT.2020.2996081

Nasution, I. S., Iskandar, M. R., & Jayanti, D. S. (2020). Internet of things: Automatic sprinklers in prototyping greenhouse using smartphone-based android. *IOP Conference Series. Earth and Environmental Science, 425*(1), 012069. doi:10.1088/1755-1315/425/1/012069

Ojha, T., Misra, S., & Raghuwanshi, N. S. (2015). Wireless sensor networks for agriculture: The state-of-the-art in practice and future challenges. *Computers and Electronics in Agriculture, 118*, 66–84. doi:10.1016/j.compag.2015.08.011

Paradiso, R., & Proietti, S. (2022). Light-quality manipulation to control plant growth and photomorphogenesis in greenhouse horticulture: The state of the art and the opportunities of modern LED systems. *Journal of Plant Growth Regulation, 41*(2), 742–780. doi:10.100700344-021-10337-y

SAgric-IoT. (1961). An IoT-Based Platform and Deep Learning for Greenhouse Monitoring. *Applied Sciences (Basel, Switzerland), 13*(3).

Saha, T., Jewel, M. K. H., Mostakim, M. N., Bhuiyan, N. H., Ali, M. S., Rahman, M. K., ... Hossain, M. K. (2017). Construction and development of an automated greenhouse system using Arduino Uno. *International Journal of Information Engineering and Electronic Business, 9*(3), 1. doi:10.5815/ijieeb.2017.03.01

Santoshkumar, K. C., & Chavhan, S. (2012). Development of wireless sensor node to monitor poultry farm. *Communications in Computer and Information Science, 296*, 27–32. doi:10.1007/978-3-642-35864-7_4

Sari, I. A., Handayani, A. N., & Lestari, D. (2018, October). Smart greenhouse sebagai media pembibitan kentang granola kembang berbasis mikrokontroler. In Prosiding Seminar Nasional Teknologi Elektro Terapan (Vol. 2, No. 1, pp. 105-110). Academic Press.

Sellami, M. H., & Sahbani, F. (2023). A Modeling Approach to Automate the Functioning of Tomato Greenhouses. *International Journal of Horticultural Science and Technology, 10*(4), 387–396.

Seo, D., Cho, B. H., & Kim, K. C. (2021). Development of monitoring robot system for tomato fruits in hydroponic greenhouses. *Agronomy (Basel), 11*(11), 2211. doi:10.3390/agronomy11112211

Siskandar, R., Santosa, S. H., Wiyoto, W., Kusumah, B. R., & Hidayat, A. P. (2022). Control and automation: Insmoaf (Integrated Smart Modern Agriculture and Fisheries) on the greenhouse model. *Jurnal Ilmu Pertanian Indonesia, 27*(1), 141–152. doi:10.18343/jipi.27.1.141

Soussi, M., Chaibi, M. T., Buchholz, M., & Saghrouni, Z. (2022). Comprehensive review on climate control and cooling systems in greenhouses under hot and arid conditions. *Agronomy (Basel), 12*(3), 626. doi:10.3390/agronomy12030626

SumalanR. L.StroiaN.MogaD.MuresanV.LodinA.VintilaT., & Popescu,

Waller, R., Kacira, M., Magadley, E., Teitel, M., & Yehia, I. (2021). Semi-transparent organic photovoltaics applied as greenhouse shade for spring and summer tomato production in arid climate. *Agronomy (Basel), 11*(6), 1152. doi:10.3390/agronomy11061152

Zhang, S., Guo, Y., Li, S., Ke, Z., Zhao, H., Yang, J., Wang, Y., Li, D., Wang, L., Yang, W., & Zhang, Z. (2022). Investigation on environment monitoring system for a combination of hydroponics and aquaculture in greenhouse. *Information Processing in Agriculture*, *9*(1), 123–134. doi:10.1016/j.inpa.2021.06.006

Chapter 14
Machine Learning Time Series Models for Tea Pest Looper Infestation in Assam, India

Dwijendranath Dwivedi
https://orcid.org/0000-0001-7662-415X
Krakow University of Economics, Poland

Pradish N. Kapur
PES University, India

Nipun N. Kapur
Indian Institute of Technology, Roorkee, India

ABSTRACT

In the agriculture industry, pest infestation is a significant challenge that is complicated by the nonlinear relationship with environmental factors. Given the effectiveness of machine learning models in simulating such complex nonlinear phenomena, the authors opted to employ them in the modelling of the life cycle of tea pests, which impact several other crops as well. Accordingly, multiple machine learning models were developed to forecast the occurrence of tea pest looper infestations. They utilized data for just two readily available parameters—temperature and rainfall—to investigate whether predictive models of good quality can be created even with limited data, particularly for small tea growers. After analyzing the various models generated, they discovered that neural network models can produce accurate predictions even with a restricted data set. Therefore, they are optimistic that new age technologies such as machine learning can benefit many small farmers in India who lack access to various technologies and, as a result, have limited data.

1. INTRODUCTION

Despite the recent hype surrounding machine learning, it has existed in some capacity for over fifty

DOI: 10.4018/979-8-3693-0200-2.ch014

years, ever since the advent of programmable computers. The quest for humans to impart instructions to machines (via written, verbal, or visual means) and for machines to comprehend these instructions intelligently and react to them like humans has been a captivating pursuit. One of the initial efforts toward this objective was labeled Cybernetics. Way back in 1948 Norbert Weiner defines Cybernetics as "Control and Communication" in engineering parlance while "thinking" in human context (Wiener, 1948). The concept was to establish a principle of regulation and communication applicable to both machines and living organisms. Scientists' comprehension of the term "thinking" has progressed considerably over the years. Initially, it was thought that thinking involved analyzing multiple scenarios related to a situation and choosing the best outcome. However, this perception limited the effectiveness of various mathematical machines, especially when faced with complex situations that resulted in a vast and nonlinear increase in the number of possible scenarios. The np versus p problems highlight this nonlinear growth in computations for np complex problems (Reference). These computations were overwhelming for computing machines of that era, even when performing relatively simple tasks (Reference). This necessitated a shift in scientists' thinking, resulting in the emergence of Knowledge-Based Expert Systems. The concept involves employing previous experience data to make decisions instead of analyzing all conceivable scenarios related to a decision-making situation. This approach can be summarized as the "If-Then" method of decision-making. Essentially, if a successful decision "x" was made in the past for a given situation "alpha," then the same decision "x" should be made in a comparable present scenario. Naturally, there are feedback loops and opportunities for refining decisions in subsequent real-world scenarios that are not binary. As one can make out that in such expert systems the availability of data of past experience becomes important in taking decisions for future. As technologies (hardware, software) improved in the last few decades, memory costs decreased and capturing of data became much easier and faster leading of explosion of data in last few years. This provided solid grounding for expert systems and the like to take a strong hold in the field of artificial intelligence / machine learning (Figure 1)

The figure above picturizes the evolution in the field of Machine Learning and how data and technology growth has accelerated it. Earlier approach to take data from a particular domain / phenomenon, analyze it for various causative factors pick up the few relevant ones to build analytical decisioning models. More recent approach with huge amounts of data available as well as huge computing power is available. The recent approach more aligned to the knowledge based expert systems described above where huge amount of data and computing power is used to train the models. There are some pros and cons associated with the currently more popular ML approach. The benefits include much faster model development, recalibration process compared to traditional analytical approach. Further with recent ML models the ability to model highly nonlinear phenomena is becoming more easier and easier. To the extent some people have started saying that the boundary between the p – np problem might be getting blur or at least shifting (Fifty Years of P vs. NP and the Possibility of the Impossible; By Lance Fortnow Communications of the ACM, January 2022, Vol. 65 No. 1, Pages 76-85). This in business sense means we can develop highly customized models. On the con side of such process / approach is that there is less of focus on identifying causal relation amongst the dependent independent variables (References). Further, with highly non-linear complex problems, the stability of solutions is an issue many a times. Last few years have been difficult for the tea industry due several factors such as cost of production, climate change and pest infestations (Kramer & Ware, 2021; Duncan et al., 2018; Brouder et al., 2017). Pest Infestations results in huge losses to the production of tea (Kachhawa and Kumawat, 2018; Ahuja et al., 2013; Samanta and Gosh, 2012). Some of the tea pests like Looper and Loopers results in losses of

Figure 1. The evolution in the field of machine learning
Source: Author

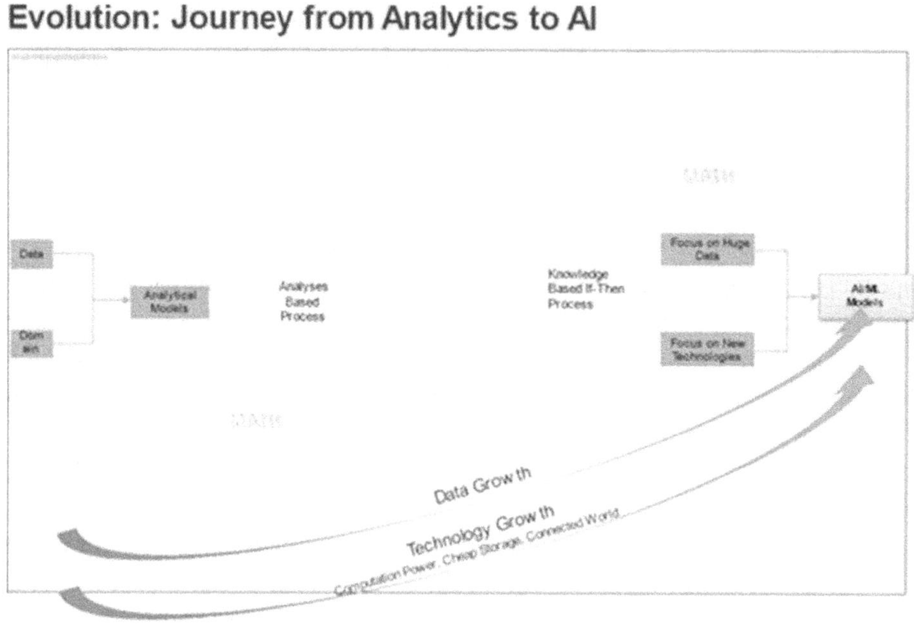

40% of even more. To add to the challenge, strict guideline has been put up in place by several countries on the amount of pesticide residue in teas.

The challenge is grave for small tea growers who constitute more than half of the tea production and have limited means to carry out non-traditional and even traditional means to control pest infestations (Dwivedi et al, 2022; Voora et al., 2019; Roy et al., 2020; Khanali et al., 2021; Gurusubramanian, 2008). Even large growers of tea are not unaffected by these challenges. They are trying to control their operations costs by controlling employed people's wages or reducing the labour force. This has resulted in new challenges especially countries and regions where alternative job opportunities are not many (Mulinya, 2021; Gebre, 2017). With tea industry employing around 13 Mn people it is resulting in socio economic challenges.

2. LITERATURE STUDIES

Over last few decades several attempts have been made by researchers to use traditional analytics to develop models to predict pest forecasting. Though many of those initiatives had limited implementation in the field. Some of the key reason for limitations of such studies include the fact that pest cycles are highly complex natural phenomena. With traditional analytics mostly limited to linear or less complex models were unable to predict pest life cycles accurately.

Further, such analytical models remained valid to a restricted geographical area considering several forms of agri-hetroginites. These Agri heterogeneities are the variation in farmland size, method of farming, use of manure and chemicals as pesticides across the country. Which means we need to have hundreds of models to cover a significant area of the country.

Now with the advent of new age technologies, as described in the Introduction of Machine Learning, we have now got the capabilities of quickly build several models in short time. Thus, the hope to positively impact people across regions and even small farmers which constitutes majority of the growers not only in teas but also other crops. Last few years has seen significant research being done in the area of pest life cycle using machine learning models (Zhang et al., 2017). Zhang et al (2017) used multilayer feed-forward neural networks (MLFN), general regression neural network (GRNN), and support vector machine (SVM) to predict the *D. superans* occurrence areas, It is concluded that Support Vector Machine Model for accurate and effective predictions of insect pest occurrence areas as a reliable predictive tool. Another study by Xiao et al.(2018) used rich data of eight different parameters viz max temperature, minimum temperatures, relative humidity in morning, relative humidity in evening, rainfall, wind speed, sunshine hours and evapotranspiration, to model cotton pests. They found out that Long Short Term Memory (LSTM) Neural Nets provide excellent results (AUC of 0.97). With above studies one can see that different modelling approach gives good results for different types of pests. Moreover, the data used also play important role which modelling approach would give good results. Though as mentioned above majority of tea production comes from small farmers who have limited means to get rich data of various parameters (Parasar et al., 2019).

With above consideration, we use different Neural Network models towards forecasting of tea pests Looper. Tea Looper is one of the most destructive pests (Reference). Further we use the limited data of just the temperature and rainfall which is easily available to even small tea growers.

Several tools are available in the market with their own pros and cons. We decided to use SAS Deep Learning Python (DLPy) and stacked model for the same. We decided to use SAS Visual Forecasting for time series analyses of the data for this research work. It automatically produces forecasts without human involvement using different approaches viz Stacked Model, Panel Series Neural Network, Seasonal Model, Temporal Aggregation Model, RNN Forecasting and Auto-Forecasting. Automation implies several benefits starting from consistency of the analyses / results to reduced need to costly experienced resources. Further the entire forecast model development and monitoring process takes much less time.

3. DATA SOURCE

The data are from Teok Tea gardens of Assam in India covering around 250 hectares of tea gardens. It consists of diurnal data of rainfall, minimum and maximum temperatures and estimated value of the area infested with pests. The duration of the data is from 2013 through 2016.

Data Exploration

Pest control data has a unique random behavior as this data is not continuous in time series. Pest control is performed only certain number of days and that is reflected in the data as per below chart. The distribution of the data is a bit not so continuous as most frequent is the day when there is no pest control performed in the tea gardens. We also tried to understand the behavior of the input datasets that are maximum temperature, Minimum Temperature and rainfall.

Figure 2. Data for pest control (area) over time series
Source: Author

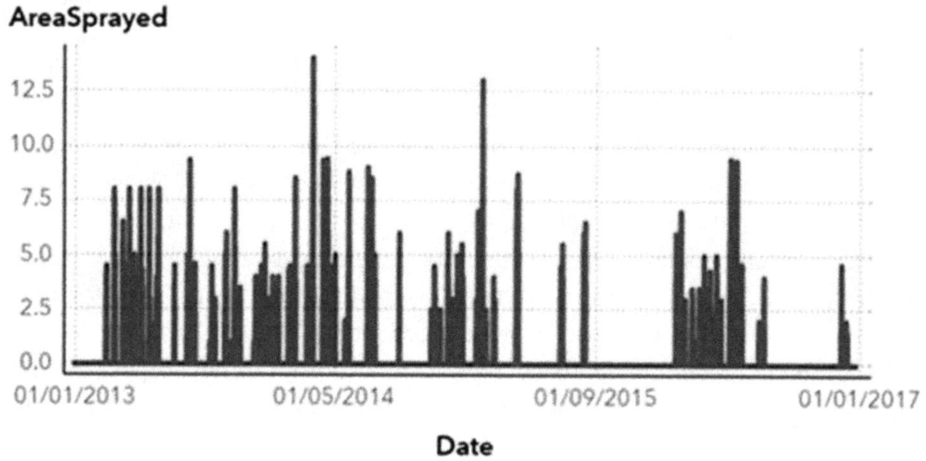

Figure 3. Cross Series plots with independent variables (minimum temperature, maximum temperature and rainfall) as well as dependent variables
Source: Author

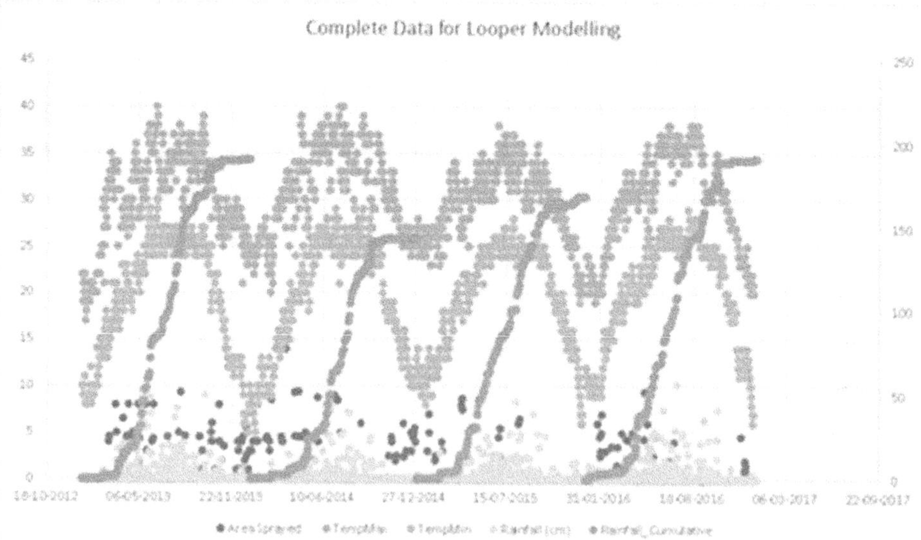

4. MODELS AND RESULTS

SARIMA Analysis: We employ a SARIMA design to the time series data. We choose the multiplied seasonal best fit model using the Akaike Information Criterion to select the most preferred model with the least value. The SARIMA is an extension of ARIMA that explicitly models the seasonal element in univariate data. A seasonal ARIMA model is an extension of ARIMA and is formed by including additional seasonal terms in the ARIMA models. The seasonal part of the model consists of terms that are like the non-seasonal components of the model, but involve back shifts of the seasonal period.

Stacked Model: The stacked modeling strategy generates forecasts using stacked models that consist of a neural network model (NN) and a time series model (TS). The time series model applies to the residuals of the neural network forecasts. This modeling strategy captures the nonlinear relationship between the dependent and independent variables as well as time series characteristics in the data, such as seasonality and trend.

Model Comparison uses weighted MAPE: The weighted version of a statistic of fit is calculated using the input data to assign weights to each series, with higher weights being assigned to series with higher volume.

Figure 4. Series weight computation

$$\text{Series weight} = \left|\frac{1}{T}\Sigma_{t=1}^{T} y_t\right|$$

Here is the used formula for the WMAPE computation.

Figure 5. Weighted mean absolute percent error

Weighted mean absolute percent error (WMAPE)

The weighted mean of the MAPE values for each time series. $\frac{\Sigma \text{MAPE} \times \text{SeriesWeight}}{\Sigma \text{SeriesWeight}}$

DISCUSSION AND CONCLUSIONS

A comparative analyses of the various Neural Net models towards simulating area infested by Looper in tea gardens highlight that Stacked Models perform better than most of the other models. The Stacked modeling strategy captures the nonlinear relationship between the area infested by pests and independent variables of temperature as well as time series characteristics in the data, such as seasonality (season dependent growth of pest Looper). This feature of modeling the season dependent variation of growth of Looper (Kalloor et al., 2020) as well as the nonlinear relation between temperatures and area infested is in line with the observed real phenomena (Figure 7). Thus, giving better results.

Figure 6. Model comparison using Weight MAPE

Champion	Model Name	Status	Weighted Mean Absolute Error
true	Stacked Model (NN + TS)	Successful	0.4555
false	Regression for Time Series	Successful	0.7739
false	Non-seasonal Model	Successful	0.7639
false	Auto-forecasting	Successful	0.7187
false	Multistage Model	Successful	0.7613
false	Hierarchical Forecasting	Successful	0.7662
false	Seasonal Model	Successful	0.7406

Figure 7. Non-linear relationship
Source: Author

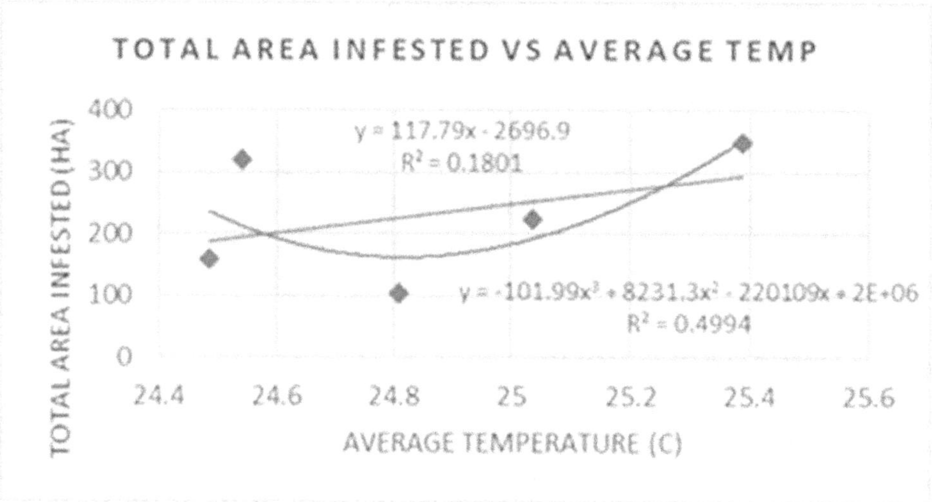

5. ACKNOWLEDGMENT

Acknowledge Mr. Jagjeet Kandhal and Mr. Kaiylanjeet Bohra for thankfully sharing the data. SAS software has been used for the analysis and modeling.

CONFLICT OF INTEREST

The authors whose names are listed immediately below certify that they have NO affiliations with or involvement in any organization or entity with any financial interest (such as honoraria; educational grants; participation in speakers' bureaus; membership, employment, consultancies, stock ownership, or other equity interest; and expert testimony or patent-licensing arrangements), or non-financial interest (such as personal or professional relationships, affiliations, knowledge or beliefs) in the subject matter or materials discussed in this manuscript.

REFERENCES

Ariyani, M., Pitoi, M. M., Koesmawati, T. A., Maulana, H., & Endah, E. S. (2020). Pyrethroid residues on tropical soil of an Indonesian tea plantation: analytical method development, monitoring, and risk assessment (Vol. 7). Academic Press.

Ayouba, K., & Vigeant, S. (2020). *Can We Really Use Prices to Control Pesticide Use? Results from a Nonparametric Model.* doi:10.100710666-020-09714-w

Bengal, Dutta, & Nayak. (2019). *S-Transferase Enzyme Activities and Their Correlation with Genotypic Variations Based on GST M1 and GST T1 Loci in Long Term-Pesticide-Exposed Tea Garden Workers of.* . doi:10.1007/s13530-019-0389-1

Bondori, Bagheri, Sookhtanlou, & Damalas. (2021). Modeling farmers ' intention for safe pesticide use: the role of risk perception and use of information sources. Academic Press.

Bonvoisin, T., Utyasheva, L., Knipe, D., Gunnell, D., & Eddleston, M. (2020). Suicide by pesticide poisoning in India: a review of pesticide regulations and their impact on suicide trends. Academic Press.

Damalas, C. A. (2021). Farmers' intention to reduce pesticide use: the role of perceived risk of loss in the model of the planned behavior theory. Academic Press.

Dwijendra, N. (2022). Machine Learning Time Series Models For Tea Pest Helopeltis Infestation In India. *Webology, 19*(2). Available at: https://www.webology.org/abstract.php?id=1625

Dwivedi, D. N., & ... Benchmarking of traditional and Deep Learning time series modelling techniques for prediction of rainfall in United Arab Emirate. *8th International conference on Time Series and Forecasting*, Spain.

Gallia, G. K. N., & Kephaliacos, C. (2021). Ecological - economic modeling of pollination complexity and pesticide use in agricultural crops. *Journal of Bioeconomics, 23*(3), 297–323. doi:10.100710818-021-09317-9

Guo, L., Cao, A., Huang, M., Li, H., & Cao, A. (2021). Effects of haze pollution on pes- ticide use by rice farmers: fresh evidence from rural areas of China. Academic Press.

Heshmati, Mehri, & Khaneghah. (2021). Simultaneous multi- determination of pesticide residues in black tea leaves and infusion: A risk assessment study. Academic Press.

Kramer & Ware. (2020). Seasonal abundance of tea mosquito bug, Looper Antoni Signoret infesting name. *Journal of Entomology and Zoology Studies*, *8*(6), 2006–2009. doi:10.22271/j.ento.2020.v8.i6aa.8118

Kwakye, M. O., Mengistie, B., & Ofosu, J. (2019). Pesticide registration, distribution and use practices. *Environment, Development and Sustainability*, *21*(6), 2647–2671. doi:10.100710668-018-0154-7

Le, V. S., Lesueur, D., Herrmann, L., Hudek, L., Ngoc, L., & Lambert, Q. (2021). Sus- tainable tea production through agroecological management practices in Vietnam: A review. *Environ. Sustain*, *4*, 589–604. Advance online publication. doi:10.100742398-021-00182-w

Machekano, Masamba, Mvumi, & Nyamukondiwa. (2019). *Cabbage or 'pesticide' on the platter? Chemical analysis reveals multiple and excessive residues in African vegetable markets*. Academic Press.

Mallik, P., & Ghosh, T. (2008). Impact of climate on tea production: A study of the Dooars region in India. *Theor. Apple. Climatol., no. Cline*. Advance online publication. doi:10.100700704-021-03848-x

Niwa, H. (2021). Detection of organic tea farms based on the density of spider webs using aerial photography with an unmanned aerial vehicle (UAV). *Landscape and Ecological Engineering*, *17*(4), 541–546. doi:10.100711355-021-00454-x

Pal, K. (2019). Bio-control of Pests in Tea: Effect of Environmental. *International Journal of Applied and Computational Mathematics*, *5*(3), 1–9. doi:10.100740819-019-0666-3

Pardo, L. A., & ... *Pesticide exposure and risk of aggressive prostate cancer among private pesticide applicators*. doi:10.118612940-020-00583-0

Prasad, K., Roy, S., Sen, S., Neave, S., Nagpal, A., & Pandit, V. (2020). *Impact of different pest management practices on natural enemy population in tea plantations of Assam special emphasis on spider fauna*. doi:10.100742690-020-00111-0

Qianjiang. (2015). *Demand for Pesticides in China to Reach*. Academic Press.

Soydan, D. K., Turgut, N., Yalç, M., & Turgut, C. (2021). Evaluation of pesticide residues in fruits and vegetables from the Aegean region of Turkey and assessment of risk to consumers. Academic Press.

Suganthi, M., Event, S., & Senthilkumar, P. (2020). Comparative bioefficacy of Bacil- lus and Pseudomonas continues against Looper severe in tea (Camellia sinensis (L.) O. Kuntze. *Physiology and Molecular Biology of Plants*, *26*(10), 2053–2060. doi:10.100712298-020-00875-2 PMID:33088049

Sun, R., Yang, W., Li, Y., & Sun, C. (2021). Multi - residue analytical methods for pesticides in teas: A review. *European Food Research and Technology*, *247*(8), 1839–1858. doi:10.100700217-021-03765-3

Tambe, A. B., Mbanga, B. M. R., Nzefa, D. L., & Name, M. G. (2019). *Pesticide usage and occupational hazards among farmers working in small-scale tomato farms in Cameroon*. doi:10.118642506-019-0021-x

Xie, S. (2019). *Does a dual reduction in chemical fertilizer and pesticides, improve nutrient loss and tea yield and quality? A pilot study in a green tea garden in Shaoxing*. Zhejiang Province. doi:10.100711356-018-3732-1

Zhang, D., & ... *Detection of systemic pesticide residues in tea products at trace level based on SERS and verified by GC – MS*. doi:10.100700216-019-02103-7

Zhang, W. Y., Jing, T. Z., & Yan, S. C. (2017). Studies on prediction models of Den- drolimus superans occurrence area based on machine learning. *Journal of Beijing Forestry University*, *39*(1), 85–93.

Chapter 15
Crop Disease Prediction Using Deep Learning Algorithms

Pancham Singh
Ajay Kumar Garg Engineering College, Ghaziabad, India

Mrignainy Kansal
Ajay Kumar Garg Engineering College, Ghaziabad, India

Manoj Kumar Singh
IMS Engineering College, Ghaziabad, India

Sachin Kumar
Galgotias College of Engineering and Technology, Greater Noida, India

Anupam Dwivedi
Ajay Kumar Garg Engineering College, Ghaziabad, India

ABSTRACT

Good crops yield good food which in turn nourishes the human body and mind. But these crops across the globe face the threat of various diseases that remain unidentified, leading to poorer quality and quantity of crops. But, in recent times, the increasing adoption of smartphones worldwide and current developments in image processing of computers enabled by deep models of learning have made smartphone-based disease detection possible. In this chapter, the authors train a deep convolutional neural network (CNN) model to recognize 18 crop species and 28 diseases by feeding it a pre-available dataset of 70,296 photographs of unhealthy and healthy crop leaves taken under control. On a sustained test set, the trained model shows up to 99% accuracy, proving the feasibility of the method.

1. INTRODUCTION

Crop diseases pose a significant threat to global food security, exerting adverse effects on agricultural productivity and food sustainability. Ensuring timely and accurate detection of these diseases becomes

DOI: 10.4018/979-8-3693-0200-2.ch015

imperative for effective disease management and the mitigation of potential losses. In recent years, the emergence of deep learning, specifically Convolutional Neural Networks (CNNs) employing the Residual Network (ResNet) architecture, has shown remarkable potential in automating crop disease detection through computer vision.

This research paper endeavours to explore the application of deep learning techniques in crop disease detection, specifically focusing on CNNs with the ResNet architecture. Deep learning algorithms have demonstrated exceptional proficiency in recognizing intricate patterns and features within images, rendering them highly suitable for identifying disease symptoms present in crops (Patil & Manohar, 2023). The primary objectives of this study encompass the development of a precise and efficient crop disease detection model using deep learning, along with a thorough evaluation of its performance when compared against conventional methods and established benchmark datasets (Sumathi et al., 2022). By harnessing the capabilities of deep learning, this research aims to streamline the disease detection process, thereby facilitating early intervention and minimizing potential crop damage.

Through its contributions to precision agriculture, this research seeks to empower farmers, researchers, and policymakers with an advanced tool for effective disease monitoring and management (Vaishali et al., 2022). By bridging the gap between deep learning methodologies and crop disease detection, this investigation anticipates substantial advancements in safeguarding global food security against the emergence of crop diseases and potential environmental challenges.

2. RELATED WORK

Many initiatives have been established to combat crop loss caused by diseases. Integrated pest management (IPM) methods have gained prominence over traditional insecticide usage in the past decade (Sumathi et al., 2022). Regardless of the approach, accurate disease identification at an early stage is crucial for effective plant disease management. Indigenous crop clinics, agricultural institutions, and associated agencies have historically played a role in disease identification. In recent times, the progress of such initiatives has been further augmented by the abundance of online resources and the easy accessibility of data for disease detection, capitalizing on the global surge in data consumption. Additionally, Android application-based tools have flourished, leveraging the widespread adoption of Android applications and their associated technologies (Sharma et al., 2020).

Smartphones, equipped with powerful high-resolution cameras and supported by efficient computing capabilities and fast processing speeds, present a distinctive platform for disease detection technology. Their inherent features, including high-resolution displays and built-in functionalities, enable innovative applications in the field of disease detection. The widespread adoption of smartphones worldwide has facilitated the potential reach of automated picture identification for disease diagnosis (S. Das et al., 2021). High-definition cameras, high-performance processors, and broad smartphone adoption are factors that contribute to the technical feasibility of such disease detection methods. In recent years, object identification and computer vision have made significant strides, with large-scale image recognition challenges based on the ImageNet dataset serving as a standard for various image recognition-related tasks, including object classification.

Advancements in deep convolutional neural networks have significantly reduced the error rate in image recognition tasks, reaching nearly 3.58 percent within the next three years (Saeed et al., 2021). Deep neural networks have been extensively employed for one-end-to-another-end learning across various domains. In the context of plant disease detection, neural networks establish a key-pair relation between input images, representing sick crops, and the corresponding outputs, indicating the specific crop and disease. Building a deep network requires careful consideration of its structure, operating parameters (nodes), and edge distortions (weights) to accurately map input values to desired outputs. The study of neural networks has seen substantial growth, driven by both logical and technical advancements (Nennuri et al., 2023).

Creating an accurate image classifier for diagnosing crop diseases necessitates a large dataset of images containing both distorted and healthy crops. Such datasets were limited or unavailable until recently, posing challenges for researchers. To address this limitation, the New-crop-disease project was initiated, compiling thousands of images depicting both healthy and diseased crop (Wang et al., 2019). By making these images freely available to the public, the project aims to facilitate research and development in disease detection techniques. In this study, we employed a convolutional neural network to analyse 70,296 images and classify 28 illnesses across 18 crop species. Evaluating the model's performance involved judging and predicting the correct disease of the crop, among a total of 39 different classes (Yatoo & Sharma, 2021). The study concludes with the successful completion of the most useful and effective model for crop disease identification.

3. METHODOLOGY

3.1 Dataset Description

The New-crop-disease dataset experiment set, consisting of 70,289 images representing 28 different diseases, represents a substantial and diverse dataset for training and evaluating deep learning models. The inclusion of multiple iterations with variations in color, grayscale, and segmented versions adds depth and robustness to the analysis, enabling a comprehensive exploration of the neural network's learning capabilities (Malathi & Gopinath, 2023).

- **Color Version:** The original dataset was used, containing color images of crop leaves. The images were downscaled to 256x256 pixels for model optimization and prediction. This version allowed the model to leverage rich color information, akin to the human visual system's perception of crop diseases (Luo et al., 2009). The color cues in the images provided additional discriminative features that the model could utilize to identify various disease patterns.
- **Grayscale Version:** The dataset was converted to grayscale images, reducing each image to a single-channel representation. The downscaled grayscale images were used for model training and evaluation. By reducing each image to a single channel, this version challenged the model to rely solely on texture and structural features without the aid of color cues (Hu et al., 2019). It provided valuable insights into the model's ability to discern disease patterns based solely on grayscale intensity variations.
- **Single Version:** A single standardized version of the dataset was used across all experiments, irrespective of color or grayscale. The segmentation process was applied to remove background information and any biases introduced during data collection. The images were then downscaled to 256x256 pixels (Zambare et al., 2022). This version ensured that the model focused exclusively on the disease characteristics without being influenced by background noise or irrelevant elements.

By removing background information, the segmentation process created a cleaner and more homogeneous dataset, reducing the risk of overfitting to specific dataset characteristics.

The outcomes of these experiments play a vital role in addressing a pivotal concern in deep learning for agricultural applications. The ability of the neural network to maintain consistent performance across different dataset versions suggests that it has learned the fundamental concepts of crop diseases, indicating robust generalization capabilities (Boote et al., 2014). This discovery instils confidence in deploying the model in real-world scenarios, where the neural network can accurately diagnose and manage crop diseases, leading to more effective agricultural practices and increased crop productivity.

The implications of this research extend beyond crop disease detection, as the methodology and insights gained can be applied to other domains where deep learning models are deployed (Jenipher & Radhika, 2022). By understanding the model's learning process and its susceptibility to biases and artifacts, researchers can develop strategies to enhance the model's interpretability, fairness, and generalization across diverse datasets. Overall, the New-crop-disease experiment set represents a significant contribution to the field of precision agriculture and deep learning applications in agriculture. It highlights the importance of thoughtful dataset curation, model evaluation, and the potential for creating more reliable and efficient systems for crop disease detection and management, ultimately promoting sustainable and resilient agriculture for the future (Nirgude & Rathi, 2021).

Figure 1 showcases a selection of leaf images from the New-crop-disease dataset, representing various crop-disease pairs used in the study. Each image captures the visual characteristics of different plant ailments, providing a diverse and representative collection for model training and evaluation (Chen et al., 2022). This dataset's rich diversity enables deep learning models to learn and recognize the unique patterns and features associated with specific crop diseases, making it a valuable resource for advancing precision agriculture and crop management practices.

3.2 Approach

The experimental flowchart Figure 2, the New-crop-disease dataset is used, consisting of diverse crop leaf images with various diseases. The dataset is split into training and testing sets using different proportions, such as 80%-20% or 60%-40%. Images are processed with colored and grayscale filters, representing two versions of the dataset. Three deep learning models, CNN, ResNet, and Model Inception V3, are fine-tuned and trained on the dataset, with the ResNet model emerging as the most efficient. The aim is to achieve the most optimum result in disease detection accuracy and precision (Fenu & Malloci, 2020). Once the desired outcome is reached, the flowchart concludes, signifying the potential deployment of the ResNet model for real-world agricultural applications, aiding in early disease detection and improving crop management.

The ResNet architecture, Model Inception v3 Finetuned, and simple convolutional networks share similarities in their design. Stacks composed of convolutional layers, single or multiple fully connected layers, and additional layers are commonly utilized in various ResNet architectural modifications. Optional normalization and pooling layers follow the convolution layer (Kaur et al., 2016). Each layer is typically equipped with a ReLu non-linear activation unit. The Simple Convolutional Network and Model Inception v3 Finetuned consist of a SoftMax layer, five convolutional layers, and fully connected layers. After the first two layers, a normalization and pooling layer are added, while a single pooling layer follows the last convolutional layer. The SoftMax layer normalizes the input and yields a distribution of values

Figure 1. Leaf images from the New-crop-disease dataset, showing every crop disease pair used

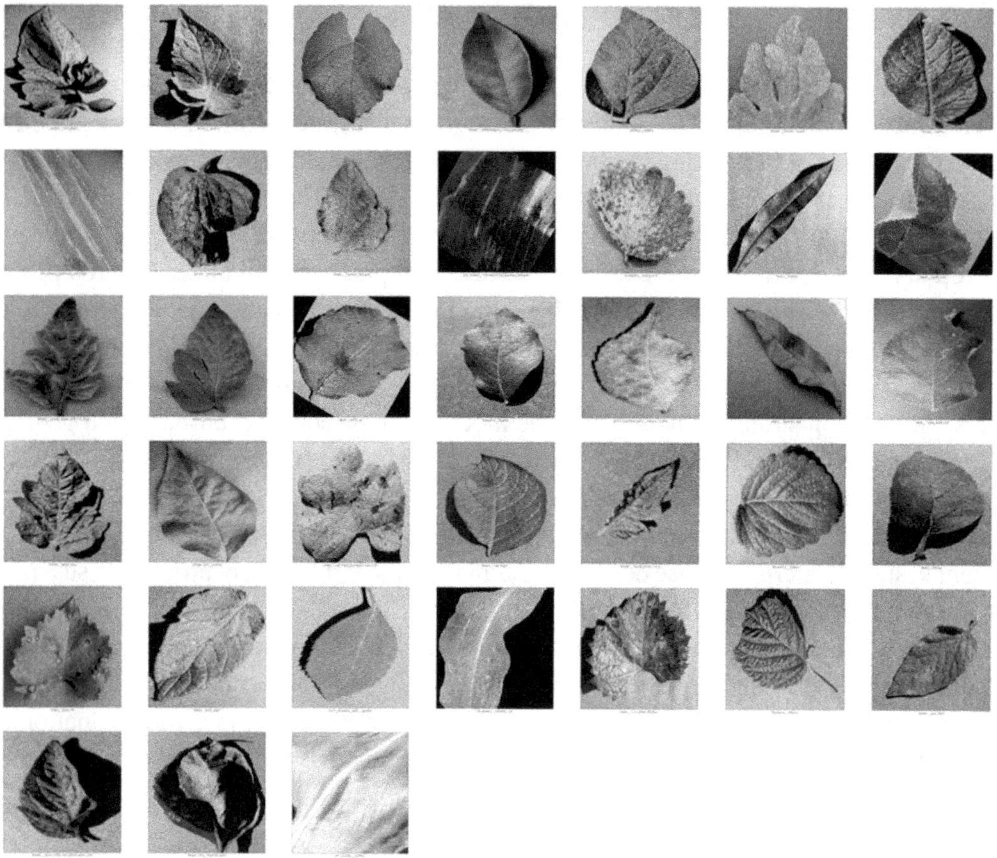

among the 38 classes, representing the network's confidence in predicting the appropriate classes for a given input image (Chopda et al., 2018).

In contrast, the ResNet architecture is characterized by its depth and width, with multiple layers and significantly fewer parameters compared to the Simple Convolutional Network and Model Inception v3 Finetuned. A key aspect of the ResNet architecture is the inclusion of inception modules, implementing the "network-in-network" concept. The inception model captures a diverse range of features through parallel convolutions and a max-pooling layer. To maintain viable processing requirements, 1x1 convolutions are used before the inception modules (Vishnoi et al., 2021).

Regarding training approaches, in some cases, the model is trained from scratch, while in others, pre-trained models initially trained on the ImageNet dataset are fine-tuned using transfer learning. In

Figure 2. Flow explains the execution of the approach

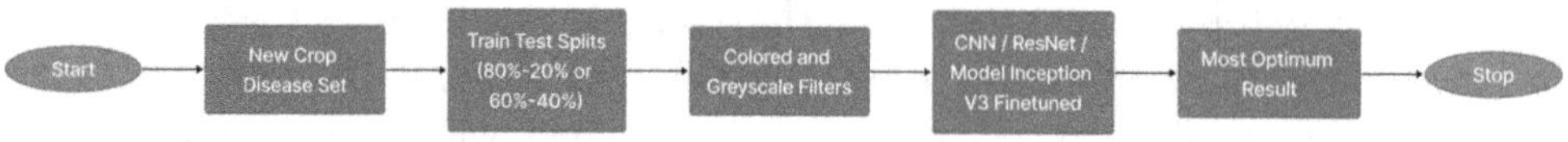

Figure 3. Residual block

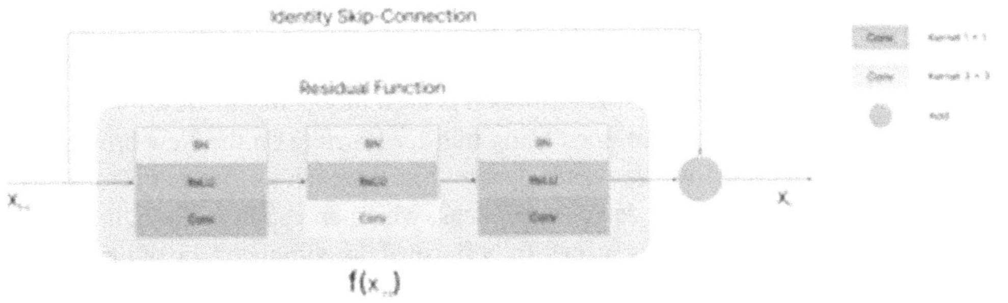

the transfer learning approach, the weights of the layers for the Simple Convolutional Network and Model Inception v3 Finetuned, and the classifier layers for ResNet, are reset. Unlike learning transfer, no restrictions or compromises are imposed on the training of any layers (Pinto et al., 2016). The key distinction between these two training approaches lies in the initialized state of biasing for some layers, allowing the transfer learning strategy to leverage the substantial visual experience gained from the pre-trained models.

The architecture of the deep learning models for crop disease detection incorporates a fundamental component known as the residual block Figure 3. This block consists of three layers, each featuring essential operations crucial for effective feature extraction and representation. The first and third layers are equipped with 1x1 convolution kernels, contributing to dimensionality reduction and channel-wise transformations. Meanwhile, the second layer employs a 3x3 convolution kernel, allowing the model to capture spatial features more effectively. Within the residual block, Batch Normalization (BN) is employed to stabilize the training process by normalizing intermediate activations, mitigating the issues of vanishing or exploding gradients (Kulkarni, 2018). Furthermore, Rectified Linear Unit (ReLU) activation introduces non-linearity, enabling the model to learn complex patterns and enhance its representational capacity. This well-designed architecture of the residual block plays a pivotal role in the deep learning models' ability to learn intricate patterns and optimize performance in crop disease detection tasks.

In summary, we have various experimental configurations with different combinations of the following variables:

1. Deep learning architecture:
    - Convolutional Neural Network (CNN)
    - Resnet
    - model inception v3 finetuned
2. Training mechanism:
    - Learning by Transfer,
    - Scratch Training.
3. Options of dataset type:
    - Color,
    - Greyscale,
    - Segmented (leaf).
4. Distribution of training testing set:

  - Train: 80%, Test: 20%,
  - Train: 60%, Test: 40%,

We have used the designation defined "Architecture" throughout the New-crop-disease research paper. Training Mechanism: Dataset: Train-Test Distribution is used to identify the results of our model. For example, in the case of Resnet models trained using transfer learning on the New-crop-disease dataset with an 80-20 train-test distribution, we would denote this approach as "Resnet: Transfer Learning." Every experiment lasts for the sum of decreased epochs, where an epoch is the number of modelling loops during which the specific CNN has successfully finished the complete loop of the entire modelling dataset. The decision to choose multiple epochs was dependent on the earlier finding that training can consistently degrade internal multiple epochs in every trial (Morbekar et al., 2020).

We endeavored to establish a standardized set of hyperparameters for all New-crop-disease experiments, ensuring a fair and justifiable comparison between different configurations. The selected hyperparameters for each experiment are as follows:

- Learning rate at the base: 0.01
- Learning rate policy: Step (decrease by some factor of epochs)
- Speed: 0.8
- Weight loss: 0.001
- Gamma: 0.15
- Batch sizes: 32 for ResNet and 16 for Simple convolutional network and fine-tuned model inception v3.

By employing these consistent hyperparameters across all experiments, we aim to establish a meaningful and coherent comparison of outcomes in each configuration. In (Haider et al., 2020), which is also known for its fast and open-source framework for deep learning. was used for all required experiments. We can also reproduce the fundamental outcomes, such as the overall accuracy, by running it in its default configuration.

3.3 Performance Measurement

In our experiment, we conduct a thorough evaluation of our approach by employing various train-test splits to assess its performance on unseen datasets and identify any potential issues. The train-test splits used include:

- 70-10: 70% of the dataset is used for training, and 10% for testing. This split allows us to gain insights into how our approach generalizes to new, unseen data and identifies any instances where our models might produce incorrect predictions (Meshram & Alvi, 2022).
- 60-20: Here, 60% of the dataset is used for testing, and 20% for training. This split further helps us understand the behaviour of our approach when a significant portion of the data is withheld for testing, giving us a measure of its robustness.
- 50-50: An equal 50% of the dataset is used for both training and testing. This balanced split allows us to gauge the performance of our models in a scenario where training and testing data are equally distributed.

Furthermore, we utilize additional train-test splits to gain a comprehensive understanding of our approach:

- 80-20: In this split, 80% of the dataset is used for training, and 20% for testing. This setting helps us investigate how our approach performs when a larger portion of data is used for training.
- 40-60: Here, 40% of the dataset is used for training, and 60% for testing. This split allows us to assess the impact of a smaller training set on the overall performance of our approach.

To ensure the integrity of the train-test splits, we take care to include every image of a single leaf in either the training set or the testing set. This ensures that each leaf's various images, captured from different angles, are correctly represented in the training and testing data. Throughout the training process, we continuously monitor the performance of our models by calculating the average precision, average recall, average F1 score, and overall accuracy at regular intervals, such as at the end of each epoch. These metrics provide valuable insights into the models' learning progress and their ability to accurately identify and classify crop diseases (Roy et al., 2023).

The final mean F1 score is used as a key performance metric to compare the outcomes of the various experimental settings. By utilizing the F1 score, which considers both precision and recall, we obtain a balanced measure of the models' performance and robustness across different train-test splits.

Additionally, we address the scenario where multiple images of a single leaf are present in the dataset. From a total of 70,289 photos, we have identified and accounted for 3696 images with mappings to ensure their proper inclusion in the training and testing sets (Bhatia et al., 2021). By conducting an in-depth evaluation using these diverse train-test splits and performance metrics, we gain valuable insights into the effectiveness and applicability of our approach to crop disease detection. This rigorous analysis enables us to develop more accurate and robust models, contributing to improved agricultural practices and crop management.

Figure 4 provides a visual representation of one of the three versions of the New-crop-disease dataset that were utilized in multiple experimental setups. This illustration showcases a sample image from the dataset, demonstrating its color version, which includes rich color information capturing the distinct characteristics of the crop leaf and disease (Iniyan & Jebakumar, 2022). The color version is an essential representation of the original dataset and serves as a basis for comparison with other versions.

In addition to the color version, the experiment also involves the grayscale version, where the same image is converted to a single-channel representation, effectively removing the color information. The grayscale version challenges the models to rely solely on texture and structural features without the aid of color cues, providing insights into the models' ability to discern disease patterns based solely on grayscale intensity variations.

The third version is the single version, where a standardized dataset is used across all experiments, regardless of color or grayscale. The segmentation process is applied to remove background information and biases introduced during data collection, ensuring a cleaner and more homogeneous dataset (Sowmiya & Krishnaveni, 2021). This version enables the models to focus exclusively on the disease characteristics without being influenced by irrelevant elements.

The inclusion of these three distinct versions of the New-crop-disease dataset in the experimental setups is instrumental in understanding the models' learning process and evaluating their performance in different scenarios. By comparing and analysing results from these variations, we gain valuable insights into the models' ability to generalize to unseen data and identify whether they truly understand

Figure 4. illustration of one of the three versions of the new-crop-disease dataset that were utilized in multiple experimental setups

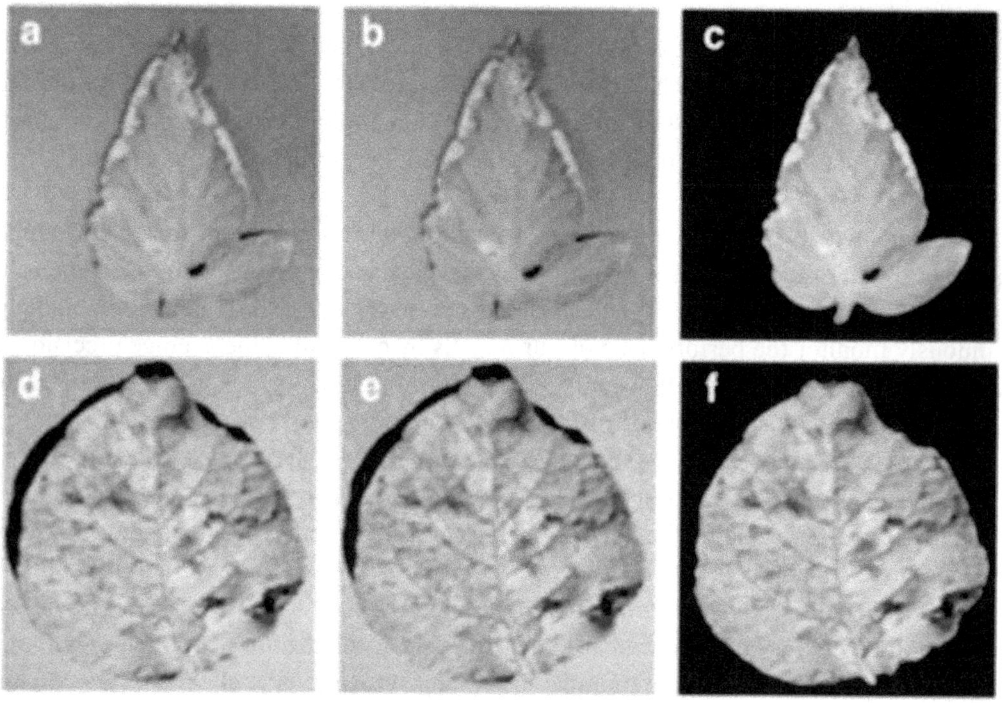

the underlying concepts of crop diseases or merely rely on superficial features or biases present in the dataset (Zhang et al., 2023). Such comprehensive evaluations contribute to the development of more reliable and effective deep learning approaches for agricultural crop disease detection and management.

4. RESULTS

We first remark that any guessing results in a precision of neglection of a dataset with 28 class labels. The complete precision we derived on the crop disease detection dataset lies between 95+% (in the case of ResNet), 92% (in the case of model inception v3 finetuned) and 90+% (in the case of simple Convolutional Neural Network(CNN)) all of our experimental configurations, which contain three visual and figurative representations of the image data, demonstrating the big promise of learning for same problems. The average f1 score, average accuracy, average recall, and all-over accuracy for completing our experimental combinations are evaluated thoroughly. The first step finishes and then decreases the learning rate, and all experimental setups almost uniformly converge after running for a total of 2 epochs apiece.

Figure 5 displays the graph showing the accuracy in the training and testing steps throughout the model's training process, providing insights into its learning progress and performance on both datasets. As the training progresses, the accuracy on the training data gradually increases, indicating effective learning, while the accuracy on the testing data improves, reflecting the model's ability to generalize to unseen data. Similarly, Figure 6 illustrates the graph showing the loss in the training and testing steps, where the decreasing loss signifies the model's improved accuracy and ability to minimize errors. The

Figure 5. Graph shows the accuracy in the training and testing step

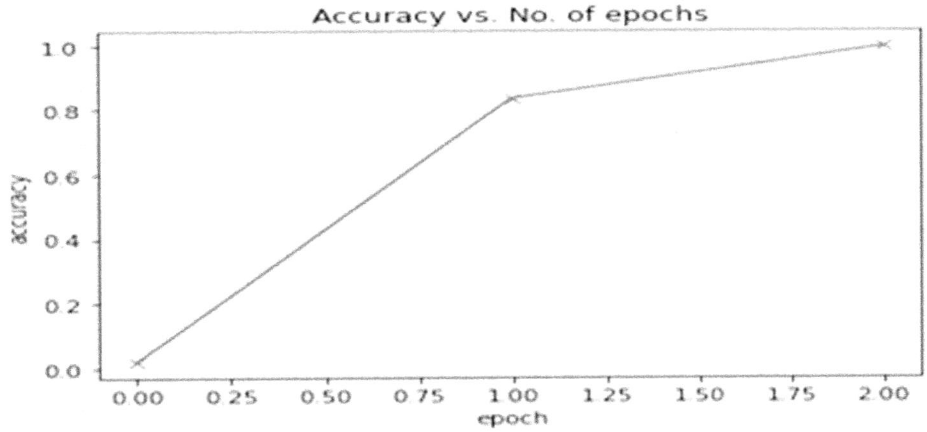

close alignment between accuracy and loss on both datasets indicates that the model is not overfitting and can generalize well. These figures are valuable tools in evaluating the model's training progress, guiding hyperparameter tuning, and ensuring optimal performance for crop disease detection in real-world agricultural applications.

We experimented with several test sets to train set ratios to give location over "over-fitting" and found that the model, in the case of ResNet Transfer Learning: completes its complete precision 99% (average f1 score of 0.99). As anticipated, if the increase test set train ratio increases, both the Simple convolutional network and model inception v3 are finetuned and ResNet's overall performance starts to decline. However, the performance decline was not as severe as we would anticipate if the model was truly overfull. Additionally, figure 5 demonstrates there is not a single split among training and the validation loss. ResNet consistently outperforms Simple convolutional network and model inception v3 finetuned among the two architectures, strategy, transfer learning consistently produces good results as was anticipated.

Figure 6. Graph shows the loss in the training and testing step

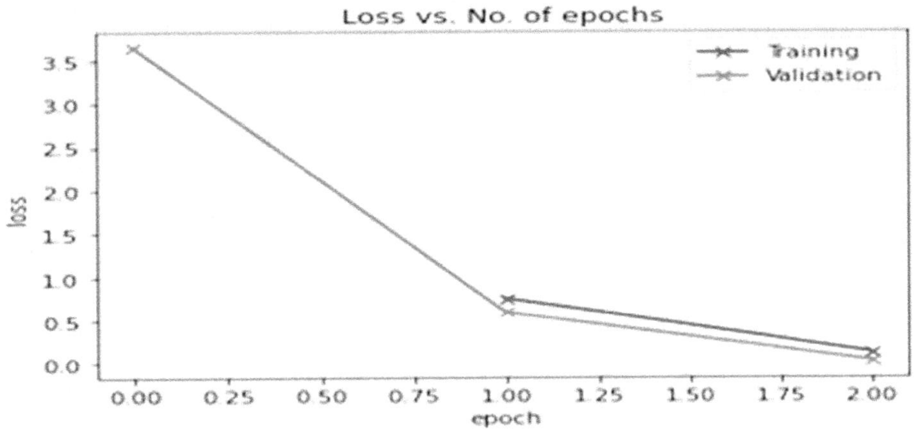

Figure 7. Graph shows the learning rate vs batch no.

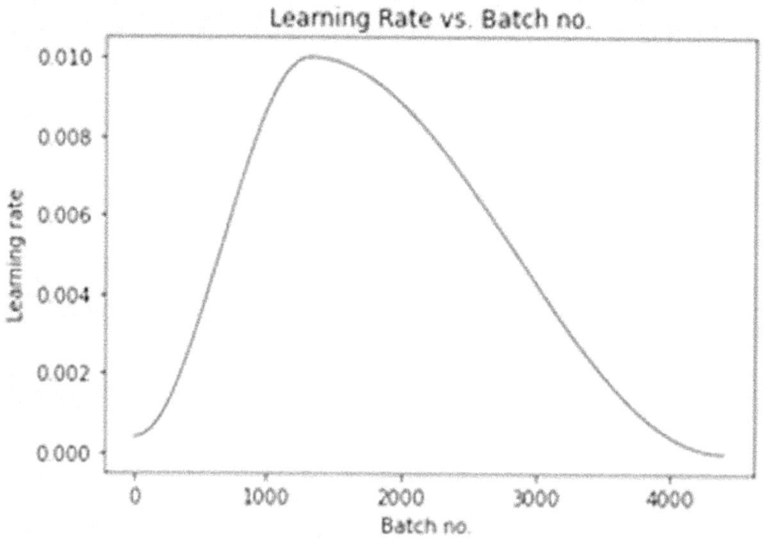

In Figure 7, representing the learning rate vs. batch number, provides valuable insights into the ResNet model's learning process and optimization during training. By analysing this graph, we understand how the model's learning rate evolves over time .When we maintain the remainder of the experimental setup constant, the three varieties of set data (color, segmented, grayscale) exhibit a distinctive variance in performance among the overall practical. When applied to the dataset with colored filters, the models perform best. The main concern when planning the tests was that the neural networks may be simply taking on innate biases related to the various states of light, the procedure, and the instrument of data collecting. To test the model's flexibility in the absence of any color and the capacity to acquire higher-level structural features, we tested the grayscale variety version of a similar dataset. Even with the below-average performance seen, mean F1 score was 0.85324 (complete accuracy of 86.53%), which was expectedly lower than the performance of the experiments when done on the colored model of the dataset. The model's performance, utilizing segmented copies of the entire dataset to investigate the influence of image backgrounds on overall performance, outperforms using a grayscale model of images, but falls slightly behind the performance achieved by using images with all colors. Even though these methods produce good results on the New-crop-disease dataset, which was gathered in a looked-after setting, we also evaluated the performance of the model on photos taken from reliable internet sources and academic extensions. These pictures were not accessible.

We derived 2 little small, validated datasets (dataset-1) as well as (dataset-2), using a mixed automatic download from Image verification phase (see Material Supplementary for describing the procedure completely). If the model with the best performance was used on the datasets, then we were able to correctly predict the class label (i.e., the information of crop and its diseases) from among 38 potential class labels showing 85.40% accuracy in the 1st dataset and 90.69% in the 2nd dataset. The average accuracy of a random classifier is only 2.63%, as we note. In dataset 1's case, 55.75% of all photos had the proper class in the top five predictions, whereas in dataset 2's case, 69.61% did.

Figure 8. Graph shows the train and validation accuracy of Dataset 1 and Dataset 2 after mixing the images

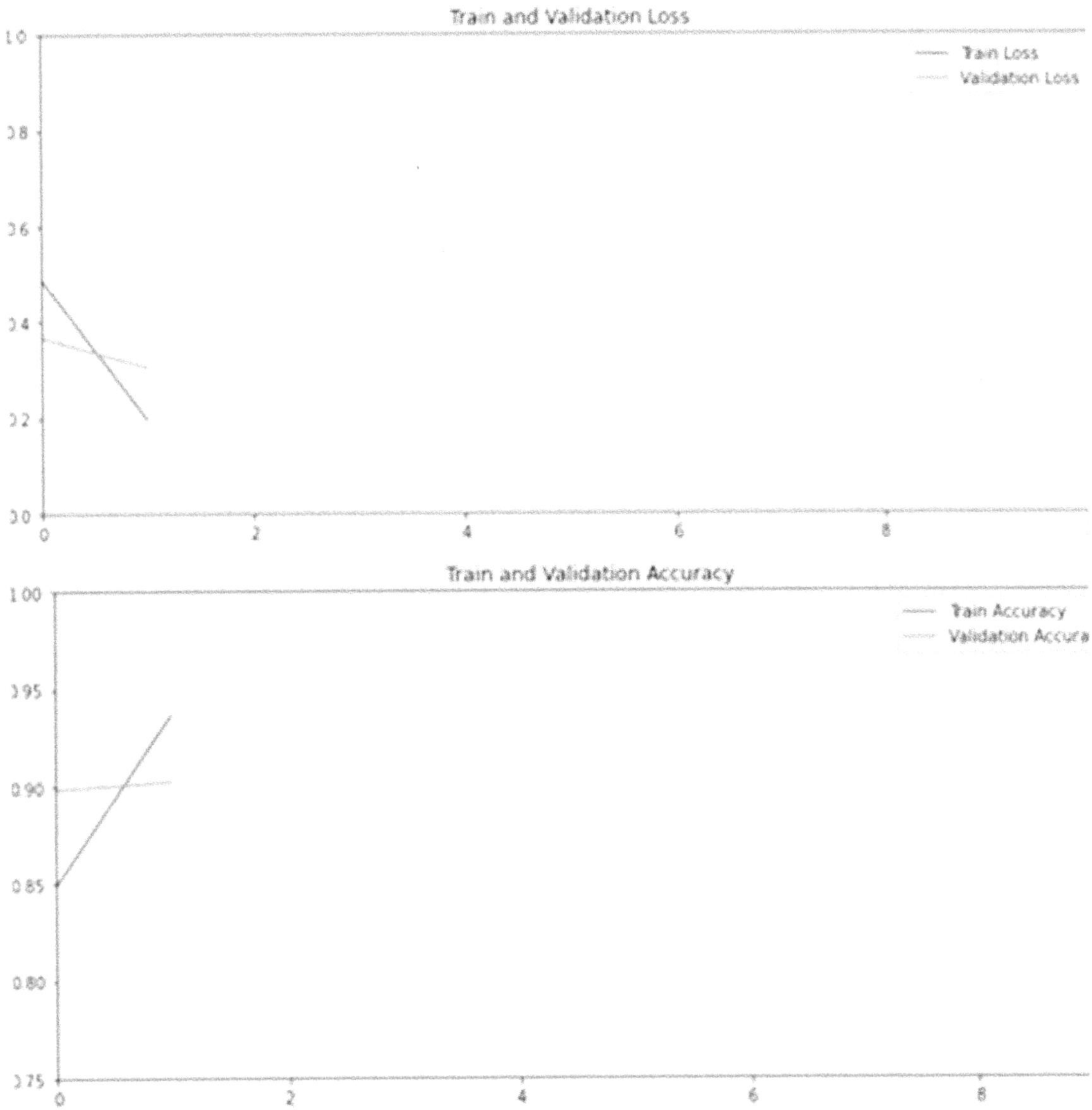

Furthermore, Figure 8 depicts the graph showing the train and validation accuracy of dataset-1 and dataset-2 after mixing the images. This figure illustrates the model's performance on these external datasets, which were not part of the original New-crop-disease dataset.

All the results yet presented before the premise that the model must be able to determine the disease of the crop state. To make the scenario more plausible, we would assume that the species of the crop is given, as it can be reasonably assumed that the person cultivating the crop would be aware of its species. This simplification allows us to focus solely on the disease detection aspect, leveraging the knowledge of the specific crop species to enhance the accuracy and relevance of the detection model. To avoid trivial categorization, we restrict ourselves to crops with the least classes per crop to evaluate the model's performance in this scenario. In the instance where dataset 1 has classes spread among crops. In such a dataset, random guessing would have an accuracy of dataset containing random files and that of random noise. While the accuracy of our model is higher. When there are classes spread among crops in the dataset. In such a dataset, random guessing would have an accuracy of 0.00-0.02, whereas our model has a precision of 0.9100.

Data set 2 in the example also has more classes spread across more types of crops. A random prediction in this dataset would be accurate to within 0.2, whereas our model is accurate to within 0.9. When there are more classes spread among crops in the sample. In a dataset like this, random guessing would be accurate to tune as opposed to our model's accuracy of 95+%.

5. CONCLUSION

Based on the observations and working techniques described above, we can draw several key conclusions. Firstly, the Resnet model proves to be the most efficient among the tested models for crop disease detection. The practical accuracy achieved using the colored filter is impressively high at 99%. This indicates that the model performs exceptionally well when leveraging rich color information to identify and classify crop diseases accurately. Additionally, the consideration of a wide variety of crop classes in datasets 1 and 2 leads to more precise results compared to other models. The extensive dataset with diverse crop-disease pairs provides the model with a comprehensive understanding of various diseases across different crops, enhancing its overall accuracy. Furthermore, the experiments demonstrate that the model's performance is consistent even under the scenario of pre-known species of crops and their associated diseases. This robustness and reliability make the model highly suitable for practical applications in real-world agricultural settings.

6. FUTURE SCOPE

While the grayscale filter yields a slightly lower accuracy of 86.53%, there is room for further improvement in fine-tuning the model's performance in this aspect. Researchers can explore various techniques, such as data augmentation, transfer learning, and hyperparameter optimization, to enhance the grayscale model's accuracy. Fine-tuning the grayscale model can lead to a more balanced and comprehensive approach, combining the strengths of both color and texture-based features.

Moreover, the success of the model with pre-known species of crops opens up exciting possibilities for extending its usability to scenarios where crop species are not known in advance. Implementing techniques for species and disease identification in the absence of prior knowledge can enhance the model's adaptability and make it suitable for broader agricultural applications.

Additionally, the model's easy-to-use nature suggests its potential for practical deployment in agricultural fields. Further research can focus on developing user-friendly interfaces and integrating the model into smart farming technologies, making it accessible to farmers and agronomists for real-time crop disease detection and management.

In conclusion, the Resnet model's efficiency, high accuracy with colored filter data, and robustness with pre-known crop species make it a promising tool for crop disease detection. With further improvements in grayscale performance and usability enhancements, the model holds great potential to revolutionize precision agriculture and contribute to sustainable crop management practices.

REFERENCES

Bhatia, G., Joshi, N., Iyengar, S., Rajpal, S., & Mahadevan, K. (2021). Crop Prediction Based on Environmental Conditions and Disease Prediction. In T. Senjyu, P. N. Mahalle, T. Perumal, & A. Joshi (Eds.), *Information and Communication Technology for Intelligent Systems* (pp. 335–344). Springer. doi:10.1007/978-981-15-7078-0_31

Boote, K. J., Porter, C. H., Hargreaves, J., Hoogenboom, G., Thorburn, P., & Mutter, C. (2014). AgMIP Training in Multiple Crop Models and Tools. In *Handbook of Climate Change and Agroecosystems* (Vol. 3, pp. 393–410). Imperial College Press. doi:10.1142/9781783265640_0025

Chen, Q., Xuan, Q., & Xiang, Y. (2022). Application of Deep Learning in Crop Stress. In *Deep Learning Applications* (pp. 61–77). World Scientific. doi:10.1142/9789811266911_0004

Chopda, J., Raveshiya, H., Nakum, S., & Nakrani, V. (2018). Cotton Crop Disease Detection using Decision Tree Classifier. *2018 International Conference on Smart City and Emerging Technology (ICSCET),* 1–5. 10.1109/ICSCET.2018.8537336

Das, S., Sharma, R., Gourisaria, M. K., Rautaray, S. S., & Pandey, M. (2021). A Model for Probabilistic Prediction of Paddy Crop Disease Using Convolutional Neural Network. In D. Mishra, R. Buyya, P. Mohapatra, & S. Patnaik (Eds.), *Intelligent and Cloud Computing* (pp. 125–134). Springer. doi:10.1007/978-981-15-5971-6_12

Fenu, G., & Malloci, F. M. (2020). Artificial Intelligence Technique in Crop Disease Forecasting: A Case Study on Potato Late Blight Prediction. In I. Czarnowski, R. J. Howlett, & L. C. Jain (Eds.), *Intelligent Decision Technologies* (pp. 79–89). Springer. doi:10.1007/978-981-15-5925-9_7

Haider, W., Ur Rehman, A., Maqsood, A., & Javed, S. Z. (2020). Crop Disease Diagnosis using Deep Learning Models. *2020 Global Conference on Wireless and Optical Technologies (GCWOT),* 1–6. 10.1109/GCWOT49901.2020.9391605

Hu, H., Su, C., Yang, C., & Yu, P. (2019). A Study on Intelligent Crop Disease and Pest Diagnosis System Based on Hadoop. *2019 International Conference on Robots & Intelligent System (ICRIS),* 171–173. 10.1109/ICRIS.2019.00052

Iniyan, S., & Jebakumar, R. (2022). Crop Yield Prediction on Soybean Crop Applying Multi-layer Stacked Ensemble Learning Technique. In G. Manogaran, A. Shanthini, & G. Vadivu (Eds.), *Proceedings of International Conference on Deep Learning, Computing and Intelligence* (pp. 335–348). Springer Nature. 10.1007/978-981-16-5652-1_29

Jenipher, V. N., & Radhika, S. (2022). An Automated System for Detecting Rice Crop Disease using CNN Inception V3 Transfer Learning Algorithm. *2022 Second International Conference on Artificial Intelligence and Smart Energy (ICAIS),* 88–94. 10.1109/ICAIS53314.2022.9742999

Kaur, R., Garg, R., & Aggarwal, H. (2016). Big data analytics framework to identify crop disease and recommendation a solution. *2016 International Conference on Inventive Computation Technologies (ICICT), 2*, 1–5. 10.1109/INVENTIVE.2016.7824791

Kulkarni, O. (2018). Crop Disease Detection Using Deep Learning. *2018 Fourth International Conference on Computing Communication Control and Automation (ICCUBEA),* 1–4. 10.1109/ICCUBEA.2018.8697390

Luo, J., Huang, W., Wang, J., & Wei, C. (2009). The crop disease and pest warning and prediction system. In D. Li & C. Zhao (Eds.), *Computer and Computing Technologies in Agriculture II* (Vol. 2, pp. 937–945). Springer US. doi:10.1007/978-1-4419-0211-5_17

Malathi, V., & Gopinath, M. P. (2023). A Review on Rice Crop Disease Classification Using Computational Approach. *International Journal of Image and Graphics, 23*(3), 2240007. doi:10.1142/S0219467822400071

Meshram, R. A., & Alvi, A. S. (2022). Crop Disease Prediction Using Computational Machine Learning Model. In J. S. Raj, Y. Shi, D. Pelusi, & V. E. Balas (Eds.), *Intelligent Sustainable Systems* (pp. 541–550). Springer Nature. doi:10.1007/978-981-19-2894-9_40

Morbekar, A., Parihar, A., & Jadhav, R. (2020). Crop Disease Detection Using YOLO. *2020 International Conference for Emerging Technology (INCET),* 1–5. 10.1109/INCET49848.2020.9153986

Nennuri, R., Kumar, R. H., Prathyusha, G., Tejaswini, K., Kanishka, G., & Sunitha, G. (2023). A Multistage Deep Model for Crop Variety and Disease Prediction. In A. Abraham, T. Hanne, N. Gandhi, P. Manghirmalani Mishra, A. Bajaj, & P. Siarry (Eds.), *Proceedings of the 14th International Conference on Soft Computing and Pattern Recognition (SoCPaR 2022)* (pp. 52–59). Springer Nature Switzerland. 10.1007/978-3-031-27524-1_6

Nirgude, V., & Rathi, S. (2021). An Extensive Survey on Deep Learning Approach for Crop Diseases Management. In A. Abraham, T. Hanne, O. Castillo, N. Gandhi, T. Nogueira Rios, & T.-P. Hong (Eds.), *Hybrid Intelligent Systems* (pp. 638–648). Springer International Publishing. doi:10.1007/978-3-030-73050-5_62

Patil, M. A., & Manohar, M. (2023). A Comprehensive Review on Crop Disease Prediction Based on Machine Learning and Deep Learning Techniques. In S. Kumar, H. Sharma, K. Balachandran, J. H. Kim, & J. C. Bansal (Eds.), *Third Congress on Intelligent Systems* (pp. 481–503). Springer Nature. 10.1007/978-981-19-9225-4_36

Pinto, L. S., Ray, A., Reddy, M. U., Perumal, P., & Aishwarya, P. (2016). Crop disease classification using texture analysis. *2016 IEEE International Conference on Recent Trends in Electronics, Information & Communication Technology (RTEICT),* 825–828. 10.1109/RTEICT.2016.7807942

Roy, A., Sehrawat, D., & Singh, J. (2023). Crop Disease Prediction Using Deep ConvNet Architecture Technique. In D. Goyal, A. Kumar, V. Piuri, & M. Paprzycki (Eds.), *Proceedings of the Third International Conference on Information Management and Machine Intelligence* (pp. 187–195). Springer Nature. 10.1007/978-981-19-2065-3_22

Saeed, Z., Raza, A., Qureshi, A. H., & Haroon Yousaf, M. (2021). A Multi-Crop Disease Detection and Classification Approach using CNN. *2021 International Conference on Robotics and Automation in Industry (ICRAI),* 1–6. 10.1109/ICRAI54018.2021.9651409

Sharma, R., Das, S., Gourisaria, M. K., Rautaray, S. S., & Pandey, M. (2020). A Model for Prediction of Paddy Crop Disease Using CNN. In H. Das, P. K. Pattnaik, S. S. Rautaray, & K.-C. Li (Eds.), *Progress in Computing, Analytics and Networking* (pp. 533–543). Springer. doi:10.1007/978-981-15-2414-1_54

Sowmiya, M., & Krishnaveni, S. (2021). Deep Learning Techniques to Detect Crop Disease and Nutrient Deficiency -A Survey. *2021 International Conference on System, Computation, Automation and Networking (ICSCAN),* 1–5. 10.1109/ICSCAN53069.2021.9526442

Sumathi, M., Rajkamal, M., Raja, S. P., Venkatachalapathy, M., & Vijayaraj, N. (2022). A crop yield prediction model based on an improved artificial neural network and yield monitoring using a blockchain technique. *International Journal of Wavelets, Multresolution, and Information Processing*, *20*(06), 2250030. doi:10.1142/S0219691322500308

Vishnoi, V. K., Kumar, K., & Kumar, B. (2021). Crop Disease Classification Through Image Processing and Machine Learning Techniques Using Leaf Images. *2021 First International Conference on Advances in Computing and Future Communication Technologies (ICACFCT),* 27–32. 10.1109/ICACFCT53978.2021.9837353

Wang, H., Li, C., Zhang, J., Wang, J., Ma, Y., & Lian, Y. (2019). A new LSTM-based gene expression prediction model: L-GEPM. *Journal of Bioinformatics and Computational Biology*, *17*(04), 1950022. doi:10.1142/S0219720019500227 PMID:31617459

Yatoo, A. A., & Sharma, A. (2021). A Novel Model for Automatic Crop Disease Detection. *2021 Sixth International Conference on Image Information Processing (ICIIP), 6,* 310–313. 10.1109/ICIIP53038.2021.9702553

Zambare, R., Awati, C., Deshmukh, R., & Shirgave, S. (2022). A Theoretical Study on CNN based Cotton Crop Disease Detection. *2022 6th International Conference on Trends in Electronics and Informatics (ICOEI),* 1062–1066. 10.1109/ICOEI53556.2022.9777219

Zhang, T., Cai, Y., Zhuang, P., & Li, J. (2023). Remotely Sensed Crop Disease Monitoring by Machine Learning Algorithms: A Review. *Unmanned Systems*, 1–11. doi:10.1142/S2301385024500237

Chapter 16
Identification of Tomato Plant Disease Using Faster R-CNN and RetinaNet

Indrajeet Kumar
https://orcid.org/0000-0003-2814-2900
Graphic Era Hill University, Dehradun, India

Aman Bisht
DIT University, Dehradun, India

Jyoti Rawat
Vertex Plus Technologies Limited, India

ABSTRACT

Plant disease plays a significant role in the low productivity of tomatoes which leads to huge loss to the farmer and the country's economy. Identification of plant disease at an early stage can play a major role in producing good amounts and even good quality tomatoes. Identification of the disease sometimes gets difficult because of lack of knowledge or having multiple diseases or even wrong prediction of disease. The chapter contains implementations of two classification algorithms Faster R-CNN and RetinaNet. Both the algorithms are initiated from transfer learning and tested on different hyperparameters for better results. The work produces impressive results with average precision (AP)-50 as FR 93.11%, RN 95.54%. The preliminary results look promising and can be helpful for harvest quality and precision agriculture.

1 INTRODUCTION

Every year, India produces a significant variety of crops and vegetables. After China, India is the world's second-largest tomato grower, with 19.75 million tons produced from 7.89 lakh hectares with an average output of 25.05 tons per hectare. Andhra Pradesh is the world's greatest tomato grower, with the commodity presently retailing for Rs 100 per kg and anticipated to soar much higher (Vikaspedia, 2022).

DOI: 10.4018/979-8-3693-0200-2.ch016

At the same time, India is dealing with a tomato disease that is destroying yields and costing farmers a lot of money.

It is found that these diseases which are very common to the tomato plants and can affect the growth and development of the crop which leads to reduced production of tomatoes which is the main cause of economic losses for a country. Mosaic virus is a virus that is mostly found in the leaves of plants, and it can be easily identified as it comes with a spotty appearance on the infected leaves of plants. It affects the growth of plants and also affects the leaves of the plants which can have deformities in some or other ways (Asif et al., 2022). Their leaves are crinkled or in a wavy pattern. It may exhibit shoestring syndrome which results in the edges of the leaves failing their ability to develop, with the leaf appearing as long narrow strips. Diseases causing fungus which is known as Septoria lycopersici produces a disease which is known as Septoria leaf spot. It is found on the leaves of plants severely where wet, warm (60-80 F) and humid weather is present for a prolonged period of time and is considered as a most harmful disease which brings a lot of destruction to the leaves of plants. This spreads rapidly and will result in removal of leaves and the weakening of plants resulting in being unable to bear fruit. Late Blight is a disease that is mostly present in potato and tomato plants under the areas where there is a high relative humidity with temperatures ranging from 4 degree to 29 degree (40 - 80 F) and mainly caused by the water mold Phytophthora infectants. It is a harmful disease which rots the plants within a week or two weeks. Lesions will appear on the stems, petioles and leaves. At the margin of the lesions under leaf surface we may find a whitish growth which is a spore producing structure.

In this technological era, people are surrounded with many tools and technologies that make our daily life easier. As in agriculture there have been a lot of enhancements on the technological side which has made the life of farmers easier. As it is known that the mainstream source of economy for all the countries is through agriculture. Artificial Intelligence solutions will help in the agriculture industry's evolution to increase efficiency as well ensure good quality and quantity for crops. AI can be used in correctly identifying the output yield of crops which can be further used to forecast the prices of crops for the upcoming weeks yielding maximum profit for farmers (Udawant. & Srinath, 2022). AI sensors can also be used for the detection of the areas on the crop that are generally affected by weed and then the herbicides are sprayed on the region that remained unaffected by the weeds which in turn reduces the usage of herbicides. Using mechanical robotic hands that are powered with AI can be used for harvesting large amounts of crops at higher volume reducing manual efforts. AI can be used in diagnosis and classification of disease which will help farmers to control and prevent the disease. It can also be used in monitoring crop health which forms the basis for diagnosis of pests/soil defects, nutrients deficiencies in soil (Udawant. & Srinath, 2022).

In this work, a computer assisted deep learning based model is designed to detect and to classify the tomato plant leaf diseases into three categories as Late Blight, Mosaic virus, Tomato Septoria leaf spot. The used samples are collected from the public dataset, which is open for research purpose. The research objectives have been achieved by implementing faster RCNN and RetinaNet deep learning model. After the implementation, the output of the system has been evaluated by the help of average precision (AP)-50, FR, and RN. The preliminary results look promising and can be helpful for harvest quality and precision agriculture.

This work is organized as started with the introduction where some basics of the plant disease is discussed and their correlation with the AI technologies is explained how AI can be helpful for detecting the plant diseases, next part is the related studies, where existing literature and previous methods are discussed for plant leaf diseases detection. In the next portion of the paper, the methodology of the work is explained which includes the details of all the steps needed to be followed for fulfillment of the objective of the work. Then the implementation part is there which consists of a list of experiments and their results, at last the paper is concluded with the discussion and conclusion part.

2 LITERATURE SURVEY

To understand the disease in plants and to identify them, a few research papers need to be inferred which are based on deep Learning as well as machine learning. The study (Asif Iqbal Khan et al. 2022) performed a real-time apple leaf disease detection system. In this experiment, 9000 high-quality RGB image datasets have been used for expert-annotated apple disease symptoms. They have divided the disease detection into two stages. The Primary stage classifies the input images into diseased as healthy or damaged, and at the secondary stage (detection stage) processing starts only if any disease is being detected in the first stage. In 2012, Dubey et al. introduced a technique for disease classification in apple fruit. In this experiment K-means clustering methods were used for segmentation and different features extracted as GCH, CCV, LBP and CLBP and support vector machine (SVM) were used for classification.

Article (Udawant. & Srinath, 2022) had performed disease detection on cotton leaf using instance segmentation and proposing an accurate classification of disease on the leaf but had to face the problem of detecting the diseases because of images which were being taken. As the training of models was done on isolated environment photos, many diseases could not be identified such as images with the black background, images with the single leaf disease. Even if it was not practical in real life, they would stop working when it is exposed to real time detection because there could be many diseases on a single leaf or numerous leaves in the background. In order to make this into reality, a good quality camera phone was used to capture around 2500 pictures of cotton leaves, photos were taken in a way that it would depict the real time detection as the datasets would have several leaves in the background or picture of affected leaves by various diseases and pests. When the background of the pictures was complicated, deep neural networks were utilized for object recognition algorithms for identification of multiple diseases in one leaf. Two detection approaches were used to separate the objects. Object detectors with one and two stages. Using single state detectors, the object detection is considered a UNI regression issue (that is, bounding box coordinates from an input image). Then, in the initial stage of the detection, region proposal networks (RPN) were deployed. It provides recommendations for object detection and bounding box regression in regions of interest. Later, a single shot detector (SSD) and a Mask R-CNN (two stage) were used on the batch having size as 52 and a rate of learning of 0.0001 datasets, then after that model was being trained (Udawant. & Srinath, 2022).

V. Tsironis et al. (Tsironis et al., 2020) worked on object detection algorithms on real world tomato datasets. This experiment introduces a task of selective treating and cropping. By using the deep learning algorithm, it was able to recognize the growth of the crop and disease, fruit detection and recognition of the species (Tsironis et al., 2020). G. Sambasivam et al. published research (Sambasivam, & Opiyo, 2021) paper on cassava disease detection and classification. Cassava is the common staple food in Africa, as it is being consumed by several local people and it is highly enriched in protein. Cassava plants are highly at risk to a wide range of diseases that includes cassava brown streak virus and cassava mosaic disease. In Tanzania first CMD infections were reported, and it spread like a pandemic, which threatened Africa's food security in the 19th century. So, the model performance showed promising results on classification of cassava disease detection by training CNN through an imbalanced dataset which achieved 93% accuracy.

Bin Yan et al. (Yan et al., 2021) tried to identify the apples in the apple tree image which should be picked by the robots, and which one should not be picked. Detection method have been used to targets lightweight apple which in turn used by picking robots using improvised YOLOv5s. In backbone architecture, replaced Bottlleneck CSP module with BottlleneckCSP-2. SE section was introduced a better performing backbone network for to the visual attention mechanism network which can be used for acquiring the feature of target apples in different conditions. To avoid the wrong prediction of the low size instance that are present in the related frame, the primary anchor box circumference is increased. Improved network models were used to recognize the fruits effectively with total precision, recall, MAP, and F1 of recognition as, 88.8%, 91.4%, 86.7%, and 87.5% respectively with an average detection rate of 0.015 second per sample.

The study reported by (Zhang C. et al, 2017) published a research paper on apple lead disease recognition method and it was purely based on techniques that involved image processing and recognition of patterns. For the input image, an RGB model designed after that was turned to HSI (Hue, Saturation, and Intensity), YUV, and gray models. To remove the background of the image a specific threshold value is used, and after that image was segmented using a region-growing algorithm (RGA). 38 features that are used to classify the diseased images such as shape, color and texture, which were then extracted. These features were combined with genetic algorithms (GA) to first increase the accuracy of identification, combining genetic algorithms (GA) and correlation-based feature selection yielded the most valuable attributes (CFS). Finally, the SVM classifier identified the disorders. Using the dataset, which comprises 90 disease images for only three types of apple leaf diseases like rust, powdery mildew, and mosaic. The proposed model achieved a 90% rate of correct recognition.

Zhaoyi Chen with their research team (Zhaoyi et al. 2018) worked on the identification of the diseases that are present in plants by making improvisation on an existing model based on the YOLOv5 network model making use of deep-learning methods. To detect the rubber tree, anthracnose, and powdery mildew, made use of convolutional neural networks (CNN). To reduce the number of convolutional neural network parameters and achieve channel sharing within the group, instead of using the Bottleneck module in the backbone network, the Involution Bottleneck module was employed. To combine the disease's features, the SE module was added to the last tier of the backbone network. Instead of using the loss function, Efficient Intersection Over Union (EIOU) was employed. Generalized IOU (GIOU), which was present in YOLOv5, was used to adjust for changes in target frame width, height, and confidence, which aided in the network model's convergence. The model had an accuracy of 86.5% for powdery mildew detection and 86.8% for anthracnose detection (Zhaoyi et al. 2018). Konstantinos P et al. performed the identification of plant disease and performed diagnosis using the dataset containing the healthy and severely affected plants, by developing convolutional neural network models. The dataset contains 87,848 images of 58 different classes in the form of [plant, disease], 25 different sets of plants are used to train the model. The final model achieved 99.53% accuracy (Ferentinos, 2018).

Hu Gensheng W. H. et al. identified and detected the diseases that are present in tea leaves and also took measures to control and prevent the diseases (Hu et al., 2019). By making use of the support vector machine the images of tea leaf's disease spots are segmented by extracting the features such as color and texture. By taking input as these segmented images, identification of disease is performed using conditional deep convolutional generative adversarial networks (C-DCGAN), and VGG16 networks. To improve the accuracy in identifying diseases VGG16 classification networks are trained on augmented images using C-DCGAN which can generate new disease spot images and the proposed method reaches 90% identification accuracy.

Yong Zhong et al. worked on the identification of apple leaf diseases using the 3 methods which are based on regression, which is then followed by multi-label classification and loss function which forms the basis on DenseNet-121 deep convolution network (Dhiman & Tamsir, 2018). For the modeling of data and method evaluation a dataset which is used to contain the 2462 different images of apple leaf which has six leaf diseases. The proposed methods concluded 93.51%, 93.31% and 93.71% accuracy on the dataset (Zhong. & Zhao, 2020).

Guichao Lin et al. published a research paper with an objective to provide a comprehensive and powerful algorithm that will be used for identification of spherical or cylindrical fruit under natural conditions which is based on shape, color, and depth information. To test, train and validate the algorithm, three datasets were used and each of the dataset comprises images of peppers, guavas, and eggplants in the form of [RGB, depth]. To exclude the pixels in the RGB image that are of no use, a probabilistic image segmentation method is used which creates a binary mask using the RGB image. The depth image which is filtered is obtained by obtaining the product of the depth image and binary mask. After that a clustering method which is based on region - growing algorithm which is then used to obtain a cluster set from the filtered depth image which is then converted to point cloud. After that to detect the fruits which are similar in a point cloud a 3D shape detection algorithm is used which is based on a novel M-estimator sample consensus (MSAC). Also, to identify true fruits an SVM classifier is used which is trained on angle, color, and shape-based features (Guichao Lin et al., 2020).

Rakesh C. Joshi et al. worked on the method which is purely based on deep learning to detect the viral infection disease which are present in leguminous plants. The dataset used contains the image of Vigna Mungo leaves which are segmented and augmented. The CNN is VirLeafNet which is trained on the dataset containing different images of either healthy, moderately affected and severely affected leaves for epochs. The proposed models achieved testing accuracy as thereafter 91.234%, 96.429%, and 97.403% for VirLeafNet-2, VirLeafNet-1 and VirLeafNet-3 respectively using the dataset (Joshi et al., 2021).

S. Appeltans et al. identified and detected white tip disease that are in either early or pre-symptomatic stage under normal conditions that are there in the field. The dataset contains 29744 images of different varieties which are as follows: weed plants, healthy leek and diseased leek and soil which is used to construct the hyperspectral training library. LDA classification is used to remove soil pixels from hyperspectral images after which a noise filter-algorithm is used which is customized. Then a model is used to classify the plant leaf disease into five different classes such as healthy, early-stage disease, moderate disease, severe disease, and disease which is fully developed by training a model using a classifier which uses logistic regression supervised machine learning (Appeltans et al., 2021).

U. Atila et al. classified the plant leaf disease. This method was proposed to identify and classify the plant disease using the deep learning architecture of EfficientNet. The Plant Village dataset having 55448 and 61486 original and augmented images were used to train the models by making use of the transfer learning approach. The models achieved the highest accuracy values 99.91% and 99.91% and highest precision values 98.42% and 99.39% with respect to original and augmented datasets (Atila et al., 2021). Juncheng Ma et al. worked on cucumber disease recognition methods using leaf symptom images. The sole purpose for this work was to reduce the time it takes to detect the diseases manually, consuming more time and labor. To recognize four cucumber diseases (downy mildew, anthracnose, powdery mildew, target leaf spots), a DCNN is shown. The DCNN is being used to perform symptom-wise disease recognition using symptom images as input. On illness photos captured in the field, the approach produced good recognition results. Cucumber leaf pictures obtained in the field were segmented to create the symptom images. Data augmentation methods were used to extend the datasets created by the segmented symptom photos in order to reduce the risk of overfitting. The DCNN showed good recognition results with the supplemented datasets containing 14,200 symptom photos, with an accuracy of 93.4%. Comparative tests were conducted utilizing traditional classifiers (SVM and Random Forest), as well as AlexNet, to compare the DCNN outcomes. The DCNN proved to be a reliable method for detecting cucumber infections in field situations (Ma et al., 2018).

Jun Sun et al. tried to detect the primary organs (flower and fruit) of tomatoes from photos of plants is a common object recognition challenge. Traditionally, tomatoes are recognized based on their physical appearances like color, shape, and size. In this work, a combination of the gray level of the input image and the pixel difference to calculate through the module were used with the Adaboost classifier. The success rate for detecting the mature tomatoes was 93.3%. A framework constructed for image segmentation by using the image data extracted from the orchard. CNN and Multi-scale multilayer perception both were included in this framework. The VGG16-based feature extraction network was trained by combining RGB and grayscale images. Fast R-CNN was then initialized with the parameters to identify feature photos of the tomato (Jun Sun et al., 2018).

Ahmad Almadhor et al. worked on a machine learning based AI-driven framework for recognizing guava plant diseases with the help of a DSLR Camera as guava is one of the tropical fruits growing in southern Asia in vast quantities. Plant diseases could cause the reduction of the quality of the fruits and quality. These illnesses will have a significant influence on the future generation of guavas if they are not adequately detected and controlled. Study of this paper shows that plants are attacked by 167 fungal and 177 pathogens and others such as nematodes, bacteria, and algae. To overcome this problem an AI driven framework is designed to detect plant disease beforehand through Machine learning from DSLR cameras. To separate the disease-infected areas, the suggested framework uses Delta E color difference image segmentation. RGB color, HSV and textural LBP histogram feature vectors are also extracted. Machine-learning algorithm like Bagged, Complex Tree, Fine KNN, Boosted Tree, Cubic SVM, Boosted Tree have been used to achieve similar outcomes compared to individual channels using a combination of textural and color features, disease recognition is performed using machine-learning classifiers. An 18 MP picture dataset of guava fruit and leaves is used to test the suggested framework. The feature vector set comprised of HSV, RGB and LBP extracted features and the Bagged Tree classifier produced the best recognition results (99% in distinguishing 4 guava fruit diseases (Mummification, Dot, Canker and Rust) (Almadhor et al., 2021). Table 1 explained all the existing literature for the detection of disease for plants and fruits.

Table 1 briefly describes the previously published work for the detection of disease for plants and fruits. It also gives the information about the prominently used methodologies and their performances. From the literature table found that machine learning techniques (SVM, KNN, LDA, GA-SVM, PCA-SVM), CNN, RCNN, Faster RCNN, densenet, VGG16, RetinaNet, DCNN, modified CNN models are frequently used and yields promising results. The obtained results using Faster RCNN and Retina-Net is very poor with respect to others. The reported accuracy for faster RCNN is 90.7% and for Retina-Net 74.5% only on self-collected dataset. Therefore, there is a possibility to increase the performance of the system using different parameters tunning process or via other methods so that the precise agriculture is adopted globally.

Table 1. Literature study for plants and fruit disease detection and classification

Author(s), Year	Title	Dataset	Methods	Accuracy (%)
Prashant Udawant et al., 2022	Cotton leaf disease detection using instance segmentation	Self-Collected	R-CNN	94.0
Srivastava et al., 2022	Plant Disease Detection using SVM Algorithm and Neural Network Approach	Village Plant Dataset	PCA-SVM	94.5
Zhaoyi Chen et al., 2022	Plant Disease Recognition Model Based on Improved YOLO	Self-Collected	Faster R-CNN	86.5
Jia Yao et al., 2021	A Real-Time Detection Algorithm for Kiwifruit Defects Based on YOLOv5	Self-Collected	Yolo V5	94.7
Sk Mahmudul Hassan et al., 2022	Identification of Plant-Leaf Diseases Using CNN and Transfer-Learning Approach	Village Plant Dataset	InceptionV3	98.6
Asif Iqbal Khan et al., 2022	Deep Diagnosis: A Real-Time Apple Leaf Disease Detection System Based on Deep Learning	Self-Collected	CNN	88.0
V-Tsirons et al., 2020	Evaluation of object detection algorithms on a new real world Tomato dataset	Self-Collected	Retina-net	74.5
G. Sambosivam et al., 2021	Cassava Disease detection and classification with imbalanced dataset using CNN	Public dataset	DCNN	93.0
B. Yan et al., 2021	A Real-Time Apple Targets Detection Method using YOLOv5	Self-Collected	Faster R-CNN	--
Zhang Chuanlei et al., 2017	Apple leaf disease identification using GA	--	GA-SVM	90.0
Umit Atila et al., 2021	Plant leaf disease classification using EfficientNet deep learning model	Public dataset	CNN	99.9
KP. Ferentinos, 2018	Deep learning models for plant disease detection and diagnosis	Public dataset	CNN	99.5
Hu Gensheng et al., 2019	A low shot learning method for tea leaf's disease identification	Self-Collected	VGG16	90.0
Yong Zhong et al., 2020	Research on deep learning in apple leaf disease recognition	Public dataset	DenseNet-121	93.7
Guichao Lin et al., 2019	Color, depth, and shape-based 3D fruit detection	Self-Collected	SVM	97.1
Juncheng Ma et al., 2018	A cucumber diseases recognition using DCNN.	PlantVillage	DCNN	93.4
Appeltans et al., 2021	Detection of leek white tip disease under field conditions using hyperspectral proximal sensing and supervised machine learning.	Self-Collected	LDA	96.7
Joshi et al., 2021	Automatic analysis and viral disease diagnosis using deep learning in the Vigna mungo plant. Ecological Informatics	Self-Collected	Virleaf-Net	97.4
Sun, J. et al., 2018	Detection of tomato from complex background using deep learning	Self-Collected	Faster R-CNN	90.7
Almadhor et al., 2021,	AI-driven framework for recognition of guava plant diseases.	Public dataset	Cubic SVM	99.0
Peng Jiang et al., 2022	Real-time detection of apple leaf diseases using improved DCNN.	Self-Collected	CNN	78.8

Figure 1. Flow chart of proposed work

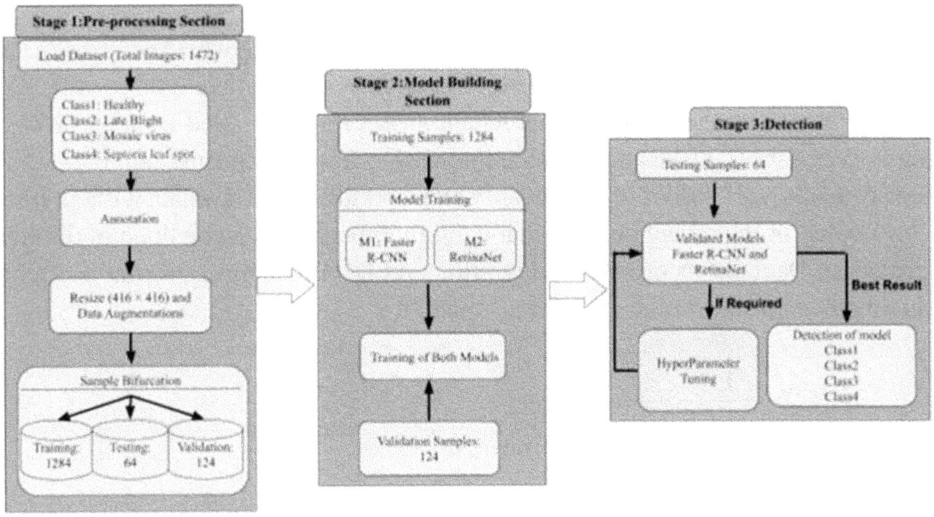

3 METHODOLOGY

In this section, a detailed description of the dataset, steps involved in methodology of the work is described. Initially an Annotated Tomato Diseases dataset of appropriate size is created that consists of roughly 1472 high-quality RGB photos encompassing three disease and symptom types. Following that, a deep learning Tomato disease model is developed that can effectively and efficiently identify symptoms. For implementation of proposed work Detectron 2 is used. Detectron2 (Loudon, J. S. 2018). is Facebook AI Research's (FAIR) object detection library written in PyTorch. Detectron2 is flexible and extensible, allowing high-quality implementations of state-of-the-art object detection algorithms. Detectron2's is used for implementation of Faster R-CNN and the research compares two separate models as Faster R-CNN (Ren et al., 2017) and RetinaNet (Ale et al., 2018). Both the models yielded adequate outcomes. The annotation of the dataset is done by the annotating tool Roboflow. The methodology of the work is explained in the Figure 1.

3.1 Dataset Description

The dataset is an open-source data collection that comprises photos of several plant diseases (dataset, Ref. 26). There are 15 distinct diseases included in the dataset. This work is focused on the tomato disease data after sifting the data from the dataset. Three distinct tomato diseases are detected and classified in the proposed model. Dataset is splitted into 3 different sets (Training, Testing, and Validation). Figure 2 contains the sample images from the dataset.

Training set: Training contains the data which will be fed into the model for the training of the model.
Validation set: While tuning model hyperparameters, a sample of data is employed to offer an unbiased evaluation of a model fit on the training dataset.
Test set: The data set used to offer an unbiased assessment of a final model's fit to the training dataset.

3.2 Pre-Processing (Augmentation, Annotation)

Preprocessing: Preprocessing is nothing but a technique of data mining which reforms the raw data in understandable format. Real world data is mostly inconsistent or incomplete or consists of lots of errors. Rearranging the data in the proper format which makes sense is called Data preprocessing. Data preprocessing is necessary to make any valuable knowledge from the inconsistent data. Data which is being collected from the real world is of poor quality and working on the raw data without filtering could lead to the low quality of the model built on that data. Preprocessing can be done in different ways as Data transformation, Data cleansing, Data integration and Data reduction. Normalized image can be obtained using mathematical expression given in equation 1.

$$O_N = (Max_{new} - Min_{new})\frac{1}{1+e^{-\frac{F-\beta}{\alpha}}} + Min_{new} \tag{1}$$

Where, O(I, J) is resultant value, α is input image intensity range value and β is centroid.

3.2.1 Annotation Tools

Label-Img: Label-Img is a tool for annotating graphical images. Image labelling is the process of recognizing and labelling unique components of an image. Image labelling comes particularly handy when it comes to automating the process of creating metadata or making recommendations to customers based on details in their photos.

RoboFlow: Roboflow (Alexandrova et al., 2015) is an augmentation tool that provides users with access to public datasets as well as the ability to submit their own bespoke data. Roboflow supports several annotation formats. We chose roboflow because it allows us to see the image annotation in real time and requires no code.

MakeSense: MakeSense is a free annotation tool, it is very simple to use, and it does not require any prerequisite to install. MakeSense is platform independent, it runs on any operating system, it is freely available on the web, annotating the images is very fast and could be downloaded in any format.

3.2.2 Augmentation

Data Augmentation is nothing, but it is a technique to increase the amount of data artificially by making new data points in existing data, this method increases the amount of data by making small changes to the existing data or generating new data points using machine learning techniques. Figure 3 shoes the annotated sample images from the dataset.

3.3 Classification

Classification is a predictive modelling task in ML where a class label is predicted for an example of input data. The ability to recognize items and categorize them is a typical task for ML systems. This method is called classification, and it helps us to divide large amounts of data into smaller values, such as 0 or 1 True or false or a predefined output label. Based on a labelled training set, in a set of catego-

Figure 2. Sample Images for each class from the dataset

Class type	Sample Image1	Sample Image2
Healthy		
Late Blight disease	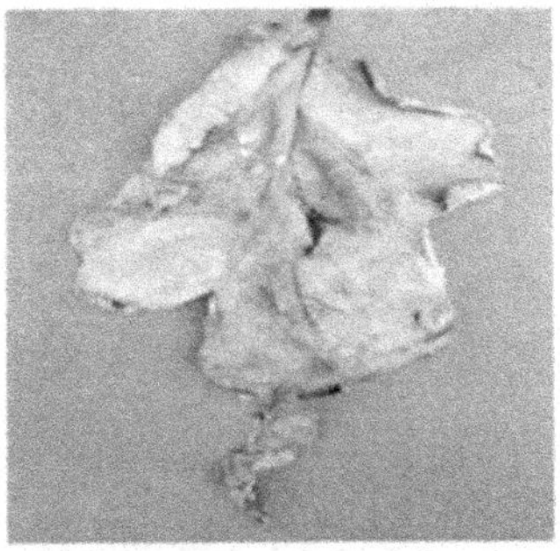	
Late Mosaic virus		
Septoria Leaf Spot.		

ries, classification algorithms attempt to identify the category of a new observation. The accuracy of categorization varies depending on anatomical anatomy, task, features, and tissue preparation. In CNN, convolution operation performed according to the given equation 2.

Figure 3. Example of annotation and augmentation

Class type	Sample Image1	Sample Image2
Healthy		
Late Blight disease	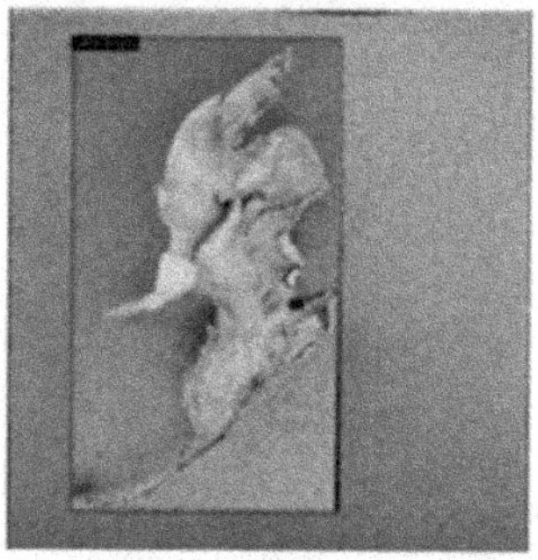	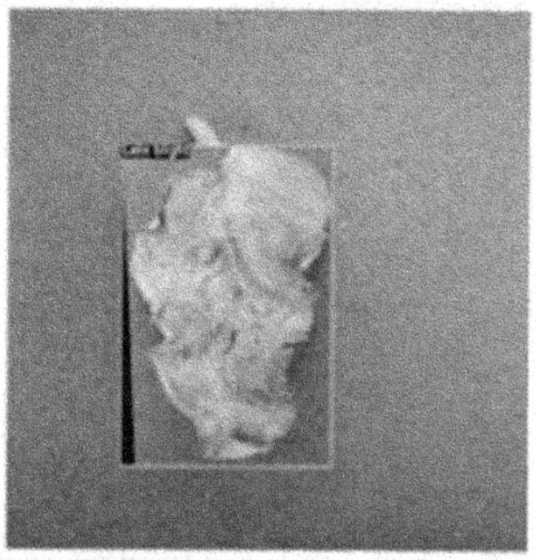
Late Mosaic virus		
Septoria Leaf Spot.	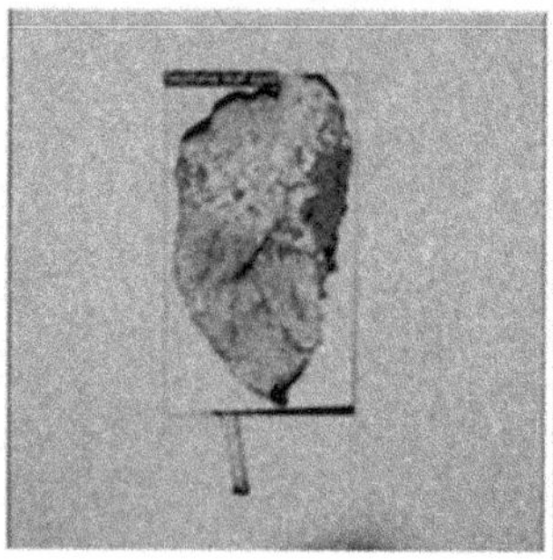	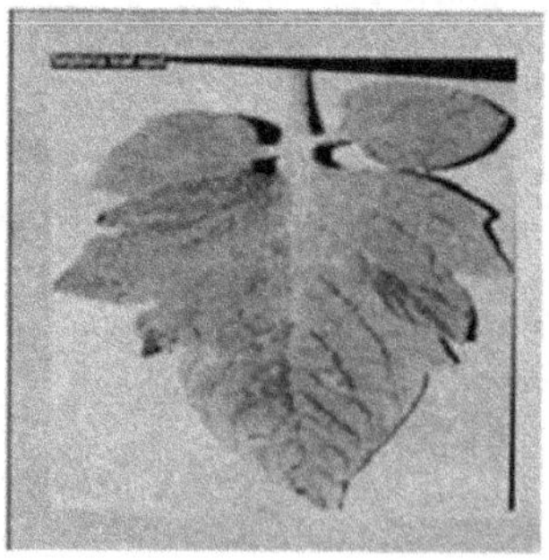

$$G(i,j) = \sum_{k=1}^{m}\sum_{l=1}^{n} f(i+k-1, j+l-1)K(k,l) \tag{2}$$

Figure 4. Faster R-CNN architecture

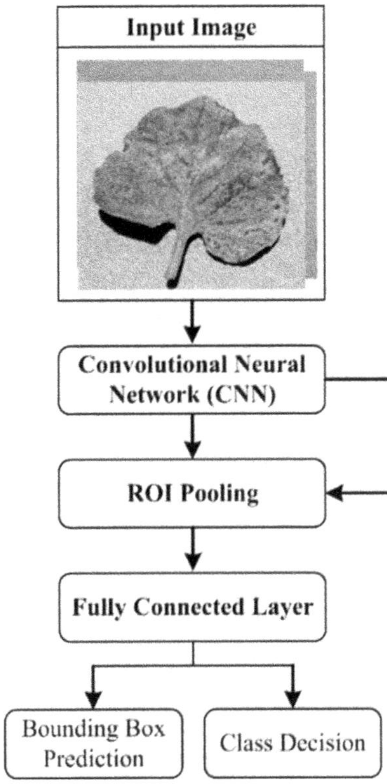

Where, *G(i, j)* is resultant image, *F(i, j)* is input image and *K(i, j)* is kernel filter. The size of the resultant image is M-m+1 rows and N-n+1 column where input image has M × N and kernel filter size is m n.

3.3.1 Faster R-CNN

Faster R-CNN is a deep convolutional network, and it was proposed by Shaoqing Ren, Ross Girshick, Jian Sun, and Kaiming in 2015 (Kaiming et al., 2015) and it appears to the user as a single and end-to-end, unified network for object detection as shown in Figure 4. The network can predict the positions of various items correctly and fast. Faster R-CNN is a better version of Fast R-CNN. State-of-the-art object detection networks use region proposal algorithms to hypothesize object placements. Detection network's operating time has been reduced thanks to SPPNet and Fast R-CNN, exposing region proposal calculation as a bottleneck. Two strategies for training the network are described in the Faster R-CNN research article. End-to-end training uses all four-loss functions to train the whole network in a single session (RPN objectness loss, detector class loss detector regression loss and RPN regression loss).

3.3.2 RetinaNet

RetinaNet (Ale et al., 2018) is a one-stage object detection model that uses a focal loss function to correct class imbalance during training as show in Figure 5. RetinaNet is a unified network made up of two task-specific subnetworks and a backbone. A commercially available convolutional network serves as the backbone, generating a convolutional feature map over the full input image. On the backbone's output, the first subnet does convolutional object classification, while the second subnet performs convolutional bounding box regression.

4 RESULTS AND DISCUSSIONS

The results from the classification are reflected in the term of mAP value and Ap value. Object detectors like the Faster R-CNN, SSD, and others are frequently evaluated using the AP (Average Precision) measure. For recall values ranging from 0 to 1, the average accuracy value is calculated. In this work, the models are compared on the bases of different types of AP values. AP (Average Precision) with IOU (Intersection over union) lies between the range from 0.5 to 0.95. For AP-50, average precision with intersection over union IOU value was 0.5, similarly for Ap-75, the average precision with intersection over union IOU value was 0.75. mAP is the average of all the AP and the range on the mAP lies between 0 to 1. The higher the mAP value, the better the model. The classification is done on the individual classes and the results are in the form of Average Precision-50 (where IOU is set as 50%) and Mean Average Precision. The average precision is calculated using the formula given in equation 3. And the mean of average precision is called mean average precision.

$$AP = \frac{1}{N} \sum_{r \in \{0, 0.1, \ldots, 1\}} \rho\, \mathrm{int}\, erp(r)$$
$$\rho\, \mathrm{int}\, erp(r) = \max_{\tilde{r}: \tilde{r} \geq r} \rho(\tilde{r}) \quad (3)$$

4.1 Hyper Parameter Tuning

Hyper Parameter tuning is an advantageous technique to improve the performance of the machine learning model. Hyper Parameters are different parameters that affect different aspects of the model. Different hyperparameters of machine learning are Momentum, Learning Rate, weight decay, warmup epochs, warmup_momentum, warmup_bias_lr, box, cls, cls_pw, obj, obj_pw, iou_t, anchor_t, anchors. At the time of training the model, different values can be set for these parameters to gain a good accuracy. The method of doing experiments on different values of different parameters is known as hyperparameter tuning.

Following experiments performed for implementation of model for the classification of leaf disease as shown in Table 2.

Experiment 1 (Classification of leaf disease using Faster R-CNN mode): In the proposed work, the different parameters which are used as hyperparameters are learning rate, Gamma, and Steps. While working on the Steps the initial count was considered as 1000 but later it was increased to 1500 for better performance. The results from Learning Rate and Gamma Tuning were observed and recorded as shown in Table 3.

Figure 5. RetinaNet architecture

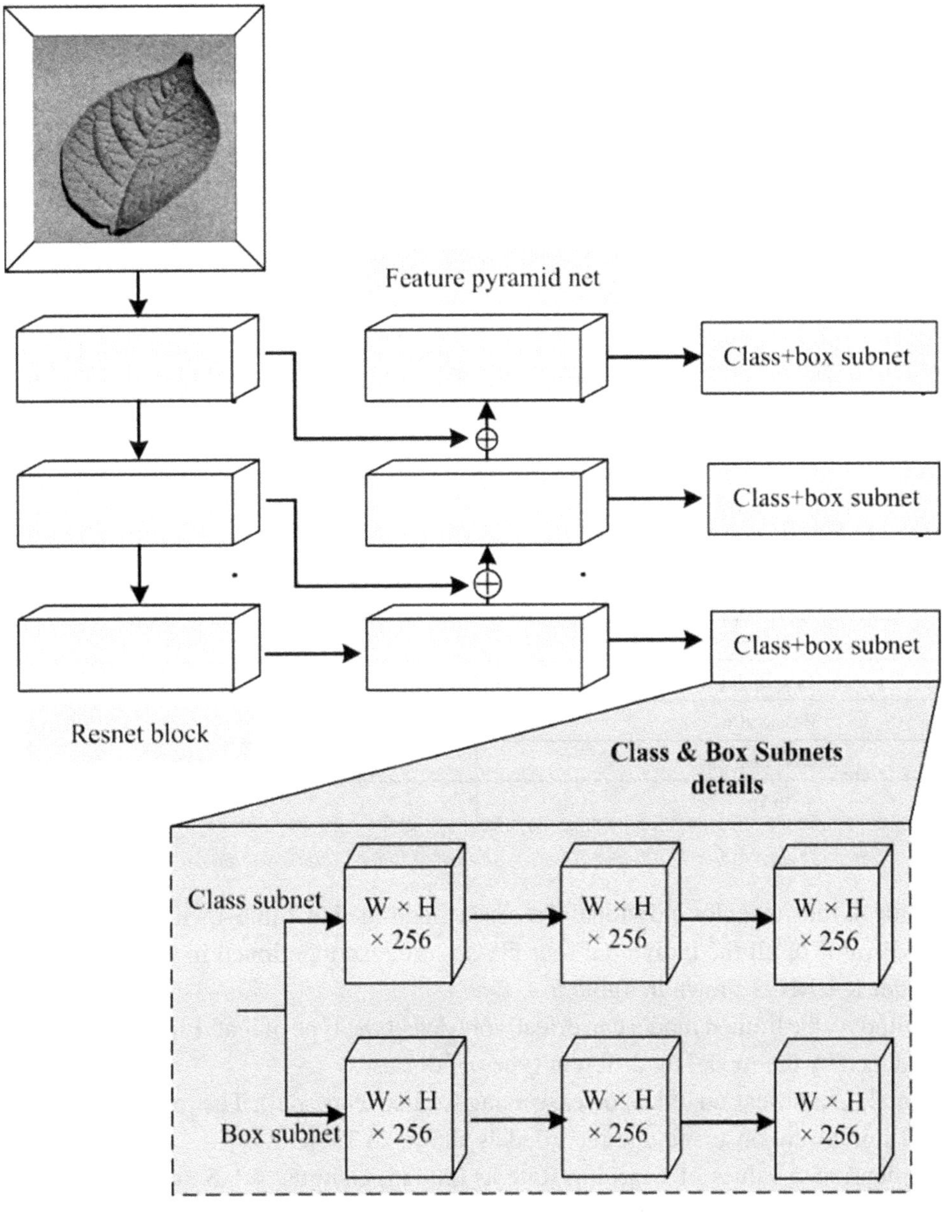

Table 2. Number of experiments

Experiment #	Description
Experiment 1	Classification of leaf disease using Faster R-CNN model
Experiment 2	Classification of leaf disease using RetinaNet model

Table 3. Hyperparameter tuning table for faster R-CNN

Hyper Parameters	Learning Rate	Gamma	Steps	Faster R-CNN AP-50
1	0.001	0.05	1500	73.13
2	0.001	0.01	1500	76.81
3	0.0025	0.1	1500	93.04
4	0.0025	0.5	1500	93.11

4.2 Results From Faster R-CNN

In the quantitative analysis, the performance of the experiments was noted in table1 and from the produced values of different experiments. The results and conclusion are narrowed down into the final best output of the Faster R-CNN.

Table 4. AP value for each class

Category	AP (Faster R-CNN)
Healthy	56.04
Late Blight	73.56
Mosaic virus	70.28
Septoria Leaf Spot	79.77
mAp	69%

The outcome of the model for individual classes is mentioned in Table 4 with an achieved mAp value of 69% and AP value of all the individual four classes are also mentioned in the same table. Also, AP values for Faster R-CNN is shown in Table 5.

In the qualitative, the trained model can effectively detect the type of leaf. Figure 6 shows the example of results produced by the model on different types of diseases.

Experiment 2 (Classification of leaf disease using RetinaNet model): The results from Learning Rate and Gamma Tuning were observed and recorded as shown in Table 6.

From the mentioned values of Learning Rate as 0.0025, Gamma as 0.5 and the Steps as 1500, the Faster R-CNN model obtained Average-Precision of 93.11% and RetinaNet obtained A1verage-Precision of 95.54% as depicted in Table.

AP values are depicted in Table 8 for Faster R-CNN.

Figure 7 shows the output for RetinaNet.

4.3 Comparative Results

After exhaustive experiments on the given dataset both the models performed well after tuning the hyperparameters. The performance of these models and the comparison of different AP values as AP, AP-50, AP-75 for Faster R-CNN and RetinaNet is given in Table 9.

Figure 6. Output results after applying Faster R-CNN

Class type	Sample Image1	Sample Image2
Healthy		
Late Blight disease	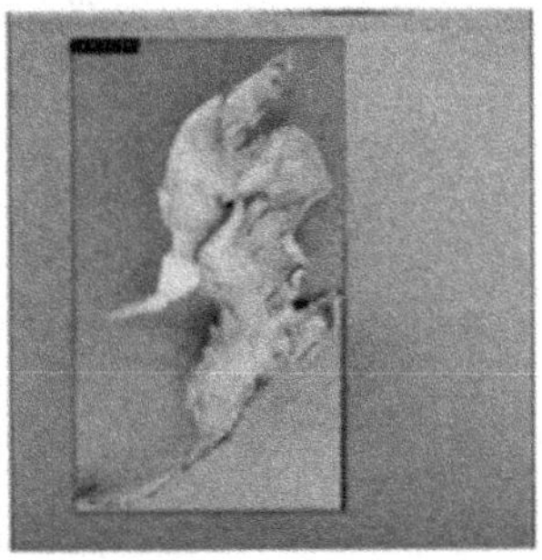	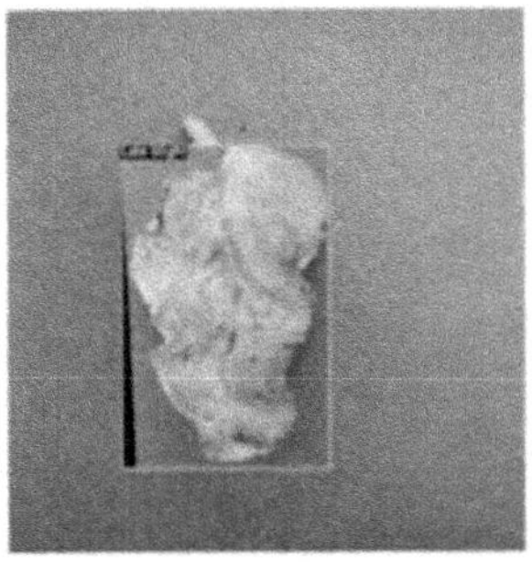
Late Mosaic virus		
Septoria Leaf Spot.	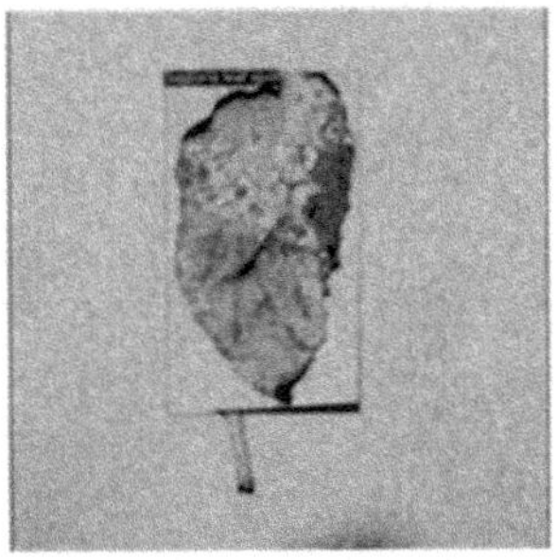	

All of the studies concluded that RetinaNet outperformed Faster R-CNN and achieved the AP-50 value of 95.54%.

Table 5. AP values for faster R-CNN

Ap	Ap-50	Ap-75
69.91	93.11	85.168

Table 6. Hyper-parameter tuning results for RetinaNet

Hyper Parameters	Learning Rate	Gamma	Steps	RetinaNet AP-50
1	0.001	0.05	1500	94.04
2	0.001	0.01	1500	92.18
3	0.0025	0.1	1500	94.50
4	0.0025	0.5	1500	95.54

Table 7. AP value for each class (results from RetinaNet)

Category	AP
Healthy	58.24
Late Blight	76.42
Mosaic virus	74.25
Septoria Leaf Spot	83.33
mAP	73.06

Figure 8 depicted the bar chart that shows the comparison between fast R-CNN and RatinaNet.

The comparison findings for both models on various classes is shown in Table 10. When compared to the Faster R-CNN model, the RetinaNet model performs well. In each particular class, the RetinaNet model has performed better than the Faster R-CNN.

Table 8. AP values for faster R-CNN

Ap	Ap-50	Ap-75
73.06	95.54	85.713

Table 9. Comparison between AP-50 of both the models

Model	Ap	Ap-50	Ap-75
Faster R-CNN	69.91	93.11	85.16
RetinaNet	73.06	95.54	85.71

Figure 7. Detection output for RetinaNet

Class type	Sample Image1	Sample Image2
Healthy	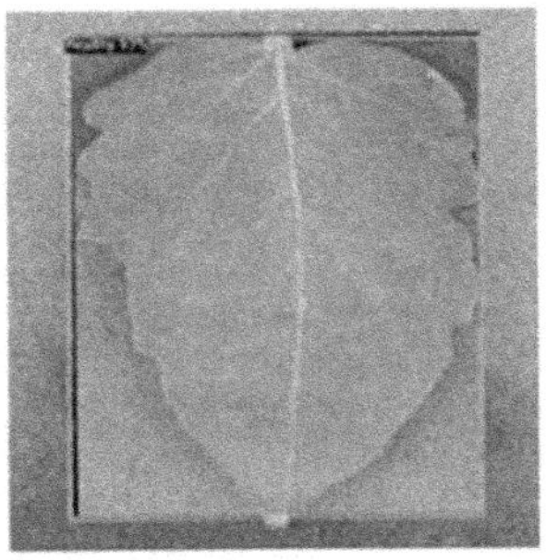	
Late Blight disease	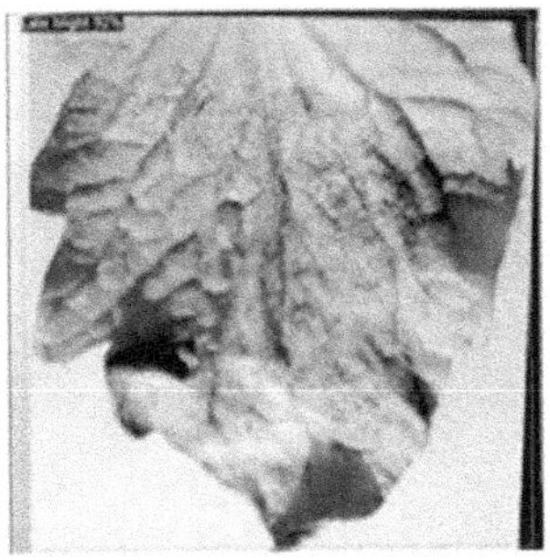	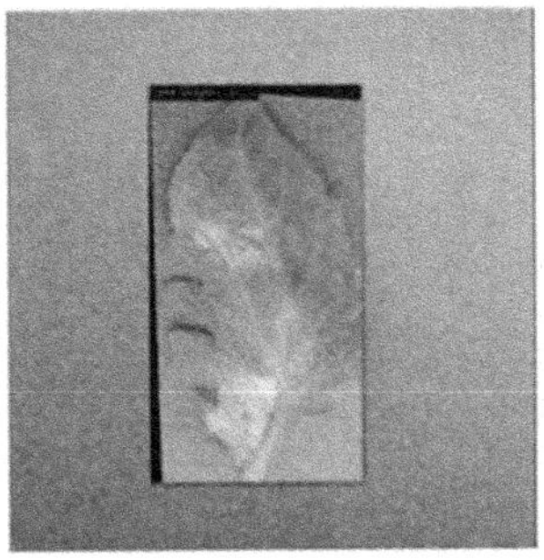
Late Mosaic virus		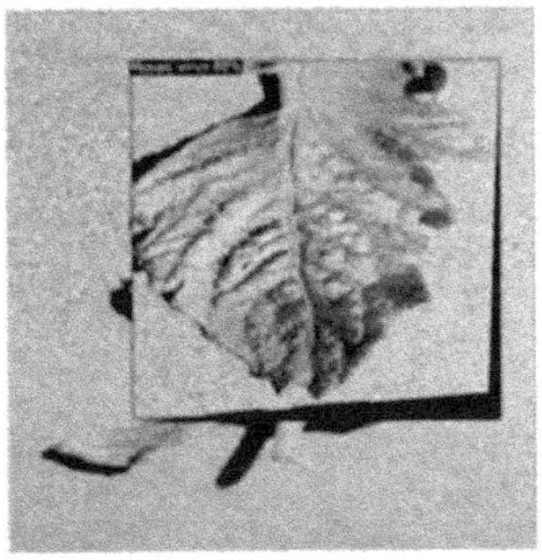
Septoria Leaf Spot.	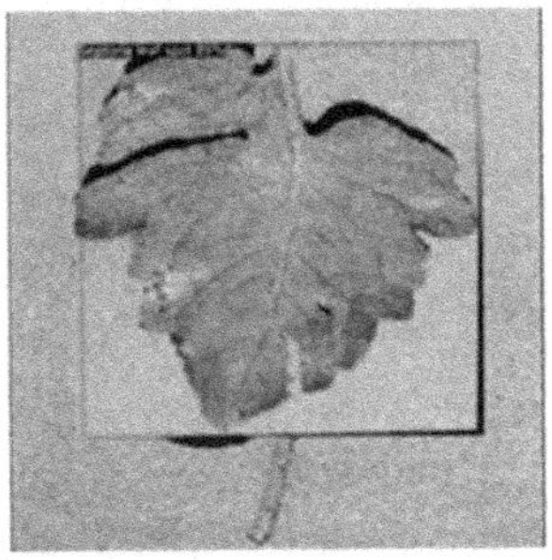	

Figure 8. Bar Chart showing the comparison between AP-50 for fast R-CNN and RatinaNet

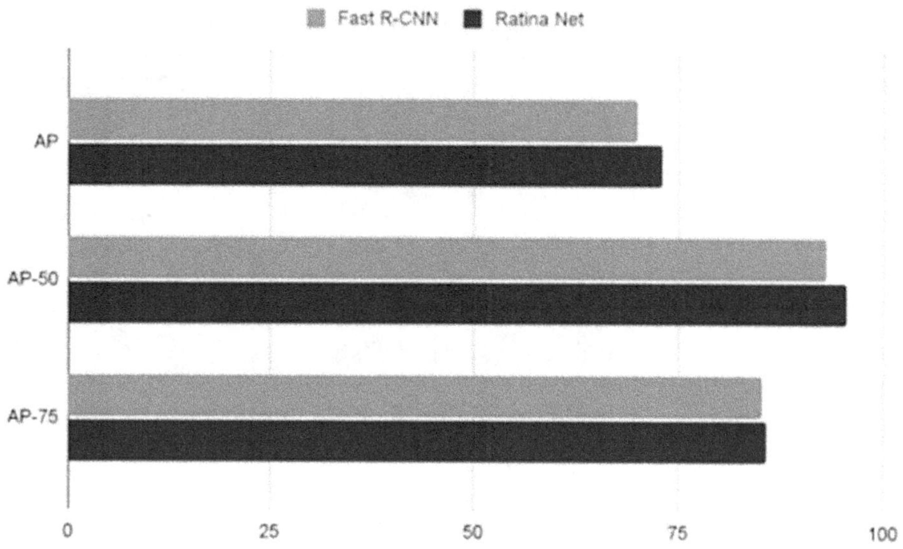

Table 10. Comparison of the two models for each class

Category	RetinaNet AP (%)	Faster R-CNN AP (%)
Healthy	58.24	56.04
Late Blight	76.42	73.56
Mosaic virus	74.25	70.28
Septoria Leaf Spot	83.33	79.77

Figure 9 shows the column chart for comparative analysis of models (Fast R-CNN and Retina Net) for different classes of plant disease.

5. CONCLUSION

Working with Plant disease is a very typical problem due to the different diseases and the amount of data that is present. While annotating the images it was difficult to keep track of the annotation and most of the leaves resemble each other. In this paper, the working of Faster R-CNN and RetinaNet algorithms were taken into consideration. The proposed work also focuses on transfer learning and learning from scratch to study the working of algorithms along with the plant disease detection problem. Data augmentation techniques like shear and rotation were applied to save the model from overfitting. The above research can be helpful for Indian farmers to prevent the crop from disease. In the future, these AI techniques

Figure 9. Column chart for comparative analysis of models (fast R-CNN and Retina Net) for different classes of plant disease

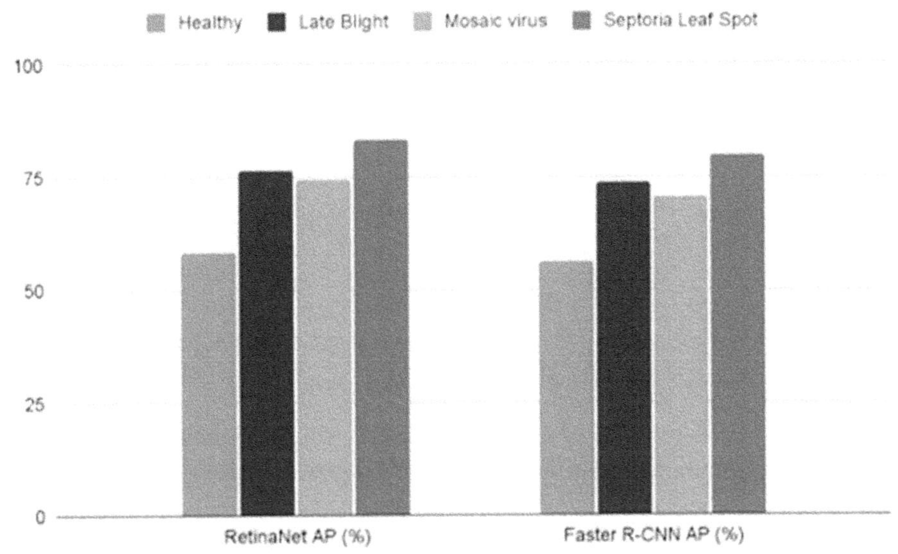

like RetinaNet and Fast R-CNN can be useful to move forward towards all the diseases of tomatoes or even different vegetables or fruits.

REFERENCES

Ale, L., Zhang, N., & Li, L. (2018, December). Road damage detection using RetinaNet. In *2018 IEEE International Conference on Big Data (Big Data)* (pp. 5197-5200). IEEE.

Alexandrova, S., Tatlock, Z., & Cakmak, M. (2015, May). RoboFlow: A flow-based visual programming language for mobile manipulation tasks. In *2015 IEEE International Conference on Robotics and Automation (ICRA)* (pp. 5537-5544). IEEE.

Almadhor, A., Rauf, H. T., Lali, M. I. U., Damaševičius, R., Alouffi, B., & Alharbi, A. (2021). Ai-driven framework for recognition of guava plant diseases through machine learning from dslr camera sensor-based high resolution imagery. *Sensors (Basel)*, *21*(11), 3830.

Appeltans, S., Pieters, J. G., & Mouazen, A. M. (2021). Detection of leek white tip disease under field conditions using hyperspectral proximal sensing and supervised machine learning. *Computers and Electronics in Agriculture*, *190*, 106453.

Asif Iqbal Khan, S. M. K. (2022). Quadri, Saba Banday, Junaid Latief Shah, Deep diagnosis: A real-time apple leaf disease detection system based on deep learning. *Computers and Electronics in Agriculture*, *198*, 107093.

Atila, Ü., Uçar, M., Akyol, K., & Uçar, E. (2021). Plant leaf disease classification using EfficientNet deep learning model. *Ecological Informatics*, *61*, 101182.

Chen, Z., Wu, R., Lin, Y., Li, C., Chen, S., Yuan, Z., Chen, S., & Zou, X. (2022). Plant Disease Recognition Model Based on Improved YOLOv5. *Agronomy (Basel), 12*(2), 365.

Chuanlei, Z., Shanwen, Z., Jucheng, Y., Yancui, S., & Jia, C. (2017). Apple leaf disease identification using genetic algorithm and correlation based feature selection method. *International Journal of Agricultural and Biological Engineering, 10*(2), 74–83.

Dataset. (n.d.). https://www.kaggle.com/datasets/arjuntejaswi/plant-village

Dhiman, N., & Tamsir, M. (2018). A collocation technique based on modified form of trigonometric cubic B-spline basis functions for Fisher's reaction-diffusion equation. Multidiscipline Modeling in Materials and Structures, 14(5), 923-939.

Ferentinos, K. P. (2018). Deep learning models for plant disease detection and diagnosis. *Computers and Electronics in Agriculture, 145*, 311–318.

Hassan, S. M., Maji, A. K., Jasiński, M., Leonowicz, Z., & Jasińska, E. (2021). Identification of plant-leaf diseases using CNN and transfer-learning approach. *Electronics (Basel), 10*(12), 1388.

Hu, G., Wu, H., Zhang, Y., & Wan, M. (2019). A low shot learning method for tea leaf's disease identification. *Computers and Electronics in Agriculture, 163*, 104852.

Jiang, P., Chen, Y., Liu, B., He, D., & Liang, C. (2019). Real-time detection of apple leaf diseases using deep learning approach based on improved convolutional neural networks. *IEEE Access : Practical Innovations, Open Solutions, 7*, 59069–59080.

Joshi, R. C., Kaushik, M., Dutta, M. K., Srivastava, A., & Choudhary, N. (2021). VirLeafNet: Automatic analysis and viral disease diagnosis using deep-learning in Vigna mungo plant. *Ecological Informatics, 61*, 101197.

Kim, D. G., Burks, T. F., Qin, J., & Bulanon, D. M. (2009). Classification of grapefruit peel diseases using color texture feature analysis. *International Journal of Agricultural and Biological Engineering, 2*(3), 41–50.

Lin, G., Tang, Y., Zou, X., Xiong, J., & Fang, Y. (2020). Color-, depth-, and shape-based 3D fruit detection. *Precision Agriculture, 21*(1), 1–17.

Loudon, J. S. (2018). *Detecting and Localizing Cell Nuclei in Medical Images* [Master's thesis]. NTNU.

Ma, J., Du, K., Zheng, F., Zhang, L., Gong, Z., & Sun, Z. (2018). A recognition method for cucumber diseases using leaf symptom images based on deep convolutional neural networks. *Computers and Electronics in Agriculture, 154*, 18–24.

Ren, S., He, K., Girshick, R., & Sun, J. (2017). Faster r-cnn: Towards real-time object detection with region proposal networks. *IEEE Transactions on Pattern Analysis and Machine Intelligence, 39*(6), 1137 – 1149.

Sambasivam, G., & Opiyo, G. D. (2021). A predictive machine learning application in agriculture: Cassava disease detection and classification with imbalanced dataset using convolutional neural networks. *Egyptian Informatics Journal, 22*(1), 27–34.

Srivastava, A., Pandey, A., & Shukla, S.K. (n.d.). *Plant Disease Detection using SVM Algorithm and Neural Network Approach*. Academic Press.

Sun, J., He, X., Ge, X., Wu, X., Shen, J., & Song, Y. (2018). Detection of key organs in tomatoes based on deep migration learning in a complex background. *Agriculture*, *8*(12), 196.

Tsironis, V., Bourou, S., & Stentoumis, C. (2020). Tomato d: Evaluation of object detection algorithms on a new real-world tomato dataset. *The International Archives of the Photogrammetry, Remote Sensing and Spatial Information Sciences*, *43*, 1077–1084.

Udawant, P., & Srinath, P. (2022). Cotton Leaf Disease Detection Using Instance Segmentation. *Journal of Cases on Information Technology*, *24*(4), 1–10.

Vikaspedia. (n.d.). Available at https://vikaspedia.in/agriculture/crop-production/integrated -pest-managment/ipm-for-vegetables/ipm-strategies-for-tomato /tomato-diseases-and-symptoms

Yan, B., Fan, P., Lei, X., Liu, Z., & Yang, F. (2021). A real-time apple targets detection method for picking robot based on improved YOLOv5. *Remote Sensing*, *13*(9), 1619.

Yao, J., Qi, J., Zhang, J., Shao, H., Yang, J., & Li, X. (2021). A real-time detection algorithm for Kiwi-fruit defects based on YOLOv5. *Electronics (Basel)*, *10*(14), 1711.

Zhong, Y., & Zhao, M. (2020). Research on deep learning in apple leaf disease recognition. *Computers and Electronics in Agriculture*, *168*, 105146.

Compilation of References

Aasha Nandhini, S., Hemalatha, R., Radha, S., & Indumathi, K. (2018). Web enabled plant disease detection system for agricultural applications using WMSN. *Wireless Personal Communications*, *102*(2), 725–740. doi:10.100711277-017-5092-4

Abbasi, A. Z., Islam, N., & Shaikh, Z. A. (2014). A review of wireless sensors and networks' applications in agriculture. *Computer Standards & Interfaces*, *36*(2), 263–270. doi:10.1016/j.csi.2011.03.004

Abbasov, A., & Ibrahimov, O. (2019). *Digitalization of the world economy and social life: Global trends, achievements, and challenges to the national interests and growth perspectives of Azerbaijan.* https://mpra.ub.uni-muenchen.de/105176/

Abdalla, Z. F., & El-Ramady, H. (2022). Applications and Challenges of Smart Farming for Developing Sustainable Agriculture. Environment. *Biodiversity and Soil Security*, *6*(2022), 81–90. doi:10.21608/jenvbs.2022.135889.1175

Abied, O., Ibrahim, O., & Kamal, S. N. (2022). Adoption of Cloud Computing in E-Government: A Systematic Literature Review. *Pertanika Journal of Science & Technology*, *30*(1), 655–689. Advance online publication. doi:10.47836/pjst.30.1.36

Abrouk, M., Ahmed, H. I., Cubry, P., Šimoníková, D., Cauet, S., Pailles, Y., Bettgenhaeuser, J., Gapa, L., Scarcelli, N., Couderc, M., Zekraoui, L., Kathiresan, N., Čížková, J., Hřibová, E., Doležel, J., Arribat, S., Bergès, H., Wieringa, J. J., Gueye, M., ... Krattinger, S. G. (2020). Fonio millet genome unlocks African orphan crop diversity for agriculture in a changing climate. *Nature Communications*, *11*(1), 4488. doi:10.103841467-020-18329-4 PMID:32901040

Adegbeye, M. J., Reddy, P. R. K., Obaisi, A. I., Elghandour, M. M. M. Y., Oyebamiji, K. J., Salem, A. Z. M., ... Camacho-Díaz, L. M. (2020). Sustainable agriculture options for production, greenhouse gasses and pollution alleviation, and nutrient recycling in emerging and transitional nations-An overview. *Journal of Cleaner Production*, *242*, 118319. doi:10.1016/j.jclepro.2019.118319

Adeloye, A. J., Rustum, R., & Kariyama, I. D. (2012). Neural computing modelling of the reference crop evapotranspiration. *Environmental Modelling & Software*, *29*(1), 61–73. doi:10.1016/j.envsoft.2011.10.012

Adesina, I., Bhowmik, A., Sharma, H., & Shahbazi, A. (2020). A review on the current state of knowledge of growing conditions, agronomic soil health practices and utilities of hemp in the United States. *Agriculture*, *10*(4), 129. doi:10.3390/agriculture10040129

Adetunji, K. E., & Joseph, M. K. (2018). Development of a Cloud-Based Monitoring System Using 4Duino: Applications in Agriculture. In *2018 International Conference on Advances in Big Data, Computing and Data Communication Systems (icABCD).* 10.1109/ICABCD.2018.8465418

Adeyemi, O., Grove, I., Peets, S., Domun, Y., & Norton, T. (2018). Dynamic neural network modelling of soil moisture content for predictive irrigation scheduling. *Sensors (Basel)*, *18*(10), 3408. doi:10.339018103408 PMID:30314346

Adnan, N., Nordin, S. M., Rahman, I., & Noor, A. (2018). *The effects of knowledge transfer on farmers decision making toward sustainable agriculture practices: In view of green fertilizer technology*. World Journal of Science, Technology and Sustainable Development. doi:10.1108/WJSTSD-11-2016-0062

Agency for Toxic Substance and Disease Registry. (2014). *Medical Management Guidelines for Chlordane*. Retrieved from https://wwwn.cdc.gov/TSP/MMG/MMGDetails.aspx?mmgid=349&toxid=62

Aggarwal, N., & Singh, D. (2021). Technology assisted farming: Implications of IoT and AI. *IOP Conference Series. Materials Science and Engineering*, *1022*(1), 012080. doi:10.1088/1757-899X/1022/1/012080

Agrimonti, C., Lauro, M., & Visioli, G. (2021). Smart agriculture for food quality: Facing climate change in the 21st century. *Critical Reviews in Food Science and Nutrition*, *61*(6), 971–981. doi:10.1080/10408398.2020.1749555 PMID:32270688

Agrovision. (n.d.). Retrieved from www.agrovision.com

Ahmad, B., Ahmed, R., Masroor, S., Mahmood, B., Hasan, S. Z. U., Jamil, M., Khan, M. T., Younas, M. T., Wahab, A., Haydar, B., Subhani, M., Khan, M. A., & Tariq, S. (2023). Evaluation of Smart Greenhouse Monitoring System using Raspberry-Pi Microcontroller for the Production of Tomato Crop. *Journal of Applied Research in Plant Sciences*, *4*(01), 452–458. doi:10.38211/joarps.2023.04.01.54

Ahmad, S., Shakeel, I., Mehfuz, S., & Ahmad, J. (2023). Deep learning models for cloud, edge, fog, and IoT computing paradigms: Survey, recent advances, and future directions. *Computer Science Review*, *49*, 100568. doi:10.1016/j.cosrev.2023.100568

Ahumada, O., & Villalobos, J. R. (2009). Application of planning models in the agri-food supply chain: A review. *European Journal of Operational Research*, *196*(1), 1–20. doi:10.1016/j.ejor.2008.02.014

Akhter, R., & Sofi, S. A. (2022). Precision agriculture using IoT data analytics and machine learning. *Journal of King Saud University-Computer and Information Sciences*, *34*(8), 5602–5618. doi:10.1016/j.jksuci.2021.05.013

Alakbarov, O. (2019). Research of architectural-technological principles of mobile cloud computing. *Problems of Information Technology*, *2*, 92–104.

Alakbarov, R.G. & Hashimov, M.A. (2020) Security issues in cloud-based SCADA systems. *Information Technology Problems, 2*, 3–12.

Alakbarov, R.G. (2021). Cloud-based electronic government system: current situation, problems and security issues. *Information Society Problems*, *1*, 18–31.

Alam, A. (2022). Employing Adaptive Learning and Intelligent Tutoring Robots for Virtual Classrooms and Smart Campuses: Reforming Education in the Age of Artificial Intelligence. In R. N. Shaw, S. Das, V. Piuri, & M. Bianchini (Eds.), *Advanced Computing and Intelligent Technologies. Lecture Notes in Electrical Engineering* (Vol. 914). Springer. doi:10.1007/978-981-19-2980-9_32

Alam, M. M., Le Moullec, Y., Ahmad, R., Magarini, M., & Reggiani, L. (2020). A Primer On Public Safety Communication in the Context of Terror Attacks: The NATO SPS "COUNTER-TERROR" Project. In C. Palestini (Ed.), *Advanced Technologies for Security Applications* (pp. 19–34). Springer Netherlands. doi:10.1007/978-94-024-2021-0_3

Albright, L. D., Both, A. J., & Chiu, A. J. (2000). Controlling greenhouse light to a consistent daily integral. *Transactions of the ASAE. American Society of Agricultural Engineers*, *43*(2), 421–431. doi:10.13031/2013.2721

Alejandrino, J., Concepcion, R., Almero, V. J., Palconit, M. G., Bandala, A., & Dadios, E. (2020). A hybrid data acquisition model using artificial intelligence and IoT messaging protocol for precision farming. *2020 IEEE 12th International Conference on Humanoid, Nanotechnology, Information Technology, Communication and Control, Environment, and Management (HNICEM),* 1–6.

Alekbarov, R.G. & Hasanli, A.Sh. (2020). Serverless computing service in cloud technologies. *Information Technology Problems, 2,* 72–79.

Alekbarov, R.G. (2021). Mobile cloud technologies: current situation, problems and security issues. *Information Technology Problems, 1,* 15–32.

Ale, L., Zhang, N., & Li, L. (2018, December). Road damage detection using RetinaNet. In *2018 IEEE International Conference on Big Data (Big Data)* (pp. 5197-5200). IEEE.

Alessandro, A., Barbara, R., & Alberto, P. (2017). E-government and cloud: Security implementation for services. *Proceedings of the 4th International Conference on eDemocracy & eGovernment (ICEDEG)*, 79-85.

Alexandratos, N., & Bruinsma, J. (2012). *World agriculture towards 2030/2050: the 2012 revision.* Academic Press.

Alexandrova, S., Tatlock, Z., & Cakmak, M. (2015, May). RoboFlow: A flow-based visual programming language for mobile manipulation tasks. In *2015 IEEE International Conference on Robotics and Automation (ICRA)* (pp. 5537-5544). IEEE.

Aliguliyev, R.M. & Abdullayeva, F.C. (2013). Research and analysis of cloud technology security issues. *Information Technology Issues, 1*(7), 3-14.

Aliguliyev, R.M., Alakbarov, R.G., & Hashimov, M.A. (2015). Issues of application of cloud technologies in the electronic government system. *Problems of Information Society, 1,* 3-11.

Ali, K. E., Mazen, S. A., & Hassanein, E. E. (2018). A proposed hybrid model for adopting cloud computing in e-government. *Future Computing and Informatics Journal, 3*(2), 286–295. doi:10.1016/j.fcij.2018.09.001

Ali, O., Shrestha, A., Chatfield, A., & Murray, P. (2020). Assessing information security risks in the cloud: A case study of Australian local government authorities. *Government Information Quarterly, 37*(1), 101419. Advance online publication. doi:10.1016/j.giq.2019.101419

Ali, O., Shrestha, A., Osmanaj, V., & Muhammed, S. (2021). Cloud computing technology adoption: An evaluation of key factors in local governments. *Information Technology & People, 34*(2), 666–703. doi:10.1108/ITP-03-2019-0119

Al-Maatouk, Q., Othman, M., Rana, M., & Al-Rahmi, W. (2018). A cloud based framework for e-government implementation in developing countries. *IACSIT International Journal of Engineering and Technology, 7*(4), 3018–3021. doi:10.14419/ijet.v7i4.21532

Almadhor, A., Rauf, H. T., Lali, M. I. U., Damaševičius, R., Alouffi, B., & Alharbi, A. (2021). Ai-driven framework for recognition of guava plant diseases through machine learning from dslr camera sensor-based high resolution imagery. *Sensors (Basel), 21*(11), 3830.

Almutairi, N. N., & Thuwaini, Sh. F. (2015). Cloud Computing Uses for E-Government in the Middle East Region Opportunities and Challenges. *International Journal of Business and Management, 10*(4), 60–69. doi:10.5539/ijbm.v10n4p60

Alreshidi, E. (2019). *Smart sustainable agriculture (SSA) solution underpinned by internet of things (IoT) and artificial intelligence (AI).* ArXiv Preprint ArXiv:1906.03106.

Al-Ruithe, M., Benkhelifa, E., & Hameed, K. (2019). A systematic literature review of data governance and cloud data governance. *Personal and Ubiquitous Computing*, *23*(5-6), 839–859. doi:10.100700779-017-1104-3

Alshehri, M. (2022). Fuzzy Logic Based Explainable AI Approach for the Easy Calibration of AI Models in IoT Environments. In J.-H. Kim, M. Singh, J. Khan, U. S. Tiwary, M. Sur, & D. Singh (Eds.), *Intelligent Human Computer Interaction* (pp. 621–633). Springer International Publishing. doi:10.1007/978-3-030-98404-5_57

Alzaghir, A., Abdellah, A. R., & Koucheryavy, A. (2022). Predicting Energy Consumption for UAV-Enabled MEC Using Machine Learning Algorithm. In Y. Koucheryavy, S. Balandin, & S. Andreev (Eds.), *Internet of Things, Smart Spaces, and Next Generation Networks and Systems* (pp. 297–309). Springer International Publishing. doi:10.1007/978-3-030-97777-1_25

Amanda, A. F. (2021, February). Acute and Chronic Toxicity of Indoxacarb in Two Populations of *Plutella xylostella* (Lepidoptera: Plutellidae). *Journal of Economic Entomology*, *114*(1), 298–306. doi:10.1093/jee/toaa260 PMID:33179745

Amara, J., Bouaziz, B., & Algergawy, A. (2017). *A Deep Learning-based approach for Banana Leaf Diseases Classification*. Academic Press.

Aminulloh, L., Sesulihatien, W. T., & Pramadihanto, D. (2019, September). Feature extraction of tomato growth model using greenhouse monitoring system. In *2019 International Electronics Symposium (IES)* (pp. 370-375). IEEE. 10.1109/ELECSYM.2019.8901534

Anand, R., Singh, J., Pandey, D., Pandey, B. K., Nassa, V. K., & Pramanik, S. (2022). Modern Technique for Interactive Communication in LEACH-Based Ad Hoc Wireless Sensor Network. In M. M. Ghonge, S. Pramanik, & A. D. Potgantwar (Eds.), *Software Defined Networking for Ad Hoc Networks*. Springer. doi:10.1007/978-3-030-91149-2_3

Andono, P. N., Saputra, F. O., Shidik, G. F., & Hasibuan, Z. A. (2022). End-to-end circular economy in onion farming with the application of artificial intelligence and internet of things. *2022 International Seminar on Application for Technology of Information and Communication (ISemantic)*, 459–462.

Ansari, M., & Alam, M. (2023). IoT-Cloud-Enabled Smart Framework for Real-World Applications. In A. J. Kulkarni, S. Mirjalili, & S. K. Udgata (Eds.), *Intelligent Systems and Applications* (pp. 87–105). Springer Nature. doi:10.1007/978-981-19-6581-4_8

Appeltans, S., Pieters, J. G., & Mouazen, A. M. (2021). Detection of leek white tip disease under field conditions using hyperspectral proximal sensing and supervised machine learning. *Computers and Electronics in Agriculture*, *190*, 106453.

Araus, J. L., Li, J., Parry, M. A., & Wang, J. (2014). Phenotyping and other breeding approaches for a New Green Revolution. *Journal of Integrative Plant Biology*, *56*(5), 422–424. doi:10.1111/jipb.12202 PMID:24703237

Ari, A. A. A., Gueroui, A., Labraoui, N., & Yenke, B. O. (2015). Concepts and evolution of research in the field of wireless sensor networks. arXiv preprint arXiv:1502.03561.

Ariyani, M., Pitoi, M. M., Koesmawati, T. A., Maulana, H., & Endah, E. S. (2020). Pyrethroid residues on tropical soil of an Indonesian tea plantation: analytical method development, monitoring, and risk assessment (Vol. 7). Academic Press.

Arnold, N., Brunori, G., Dessein, J., Galli, F., Ghosh, R., Loconto, A. M., & Maye, D. (2022). Governing food futures: Towards a 'responsibility turn' in food and agriculture. *Journal of Rural Studies*, *89*, 82–86. doi:10.1016/j.jrurstud.2021.11.017

Artmann, M., Breuste, J., Breuste, J., Artmann, M., Liu, Y., Breuste, J., & Dilini, M. M. G. S. (2020). Urban agriculture—more than food production. In *Making Green Cities: Concepts, Challenges and Practice* (pp. 75–176). Springer International Publishing. doi:10.1007/978-3-030-37716-8_4

Arulkumaran, G., Jayagopalan, S. L, T. M., & Balamurugan, P. (2023). An Effective Analysis of Proficient Two Level Security Contraptions For Loading Data In Cloud. *2023 1st International Conference on Innovations in High Speed Communication and Signal Processing (IHCSP)*, 536–540. 10.1109/IHCSP56702.2023.10127217

Aryal, J. P., Sapkota, T. B., Khurana, R., Khatri-Chhetri, A., Rahut, D. B., & Jat, M. L. (2020). Climate change and agriculture in South Asia: Adaptation options in smallholder production systems. *Environment, Development and Sustainability*, *22*(6), 5045–5075. doi:10.100710668-019-00414-4

Ashraf, S. A., Siddiqui, A. J., Abd Elmoneim, O. E., Khan, M. I., Patel, M., Alreshidi, M., ... Adnan, M. (2021). Innovations in nanoscience for the sustainable development of food and agriculture with implications on health and environment. *The Science of the Total Environment*, *768*, 144990. doi:10.1016/j.scitotenv.2021.144990 PMID:33736303

Asif Iqbal Khan, S. M. K. (2022). Quadri, Saba Banday, Junaid Latief Shah, Deep diagnosis: A real-time apple leaf disease detection system based on deep learning. *Computers and Electronics in Agriculture*, *198*, 107093.

Asrat, D., & Anteneh, A. (2020). Status of food insecurity in dryland areas of Ethiopia: A review. *Cogent Food & Agriculture*, *6*(1), 1853868. doi:10.1080/23311932.2020.1853868

Assaf, A., Hamsir, A.W.I., & Muhammad, M. (2021). Benefits and Risks of Cloud Computing in E-Government Tasks: A Systematic Review. *E3S Web of Conferences, 328.* doi:10.1051/e3sconf/202132804005

Atila, Ü., Uçar, M., Akyol, K., & Uçar, E. (2021). Plant leaf disease classification using EfficientNet deep learning model. *Ecological Informatics*, *61*, 101182.

Augustin, A., Yi, J., Clausen, T., & Townsley, W. M. (2016). A study of LoRa: Long range & low power networks for the internet of things. *Sensors (Basel)*, *16*(9), 1466. doi:10.339016091466 PMID:27618064

Avşar, E., & Mowla, M. N. (2022). Wireless communication protocols in smart agriculture: A review on applications, challenges and future trends. *Ad Hoc Networks*, *136*, 102982. doi:10.1016/j.adhoc.2022.102982

Ayouba, K., & Vigeant, S. (2020). *Can We Really Use Prices to Control Pesticide Use? Results from a Nonparametric Model*. doi:10.100710666-020-09714-w

Azaza, M., Tanougast, C., Fabrizio, E., & Mami, A. (2016). Smart greenhouse fuzzy logic-based control system enhanced with wireless data monitoring. *ISA Transactions*, *61*, 297–307. doi:10.1016/j.isatra.2015.12.006 PMID:26749556

AzCloud computing. (n.d.). www.azcloud.az

Azercell. (n.d.). www.azercell.com/az/corporate/cloud-services.html

Aznar-Sánchez, J. A., Piquer-Rodríguez, M., Velasco-Muñoz, J. F., & Manzano-Agugliaro, F. (2019). Worldwide research trends on sustainable land use in agriculture. *Land Use Policy*, *87*, 104069. doi:10.1016/j.landusepol.2019.104069

AzScienceNet cloud service. (n.d.). https://azsciencenet.az/az/azcloud

B. (1993). Accuracy of quantum sensors measuring yield photon flux and photosynthetic photon flux. *HortScience, 28*(12), 1197-1200.

Bac, C. W., van Henten, E. J., Hemming, J., & Edan, Y. (2014). Robots for high-value crops: State-of-the-art review and challenges. *Journal of Field Robotics*, *31*(6), 888–911. doi:10.1002/rob.21525

Bannister, K., Giorgetti, G., & Gupta, S. K. (2008, June). Wireless sensor networking for hot applications: Effects of temperature on signal strength, data collection and localization. In *Proceedings of the 5th workshop on embedded networked sensors (HotEmNets' 08)* (pp. 1-5). Academic Press.

Bansal, R., Obaid, A. J., Gupta, A., Singh, R., & Pramanik, S. (2021). Impact of Big Data on Digital Transformation in 5G Era. *2nd International Conference on Physics and Applied Sciences.* 10.1088/1742-6596/1963/1/012170

Barichello, R. (2020). The COVID-19 pandemic: Anticipating its effects on Canada's agricultural trade. *Canadian Journal of Agricultural Economics/Revue canadienne d'agroeconomie, 68*(2), 219-224.

Beckman, J., Ivanic, M., Jelliffe, J. L., Baquedano, F. G., & Scott, S. G. (2020). *Economic and food security impacts of agricultural input reduction under the European Union Green Deal's Farm to Fork and biodiversity strategies* (No. 1473-2020-1039). Academic Press.

Belsare, K. S., & Singh, M. (2022). Various Frameworks for IoT-Enabled Intelligent Waste Management System Using ML for Smart Cities. In S. Shakya, K. Ntalianis, & K. A. Kamel (Eds.), *Mobile Computing and Sustainable Informatics* (pp. 797–817). Springer Nature. doi:10.1007/978-981-19-2069-1_55

Ben Ayed, R., & Hanana, M. (2021). Artificial intelligence to improve the food and agriculture sector. *Journal of Food Quality, 2021*, 1–7. doi:10.1155/2021/5584754

Béné, C., Barange, M., Subasinghe, R., Pinstrup-Andersen, P., Merino, G., Hemre, G. I., & Williams, M. (2015). Feeding 9 billion by 2050–Putting fish back on the menu. *Food Security, 7*(2), 261–274. doi:10.100712571-015-0427-z

Bengal, Dutta, & Nayak. (2019). *S-Transferase Enzyme Activities and Their Correlation with Genotypic Variations Based on GST M1 and GST T1 Loci in Long Term-Pesticide-Exposed Tea Garden Workers of.* . doi:10.1007/s13530-019-0389-1

Benincasa, P., Antognelli, S., Brunetti, L., Fabbri, C. A., Natale, A., Sartoretti, V., Modeo, G., Guiducci, M., Tei, F., & Vizzari, M. (2018). Reliability of NDVI derived by high resolution satellite and UAV compared to in-field methods for the evaluation of early crop N status and grain yield in wheat. *Experimental Agriculture, 54*(4), 604–622. doi:10.1017/S0014479717000278

Benson Hill. (n.d.). Retrieved from www.bensonhill.com

Bersani, C., Ruggiero, C., Sacile, R., Soussi, A., & Zero, E. (2022). Internet of Things Approaches for Monitoring and Control of Smart Greenhouses in Industry 4.0. *Energies, 15*(10), 3834. doi:10.3390/en15103834

Bhargav, A., & Manhar, A. (2020). A Review on Cryptography in Cloud Computing. *International Journal of Scientific Research in Computer Science, Engineering and Information Technology*, 225-230. doi:10.32628/CSEIT206639

Bhatia, G., Joshi, N., Iyengar, S., Rajpal, S., & Mahadevan, K. (2021). Crop Prediction Based on Environmental Conditions and Disease Prediction. In T. Senjyu, P. N. Mahalle, T. Perumal, & A. Joshi (Eds.), *Information and Communication Technology for Intelligent Systems* (pp. 335–344). Springer. doi:10.1007/978-981-15-7078-0_31

Bhat, S. A., & Huang, N.-F. (2021). Big data and ai revolution in precision agriculture: Survey and challenges. *IEEE Access : Practical Innovations, Open Solutions, 9*, 110209–110222. doi:10.1109/ACCESS.2021.3102227

Bhutta, M. N. M., & Ahmad, M. (2021). Secure identification, traceability and real-time tracking of agricultural food supply during transportation using internet of things. *IEEE Access : Practical Innovations, Open Solutions, 9*, 65660–65675. doi:10.1109/ACCESS.2021.3076373

Binns, C. W., Lee, M. K., Maycock, B., Torheim, L. E., Nanishi, K., & Duong, D. T. T. (2021). Climate change, food supply, and dietary guidelines. *Annual Review of Public Health, 42*(1), 233–255. doi:10.1146/annurev-publhealth-012420-105044 PMID:33497266

Blancaflor, E., Saunar, B. Y. P., Darrel, C., Bilbao, T., Villarias, I. H. B., Paula, V., & Mapue, I. (2023). The Use of Cloud Computing and its Security Risks in a Philippine Education System: A Literature Review. *2023 11th International Conference on Information and Education Technology (ICIET)*, 66–70. 10.1109/ICIET56899.2023.10111146

Bogoviz, A. V., Suglobov, A. E., Maloletko, A. N., & Kaurova, O. V. (Eds.). (2022). *Cooperation and Sustainable Development* (Vol. 245). Springer International Publishing. doi:10.1007/978-3-030-77000-6

Bondori, Bagheri, Sookhtanlou, & Damalas. (2021). Modeling farmers ' intention for safe pesticide use: the role of risk perception and use of information sources. Academic Press.

Bonvoisin, T., Utyasheva, L., Knipe, D., Gunnell, D., & Eddleston, M. (2020). Suicide by pesticide poisoning in India: a review of pesticide regulations and their impact on suicide trends. Academic Press.

Boonyopakorn, P. (2018, October). Applying Data Analytics to Findings of User Behaviour Usage in Network Systems. In *2018 International Conference on Information Technology (InCIT)* (pp. 1-6). IEEE. 10.23919/INCIT.2018.8584865

Boote, K. J., Porter, C. H., Hargreaves, J., Hoogenboom, G., Thorburn, P., & Mutter, C. (2014). AgMIP Training in Multiple Crop Models and Tools. In *Handbook of Climate Change and Agroecosystems* (Vol. 3, pp. 393–410). Imperial College Press. doi:10.1142/9781783265640_0025

Borenstein, J., & Miller, K. (2013). Robots and the internet: Causes for concern. *IEEE Technology and Society Magazine, 32*(1), 60–65. doi:10.1109/MTS.2013.2241475

Borodin, V., Bourtembourg, J., Hnaien, F., & Labadie, N. (2016). Handling uncertainty in agricultural supply chain management: A state of the art. *European Journal of Operational Research, 254*(2), 348–359. doi:10.1016/j.ejor.2016.03.057

Bo, Y., & Wang, H. (2011). The Application of Cloud Computing and the Internet of Things in Agriculture and Forestry. In *2011 International Joint Conference on Service Sciences*. 10.1109/IJCSS.2011.40

Bronson, K., & Knezevic, I. (2016). Big Data in food and agriculture. *Big Data & Society, 3*(1), 2053951716648174. doi:10.1177/2053951716648174

Brown, P., & Roper, S. (2017). Innovation and networks in New Zealand farming. *The Australian Journal of Agricultural and Resource Economics, 61*(3), 422–442. doi:10.1111/1467-8489.12211

Bryngelsson, D., Wirsenius, S., Hedenus, F., & Sonesson, U. (2016). How can the EU climate targets be met? A combined analysis of technological and demand-side changes in food and agriculture. *Food Policy, 59*, 152–164. doi:10.1016/j.foodpol.2015.12.012

Burkle, A. (2009). Collaborating miniature drones for Surveillance &recon-renaissance. In Unmanned/Unattended Sensors & Sensor Networks (vol. 7480). International Society for Optics & Photonics.

Butt, U. A., Mehmood, M., Shah, S. B. H., Amin, R., Shaukat, M. W., Raza, S. M., Suh, D. Y., & Piran, M. J. (2020). A Review of Machine Learning Algorithms for Cloud Computing Security. *Electronics (Basel), 9*(9), 9. Advance online publication. doi:10.3390/electronics9091379

Butu, A., Brumă, I. S., Tanasă, L., Rodino, S., Dinu Vasiliu, C., Doboş, S., & Butu, M. (2020). The impact of COVID-19 crisis upon the consumer buying behavior of fresh vegetables directly from local producers. Case study: The quarantined area of Suceava County, Romania. *International Journal of Environmental Research and Public Health, 17*(15), 5485. doi:10.3390/ijerph17155485 PMID:32751368

C. A. (2020). A Cost-effective embedded platform for greenhouse environment control and remote monitoring. *Agronomy, 10*(7), 936.

Carlin, S., & Curran, K. (2011). Cloud computing security. *International Journal of Ambient Computing and Intelligence, 3*(1), 14–19. doi:10.4018/jaci.2011010102

Centenaro, M., Vangelista, L., Zanella, A., & Zorzi, M. (2016). Long-range communications in unlicensed bands: The rising stars in the IoT and smart city scenarios. *IEEE Wireless Communications*, *23*(5), 60–67. doi:10.1109/MWC.2016.7721743

Chandra, M. S., Srinivasulu, M., Yadav, P. S., Ramesh, B., Narasimha, G., & Chandrasekhar, T. (2021). Value added products from agriculture, paper and food waste: a source of bioenergy production. *Bioenergy Research: Commercial Opportunities & Challenges*, 91-126.

Chang, C. L., Chung, S. C., Fu, W. L., & Huang, C. C. (2021). Artificial intelligence approaches to predict growth, harvest day, and quality of lettuce (Lactuca sativa L.) in a IoT-enabled greenhouse system. *Biosystems Engineering*, *212*, 77–105. doi:10.1016/j.biosystemseng.2021.09.015

Chan, K. Y., Abu-Salih, B., Qaddoura, R., Al-Zoubi, A. M., Palade, V., Pham, D.-S., Ser, J. D., & Muhammad, K. (2023). Deep neural networks in the cloud: Review, applications, challenges and research directions. *Neurocomputing*, *545*, 126327. doi:10.1016/j.neucom.2023.126327

Channe, H., Kothari, S., & Kadam, D. (2015). Multidisciplinary model for smart agriculture using internet-of-things (IoT), sensors, cloud-computing, mobile-computing & big-data analysis. Int. J. *Computer Technology and Application*, *6*(3), 374–382.

Chavan, J., Patil, R., Patil, S., Gutte, V., & Karande, S. (2022). A Survey on Security Threats in Cloud Computing Service Models. *2022 6th International Conference on Intelligent Computing and Control Systems (ICICCS)*, 574–580. 10.1109/ICICCS53718.2022.9788148

Che Fauzi, A. A., Noraziah, A., Herawan, T., & Mohd. Zin, N. (2012). On Cloud Computing Security Issues. In J.-S. Pan, S.-M. Chen, & N. T. Nguyen (Eds.), *Intelligent Information and Database Systems* (pp. 560–569). Springer. doi:10.1007/978-3-642-28490-8_58

Chen, T., & Liu, H. (2022). Discussion on Network Information Security Based on Cloud Computing Environment. *2022 8th Annual International Conference on Network and Information Systems for Computers (ICNISC)*, 65–68. 10.1109/ICNISC57059.2022.00023

Cheng, E., Zhang, B., Peng, D., Zhong, L., Yu, L., Liu, Y., Xiao, C., Li, C., Li, X., Chen, Y., Ye, H., Wang, H., Yu, R., Hu, J., & Yang, S. (2022). Wheat yield estimation using remote sensing data based on machine learning approaches. *Frontiers in Plant Science*, *13*, 13. doi:10.3389/fpls.2022.1090970 PMID:36618627

Chen, Q., Xuan, Q., & Xiang, Y. (2022). Application of Deep Learning in Crop Stress. In *Deep Learning Applications* (pp. 61–77). World Scientific. doi:10.1142/9789811266911_0004

Chen, X., Shi, Q., Yang, L., & Xu, J. (2018). ThriftyEdge: Resource-efficient edge computing for intelligent IoT applications. *IEEE Network*, *32*(1), 61–65. doi:10.1109/MNET.2018.1700145

Chen, Y. (2020). A Survey on Industrial Information Integration 2016–2019. *Journal of Industrial Integration and Management*, *05*(01), 33–163. doi:10.1142/S2424862219500167

Chen, Z., Wu, R., Lin, Y., Li, C., Chen, S., Yuan, Z., Chen, S., & Zou, X. (2022). Plant Disease Recognition Model Based on Improved YOLOv5. *Agronomy (Basel)*, *12*(2), 365.

Cheshkova, A. F. (2022). A review of hyperspectral image analysis techniques for plant disease detection and identification. *Vavilovskii Zhurnal Genetiki i Selektsii*, *26*(2), 202–213. doi:10.18699/VJGB-22-25 PMID:35434482

Chopda, J., Raveshiya, H., Nakum, S., & Nakrani, V. (2018). Cotton Crop Disease Detection using Decision Tree Classifier. *2018 International Conference on Smart City and Emerging Technology (ICSCET)*, 1–5. 10.1109/ICSCET.2018.8537336

Choudhary, S. K., Jadoun, R. S., & Mandoriya, H. L. (2016). Role of cloud computing technology in agriculture fields. *Computing, 7*(3), 1–7.

Choudhary, S., Narayan, V., Faiz, M., & Pramanik, S. (2022). Fuzzy Approach-Based Stable Energy-Efficient AODV Routing Protocol in Mobile Ad hoc Networks. In M. M. Ghonge, S. Pramanik, & A. D. Potgantwar (Eds.), *Software Defined Networking for Ad Hoc Networks*. Springer. doi:10.1007/978-3-030-91149-2_6

Chuanlei, Z., Shanwen, Z., Jucheng, Y., Yancui, S., & Jia, C. (2017). Apple leaf disease identification using genetic algorithm and correlation based feature selection method. *International Journal of Agricultural and Biological Engineering, 10*(2), 74–83.

Chukkapalli, S. S. L., Mittal, S., Gupta, M., Abdelsalam, M., Joshi, A., Sandhu, R., & Joshi, K. (2020). Ontologies and artificial intelligence systems for the cooperative smart farming ecosystem. *IEEE Access : Practical Innovations, Open Solutions, 8*, 164045–164064. doi:10.1109/ACCESS.2020.3022763

Clarke, D., Al-Aqrabi, H., Hill, R., Mistry, P., & Lane, P. (2019). Implementing a Lightweight Cloud-Based Process Monitoring Solution for Smart Agriculture. *Smart City and Informatization: 7th International Conference, iSCI 2019, Guangzhou, China, November 12–15, 2019 Proceedings, 7*, 379–391.

Cloud Computing at the U.S. Department of Agriculture - GovWin IQ. (n.d.). https://iq.govwin.com/neo/marketAnalysis/view/Cloud-Computing-at-the-US-Department-of-Agriculture/6048?researchTypeId=1&researchMarket=

Cloud Computing strategy for Norway. (2016). www.regjeringen.no

Cloud ComputingI. T. U. (n.d.). www.itu.int

Coble, K. H., Mishra, A. K., Ferrell, S., & Griffin, T. (2018). Big data in agriculture: A challenge for the future. *Applied Economic Perspectives and Policy, 40*(1), 79–96. doi:10.1093/aepp/ppx056

Cook, S., Jackson, E. L., Fisher, M. J., Baker, D., & Diepeveen, D. (2022). Embedding digital agriculture into sustainable Australian food systems: Pathways and pitfalls to value creation. *International Journal of Agricultural Sustainability, 20*(3), 346–367. doi:10.1080/14735903.2021.1937881

Copa, C., Cema, F. E., Ceettar, C., Ecpa, E., & Fefac, E. S. A. (2018). *EU Code of conduct on agricultural data sharing by contractual agreement*. Academic Press.

CSIS. (n.d.). *Who Benefits from China's Cybersecurity Laws? New Perspectives on Asia*. Retrieved June 2023, from https://www.csis.org/blogs/new-perspectives-asia/who-benefits-chinas-cybersecurity-laws

Damalas, C. A. (2021). Farmers' intention to reduce pesticide use: the role of perceived risk of loss in the model of the planned behavior theory. Academic Press.

Danielsen, F., Flak, L. S., & Ronzhyn, A. (2019). Cloud Computing in eGovernment: Benefits and Challenges. *Proceedings of the Thirteenth International Conference on Digital Society and eGovernments (ICDS)*, 71-77. www.weforum.org/reports/digital-transformation-of-industries

Das, S., Sharma, R., Gourisaria, M. K., Rautaray, S. S., & Pandey, M. (2021). A Model for Probabilistic Prediction of Paddy Crop Disease Using Convolutional Neural Network. In D. Mishra, R. Buyya, P. Mohapatra, & S. Patnaik (Eds.), *Intelligent and Cloud Computing* (pp. 125–134). Springer. doi:10.1007/978-981-15-5971-6_12

Dataset. (n.d.). https://www.kaggle.com/datasets/arjuntejaswi/plant-village

DB, A. K., Doddabasappa, N., Bairwa, B., Kumar, A., & Raju, G. (2023, April). IoT-based Water Harvesting, Moisture Monitoring, and Crop Monitoring System for Precision Agriculture. In *2023 International Conference on Distributed Computing and Electrical Circuits and Electronics (ICDCECE)* (pp. 1-6). IEEE.

De Castro, P., Miglietta, P. P., & Vecchio, Y. (2020). The Common Agricultural Policy 2021–2027: A new history for European agriculture. *Ital. Rev. Agric. Econ, 75*(3), 5–12.

De Donno, M., Tange, K., & Dragoni, N. (2019). Foundations and evolution of modern computing paradigms: Cloud, iot, edge, and fog. *IEEE Access : Practical Innovations, Open Solutions, 7*, 150936–150948. doi:10.1109/ACCESS.2019.2947652

de Souza, M., Koo-Oshima, S., Kahil, T., Wada, Y., Qadir, M., Jewitt, G., ... Zhang, L. (2021). *Food and agriculture*. Academic Press.

Deaton, B. J., & Deaton, B. J. (2020). Food security and Canada's agricultural system challenged by COVID-19. *Canadian Journal of Agricultural Economics/Revue canadienne d'agroeconomie, 68*(2), 143-149.

Deebak, B. D., Al-Turjman, F., Aloqaily, M., & Alfandi, O. (2020). IoT-BSFCAN: A smart context-aware system in IoT-Cloud using mobile-fogging. *Future Generation Computer Systems, 109*, 368–381. doi:10.1016/j.future.2020.03.050

Deepika, K., R., & Dalip. (2022). Security Enabled Framework to Access Information in Cloud Environment. *2022 International Conference on Machine Learning, Big Data, Cloud and Parallel Computing (COM-IT-CON), 1*, 578–582. 10.1109/COM-IT-CON54601.2022.9850906

Dehghani, A., Setare, M. R., & Zarepour, S. (2022). Self-energy problem, vacuum polarization, and dual symmetry in Born–Infeld-type *U*(1) gauge theories. *The European Physical Journal Plus, 137*(7), 859. doi:10.1140/epjp13360-022-03066-y

Delgado, C., Rosegrant, M., Steinfeld, H., Ehui, S., & Courbois, C. (2001). Livestock to 2020: The next food revolution. *Outlook on Agriculture, 30*(1), 27–29. doi:10.5367/000000001101293427

Dharmaraj, V., & Vijayanand, C. (2018). Artificial intelligence (AI) in agriculture. *International Journal of Current Microbiology and Applied Sciences, 7*(12), 2122–2128. doi:10.20546/ijcmas.2018.712.241

Dhiman, N., & Tamsir, M. (2018). A collocation technique based on modified form of trigonometric cubic B-spline basis functions for Fisher's reaction-diffusion equation. Multidiscipline Modeling in Materials and Structures, 14(5), 923-939.

F. Dignum, J. M. Corchado, & F. De La Prieta (Eds.). (2021). Advances in Practical Applications of Agents, Multi-Agent Systems, and Social Good. In *The PAAMS Collection: 19th International Conference, PAAMS 2021*, Salamanca, Spain, October 6–8, 2021, *Proceedings* (Vol. 12946). Springer International Publishing. 10.1007/978-3-030-85739-4

Domínguez-Niño, J. M., Oliver-Manera, J., Girona, J., & Casadesús, J. (2020). Differential irrigation scheduling by an automated algorithm of water balance tuned by capacitance-type soil moisture sensors. *Agricultural Water Management, 228*, 105880. doi:10.1016/j.agwat.2019.105880

Doron, L. (2017). Flexible and precise irrigation platform to improve farm scale water productivity. *Impact, 2017*(1), 77–79. doi:10.21820/23987073.2017.1.77

Doshi, Z., Nadkarni, S., Agrawal, R., & Shah, N. K. (2018). AgroConsultant: Intelligent Crop Recommendation System Using Machine Learning Algorithms. *2018 Fourth International Conference on Computing Communication Control and Automation (ICCUBEA)*, 1-6. 10.1109/ICCUBEA.2018.8697349

Driessen, C., & Heutinck, L. F. (2015). Cows desiring to be milked? Milking robots and the co-evolution of ethics and technology on Dutch dairy farms. *Agriculture and Human Values, 32*(1), 3–20. doi:10.100710460-014-9515-5

Dubey, S. R., & Jalal, A. S. (2013). Adapted approach for fruit disease identification using images. In *Image processing: Concepts, methodologies, tools, and applications* (pp. 1395–1409). IGI Global. doi:10.4018/978-1-4666-3994-2.ch069

Du, C. J., & Sun, D. W. (2006). Learning techniques used in computer vision for food quality evaluation: A review. *Journal of Food Engineering*, *72*(1), 39–55. doi:10.1016/j.jfoodeng.2004.11.017

Dudiki, N., Sangeetha, S., Manna, A., Pokhariyal, R., Mohanaprakash, T. A., & Srivastava, A. P. (2022). A Hybrid Cryptography Algorithm to Improve Cloud Computing Security. *2022 5th International Conference on Contemporary Computing and Informatics (IC3I),* 1020–1023. 10.1109/IC3I56241.2022.10072437

Duman, G. M., Tozanli, O., Kongar, E., & Gupta, S. M. (2017). A holistic approach for performance evaluation using quantitative and qualitative data: A food industry case study. *Expert Systems with Applications*, *81*, 410–422. doi:10.1016/j.eswa.2017.03.070

Duque-Acevedo, M., Belmonte-Ureña, L. J., Plaza-Úbeda, J. A., & Camacho-Ferre, F. (2020). The management of agricultural waste biomass in the framework of circular economy and bioeconomy: An opportunity for greenhouse agriculture in Southeast Spain. *Agronomy (Basel)*, *10*(4), 489. doi:10.3390/agronomy10040489

Dutta, M., Gupta, D., Sahu, S., Limkar, S., Singh, P., Mishra, A., Kumar, M., & Mutlu, R. (2023). Evaluation of Growth Responses of Lettuce and Energy Efficiency of the Substrate and Smart Hydroponics Cropping System. *Sensors (Basel)*, *23*(4), 1875. doi:10.339023041875 PMID:36850471

Dwijendra, N. (2022). Machine Learning Time Series Models For Tea Pest Helopeltis Infestation In India. *Webology*, *19*(2). Available at: https://www.webology.org/abstract.php?id=1625

Dwivedi, D. N., & ... Benchmarking of traditional and Deep Learning time series modelling techniques for prediction of rainfall in United Arab Emirate. *8th International conference on Time Series and Forecasting*, Spain.

Ekanayake, J., & Saputhanthri, L. (2020). E-AGRO: Intelligent chat-bot. IoT and artificial intelligence to enhance farming industry. *AGRIS On-Line Papers in Economics and Informatics, 12*, 15–21.

El Bilali, H., & Allahyari, M. S. (2018). Transition towards sustainability in agriculture and food systems: Role of information and communication technologies. *Information Processing in Agriculture*, *5*(4), 456–464. doi:10.1016/j.inpa.2018.06.006

El Chami, D., Daccache, A., & El Moujabber, M. (2020). How can sustainable agriculture increase climate resilience? A systematic review. *Sustainability (Basel)*, *12*(8), 3119. doi:10.3390u12083119

Elavarasan, D., Vincent, D. R., Sharma, V., Zomaya, A. Y., & Srinivasan, K. (2018). Forecasting yield by integrating agrarian factors and machine learning models: A survey. *Computers and Electronics in Agriculture*, *155*, 257–282. doi:10.1016/j.compag.2018.10.024

Elbasi, E., Mostafa, N., Alarnaout, Z., Zreikat, A. I., Cina, E., Varghese, G., Shdefat, A., Topcu, A. E., Abdelbaki, W., Mathew, S., & Zaki, C. (2023). Artificial Intelligence Technology in the Agricultural Sector : A Systematic Literature Review. *IEEE Access : Practical Innovations, Open Solutions*, *11*, 171–202. doi:10.1109/ACCESS.2022.3232485

Elferink, M., & Schierhorn, F. (2016). Global demand for food is rising. Can we meet it. *Harvard Business Review*, *7*(04), 2016.

Elijah, O., Member, S., & Rahman, T. A. (2018). An Overview of Internet of Things (IoT) and Data Analytics in Agriculture : Benefits and Challenges. *IEEE Internet of Things Journal*, *5*(5), 3758–3773. doi:10.1109/JIOT.2018.2844296

Elvan, O. D., Birben, Ü., Özkan, U. Y., Yıldırım, H. T., & Türker, Y. Ö. (2021). Forest fire and law: An analysis of Turkish forest fire legislation based on Food and Agriculture Organization criteria. *Fire Ecology*, *17*(1), 1–15. doi:10.118642408-021-00102-7

Engler, N., & Krarti, M. (2021). Review of energy efficiency in controlled environment agriculture. *Renewable & Sustainable Energy Reviews*, *141*, 110786. doi:10.1016/j.rser.2021.110786

Esham, M., Jacobs, B., Rosairo, H. S. R., & Siddighi, B. B. (2018). Climate change and food security: A Sri Lankan perspective. *Environment, Development and Sustainability*, *20*(3), 1017–1036. doi:10.100710668-017-9945-5

Estonia's Digital Agenda 2030. (2021). www.mkm.ee

Eyhorn, F., Muller, A., Reganold, J. P., Frison, E., Herren, H. R., Luttikholt, L., & Smith, P. (2019). Sustainability in global agriculture driven by organic farming. *Nature Sustainability*, *2*(4), 253–255. doi:10.103841893-019-0266-6

Faid, A., Sadik, M., & Sabir, E. (2021). An agile AI and IoT-augmented smart farming: A cost-effective cognitive weather station. *Agriculture*, *12*(1), 35. doi:10.3390/agriculture12010035

Fairbairn, M., Kish, Z., & Guthman, J. (2022). Pitching agri-food tech: Performativity and non-disruptive disruption in Silicon Valley. *Journal of Cultural Economics*, *15*(5), 652–670. doi:10.1080/17530350.2022.2085142

Fang, W., Tang, L., Cheng, P., & Ahmad, N. (2018). Evolution decision, drivers and green innovation performance for collaborative innovation center of ecological building materials and environmental protection equipment in Jiangsu province of China. *International Journal of Environmental Research and Public Health*, *15*(11), 2365. doi:10.3390/ijerph15112365 PMID:30366457

FAO. (2009). *Press release, 19 June 2009*. http://www.fao.org/news/story/en/item/ 20568/icode/

Federal Cloud Computing Strategy. (2019). https://cloud.cio.gov

Federal Data Center Consolidation Initiative. (2011). *Data Center Consolidation Plan & Progress Report*. https://www.dhs.gov/

Fenu, G., & Malloci, F. M. (2020). Artificial Intelligence Technique in Crop Disease Forecasting: A Case Study on Potato Late Blight Prediction. In I. Czarnowski, R. J. Howlett, & L. C. Jain (Eds.), *Intelligent Decision Technologies* (pp. 79–89). Springer. doi:10.1007/978-981-15-5925-9_7

Ferentinos, K. P. (2018). Deep learning models for plant disease detection and diagnosis. *Computers and Electronics in Agriculture*, *145*, 311–318.

Fernández-Ahumada, L. M., Ramírez-Faz, J., Torres-Romero, M., & López-Luque, R. (2019). Proposal for the design of monitoring and operating irrigation networks based on IoT, cloud computing and free hardware technologies. *Sensors (Basel)*, *19*(10), 2318. doi:10.339019102318 PMID:31137517

Ferrández-Pastor, F. J., García-Chamizo, J. M., Nieto-Hidalgo, M., & Mora-Martínez, J. (2018). Precision agriculture design method using a distributed computing architecture on internet of things context. *Sensors (Basel)*, *18*(6), 1731. doi:10.339018061731 PMID:29843386

Flores, K. O., Butaslac, I. M., Gonzales, J. E. M., Dumlao, S. M. G., & Reyes, R. S. (2016, November). Precision agriculture monitoring system using wireless sensor network and Raspberry Pi local server. In *2016 IEEE region 10 conference (TENCON)* (pp. 3018- 3021). IEEE.

Foubert, B., & Mitton, N. (2020). Long-range wireless radio technologies: A survey. *Future Internet*, *12*(1), 13. doi:10.3390/fi12010013

Friedmann, H. (2021). Changes in the international division of labor: agri-food complexes and export agriculture. In *Towards a new political economy of agriculture* (pp. 65–93). Routledge. doi:10.4324/9780429269493-4

Friha, O., Ferrag, M. A., Shu, L., Maglaras, L. A., & Wang, X. (2021). Internet of Things for the Future of Smart Agriculture: A Comprehensive Survey of Emerging Technologies. *IEEE/CAA Journal of Automatica Sinica, 8*(4), 718–752. doi:10.1109/JAS.2021.1003925

Fuentes, A., Yoon, S., & Park, D. S. (2019). Deep learning-based phenotyping system with glocal description of plant anomalies and symptoms. *Frontiers in Plant Science, 10*, 1321. doi:10.3389/fpls.2019.01321 PMID:31798598

Gallia, G. K. N., & Kephaliacos, C. (2021). Ecological - economic modeling of pollination complexity and pesticide use in agricultural crops. *Journal of Bioeconomics, 23*(3), 297–323. doi:10.100710818-021-09317-9

Garcia-Mañas, F., Rodríguez, F., & Berenguel, M. (2020). Leaf area index soft sensor for tomato crops in greenhouses. *IFAC-PapersOnLine, 53*(2), 15796–15803. doi:10.1016/j.ifacol.2020.12.230

Garcia, S. N., Osburn, B. I., & Jay-Russell, M. T. (2020). One health for food safety, food security, and sustainable food production. *Frontiers in Sustainable Food Systems, 4*, 1. doi:10.3389/fsufs.2020.00001

Garske, B., Bau, A., & Ekardt, F. (2021). Digitalization and AI in European agriculture: A strategy for achieving climate and biodiversity targets? *Sustainability (Basel), 13*(9), 4652. doi:10.3390u13094652

G-Cloud framework, Guidance Applying to the G-Cloud framework. (2013). *How to apply to sell cloud hosting, software and support to public sector buyers.* https://www.gov.uk/guidance/g-cloud-suppliers-guide

Gharaei, A., Karimi, M., & Hoseini Shekarabi, S. A. (2019). Joint economic lot-sizing in multiproduct multi-level integrated supply chains: Generalized benders decomposition. *International Journal of Systems Science: Operations & Logistics*, 1-17.

Ghosh, R., Mohanty, S., & Pramanik, S. (2020). A Novel Approach towards Selection of Role Model Cluster Head for Power Management in WSN: Selection of Role Model Cluster Head for Power Management in WSN. In G. Bose & P. Pain (Eds.), *Machine Learning Applications in Non-Conventional Machining Processes* (pp. 235–249). IGI Global. doi:10.4018/978-1-7998-3624-7.ch015

Ghosh, U., & Dolui, S. (2020). Cloud Computing: It's Security and Introduction to Amazon Web Services. In S. Dawn, V. E. Balas, A. Esposito, & S. Gope (Eds.), *Intelligent Techniques and Applications in Science and Technology* (pp. 942–951). Springer International Publishing. doi:10.1007/978-3-030-42363-6_108

Giri, A., Dutta, S., & Neogy, S. (2016, October). Enabling agricultural automation to optimize utilization of water, fertilizer and insecticides by implementing Internet of Things (IoT). In *2016 International Conference on Information Technology (InCITe)-The Next Generation IT Summit on the Theme-Internet of Things: Connect your Worlds* (pp. 125-131). IEEE.

Giri, B. C., & Masanta, M. (2018). Developing a closed-loop supply chain model with price and quality dependent demand and learning in production in a stochastic environment. *International Journal of Systems Science: Operations & Logistics*, 1-17.

Giri, C., Jain, S., Zeng, X., & Bruniaux, P. (2019). A detailed review of artificial intelligence applied in the fashion and apparel industry. *IEEE Access : Practical Innovations, Open Solutions, 7*, 95376–95396. doi:10.1109/ACCESS.2019.2928979

Giusti, E., & Marsili-Libelli, S. (2015). A fuzzy decision support system for irrigation and water conservation in agriculture. *Environmental Modelling & Software, 63*, 73–86. doi:10.1016/j.envsoft.2014.09.020

Goap, A., Sharma, D., Shukla, A. K., & Krishna, C. R. (2018). An IoT based smart irrigation management system using Machine learning and open source technologies. *Computers and Electronics in Agriculture*, *155*, 41–49. doi:10.1016/j.compag.2018.09.040

Goraya, M. S., & Kaur, H. (2015). Cloud computing in agriculture. *HCTL Open International Journal of Technology Innovations and Research*, *16*, 2321–1814.

Gourisaria, M. K., Samanta, A., Saha, A., Patra, S. S., & Khilar, P. M. (2020). An Extensive Review on Cloud Computing. Data Engineering and Communication Technology. Advances in Intelligent Systems and Computing, 1079. doi:10.1007/978-981-15-1097-7_6

Government Cloud Concept. (2019). *Decree of the President of the Republic of Azerbaijan on the Concept of "Government Cloud" (G-cloud), creation of "Government Cloud" (G-cloud) and provision of "cloud" services*. https://e-qanun.az/framework/42560

Government Cloud Strategy. (2011). https://service.gov.uk

Guesbaya, M., García-Mañas, F., Rodríguez, F., & Megherbi, H. (2023). A Soft Sensor to Estimate the Opening of Greenhouse Vents Based on an LSTM-RNN Neural Network. *Sensors (Basel)*, *23*(3), 1250. doi:10.339023031250 PMID:36772289

Guggisberg, S. (2022). Transparency in the activities of the Food and Agriculture Organization for sustainable fisheries. *Marine Policy*, *136*, 104498. doi:10.1016/j.marpol.2021.104498

Guiné, R. P., Florença, S. G., Barroca, M. J., & Anjos, O. (2020). The link between the consumer and the innovations in food product development. *Foods*, *9*(9), 1317. doi:10.3390/foods9091317 PMID:32962007

Guo, L., Cao, A., Huang, M., Li, H., & Cao, A. (2021). Effects of haze pollution on pes- ticide use by rice farmers: fresh evidence from rural areas of China. Academic Press.

Guo, Y., Yin, Q., Wang, Y., Xu, J., & Zhu, L. (2023). Efficiency and optimization of government service resource allocation in a cloud computing environment. *Journal of Cloud Computing (Heidelberg, Germany)*, *12*(1), 18. doi:10.118613677-023-00400-2 PMID:36789367

Gupta, A., Asad, A., Meena, L., & Anand, R. (2023). IoT and RFID-Based Smart Card System Integrated with Health Care, Electricity, QR and Banking Sectors. In M. Gupta, S. Ghatak, A. Gupta, & A. L. Mukherjee (Eds.), *Artificial Intelligence on Medical Data. Lecture Notes in Computational Vision and Biomechanics* (Vol. 37). Springer. doi:10.1007/978-981-19-0151-5_22

Gupta, M., Abdelsalam, M., & Mittal, S. (2020). Security and Privacy in Smart Farming : Challenges and Opportunities. *IEEE Access : Practical Innovations, Open Solutions*, *8*, 34564–34584. doi:10.1109/ACCESS.2020.2975142

Gupta, S., Tyagi, S., & Kishor, K. (2022). Study and Development of Self Sanitizing Smart Elevator. In D. Gupta, Z. Polkowski, A. Khanna, S. Bhattacharyya, & O. Castillo (Eds.), *Proceedings of Data Analytics and Management. Lecture Notes on Data Engineering and Communications Technologies* (Vol. 90). Springer. doi:10.1007/978-981-16-6289-8_15

Haider, W., Ur Rehman, A., Maqsood, A., & Javed, S. Z. (2020). Crop Disease Diagnosis using Deep Learning Models. *2020 Global Conference on Wireless and Optical Technologies (GCWOT)*, 1–6. 10.1109/GCWOT49901.2020.9391605

Harish Naik, B. M., & Kannan, J. (2023). A Research on Various Security Aware Mechanisms in Multi-Cloud Environment for Improving Data Security. *2023 International Conference on Distributed Computing and Electrical Circuits and Electronics (ICDCECE)*, 1–5. 10.1109/ICDCECE57866.2023.10151135

Haryanto, U. M., Ibadillah, A. F., Alfita, R., Aji, K., & Rizkyandi, R. (2019, April). Smart aquaponic system based Internet of Things (IoT). In Journal of Physics: Conference Series (Vol. 1211, p. 012047). IOP Publishing.

Hassan, S. I., Alam, M. M., Illahi, U., Al Ghamdi, M. A., Almotiri, S. H., & Su'ud, M. M. (2021). A systematic review on monitoring and advanced control strategies in smart agriculture. *IEEE Access : Practical Innovations, Open Solutions*, *9*, 32517–32548. doi:10.1109/ACCESS.2021.3057865

Hassan, S. M., Maji, A. K., Jasiński, M., Leonowicz, Z., & Jasińska, E. (2021). Identification of plant-leaf diseases using CNN and transfer-learning approach. *Electronics (Basel)*, *10*(12), 1388.

Hassler, S. C., & Baysal-Gurel, F. (2019). Unmanned aircraft system (UAS) technology and applications in agriculture. *Agronomy (Basel)*, *9*(10), 618. doi:10.3390/agronomy9100618

Heldreth, C., Akrong, D., Holbrook, J., & Su, N. M. (2021). What does AI mean for smallholder farmers? A proposal for farmer-centered AI research. *Interactions (New York, N.Y.)*, *28*(4), 56–60. doi:10.1145/3468004

Hemalatha, D. K. M. R., & Deepika, G. (2016). Internet of things (iot) based smart irrigation. *International Journal Of Advanced Research in Basic Engineering Sciences and Technology*, *2*(2), 128–132.

Heshmati, Mehri, & Khaneghah. (2021). Simultaneous multi- determination of pesticide residues in black tea leaves and infusion: A risk assessment study. Academic Press.

Heuvelink, E. P., & Challa, H. (1989, May). Dynamic optimization of artificial lighting in greenhouses. In *International Symposium on Growth and Yield Control in Vegetable Production 260* (pp. 401-412). 10.17660/ActaHortic.1989.260.26

Hori, M., Kawashima, E., & Yamazaki, T. (2010). Application of cloud computing to agriculture and prospects in other fields. *Fujitsu Scientific and Technical Journal*, *46*(4), 446–454.

Hu, G., Wu, H., Zhang, Y., & Wan, M. (2019). A low shot learning method for tea leaf's disease identification. *Computers and Electronics in Agriculture*, *163*, 104852.

Hu, H., Su, C., Yang, C., & Yu, P. (2019). A Study on Intelligent Crop Disease and Pest Diagnosis System Based on Hadoop. *2019 International Conference on Robots & Intelligent System (ICRIS)*, 171–173. 10.1109/ICRIS.2019.00052

Hussain, S. M., Yusof, K. M., Asuncion, R., Hussain, S. A., & Ahmad, A. (2022). An Integrated Approach of 4G LTE and DSRC (IEEE 802.11p) for Internet of Vehicles (IoV) by Using a Novel Cluster-Based Efficient Radio Interface Selection Algorithm to Improve Vehicular Network (VN) Performance. In Sustainable Advanced Computing. Lecture Notes in Electrical Engineering (vol 840). Springer. doi:10.1007/978-981-16-9012-9_46

Iftikhar, S., Gill, S. S., Song, C., Xu, M., Aslanpour, M. S., Toosi, A. N., Du, J., Wu, H., Ghosh, S., Chowdhury, D., Golec, M., Kumar, M., Abdelmoniem, A. M., Cuadrado, F., Varghese, B., Rana, O., Dustdar, S., & Uhlig, S. (2023). AI-based fog and edge computing: A systematic review, taxonomy and future directions. *Internet of Things*, *21*, 100674. doi:10.1016/j.iot.2022.100674

Imamverdiyev, Y.N. (2021). Oil and Gas 4.0 existing solutions, prospects and problems. *Information Society Problems*, *1*, 32–42.

Imamverdiyev, Y.N., & Abbasov, H.H. (2021). Current scientific-research problems of the national e-signature infrastructure. *Information Technology Problems*, *1*, 33–45.

Information Economy Report. (2013). The Cloud Economy and Developing Countries. United Nations Publication UNCTAD.

Iniyan, S., & Jebakumar, R. (2022). Crop Yield Prediction on Soybean Crop Applying Multi-layer Stacked Ensemble Learning Technique. In G. Manogaran, A. Shanthini, & G. Vadivu (Eds.), *Proceedings of International Conference on Deep Learning, Computing and Intelligence* (pp. 335–348). Springer Nature. 10.1007/978-981-16-5652-1_29

Internet of Things Monitoring System of Modern Eco-Agriculture Based on Cloud Computing. (2019). *IEEE Journals & Magazine*. https://ieeexplore.ieee.org/abstract/document/8662571

Internet, O. E. C. D. (2012). *Economy Outlook*. OECD Publishing.

Ivanus, C., & Iovan, S. (2014). Governmental cloud-part of cloud computing. *Informações Econômicas*, *18*(4), 91–100. doi:10.12948/issn14531305/18.4.2014.08

Jain, A., Sharma, Y., & Kishor, K. (2020). Financial Supervision and Management System Using Ml Algorithm. *Solid State Technology*, *63*(6), 18974–18982. http://solidstatetechnology.us/index.php/JSST/article/view/8080

JainA.SharmaY.KishorK. (2021). Prediction and Analysis of Financial Trends Using Ml Algorithm. *Proceedings of the International Conference on Innovative Computing & Communication (ICICC)*. doi:10.2139/ssrn.3884458

Jaishetty, S. A., & Patil, R. (2016). Iot sensor network based approach for agricultural field monitoring and control. *International Journal of Research in Engineering and Technology*, *5*(6).

Jakku, E., Taylor, B., Fleming, A., Mason, C., Fielke, S., Sounness, C., & Thorburn, P. (2019). "If they don't tell us what they do with it, why would we trust them?" Trust, transparency and benefit-sharing in Smart Farming. *NJAS Wageningen Journal of Life Sciences*, *90*(1), 100285. doi:10.1016/j.njas.2018.11.002

Jamil, F., Ibrahim, M., Ullah, I., Kim, S., Kahng, H. K., & Kim, D. H. (2022). Optimal smart contract for autonomous greenhouse environment based on IoT blockchain network in agriculture. *Computers and Electronics in Agriculture*, *192*, 106573. doi:10.1016/j.compag.2021.106573

Jarzebski, M. P., Ahmed, A., Boafo, Y. A., Balde, B. S., Chinangwa, L., Saito, O., von Maltitz, G., & Gasparatos, A. (2020). Food security impacts of industrial crop production in sub-Saharan Africa: A systematic review of the impact mechanisms. *Food Security*, *12*(1), 105–135. doi:10.100712571-019-00988-x

Javaid, M., Haleem, A., Singh, R. P., & Suman, R. (2022). Artificial Intelligence Applications for Industry 4.0: A Literature-Based Study. *Journal of Industrial Integration and Management*, *07*(01), 83–111. doi:10.1142/S2424862221300040

Javed, S. A., Haider, A., & Nawaz, M. (2020). How agricultural practices managing market risk get attributed to climate change? Quasi-experiment evidence. *Journal of Rural Studies*, *73*, 46–55. doi:10.1016/j.jrurstud.2019.11.020

Jenipher, V. N., & Radhika, S. (2022). An Automated System for Detecting Rice Crop Disease using CNN Inception V3 Transfer Learning Algorithm. *2022 Second International Conference on Artificial Intelligence and Smart Energy (ICAIS)*, 88–94. 10.1109/ICAIS53314.2022.9742999

Jiang, P., Chen, Y., Liu, B., He, D., & Liang, C. (2019). Real-time detection of apple leaf diseases using deep learning approach based on improved convolutional neural networks. *IEEE Access : Practical Innovations, Open Solutions*, *7*, 59069–59080.

John, K., Abraham Isong, I., Michael Kebonye, N., Okon Ayito, E., Chapman Agyeman, P., & Marcus Afu, S. (2020). Using machine learning algorithms to estimate soil organic carbon variability with environmental variables and soil nutrient indicators in an alluvial soil. *Land (Basel)*, *9*(12), 487. doi:10.3390/land9120487

Jones, S., Irani, Z., Sivarajah, U., & Love, P. E. D. (2019). Risks and rewards of cloud computing in the UK public sector: A reflection on three Organisational case studies. *Information Systems Frontiers*, *21*(2), 359–382. doi:10.100710796-017-9756-0

Jordan, M. I., & Mitchell, T. M. (2015). Machine learning: Trends, perspectives, and prospects. *Science*, *349*(6245), 255–260. doi:10.1126cience.aaa8415 PMID:26185243

Joshi, K., Singh, R., & Kumar, N. (2023, March). Cloud Based Real Time Soil Moisture Content Monitoring Using IoT and Unmanned Aerial Vehicles. In *2023 Sixth International Conference of Women in Data Science at Prince Sultan University (WiDS PSU)* (pp. 206-210). IEEE. 10.1109/WiDS-PSU57071.2023.00050

Joshi, R. C., Kaushik, M., Dutta, M. K., Srivastava, A., & Choudhary, N. (2021). VirLeafNet: Automatic analysis and viral disease diagnosis using deep-learning in Vigna mungo plant. *Ecological Informatics*, *61*, 101197.

Jouini, M., & Rabai, L. B. A. (2017). A Security Risk Management Model for Cloud Computing Systems: Infrastructure as a Service. In G. Wang, M. Atiquzzaman, Z. Yan, & K.-K. R. Choo (Eds.), *Security, Privacy, and Anonymity in Computation, Communication, and Storage* (pp. 594–608). Springer International Publishing. doi:10.1007/978-3-319-72389-1_47

Jung, E. M., Cui, Y., Lin, T. H., He, X., Eid, A., Hester, J. G., Abowd, G. D., Starner, T. E., Lee, W.-S., & Tentzeris, M. M. (2019). A wideband, quasi-isotropic, kilometer-range FM energy harvester for perpetual IoT. *IEEE Microwave and Wireless Components Letters*, *30*(2), 201–204. doi:10.1109/LMWC.2019.2957653

Kamble, S. S., Gunasekaran, A., & Sharma, R. (2020). Modeling the blockchain enabled traceability in agriculture supply chain. *International Journal of Information Management*, *52*, 101967. doi:10.1016/j.ijinfomgt.2019.05.023

Kamienski C., Soininen J. P., Taumberger M., Fernandes S., Toscano A., Cinotti T. S., ... & Neto, A. T. (2018, June). SWAMP: An IoT-based smart water management platform for precision irrigation in agriculture. In *2018 Global Internet of Things Summit (GIoTS)* (pp. 1-6). IEEE.

Kamilaris, A., Kartakoullis, A., & Prenafeta-Boldú, F. X. (2017). A review on the practice of big data analysis in agriculture. *Computers and Electronics in Agriculture*, *143*, 23–37. doi:10.1016/j.compag.2017.09.037

Kamilaris, A., & Prenafeta-Boldú, F. X. (2018). Deep learning in agriculture: A survey. *Computers and Electronics in Agriculture*, *147*, 70–90. doi:10.1016/j.compag.2018.02.016

Kamruzzaman, S. M., Pavel M. I., Hoque M. A., & Sabuj S. R. (2019). Promoting greenness with IoT-based plant growth system. *Computational Intelligence and Sustainable Systems: Intelligence and Sustainable Computing*, 235-253.

Kapoor, A., & Attri, V. (2021, December). A Review of Tools and Technologies for Big Data Analytics. In *2021 10th International Conference on System Modeling & Advancement in Research Trends (SMART)* (pp. 13-16). IEEE. 10.1109/SMART52563.2021.9676237

Karlekar, A., & Seal, A. (2020). SoyNet: Soybean leaf diseases classification. *Computers and Electronics in Agriculture*, *172*, 105342. doi:10.1016/j.compag.2020.105342

Karthikeyan, L., Chawla, I., & Mishra, A. K. (2020). A review of remote sensing applications in agriculture for food security: Crop growth and yield, irrigation, and crop losses. *Journal of Hydrology (Amsterdam)*, *586*, 124905. doi:10.1016/j.jhydrol.2020.124905

Katzin, D., van Henten, E. J., & van Mourik, S. (2022). Process-based greenhouse climate models: Genealogy, current status, and future directions. *Agricultural Systems*, *198*, 103388. doi:10.1016/j.agsy.2022.103388

Kaur Sandeep, C. S. (2016). Indoxacarb interaction alters immunotoxic and genotoxic potential of endotoxin. *Jpurnal of Pesticide Science, 41*(3), 65–70. doi:10.1584/jpestics.D16-012

Kaur, P. J., & Kaushal, S. (2011). Security Concerns in Cloud Computing. In A. Mantri, S. Nandi, G. Kumar, & S. Kumar (Eds.), *High Performance Architecture and Grid Computing* (pp. 103–112). Springer. doi:10.1007/978-3-642-22577-2_14

Kaur, R., Garg, R., & Aggarwal, H. (2016). Big data analytics framework to identify crop disease and recommendation a solution. *2016 International Conference on Inventive Computation Technologies (ICICT), 2*, 1–5. 10.1109/INVENTIVE.2016.7824791

Kayser, V., Nehrke, B., & Zubovic, D. (2018). Data science as an innovation challenge: From big data to value proposition. *Technology Innovation Management Review, 8*(3), 16–25. doi:10.22215/timreview/1143

Kazemi, Y. (2019). *AI, Big Data & Advanced Analytics In The Supply Chain*. Available at: https://www.forbes.com/sites/yasamankazemi/2019/01/29/ai-big-data-advanced-analyticsin-the-supply-chain/#75db62f4244f

Keerthi, V., & Kodandaramaiah, G. N. (2015). Cloud iot based greenhouse monitoring system. *International Journal of Engineering Research and Applications, 5*(10), 35–41.

Keswani, B., Mohapatra, A. G., Mohanty, A., Khanna, A., Rodrigues, J. J., Gupta, D., & De Albuquerque, V. H. C. (2019). Adapting weather conditions based IoT enabled smart irrigation technique in precision agriculture mechanisms. *Neural Computing & Applications, 31*(S1), 277–292. doi:10.100700521-018-3737-1

Khalifeh, A., AlQammaz, A., Darabkh, K. A., Sha'ar, B. A., & Ghatasheh, O. (2021). A Framework for Artificial Intelligence Assisted Smart Agriculture Utilizing LoRaWAN Wireless Sensor Networks. In V. E. Balas, L. C. Jain, M. M. Balas, & S. N. Shahbazova (Eds.), *Soft Computing Applications* (pp. 408–421). Springer International Publishing. doi:10.1007/978-3-030-52190-5_29

Khan, N., Ray, R. L., Sargani, G. R., Ihtisham, M., Khayyam, M., & Ismail, S. (2021). Current Progress and Future Prospects of Agriculture Technology: Gateway to Sustainable Agriculture. *Sustainability (Basel), 13*(9), 4883. doi:10.3390u13094883

Khan, N., Yaqoob, I., Hashem, I. A. T., Inayat, Z., Mahmoud Ali, W. K., Alam, M., Shiraz, M., & Gani, A. (2014). Big data: Survey, technologies, opportunities, and challenges. *TheScientificWorldJournal, 2014*, 2014. doi:10.1155/2014/712826 PMID:25136682

Khanna, A., & Kaur, S. (2019). Evolution of Internet of Things (IoT) and its significant impact in the field of Precision Agriculture. *Computers and Electronics in Agriculture, 157*, 218–231. doi:10.1016/j.compag.2018.12.039

Kim, D. G., Burks, T. F., Qin, J., & Bulanon, D. M. (2009). Classification of grapefruit peel diseases using color texture feature analysis. *International Journal of Agricultural and Biological Engineering, 2*(3), 41–50.

Kishor, K. (2023). Cloud Computing in Blockchain. In Cloud-based Intelligent Informative Engineering for Society 5.0 (pp. 79-105). Chapman and Hall/CRC. doi:10.1201/9781003213895-5

Kishor, K. (2023). Impact of Cloud Computing on Entrepreneurship, Cost, and Security. In Cloud-based Intelligent Informative Engineering for Society 5.0 (pp. 171-191). CRC Press. doi:10.1201/9781003213895-10

Kishor, K., & Nand, P. (2023). Wireless Networks Based in the Cloud That Support 5G. In Cloud-based Intelligent Informative Engineering for Society 5.0 (pp. 23-40). Chapman and Hall/CRC. doi:10.1201/9781003213895-2

Kishor, K., & Pandey, D. (2022). Study and Development of Efficient Air Quality Prediction System Embedded with Machine Learning and IoT. In *Proceeding International Conference on Innovative Computing and Communications.* Springer. 10.1007/978-981-19-2535-1_24

Kishor, K., Saxena, N., & Pandey, D. (2023). Cloud-based Intelligent Informative Engineering for Society 5.0. Chapman and Hall/CRC. doi:10.1201/9781003213895

Kishor, K., Tyagi, R., Bhati, R., & Rai, B. K. (2023). Develop Model for Recognition of Handwritten Equation Using Machine Learning. In *Proceedings of International Conference on Recent Trends in Computing. Lecture Notes in Networks and Systems* (vol. 600). Springer. 10.1007/978-981-19-8825-7_23

Kishor, K. (2022). Communication-Efficient Federated Learning. In S. P. Yadav, B. S. Bhati, D. P. Mahato, & S. Kumar (Eds.), *Federated Learning for IoT Applications. EAI/Springer Innovations in Communication and Computing*. Springer. doi:10.1007/978-3-030-85559-8_9

Kishor, K. (2022). Personalized Federated Learning. In S. P. Yadav, B. S. Bhati, D. P. Mahato, & S. Kumar (Eds.), *Federated Learning for IoT Applications. EAI/Springer Innovations in Communication and Computing*. Springer. doi:10.1007/978-3-030-85559-8_3

KishorK.AttriS.RanjanS.TyagiT. (2018). Smart cart based on IOT. *International Journal of Creative Research Thoughts, 6*(1), 1388-1392, Available at: https://www.ijcrt.org/papers/IJCRT1803295.pdf

Kishor, K., Nand, P., & Agarwal, P. (2017). Subnet based ad hoc network algorithm reducing energy consumption in manet. *International Journal of Applied Engineering Research, 12*(22), 11796–11802.

Kishor, K., Nand, P., & Agarwal, P. (2018). Notice of Retraction Design adaptive Subnetting Hybrid Gateway MANET Protocol on the basis of Dynamic TTL value adjustment. *Aptikom Journal on Computer Science and Information Technologies, 3*(2), 59–65. doi:10.11591/APTIKOM.J.CSIT.115

Kishor, K., Nand, P., & Agarwal, P. (2018). Secure and Efficient Subnet Routing Protocol for MANET. *Indian Journal of Public Health Research & Development, 9*(12), 200. doi:10.5958/0976-5506.2018.01830.2

Kishor, K., Sharma, R., & Chhabra, M. (2022). Student Performance Prediction Using Technology of Machine Learning. In D. K. Sharma, S. L. Peng, R. Sharma, & D. A. Zaitsev (Eds.), *Micro-Electronics and Telecommunication Engineering. Lecture Notes in Networks and Systems* (Vol. 373). Springer. doi:10.1007/978-981-16-8721-1_53

Kishor, K., Singh, P., & Vashishta, R. (2023). Develop Model for Malicious Traffic Detection Using Deep Learning. In D. K. Sharma, S. L. Peng, R. Sharma, & G. Jeon (Eds.), *Micro-Electronics and Telecommunication Engineering. Lecture Notes in Networks and Systems* (Vol. 617). Springer. doi:10.1007/978-981-19-9512-5_8

Kochhar, A., & Kumar, N. (2019). Wireless sensor networks for greenhouses: An end- to-end review. *Computers and Electronics in Agriculture, 163*, 104877. doi:10.1016/j.compag.2019.104877

Kolomvatsos, K., & Anagnostopoulos, C. (2021). Proactive, uncertainty-driven queries management at the edge. *Future Generation Computer Systems, 118*, 75–93. doi:10.1016/j.future.2020.12.028

Konovalenko, I., & Ludwig, A. (2019). Event processing in supply chain management–The status quo and research outlook. *Computers in Industry, 105*, 229–249. doi:10.1016/j.compind.2018.12.009

Kramer & Ware. (2020). Seasonal abundance of tea mosquito bug, Looper Antoni Signoret infesting name. *Journal of Entomology and Zoology Studies, 8*(6), 2006–2009. doi:10.22271/j.ento.2020.v8.i6aa.8118

Kuczuk, A., & Widera, K. (2021). A Greater Share of Organic Agriculture in Relation to Food Security Resulting from the Energy Demand Obtained from Food—Scenarios for Poland until 2030. *Energies, 14*(21), 6959. doi:10.3390/en14216959

Kuhnlein, H. V., & Turner, N. J. (1991). *Traditional plant foods of Canadian indigenous peoples: nutrition, botany, and use* (Vol. 8). Taylor & Francis.

Kuiken, T., Barrangou, R., & Grieger, K. (2021). (Broken) promises of sustainable food and agriculture through new biotechnologies: The CRISPR case. *The CRISPR Journal, 4*(1), 25–31. doi:10.1089/crispr.2020.0098 PMID:33570455

Kulkarni, O. (2018). Crop Disease Detection Using Deep Learning. *2018 Fourth International Conference on Computing Communication Control and Automation (ICCUBEA),* 1–4. 10.1109/ICCUBEA.2018.8697390

Kumari, S., Modani, V., Oswal, S., Sarma, S. S., & Kulkarni, P. (2019). An Android-Based Solution for Solar Power Harvesting and Irrigation Using IoT. In *International Conference on Intelligent Data Communication Technologies and Internet of Things (ICICI) 2018* (pp. 1136-1147). Springer International Publishing. 10.1007/978-3-030-03146-6_132

Kumari, T., Singh, R., & Verma, P. (2019). Agriculture, forest and environmental management trinity: towards environmental sustainability and climate change mitigation. In Sustainable Agriculture, Forest and Environmental Management. Springer.

Kumari, M., Kumar, V., Kaur, R., Kumar, S., & Sharma, R. (2022). Process Optimization for the Development of Nutritionally Enhanced Nuggets using *Ficus geniculata*: A Nutritional Approach. *Plant Foods for Human Nutrition (Dordrecht, Netherlands), 77*(2), 241–249. doi:10.100711130-022-00967-2 PMID:35583708

Kumar, R., Singh, M. P., Kumar, P., & Singh, J. P. (2015). Crop Selection Method to maximize crop yield rate using machine learning technique. *2015 International Conference on Smart Technologies and Management for Computing, Communication, Controls, Energy and Materials (ICSTM),* 138-145. 10.1109/ICSTM.2015.7225403

Kumbar, B., Galagi, B., shankar, B., & Honnalli, N. (2016). Smart irrigation system using internet of things. *Bonfring International Journal of Research in Communication Engineering, 6*(November), 2016. doi:10.9756/BIJRCE.8189

Kummu, M., De Moel, H., Porkka, M., Siebert, S., Varis, O., & Ward, P. J. (2012). Lost food, wasted resources: Global food supply chain losses and their impacts on freshwater, cropland, and fertiliser use. *The Science of the Total Environment, 438*, 477–489. doi:10.1016/j.scitotenv.2012.08.092 PMID:23032564

Kwakye, M. O., Mengistie, B., & Ofosu, J. (2019). Pesticide registration, distribution and use practices. *Environment, Development and Sustainability, 21*(6), 2647–2671. doi:10.100710668-018-0154-7

Lacko, A. M., Guilkey, D., Popkin, B., & Ng, S. W. (2022). Associations Among Select State Policies and the Nutritional Quality of Household Packaged Food Purchases in the United States from 2008 Through 2017. *Journal of the Academy of Nutrition and Dietetics, 122*(4), 731–744. doi:10.1016/j.jand.2021.09.017 PMID:34626825

Lajoie-O'Malley, A., Bronson, K., van der Burg, S., & Klerkx, L. (2020). The future (s) of digital agriculture and sustainable food systems: An analysis of high-level policy documents. *Ecosystem Services, 45*, 101183. doi:10.1016/j.ecoser.2020.101183

Lake, T. A., Briscoe Runquist, R. D., & Moeller, D. A. (2022). Deep learning detects invasive plant species across complex landscapes using Worldview-2 and Planetscope satellite imagery. *Remote Sensing in Ecology and Conservation, 8*(6), 875–889. doi:10.1002/rse2.288

Lambré, K. L. (2015). *FLONICAMID*. JMPR. WHO. Retrieved from https://apps.who.int/pesticide-residues-jmpr-database/Document/242

Lauridsen, M., Vejlgaard, B., Kovacs, I. Z., Nguyen, H., & Mogensen, P. (2017, March). Interference measurements in the European 868 MHz ISM band with focus on LoRa and SigFox. In *2017 IEEE Wireless Communications and Networking Conference (WCNC)* (pp. 1-6). IEEE. 10.1109/WCNC.2017.7925650

LaValle, S., Lesser, E., Shockley, R., Hopkins, M. S., & Kruschwitz, N. (2010). Big data, analytics and the path from insights to value. *MIT Sloan Management Review*.

Lavanya, G., & Rani, C., & GaneshKumar, P. (2020). An automated low cost IoT based Fertilizer Intimation System for smart agriculture. *Sustainable Computing: Informatics and Systems, 28,* 100300.

Lavric, A., Petrariu, A. I., & Popa, V. (2019). Long range sigfox communication protocol scalability analysis under large-scale, high-density conditions. *IEEE Access : Practical Innovations, Open Solutions, 7,* 35816–35825. doi:10.1109/ACCESS.2019.2903157

Lawal, M. O. (2021). Tomato detection based on modified YOLOv3 framework. *Scientific Reports, 11*(1), 1–11. doi:10.103841598-021-81216-5 PMID:33446897

Lee, C. P., Leng, F. T. J., Habeeb, R. A. A., Amanullah, M. A., & Rehman, M. H. (2022). Edge computing-enabled secure and energy-efficient smart parking: A review. *Microprocessors and Microsystems, 93,* 104612. doi:10.1016/j.micpro.2022.104612

Lee, H., & Moon, A. (2014). Development of yield prediction system based on real-time agricultural meteorological information.. *16th International Conference on Advanced Communication Technology,* 1292-1295. 10.1109/ICACT.2014.6779168

Lee, S., Choi, Y., & Ra, J. (2020). Impact of public cloud computing service in Korean government organizations. *ICIC Express Letters, 11*(3), 313–318. doi:10.24507/icicelb.11.03.313

Le, V. S., Lesueur, D., Herrmann, L., Hudek, L., Ngoc, L., & Lambert, Q. (2021). Sus- tainable tea production through agroecological management practices in Vietnam: A review. *Environ. Sustain, 4,* 589–604. Advance online publication. doi:10.100742398-021-00182-w

Liakos, K. G., Busato, P., Moshou, D., Pearson, S., & Bochtis, D. (2018). Machine learning in agriculture: A review. *Sensors (Basel), 18*(8), 2674. doi:10.339018082674 PMID:30110960

Liang, Y., Qi, G., Wei, K., & Chen, J. (2017). Exploring the determinant and influence mechanism of e-government cloud adoption in government agencies in China. *Government Information Quarterly, 34*(3), 481–495. doi:10.1016/j.giq.2017.06.002

Liang, Y., Qi, G., Zhang, X., & Li, G. (2019). The effects of e-Government cloud assimilation on public value creation: An empirical study of China. *Government Information Quarterly, 36*(4), 101397. Advance online publication. doi:10.1016/j.giq.2019.101397

Li, H., Tian, X., Wei, W., & Sun, C. (2012). A Deep Understanding of Cloud Computing Security. In J. Lei, F. L. Wang, M. Li, & Y. Luo (Eds.), *Network Computing and Information Security* (pp. 98–105). Springer. doi:10.1007/978-3-642-35211-9_13

Lin, G., Tang, Y., Zou, X., Xiong, J., & Fang, Y. (2020). Color-, depth-, and shape-based 3D fruit detection. *Precision Agriculture, 21*(1), 1–17.

Lin, J., Shen, Z., Zhang, A., & Chai, Y. (2018, July). Blockchain and IoT based food traceability for smart agriculture. In *Proceedings of the 3rd international conference on crowd science and engineering* (pp. 1-6). 10.1145/3265689.3265692

Lioutas, E. D., & Charatsari, C. (2018). Green innovativeness in farm enterprises: What makes farmers think green? *Sustainable Development (Bradford), 26*(4), 337–349. doi:10.1002d.1709

Lioutas, E. D., Charatsari, C., La Rocca, G., & De Rosa, M. (2019). Key questions on the use of big data in farming: An activity theory approach. *NJAS Wageningen Journal of Life Sciences, 90*(1), 100297. doi:10.1016/j.njas.2019.04.003

Liqiang, Z., Shouyi, Y., Leibo, L., Zhen, Z., & Shaojun, W. (2011). A crop monitoring system based on wireless sensor network. *Procedia Environmental Sciences, 11,* 558–565. doi:10.1016/j.proenv.2011.12.088

Liu, W., Hao, Z., Florkowski, W. J., Wu, L., & Yang, Z. (2022). A Review of the Challenges Facing Global Commercialization of the Artificial Meat Industry. *Foods*, *11*(22), 3609. doi:10.3390/foods11223609 PMID:36429201

Liu, Y., Li, D., Wan, S., Wang, F., Dou, W., Xu, X., Li, S., Ma, R., & Qi, L. (2022). A long short- term memory-based model for greenhouse climate prediction. *International Journal of Intelligent Systems*, *37*(1), 135–151. doi:10.1002/int.22620

Liu, Y., Ma, X., Shu, L., Hancke, G. P., & Abu-Mahfouz, A. M. (2020). From Industry 4.0 to Agriculture 4.0: Current status, enabling technologies, and research challenges. *IEEE Transactions on Industrial Informatics*, *17*(6), 4322–4334. doi:10.1109/TII.2020.3003910

Loudon, J. S. (2018). *Detecting and Localizing Cell Nuclei in Medical Images* [Master's thesis]. NTNU.

Luo, J., Huang, W., Wang, J., & Wei, C. (2009). The crop disease and pest warning and prediction system. In D. Li & C. Zhao (Eds.), *Computer and Computing Technologies in Agriculture II* (Vol. 2, pp. 937–945). Springer US. doi:10.1007/978-1-4419-0211-5_17

Lutu, P. E. (2018, December). Improving Classification Performance for Minority Classes through the use of Positive-Versus-Negative Ensemble Models. In *2018 International Conference on Intelligent and Innovative Computing Applications (ICONIC)* (pp. 1-6). IEEE. 10.1109/ICONIC.2018.8601260

Machekano, Masamba, Mvumi, & Nyamukondiwa. (2019). *Cabbage or 'pesticide' on the platter? Chemical analysis reveals multiple and excessive residues in African vegetable markets*. Academic Press.

Madumidha, S., Ranjani, P. S., Vandhana, U., & Venmuhilan, B. (2019, May). A theoretical implementation: Agriculture-food supply chain management using blockchain technology. In *2019 TEQIP III Sponsored International Conference on Microwave Integrated Circuits, Photonics and Wireless Networks (IMICPW)* (pp. 174-178). IEEE. 10.1109/IMICPW.2019.8933270

Madushanki, A. R., Halgamuge, M. N., Wirasagoda, W. S., & Syed, A. (2019). Adoption of the Internet of Things (IoT) in agriculture and smart farming towards urban greening: A review. *International Journal of Advanced Computer Science and Applications*, *10*(4), 11–28. doi:10.14569/IJACSA.2019.0100402

Maginga, T., Nsenga, J., Bakunzibake, P., & Masabo, E. (2022). Smallholder farmer-centric integration of IoT and Chatbot for early Maize diseases detection and management in pre-visual symptoms phase. *2022 IEEE Global Humanitarian Technology Conference (GHTC)*, 369–372. 10.1109/GHTC55712.2022.9911047

Mahama, S., Manteaw, S. A., & Decker, E. (2021). Gender perceptions on the causes of climate variation and its effects on cassava production among farmers in Ghana. *Cogent Food & Agriculture*, *7*(1), 1911398. doi:10.1080/23311932.2021.1911398

Mahlaule, S., van der Poll, J. A., & Ochola, E. O. (2022). Towards an Enhanced Framework to Facilitate Data Security in Cloud Computing. In M. Saraswat, H. Sharma, K. Balachandran, J. H. Kim, & J. C. Bansal (Eds.), *Congress on Intelligent Systems* (pp. 227–242). Springer Nature. doi:10.1007/978-981-16-9113-3_18

Mahmoud M. S., & Mohamad A. A. (2016). *A study of efficient power consumption wireless communication techniques/modules for internet of things (IoT) applications*. Academic Press.

Mahmudov, R. Sh. (2014). Analysis of the current state of the economy of cloud technologies and development prospects. *Problems of the Information Society*, *2*, 14-23. www.microtech.az

Ma, J., Du, K., Zheng, F., Zhang, L., Gong, Z., & Sun, Z. (2018). A recognition method for cucumber diseases using leaf symptom images based on deep convolutional neural networks. *Computers and Electronics in Agriculture*, *154*, 18–24.

Malathi, V., & Gopinath, M. P. (2023). A Review on Rice Crop Disease Classification Using Computational Approach. *International Journal of Image and Graphics*, *23*(3), 2240007. doi:10.1142/S0219467822400071

Mallik, P., & Ghosh, T. (2008). Impact of climate on tea production: A study of the Dooars region in India. *Theor. Apple. Climatol., no. Cline*. Advance online publication. doi:10.100700704-021-03848-x

Mall, P. K., Pramanik, S., Srivastava, S., Faiz, M., Sriramulu, S., & Kumar, M. N. (2023). FuzztNet-Based Modelling Smart Traffic System in Smart Cities Using Deep Learning Models. In *Data-Driven Mathematical Modeling in Smart Cities*. IGI Global. doi:10.4018/978-1-6684-6408-3.ch005

Malone, T., Schaefer, K. A., & Wu, F. (2021). The razor's edge of "essential" labor in food and agriculture. *Applied Economic Perspectives and Policy*, *43*(1), 368–381. doi:10.1002/aepp.13070

Manoharan, S. (2020). Supervised Learning for Microclimatic parameter Estimation in a Greenhouse environment for productive Agronomics. *Journal of Artificial Intelligence*, *2*(03), 170–176.

Marcelis, L. F. M., Maas, F. M., & Heuvelink, E. (2000, November). The latest developments in the lighting technologies in Dutch horticulture. In *IV International ISHS Symposium on Artificial Lighting 580* (pp. 35-42). Academic Press.

Marr, B. (2017, October). How Much Data Do We Create Every Day? The Mind-Blowing Stats Everyone Should Read. *Forbes*.

Masip-Bruin, X., Souza, V. B., Marín-Tordera, E., Ren, G.-J., Jukan, A., & Garcia, J. (2020). A Multidimensional Control Architecture for Combined Fog-to-Cloud Systems. In K. Arai, S. Kapoor, & R. Bhatia (Eds.), *Advances in Information and Communication* (pp. 281–299). Springer International Publishing. doi:10.1007/978-3-030-39445-5_22

Masroor, R., Naeem, M., & Ejaz, W. (2021). Resource management in UAV-assisted wireless networks: An optimization perspective. *Ad Hoc Networks*, *121*, 102596. doi:10.1016/j.adhoc.2021.102596

Mattern, F., & Floerkemeier, C. (2010). *From the Internet of Computers to the Internet of Things. In From active data management to event-based systems and more*. Springer.

Matthews, G. (2001). The Pesticide Manual. Crop Protection -. *Crop Protection (Guildford, Surrey)*, *20*(6), 539–539. doi:10.1016/S0261-2194(01)00019-9

McGreevy, S. R., Rupprecht, C. D., Niles, D., Wiek, A., Carolan, M., Kallis, G., Kantamaturapoj, K., Mangnus, A., Jehlička, P., Taherzadeh, O., Sahakian, M., Chabay, I., Colby, A., Vivero-Pol, J.-L., Chaudhuri, R., Spiegelberg, M., Kobayashi, M., Balázs, B., Tsuchiya, K., ... Tachikawa, M. (2022). Sustainable agrifood systems for a post-growth world. *Nature Sustainability*, *5*(12), 1011–1017. doi:10.103841893-022-00933-5

McMichael, P. (2021). Political economy of the global food and agriculture system. In *Rethinking food and agriculture* (pp. 53–75). Woodhead Publishing. doi:10.1016/B978-0-12-816410-5.00003-7

Mdegela, R. H., Mwakapeje, E. R., Rubegwa, B., Gebeyehu, D. T., Niyigena, S., Msambichaka, V., Nonga, H. E., Antoine-Moussiaux, N., & Fasina, F. O. (2021). Antimicrobial use, residues, resistance and governance in the food and agriculture sectors, Tanzania. *Antibiotics (Basel, Switzerland)*, *10*(4), 454. doi:10.3390/antibiotics10040454 PMID:33923689

Mdemu, M., Kissoly, L., Bjornlund, H., Kimaro, E., Christen, E. W., van Rooyen, A., ... Ramshaw, P. (2020). The role of soil water monitoring tools and agricultural innovation platforms in improving food security and income of farmers in smallholder irrigation schemes in Tanzania. *International Journal of Water Resources Development*, *36*(sup1), S148-S170.

Mehra, N., Aggarwal, S., Shokeen, A., & Bura, D. (2018). Analyzing Cloud Computing Security Issues and Challenges. In P. K. Pattnaik, S. S. Rautaray, H. Das, & J. Nayak (Eds.), *Progress in Computing, Analytics and Networking* (pp. 193–202). Springer. doi:10.1007/978-981-10-7871-2_19

Mercier-Laurent, E. (2022). Greening and Smarting IT – Case of Digital Transformation. In E. Mercier-Laurent & G. Kayakutlu (Eds.), *Artificial Intelligence for Knowledge Management, Energy, and Sustainability* (pp. 1–18). Springer International Publishing. doi:10.1007/978-3-030-96592-1_1

Merck, A. W., Grieger, K. D., Cuchiara, M., & Kuzma, J. (2022). What Role Does Regulation Play in Responsible Innovation of Nanotechnology in Food and Agriculture? Insights and Framings from US Stakeholders. *Bulletin of Science, Technology & Society*, *42*(3), 85–103. doi:10.1177/02704676221102066

Merec. (2022). *safety data sheet Indoxacarb*. Retrieved from https://www.merck.com/docs/product/safety-data-sheets/ah-sds/Indoxacarb%20Formulation_AH_CA_EN.pdf

Meshram, R. A., & Alvi, A. S. (2022). Crop Disease Prediction Using Computational Machine Learning Model. In J. S. Raj, Y. Shi, D. Pelusi, & V. E. Balas (Eds.), *Intelligent Sustainable Systems* (pp. 541–550). Springer Nature. doi:10.1007/978-981-19-2894-9_40

Min, H. (2010). Artificial intelligence in supply chain management: Theory and applications. *International Journal of Logistics*, *13*(1), 13–39. doi:10.1080/13675560902736537

Mir, H., Zaraatgari, R., & Sotoudeh, R. (2021). Improving the food and agriculture sector tehran stock exchange by using artificial intelligence. *Agricultural Marketing and Commercialization Journal*, *5*(2), 90–114.

Mishra, S., Kumar, M., Singh, N., & Dwivedi, S. (2022). A Survey on AWS Cloud Computing Security Challenges & Solutions. *2022 6th International Conference on Intelligent Computing and Control Systems (ICICCS)*, 614–617. 10.1109/ICICCS53718.2022.9788254

Mishra, S., & Sharma, S. K. (2023). Advanced contribution of IoT in agricultural production for the development of smart livestock environments. *Internet of Things*, *22*, 100724. doi:10.1016/j.iot.2023.100724

Miśniakiewicz, M., Łuczak, J., & Maruszewska, N. (2021). Improvement of organic farm assessment procedures on the example of organic farming in poland—Recommendations for organic farming in Poland. *Agronomy (Basel)*, *11*(8), 1560. doi:10.3390/agronomy11081560

Misra, N. N., Dixit, Y., Al-Mallahi, A., Bhullar, M. S., Upadhyay, R., & Martynenko, A. (2020). IoT, big data, and artificial intelligence in agriculture and food industry. *IEEE Internet of Things Journal*, *9*(9), 6305–6324. doi:10.1109/JIOT.2020.2998584

Mittelstadt, B. D., & Floridi, L. (2016). The ethics of big data: current and foreseeable issues in biomedical contexts. *The Ethics of Biomedical Big Data*, 445-480.

Mohagheghi, A., & Moallem, M. (2023). An Energy-Efficient PAR-Based Horticultural Lighting System for Greenhouse Cultivation of Lettuce. *IEEE Access : Practical Innovations, Open Solutions*, *11*, 8834–8844. doi:10.1109/ACCESS.2023.3237757

Mohammed, F., Ibrahim, O., Nilashi, M., & Alzurqa, E. (2017). Cloud computing adoption model for e-government implementation. *Information Development*, *33*(3), 303–323. doi:10.1177/0266666916656033

Mohammed, F., Olayah, F., Ali, A., & Gazem, N. A. (2020). The effect of cloud computing adoption on the sustainability of e-government services: A review. *International Journal of Advanced Science and Technology*, *29*(5), 2636–2642.

Mohanty, S. P., Hughes, D. P., & Salathé, M. (2016). Using deep learning for image-based plant disease detection. *Frontiers in Plant Science*, *7*, 1419. doi:10.3389/fpls.2016.01419 PMID:27713752

Mohri, M., Rostamizadeh, A., & Talwalkar, A. (2018). *Foundations of machine learning*. MIT press.

Mohr, S., & Kühl, R. (2021). Acceptance of artificial intelligence in German agriculture: An application of the technology acceptance model and the theory of planned behavior. *Precision Agriculture, 22*(6), 1816–1844. doi:10.100711119-021-09814-x

Morbekar, A., Parihar, A., & Jadhav, R. (2020). Crop Disease Detection Using YOLO. *2020 International Conference for Emerging Technology (INCET),* 1–5. 10.1109/INCET49848.2020.9153986

Morrow, R. C. (2008). LED lighting in horticulture. *HortScience, 43*(7), 1947–1950. doi:10.21273/HORTSCI.43.7.1947

Mueller, R. A. (2000). *Emergent e-commerce in agriculture (No. 14).* University of California, Agricultural Issues Center.

Mullachery, B., & Alismail, S. (2023). A Smart Healthcare Framework: Opportunities for Integrating Emerging Technologies (5G, IoT, AI, and GIS). In K. Arai (Ed.), *Proceedings of the Future Technologies Conference (FTC) 2022, Volume 3* (pp. 325–340). Springer International Publishing. 10.1007/978-3-031-18344-7_22

Muñoz, M., Guzmán, J. L., Sánchez-Molina, J. A., Rodríguez, F., Torres, M., & Berenguel, M. (2020). A new IoT-based platform for greenhouse crop production. *IEEE Internet of Things Journal, 9*(9), 6325–6334. doi:10.1109/JIOT.2020.2996081

Myers, S. S., Smith, M. R., Guth, S., Golden, C. D., Vaitla, B., Mueller, N. D., Dangour, A. D., & Huybers, P. (2017). Climate change and global food systems: Potential impacts on food security and undernutrition. *Annual Review of Public Health, 38*(1), 259–277. doi:10.1146/annurev-publhealth-031816-044356 PMID:28125383

N, B., K, P., S, M., S, H., M, V. K., & MRM, V. (2022). A Novel Framework for Cyber Security Attacks on Cloud-Based Services. *2022 Fourth International Conference on Cognitive Computing and Information Processing (CCIP),* 1–4. doi:10.1109/CCIP57447.2022.10058673

Na, A., Isaac, W., Varshney, S., & Khan, E. (2016, October). An IoT based system for remote monitoring of soil characteristics. In *2016 International conference on information technology (InCITe)-the next generation IT summit on the theme-internet of things: Connect your Worlds* (pp. 316-320). IEEE. 10.1109/INCITE.2016.7857638

Nacef, A., Kaci, A., Aklouf, Y., & Dutra, D. L. C. (2022). Machine learning based fast self optimized and life cycle management network. *Computer Networks, 209,* 108895. doi:10.1016/j.comnet.2022.108895

Naik, G., & Suresh, D. N. (2018). Challenges of creating sustainable agri-retail supply chains. *IIMB Management Review, 30*(3), 270–282. doi:10.1016/j.iimb.2018.04.001

Nair K. K., Abu-Mahfouz A. M., & Lefophane S. (2019, March). Analysis of the narrow band internet of things (NB-IoT) technology. In *2019 conference on information communications technology and society (ICTAS)* (pp. 1-6). IEEE.

Nasution, I. S., Iskandar, M. R., & Jayanti, D. S. (2020). Internet of things: Automatic sprinklers in prototyping greenhouse using smartphone-based android. *IOP Conference Series. Earth and Environmental Science, 425*(1), 012069. doi:10.1088/1755-1315/425/1/012069

National Artificial Intelligence Strategy 2021-2025. (2021). https://cbddo.gov.tr/en/nais

National Center for Biotechnology Information. (2023). *PubChem Compound Summary for CID 107720, Indoxacarb.* Retrieved June 14, 2023 from https://pubchem.ncbi.nlm.nih.gov/compound/Indoxacarb

National Institute for Public Health and the Environment. (2015). *CLH report for spirodiclofen.* Retrieved from https://echa.europa.eu/documents/10162/41b26907-3ca7-4059-84f2-4694826f3598

Nennuri, R., Kumar, R. H., Prathyusha, G., Tejaswini, K., Kanishka, G., & Sunitha, G. (2023). A Multi-stage Deep Model for Crop Variety and Disease Prediction. In A. Abraham, T. Hanne, N. Gandhi, P. Manghirmalani Mishra, A. Bajaj, & P. Siarry (Eds.), *Proceedings of the 14th International Conference on Soft Computing and Pattern Recognition (SoCPaR 2022)* (pp. 52–59). Springer Nature Switzerland. 10.1007/978-3-031-27524-1_6

Nguyen, N. T. T., Nguyen, L. M., Nguyen, T. T. T., Liew, R. K., Nguyen, D. T. C., & Van Tran, T. (2022). Recent advances on botanical biosynthesis of nanoparticles for catalytic, water treatment and agricultural applications: A review. *The Science of the Total Environment*, *827*, 154160. doi:10.1016/j.scitotenv.2022.154160 PMID:35231528

Nigussie, E., Olwal, T., Musumba, G., Tegegne, T., Lemma, A., & Mekuria, F. (2020). IoT-based irrigation management for smallholder farmers in rural sub-Saharan Africa. *Procedia Computer Science*, *177*, 86–93. doi:10.1016/j.procs.2020.10.015

Nirgude, V., & Rathi, S. (2021). An Extensive Survey on Deep Learning Approach for Crop Diseases Management. In A. Abraham, T. Hanne, O. Castillo, N. Gandhi, T. Nogueira Rios, & T.-P. Hong (Eds.), *Hybrid Intelligent Systems* (pp. 638–648). Springer International Publishing. doi:10.1007/978-3-030-73050-5_62

Niwa, H. (2021). Detection of organic tea farms based on the density of spider webs using aerial photography with an unmanned aerial vehicle (UAV). *Landscape and Ecological Engineering*, *17*(4), 541–546. doi:10.100711355-021-00454-x

Novick, K. A., Ficklin, D. L., Baldocchi, D., Davis, K. J., Ghezzehei, T. A., Konings, A. G., MacBean, N., Raoult, N., Scott, R. L., Shi, Y., Sulman, B. N., & Wood, J. D. (2022). Confronting the water potential information gap. *Nature Geoscience*, *15*(3), 158–164. doi:10.103841561-022-00909-2 PMID:35300262

Nwobodo, I. (2016). Cloud Computing: Models, Services, Utility, Advantages, Security Issues, and Prototype. In Q.-A. Zeng (Ed.), *Wireless Communications, Networking and Applications* (pp. 1207–1222). Springer India. doi:10.1007/978-81-322-2580-5_110

OECD. (2017). *Going digital: Making the transformation work for growth and well-being.* Meeting of the OECD Council at Ministerial Level, Paris, France.

OECD. (2019). *Technology and digital in agriculture.* Retrieved from: https://www.oecd.org/agriculture/topics/technology-and-digital-agriculture/

Office of Environmental Health Hazard Assessment. (2011). *CIC Consultation: Flonicamid.* Retrieved from https://oehha.ca.gov/media/downloads/crnr/101211flonicamidcic.pdf

Oh, T. H., Lim, S., Choi, B. Y., Park, K.-R., Lee, H., & Choi, H. (2010). State of the Art of Network Security Perspectives in Cloud Computing. In T. Kim, A. Stoica, & R.-S. Chang (Eds.), Security-Enriched Urban Computing and Smart Grid (pp. 629–637). Springer. doi:10.1007/978-3-642-16444-6_79

Ojha, T., Misra, S., & Raghuwanshi, N. S. (2015). Wireless sensor networks for agriculture: The state-of-the-art in practice and future challenges. *Computers and Electronics in Agriculture*, *118*, 66–84. doi:10.1016/j.compag.2015.08.011

Padarian, J., Minasny, B., & McBratney, A. B. (2015). Using Google's cloud-based platform for digital soil mapping. *Computers & Geosciences*, *83*, 80–88. doi:10.1016/j.cageo.2015.06.023

Pal, K. (2019). Bio-control of Pests in Tea: Effect of Environmental. *International Journal of Applied and Computational Mathematics*, *5*(3), 1–9. doi:10.100740819-019-0666-3

Paradiso, R., & Proietti, S. (2022). Light-quality manipulation to control plant growth and photomorphogenesis in greenhouse horticulture: The state of the art and the opportunities of modern LED systems. *Journal of Plant Growth Regulation, 41*(2), 742–780. doi:10.100700344-021-10337-y

Pardo, L. A., & ... *Pesticide exposure and risk of aggressive prostate cancer among private pesticide applicators.* doi:10.118612940-020-00583-0

Park, J. S., Yi, K. J., & Park, J. H. (2011). SSP-MCloud: A Study on Security Service Protocol for Smartphone Centric Mobile Cloud Computing. In J. J. Park, H. Arabnia, H.-B. Chang, & T. Shon (Eds.), *IT Convergence and Services* (pp. 165–172). Springer Netherlands. doi:10.1007/978-94-007-2598-0_18

Patidar, R., Agrawal, S., & Pratap, S. (2018). Development of novel strategies for designing sustainable Indian agri-fresh food supply chain. *Sadhana, 43*(10), 1–16. doi:10.100712046-018-0927-6

Patil, M. A., & Manohar, M. (2023). A Comprehensive Review on Crop Disease Prediction Based on Machine Learning and Deep Learning Techniques. In S. Kumar, H. Sharma, K. Balachandran, J. H. Kim, & J. C. Bansal (Eds.), *Third Congress on Intelligent Systems* (pp. 481–503). Springer Nature. 10.1007/978-981-19-9225-4_36

Patil, V. C., Al-Gaadi, K. A., Biradar, D. P., & Rangaswamy, M. (2012). Internet of things (Iot) and cloud computing for agriculture: An overview. Proceedings of agro-informatics and precision agriculture (AIPA 2012), 292-296.

Patwa, D., & Bharat, T. V. (2023). Influence of Hydraulic Response Time of Tensiometer in Hydraulic Characteristics Estimation for Riverbank Sand. *Geotechnical and Geological Engineering, 41*(1), 413–427. doi:10.100710706-022-02283-7

Paul, M. T., Vishwakarma, S. K., & Verma, A. (2015). Analysis of Soil Behaviour and Prediction of Crop Yield Using Data Mining Approach. *2015 International Conference on Computational Intelligence and Communication Networks (CICN)*, 766-771. 10.1109/CICN.2015.156

Pelletier, N., & Tyedmers, P. (2010). Forecasting potential global environmental costs of livestock production 2000–2050. *Proceedings of the National Academy of Sciences of the United States of America, 107*(43), 18371–18374. doi:10.1073/pnas.1004659107 PMID:20921375

Peng, J. (2022). Research on E-government Information Security Based on Cloud Computing. *2022 IEEE 10th Joint International Information Technology and Artificial Intelligence Conference (ITAIC), 10*, 312–316. 10.1109/ITAIC54216.2022.9836548

Pesticide and You. (2012). *Chemical Watch Factsheet.* Retrieved 2023, from https://www.beyondpesticides.org/assets/media/documents/pesticides/factsheets/Indoxacarb.pdf

Pesticide Directory. (n.d.). *Adverse Effects Flonicamid.* Retrieved from https://www.fluoridealert.org/wp-content/pesticides/epage.flonicamid.effects.htm

Phasinam, K., Kassanuk, T., Shinde, P. P., Thakar, C. M., Sharma, D. K., Mohiddin, S. K., & Rahmani, A. W. (2022). Application of IoT and Cloud Computing in Automation of Agriculture Irrigation. *Journal of Food Quality, 2022*, 1–8. doi:10.1155/2022/8285969

Phillips-Wren, G., & Hoskisson, A. (2015). An analytical journey towards big data. *Journal of Decision Systems, 24*(1), 87–102. doi:10.1080/12460125.2015.994333

Phua, K. h., Wang, C. m., Yeong, W. y., Leong, T. y., Loh, H. t., Tan, K. c., & Chou, F. s. (1992). Optimization Techniques and Applications. World Scientific. doi:10.1142/9789814537285

Pietrzyck, K., Berke, N., Wendel, V., Steinhoff-Wagner, J., Jarzębowski, S., & Petersen, B. (2021). Understanding the Importance of International Quality Standards Regarding Global Trade in Food and Agricultural Products: Analysis of the German Media. *Agriculture*, *11*(4), 328. doi:10.3390/agriculture11040328

Pingli, G., Yanlei, S., Junliang, C., Miaoting, D., & Bojia, L. (2011). Enterprise- oriented communication among multiple esbs based on wsnotification and cloud queue model. *International Journal of Advancements in Computing Technology*, *3*(7), 255–263. doi:10.4156/ijact.vol3.issue7.30

Pinto, L. S., Ray, A., Reddy, M. U., Perumal, P., & Aishwarya, P. (2016). Crop disease classification using texture analysis. *2016 IEEE International Conference on Recent Trends in Electronics, Information & Communication Technology (RTEICT)*, 825–828. 10.1109/RTEICT.2016.7807942

Poggio, L., De Sousa, L. M., Batjes, N. H., Heuvelink, G., Kempen, B., Ribeiro, E., & Rossiter, D. (2021). SoilGrids 2.0: Producing soil information for the globe with quantified spatial uncertainty. *Soil (Göttingen)*, *7*(1), 217–240. doi:10.5194oil-7-217-2021

Popescu, D., Stoican, F., Stamatescu, G., Ichim, L., & Dragana, C. (2020). Advanced UAV–WSN system for intelligent monitoring in precision agriculture. *Sensors (Basel)*, *20*(3), 817. doi:10.339020030817 PMID:32028736

Prabha, K. (2021). Disease sniffing robots to apps fixing plant diseases: Applications of artificial intelligence in plant pathology—a mini review. *Indian Phytopathology*, *74*(1), 13–20. doi:10.100742360-020-00290-3

Pramanik, S. (2022). An Effective Secured Privacy-Protecting Data Aggregation Method in IoT. In M. O. Odhiambo, W. Mwashita, & I. G. I. Global (Eds.), *Achieving Full Realization and Mitigating the Challenges of the Internet of Things*. doi:10.4018/978-1-7998-9312-7.ch008

Prasad, K., Roy, S., Sen, S., Neave, S., Nagpal, A., & Pandit, V. (2020). *Impact of different pest management practices on natural enemy population in tea plantations of Assam special emphasis on spider fauna*. doi:10.100742690-020-00111-0

Prasath, B., & Akila, M. (2023). IoT-based pest detection and classification using deep features with enhanced deep learning strategies. *Engineering Applications of Artificial Intelligence*, *121*, 105985. doi:10.1016/j.engappai.2023.105985

Preti, M., Verheggen, F., & Angeli, S. (2021). Insect pest monitoring with camera-equipped traps: Strengths and limitations. *Journal of Pest Science*, *94*(2), 203–217. doi:10.100710340-020-01309-4

Pretty, J. N. (1998). Supportive policies and practice for scaling up sustainable agriculture. *Facilitating sustainable agriculture: Participatory learning and adaptive management in times of environmental uncertainty*, 23-45.

Public Service ICT Strategy. (2015). https://ictstrategy.per.gov.ie

Pudumalar, S., Ramanujam, E., Rajashree, R. H., Kavya, C., Kiruthika, T., & Nisha, J. (2017). Crop recommendation system for precision agriculture. *2016 Eighth International Conference on Advanced Computing (ICoAC)*, 32-36. 10.1109/ICoAC.2017.7951740

Qazi, S., Khawaja, B. A., & Farooq, Q. U. (2022). IoT-equipped and AI-enabled next generation smart agriculture: A critical review, current challenges and future trends. *IEEE Access : Practical Innovations, Open Solutions*, *10*, 21219–21235. doi:10.1109/ACCESS.2022.3152544

Qianjiang. (2015). *Demand for Pesticides in China to Reach*. Academic Press.

Qian, L., Luo, Z., Du, Y., & Guo, L. (2009). Cloud Computing: An Overview. In M. G. Jaatun, G. Zhao, & C. Rong (Eds.), Lecture Notes in Computer Science: Vol. 5931. *Cloud Computing. CloudCom 2009*. Springer. doi:10.1007/978-3-642-10665-1_63

Qureshi, I.Z., Bibi, A., Shahid, S., & Ghazanfar, M. (2016). Exposure to sub-acute doses of fipronil and buprofezin in combination or alone induces biochemical, hematological, histopathological and genotoxic damage in common carp (Cyprinus carpio L.). *Aquat Toxicol., 179*, 103-14. doi:10.1016/j.aquatox.2016.08.012

Quy, V. K., Hau, N. V., Anh, D. V., Quy, N. M., Ban, N. T., Lanza, S., Randazzo, G., & Muzirafuti, A. (2022). IoT-enabled smart agriculture: Architecture, applications, and challenges. *Applied Sciences (Basel, Switzerland), 12*(7), 3396. doi:10.3390/app12073396

Rabbani, M., Foroozesh, N., Mousavi, S. M., & Farrokhi-Asl, H. (2019). Sustainable supplier selection by a new decision model based on interval-valued fuzzy sets and possibilistic statistical reference point systems under uncertainty. *International Journal of Systems Science: Operations & Logistics, 6*(2), 162–178.

Rabbani, M., Hosseini-Mokhallesun, S. A. A., Ordibazar, A. H., & Farrokhi-Asl, H. (2020). A hybrid robust possibilistic approach for a sustainable supply chain location-allocation network design. *International Journal of Systems Science: Operations & Logistics, 7*(1), 60–75.

Rai, B. K., Sharma, S., Kumar, G., & Kishor, K. (2022). Recognition of Different Bird Category Using Image Processing. *International Journal of Online and Biomedical Engineering, 18*(07), 101–114. doi:10.3991/ijoe.v18i07.29639

Raj, E. F. I., Appadurai, M., & Athiappan, K. (2022). Precision farming in modern agriculture. In Smart Agriculture Automation Using Advanced Technologies: Data Analytics and Machine Learning, Cloud Architecture, Automation and IoT (pp. 61–87). Springer.

Raj, A., Jain, N., & Chauhan, S. S. (2021). Mapping of Security Issues and Concerns in Cloud Computing with Compromised Security Attributes. In R. Agrawal, G. Sanyal, K. Curran, V. E. Balas, & M. S. Gaur (Eds.), *Cybersecurity in Emerging Digital Era* (pp. 24–40). Springer International Publishing. doi:10.1007/978-3-030-84842-2_2

Rajagopal G., Lodd V. M., Vignesh A., Rajesh, R. & Vijayaraghavan V. (2014, August). Low cost cloud based intelligent farm automation system using Bluetooth low energy. In *2014 IEEE Region 10 Humanitarian Technology Conference (R10 HTC)* (pp. 127-132). IEEE.

Ramakrishnam Raju, S. V. S., Dappuri, B., Ravi Kiran Varma, P., Yachamaneni, M., Verghese, D. M. G., & Mishra, M. K. (2022). Design and implementation of smart hydroponics farming using iot-based ai controller with mobile application system. *Journal of Nanomaterials, 2022*, 1–12. doi:10.1155/2022/4435591

Rastogi, A. (2010). Application of cloud computing for the better e-governance in developing countries. *Oriental Journal of Computer Science & Technology, 3*(2), 343–348.

Rathod, M. L., Shivaputra, A., Umadevi, H., Nagamani, K., & Periyasamy, S. (2022). Cloud Computing and Networking for SmartFarm AgriTech. *Journal of Nanomaterials, 6491747*, 1–7. Advance online publication. doi:10.1155/2022/6491747

Rathor, S., & Kumari, S. (2021, October). Smart agriculture system using iot and cloud computing. In *2021 5th International Conference on Information Systems and Computer Networks (ISCON)* (pp. 1-4). IEEE. 10.1109/ISCON52037.2021.9702345

Reepu, K. S., Chaudhary, M. G., Gupta, K. G., Pramanik, S., & Gupta, A. (2023). Information Security and Privacy on IoT. In Handbook of Research in Advancements in AI and IoT Convergence Technologies. IGI Global.

Reese, D. (2011). *Cloud Computing*. BHV-Petersburg.

Rehab, A., & Azouz, H. O.-Z. (2020). Buprofezin toxication implicates health hazards in Nile tilapia (Oreochromis niloticus). *Aquaculture Research*. doi:10.1111/are.14884

Rehman, H. U. (2022, December 19). *5 Most Advanced Countries in Agriculture Technology*. Insider Monkey. https://insidermonkey.com/blog/5-most-advanced-countries-in-agriculture-technology-1098530/

Rejeb, A., Keogh, J. G., Zailani, S., Treiblmaier, H., & Rejeb, K. (2020). Blockchain technology in the food industry: A review of potentials, challenges and future research directions. *Logistics*, *4*(4), 27. doi:10.3390/logistics4040027

Ren, S., He, K., Girshick, R., & Sun, J. (2017). Faster r-cnn: Towards real-time object detection with region proposal networks. *IEEE Transactions on Pattern Analysis and Machine Intelligence, 39*(6), 1137 – 1149.

Reporter, S. (2022, September 6). *Cropin launches the world's first purpose-built industry cloud for agriculture*. Farmers Review Africa. https://farmersreviewafrica.com/cropin-launches-the-worlds-first-purpose-built-industry-cloud-for-agriculture-cropin-cloud/

Reshma, R., Sathiyavathi, V., Sindhu, T., & Selvakumar, K., & SaiRamesh, L. (2020, October). IoT based classification techniques for soil content analysis and crop yield prediction. In *2020 Fourth International Conference on I-SMAC (IoT in Social, Mobile, Analytics and Cloud)(I-SMAC)* (pp. 156-160). IEEE. 10.1109/I-SMAC49090.2020.9243600

Reynolds, J. (2003). *Going Wi-Fi: Networks Untethered with 802.11 Wireless Technology*. CRC Press. doi:10.1201/9781482280869

Ribi Forclaz, A. (2019). From reconstruction to development: The early years of the Food and Agriculture Organization (FAO) and the conceptualization of rural welfare, 1945–1955. *The International History Review*, *41*(2), 351–371. doi:10.1080/07075332.2018.1478873

Roadmap, I. C. T. (2016). *Strategic Roadmap for the development of telecommunications and information technologies.* Decree of the President of the Republic of Azerbaijan. https://president.az/az/articles/view/22382

Rockström, J., Williams, J., Daily, G., Noble, A., Matthews, N., Gordon, L., Wetterstrand, H., DeClerck, F., Shah, M., Steduto, P., de Fraiture, C., Hatibu, N., Unver, O., Bird, J., Sibanda, L., & Smith, J. (2017). Sustainable intensification of agriculture for human prosperity and global sustainability. *Ambio*, *46*(1), 4–17. doi:10.100713280-016-0793-6 PMID:27405653

Rosendo, D., Costan, A., Valduriez, P., & Antoniu, G. (2022). Distributed intelligence on the Edge-to-Cloud Continuum: A systematic literature review. *Journal of Parallel and Distributed Computing*, *166*, 71–94. doi:10.1016/j.jpdc.2022.04.004

Roy, A., Sehrawat, D., & Singh, J. (2023). Crop Disease Prediction Using Deep ConvNet Architecture Technique. In D. Goyal, A. Kumar, V. Piuri, & M. Paprzycki (Eds.), *Proceedings of the Third International Conference on Information Management and Machine Intelligence* (pp. 187–195). Springer Nature. 10.1007/978-981-19-2065-3_22

Ruan, D., Abderrahim, H. A., D'hondt, P., & Kerre, E. E. (1998). Fuzzy Logic and Intelligent Technologies for Nuclear Science and Industry. World Scientific. doi:10.1142/9789814528368

Rupanagudi, S. R., Ranjani, B. S., Nagaraj, P., Bhat, V. G., & Thippeswamy, G. (2015). A novel cloud computing based smart farming system for early detection of borer insects in tomatoes. In *2015 International Conference on Communication, Information & Computing Technology (ICCICT)*. IEEE. 10.1109/ICCICT.2015.7045722

Ryan, M., van der Burg, S., & Bogaardt, M. J. (2021). Identifying key ethical debates for autonomous robots in agri-food: A research agenda. *AI and Ethics*, 1–15.

S., L. E. (2009). *Potential Health Effects of pesticides.* Pesticide Education Program in Penn State's College of Agricultural Sciences. Retrieved 2023, from https://extension.psu.edu/potential-health-effects-of-pesticides

Saadane, R., Chehri, A., & Jeon, S. (2022). AI-based modeling and data-driven evaluation for smart farming-oriented big data architecture using IoT with energy harvesting capabilities. *Sustainable Energy Technologies and Assessments*, *52*, 102093. doi:10.1016/j.seta.2022.102093

Saban, M., Aghzout, O., & Rosado-Muñoz, A. (2022). Deployment of a LoRa-based Network and Web Monitoring Application for a Smart Farm. *2022 IEEE International Workshop on Metrology for Industry 4.0 & IoT (MetroInd4.0&IoT)*, 424-427. 10.1109/MetroInd4.0IoT54413.2022.9831521

Sadowski, S., & Spachos, P. (2018, November). Solar-powered smart agricultural monitoring system using internet of things devices. In *2018 IEEE 9th Annual Information Technology, Electronics and Mobile Communication Conference (IEMCON)* (pp. 18-23). IEEE. 10.1109/IEMCON.2018.8614981

Sadu, V. B., Budidha, N., Billewar, S., Pimo, S. J., Kumar, D. R., & Kumar, T. C. A. (2022, November). Digitalization of agriculture by using the application of IoT and robotics. In *2022 International Interdisciplinary Humanitarian Conference for Sustainability (IIHC)* (pp. 183-186). IEEE. 10.1109/IIHC55949.2022.10060724

Saeed, Z., Raza, A., Qureshi, A. H., & Haroon Yousaf, M. (2021). A Multi-Crop Disease Detection and Classification Approach using CNN. *2021 International Conference on Robotics and Automation in Industry (ICRAI)*, 1–6. 10.1109/ICRAI54018.2021.9651409

SAgric-IoT. (1961). An IoT-Based Platform and Deep Learning for Greenhouse Monitoring. *Applied Sciences (Basel, Switzerland)*, *13*(3).

Saha, T., Jewel, M. K. H., Mostakim, M. N., Bhuiyan, N. H., Ali, M. S., Rahman, M. K., ... Hossain, M. K. (2017). Construction and development of an automated greenhouse system using Arduino Uno. *International Journal of Information Engineering and Electronic Business*, *9*(3), 1. doi:10.5815/ijieeb.2017.03.01

Salimov, S. M. (2022). Comparative analysis and evaluation of ecological indicators of the Republic of Azerbaijan. Baku. "Science and Knowledge".

Samanta, D., Dutta, S., Galety, M. G., & Pramanik, S. (2021). A Novel Approach for Web Mining Taxonomy for High-Performance Computing. *The 4th International Conference of Computer Science and Renewable Energies (ICCSRE'2021)*. 10.1051/e3sconf/202129701073

Sambasivam, G., & Opiyo, G. D. (2021). A predictive machine learning application in agriculture: Cassava disease detection and classification with imbalanced dataset using convolutional neural networks. *Egyptian Informatics Journal*, *22*(1), 27–34.

Sanger, A. K. S., & Johari, R. (2022). Survey of Security Issues in Cloud. *2022 International Mobile and Embedded Technology Conference (MECON)*, 490–493. 10.1109/MECON53876.2022.9751959

Santoshkumar, K. C., & Chavhan, S. (2012). Development of wireless sensor node to monitor poultry farm. *Communications in Computer and Information Science*, *296*, 27–32. doi:10.1007/978-3-642-35864-7_4

Santur, Y., Karaköse, M., & Akin, E. (2016, September). Improving of personal educational content using big data approach for mooc in higher education. In *2016 15th International Conference on Information Technology Based Higher Education and Training (ITHET)* (pp. 1-4). IEEE. 10.1109/ITHET.2016.7760728

Saravanan, K., & Saraniya, S. (2017). Cloud IOT based novel livestock monitoring and identification system using UID. *Sensor Review*, *38*(1), 21–33.

Sari, I. A., Handayani, A. N., & Lestari, D. (2018, October). Smart greenhouse sebagai media pembibitan kentang granola kembang berbasis mikrokontroler. In Prosiding Seminar Nasional Teknologi Elektro Terapan (Vol. 2, No. 1, pp. 105-110). Academic Press.

Schiering, I., & Kretschmer, J. (2012). The Infrastructure Level of Cloud Computing as a Basis for Privacy and Security of Software Services. In J. Camenisch, B. Crispo, S. Fischer-Hübner, R. Leenes, & G. Russello (Eds.), *Privacy and Identity Management for Life* (pp. 88–101). Springer. doi:10.1007/978-3-642-31668-5_7

Sellami, M. H., & Sahbani, F. (2023). A Modeling Approach to Automate the Functioning of Tomato Greenhouses. *International Journal of Horticultural Science and Technology*, *10*(4), 387–396.

Semios. (n.d.). *Integrated Pest Management*. https://semios.com/ipm/

Sennaar, K. (2019). *AI in Agriculture – Present Applications and Impact*. Available at: https://emerj.com/ai-sector-overviews/ai-agriculture-present-applications-impact/

Seo, D., Cho, B. H., & Kim, K. C. (2021). Development of monitoring robot system for tomato fruits in hydroponic greenhouses. *Agronomy (Basel)*, *11*(11), 2211. doi:10.3390/agronomy11112211

Shaikh, T. A., Mir, W. A., Rasool, T., & Sofi, S. (2022). Machine learning for smart agriculture and precision farming: Towards making the fields talk. *Archives of Computational Methods in Engineering*, *29*(7), 4557–4597. doi:10.100711831-022-09761-4

Shaikh, T. A., Rasool, T., & Lone, F. R. (2022). Towards leveraging the role of machine learning and artificial intelligence in precision agriculture and smart farming. *Computers and Electronics in Agriculture*, *198*, 107119. doi:10.1016/j.compag.2022.107119

Sharif, S., Zeadally, S., & Ejaz, W. (2023). Space-aerial-ground-sea integrated networks: Resource optimization and challenges in 6G. *Journal of Network and Computer Applications*, *215*, 103647. doi:10.1016/j.jnca.2023.103647

Sharma D., Singh A., & Singhal S. (2021). The Technological Shift: AI in Big Data and IoT. *The Smart Cyber Ecosystem for Sustainable Development*, 69-90.

Sharma, A., Jain, A., Chowdary, V., & Gupta, P. (2021). *Machine Learning Applications for Precision Agriculture : A Comprehensive Review*. Academic Press.

Sharma, R., & Parhi, S. (2017, October). A review on use of big data in warehousing to enhance accessibility of food. In *2017 2nd International Conference on Communication and Electronics Systems (ICCES)* (pp. 663-667). IEEE. 10.1109/CESYS.2017.8321162

Sharma, A. K., Chaurasia, S., & Srivastava, D. K. (2018). Supervised rainfall learning model using machine learning algorithms. In *The International Conference on Advanced Machine Learning Technologies and Applications (AMLTA2018)* (pp. 275-283). 10.1007/978-3-319-74690-6_27

Sharma, A., Jha, N., & Kishor, K. (2022). Predict COVID-19 with Chest X-ray. In D. Gupta, Z. Polkowski, A. Khanna, S. Bhattacharyya, & O. Castillo (Eds.), *Proceedings of Data Analytics and Management. Lecture Notes on Data Engineering and Communications Technologies* (Vol. 90). Springer. doi:10.1007/978-981-16-6289-8_16

Sharma, R., Das, S., Gourisaria, M. K., Rautaray, S. S., & Pandey, M. (2020). A Model for Prediction of Paddy Crop Disease Using CNN. In H. Das, P. K. Pattnaik, S. S. Rautaray, & K.-C. Li (Eds.), *Progress in Computing, Analytics and Networking* (pp. 533–543). Springer. doi:10.1007/978-981-15-2414-1_54

Sharma, V., Tripathi, A. K., & Mittal, H. (2022). Technological revolutions in smart farming: Current trends, challenges & future directions. *Computers and Electronics in Agriculture*, *201*, 107217. doi:10.1016/j.compag.2022.107217

Shatnawi, M. Q., Yassein, M. B., Abuein, Q., & Nsuir, L. (2019, December). Big data analytics tools and applications: survey. In *Proceedings of the Second International Conference on Data Science, E-Learning and Information Systems* (pp. 1-4). 10.1145/3368691.3368741

Sheikh, J. A., Cheema, S. M., Ali, M., Amjad, Z., Tariq, J. Z., & Naz, A. (2021). IoT and AI in precision agriculture: Designing smart system to support illiterate farmers. *Advances in Artificial Intelligence, Software and Systems Engineering: Proceedings of the AHFE 2020 Virtual Conferences on Software and Systems Engineering, and Artificial Intelligence and Social Computing,* July 16-20, 2020, USA, 490–496.

Shei, S., Kalloniatis, C., Mouratidis, H., & Delaney, A. (2016). Modelling Secure Cloud Computing Systems from a Security Requirements Perspective. In S. Katsikas, C. Lambrinoudakis, & S. Furnell (Eds.), *Trust, Privacy and Security in Digital Business* (pp. 48–62). Springer International Publishing. doi:10.1007/978-3-319-44341-6_4

Shekhar, B. G., & Tyagi, S. (2023). Dynamic Bandwidth Allocation Using Linear Regression Model in IOT with Machine Learning Techniques. In D. K. Sharma, S.-L. Peng, R. Sharma, & G. Jeon (Eds.), Micro-Electronics and Telecommunication Engineering (pp. 597–609). Springer Nature. doi:10.1007/978-981-19-9512-5_54

Shekhawat, K., Singh, V. K., Rathore, S. S., Raj, R., & Das, T. K. (2021). Achieving sustainability in food systems: Addressing changing climate through real time nitrogen and weed management in a conservation agriculture-based maize–wheat system. *Sustainability (Basel)*, *13*(9), 5010. doi:10.3390u13095010

Shukla, M., & Jharkharia, S. (2013). Agri-fresh produce supply chain management: A state-of-the-art literature review. *International Journal of Operations & Production Management*, *33*(2), 114–158. doi:10.1108/01443571311295608

Siamabele, B. (2021). The significance of soybean production in the face of changing climates in Africa. *Cogent Food & Agriculture*, *7*(1), 1933745. doi:10.1080/23311932.2021.1933745

Sinha, M., Chacko, E., Makhija, P., & Pramanik, S. (2021). Energy Efficient Smart Cities with Green IoT. In C. Chakrabarty (Ed.), *Green Technological Innovation for Sustainable Smart Societies: Post Pandemic Era*. Springer. doi:10.1007/978-3-030-73295-0_16

Siregar, R. R. A., Seminar, K. B., Wahjuni, S., & Santosa, E. (2022). Vertical farming perspectives in support of precision agriculture using artificial intelligence: A review. *Computers*, *11*(9), 135. doi:10.3390/computers11090135

Sisinni, E., Saifullah, A., Han, S., Jennehag, U., & Gidlund, M. (2018). Industrial internet of things: Challenges, opportunities, and directions. *IEEE Transactions on Industrial Informatics*, *14*(11), 4724–4734. doi:10.1109/TII.2018.2852491

Siskandar, R., Santosa, S. H., Wiyoto, W., Kusumah, B. R., & Hidayat, A. P. (2022). Control and automation: Insmoaf (Integrated Smart Modern Agriculture and Fisheries) on the greenhouse model. *Jurnal Ilmu Pertanian Indonesia*, *27*(1), 141–152. doi:10.18343/jipi.27.1.141

Sladojevic, S., Arsenovic, M., Anderla, A., Culibrk, D., & Stefanovic, D. (2016). Deep neural networks based recognition of plant diseases by leaf image classification. *Computational Intelligence and Neuroscience*, *2016*, 2016. doi:10.1155/2016/3289801 PMID:27418923

Smith, M. (1992). CROPWAT: A computer program for irrigation planning and management (No. 46). Food & Agriculture Org.

Soni, D., & Kumar, N. (2022). Machine learning techniques in emerging cloud computing integrated paradigms: A survey and taxonomy. *Journal of Network and Computer Applications*, *205*, 103419. doi:10.1016/j.jnca.2022.103419

Soul, P. K., & Saini, S. (2017). Data Security Approach in Cloud Computing Using SHA. *International Research Journal of Engineering and Technology, 4.*

Soussi, M., Chaibi, M. T., Buchholz, M., & Saghrouni, Z. (2022). Comprehensive review on climate control and cooling systems in greenhouses under hot and arid conditions. *Agronomy (Basel), 12*(3), 626. doi:10.3390/agronomy12030626

Sowmiya, M., & Krishnaveni, S. (2021). Deep Learning Techniques to Detect Crop Disease and Nutrient Deficiency -A Survey. *2021 International Conference on System, Computation, Automation and Networking (ICSCAN),* 1–5. 10.1109/ICSCAN53069.2021.9526442

Soydan, D. K., Turgut, N., Yalç, M., & Turgut, C. (2021). Evaluation of pesticide residues in fruits and vegetables from the Aegean region of Turkey and assessment of risk to consumers. Academic Press.

Srivastava, A., Pandey, A., & Shukla, S.K. (n.d.). *Plant Disease Detection using SVM Algorithm and Neural Network Approach.* Academic Press.

Srivastava, A., & Dashora, K. (2021). A Fuzzy ISM approach for modelling electronic traceability in agri-food supply chain in India. *Annals of Operations Research,* 1–19.

Srivastava, R., Sharma, V., Jaiswal, V., & Raj, S. (2020). A research paper on smart agriculture using IoT. *International Research Journal of Engineering and Technology, 7*(07), 2708–2710.

Subeesh, A., & Mehta, C. R. (2021). Automation and digitization of agriculture using artificial intelligence and internet of things. *Artificial Intelligence in Agriculture, 5,* 278–291. doi:10.1016/j.aiia.2021.11.004

Suganthi, M., Event, S., & Senthilkumar, P. (2020). Comparative bioefficacy of Bacil- lus and Pseudomonas continues against Looper severe in tea (Camellia sinensis (L.) O. Kuntze. *Physiology and Molecular Biology of Plants, 26*(10), 2053–2060. doi:10.100712298-020-00875-2 PMID:33088049

SumalanR. L.StroiaN.MogaD.MuresanV.LodinA.VintilaT., & Popescu,

Sumathi, M., Rajkamal, M., Raja, S. P., Venkatachalapathy, M., & Vijayaraj, N. (2022). A crop yield prediction model based on an improved artificial neural network and yield monitoring using a blockchain technique. *International Journal of Wavelets, Multresolution, and Information Processing, 20*(06), 2250030. doi:10.1142/S0219691322500308

Sun, J., He, X., Ge, X., Wu, X., Shen, J., & Song, Y. (2018). Detection of key organs in tomatoes based on deep migration learning in a complex background. *Agriculture, 8*(12), 196.

Sun, R., Yang, W., Li, Y., & Sun, C. (2021). Multi - residue analytical methods for pesticides in teas: A review. *European Food Research and Technology, 247*(8), 1839–1858. doi:10.100700217-021-03765-3

Sutton, R. S., & Barto, A. G. (2018). *Reinforcement learning: An introduction.* MIT press.

Swathi, B. S., & Guruprasad, H. S. (2014). Integration of wireless sensor networks and cloud computing. *International Journal of Computational Science, 2*(5), 49–53.

Sweety, T. J. R., Arun, T.R., Sajith, K., Jose, J., & Reeha, K.R. (2023). CBCPW-Fed ADFD Shape On-Body Antenna for ISM Band and Sub-6 GHz 5G Applications. In Optical and Wireless Technologies. OWT 2021. Lecture Notes in Electrical Engineering (vol. 892). Springer. doi:10.1007/978-981-19-1645-8_36

Sykuta, M. E. (2016). Big data in agriculture: property rights, privacy and competition in ag data services. *International Food and Agribusiness Management Review, 19,* 57-74.

Symeonaki, E. G., Arvanitis, K. G., & Piromalis, D. D. (2019). Cloud computing for IoT applications in climate-smart agriculture: A review on the trends and challenges toward sustainability. In *Innovative Approaches and Applications for Sustainable Rural Development: 8th International Conference, HAICTA 2017, Chania, Crete, Greece, September 21-24, 2017, Selected Papers 8* (pp. 147-167). Springer International Publishing. 10.1007/978-3-030-02312-6_9

Symeonaki, E., Arvanitis, K. G., & Piromalis, D. D. (2017). Review on the Trends and Challenges of Cloud Computing Technology in Climate-Smart Agriculture. *HAICTA*, 66-78.

Takabi, H., Joshi, J. B. D., & Ahn, G. (2010). Security and Privacy Challenges in Cloud Computing Environments. *IEEE Security and Privacy*, *8*(6), 24–31. doi:10.1109/MSP.2010.186

Tambe, A. B., Mbanga, B. M. R., Nzefa, D. L., & Name, M. G. (2019). *Pesticide usage and occupational hazards among farmers working in small-scale tomato farms in Cameroon.* doi:10.118642506-019-0021-x

Taneja, A., Nair, G., Joshi, M., Sharma, S., Sharma, S., Jambrak, A. R., Roselló-Soto, E., Barba, F. J., Castagnini, J. M., Leksawasdi, N., & Phimolsiripol, Y. (2023). Artificial Intelligence: Implications for the Agri-Food Sector. *Agronomy (Basel)*, *13*(5), 1397. doi:10.3390/agronomy13051397

Tang, Z., Yang, J., Li, Z., & Qi, F. (2020). Grape disease image classification based on lightweight convolution neural networks and channel wise attention. *Computers and Electronics in Agriculture*, *178*, 105735. doi:10.1016/j.compag.2020.105735

Tauqeer, H. M., Turan, V., & Iqbal, M. (2022). Production of Safer Vegetables from Heavy Metals Contaminated Soils: The Current Situation, Concerns Associated with Human Health and Novel Management Strategies. In J. A. Malik (Ed.), *Advances in Bioremediation and Phytoremediation for Sustainable Soil Management*. Springer. doi:10.1007/978-3-030-89984-4_19

Taviti Naidu, G., Ganesh, K. V. B., Vidya Chellam, V., Praveenkumar, S., Dhabliya, D., Pramanik, S., & Gupta, A. (2023). Technological Innovation Driven by Big Data. In *Advanced Bioinspiration Methods for Healthcare Standards, Policies, and Reform*. IGI Global. doi:10.4018/978-1-6684-5656-9

Taylor, G. A., Torres, H. B., Ruiz, F., Marin, M. N., Chaves, D. M., Arboleda, L. T., ... Mouazen, A. M. (2019). pH measurement IoT system for precision agriculture applications. *Revista IEEE América Latina*, *17*(05), 823–832. doi:10.1109/TLA.2019.8891951

Taylor, M. (2018). Climate-smart agriculture: What is it good for? *The Journal of Peasant Studies*, *45*(1), 89–107. doi:10.1080/03066150.2017.1312355

Thilakarathne, N. N., Bakar, M. S. A., Abas, P. E., & Yassin, H. (2022). Towards making the fields talks: A real-time cloud enabled IoT crop management platform for smart agriculture. *Frontiers in Plant Science*, 13. PMID:36684733

Tomar, P., & Kaur, G. (2021). *Artificial Intelligence and IoT-based Technologies for Sustainable Farming and Smart Agriculture*. IGI Global. doi:10.4018/978-1-7998-1722-2

Tomaszewski, L., Kołakowski, R., Mesodiakaki, A., Ntontin, K., Antonopoulos, A., Pappas, N., Fiore, M., Mosahebfard, M., Watts, S., Harris, P., Lin, C.-K., Santiago, A. R., Lazarakis, F., & Chatzinotas, S. (2023). ETHER: Energy- and Cost-Efficient Framework for Seamless Connectivity over the Integrated Terrestrial and Non-terrestrial 6G Networks. In I. Maglogiannis, L. Iliadis, A. Papaleonidas, & I. Chochliouros (Eds.), *Artificial Intelligence Applications and Innovations. AIAI 2023 IFIP WG 12.5 International Workshops* (pp. 32–44). Springer Nature Switzerland. doi:10.1007/978-3-031-34171-7_2

Tomaszewski, L., Kołakowski, R., & Zagórda, M. (2022). Application of Mobile Networks (5G and Beyond) in Precision Agriculture. In I. Maglogiannis, L. Iliadis, J. Macintyre, & P. Cortez (Eds.), *Artificial Intelligence Applications and Innovations. AIAI 2022 IFIP WG 12.5 International Workshops* (pp. 71–86). Springer International Publishing. doi:10.1007/978-3-031-08341-9_7

Traore, B. B., Kamsu-Foguem, B., & Tangara, F. (2017). Data mining techniques on satellite images for discovery of risk areas. *Expert Systems with Applications*, *72*, 443–456. doi:10.1016/j.eswa.2016.10.010

Trikha, P. (2021). A Critical Review on Security Issues in Cloud Computing. In D. Goyal, V. E. Bălaş, A. Mukherjee, V. Hugo C. de Albuquerque, & A. K. Gupta (Eds.), Information Management and Machine Intelligence (pp. 643–652). Springer. doi:10.1007/978-981-15-4936-6_70

Tripathi, A. D., Mishra, R., Maurya, K. K., Singh, R. B., & Wilson, D. W. (2019). Estimates for world population and global food availability for global health. In *The role of functional food security in global health* (pp. 3–24). Academic Press. doi:10.1016/B978-0-12-813148-0.00001-3

Tripathi, A., & Parihar, B. (2011). E-Governance challenges and cloud benefits. *Proceedings of IEEE International Conference on Computer Science and Automation Engineering (CSAE 2011)*, 351-354. 10.1109/CSAE.2011.5953237

Tsironis, V., Bourou, S., & Stentoumis, C. (2020). Tomato d: Evaluation of object detection algorithms on a new real-world tomato dataset. *The International Archives of the Photogrammetry, Remote Sensing and Spatial Information Sciences*, *43*, 1077–1084.

Tubb, C., & Seba, T. (2021). Rethinking food and agriculture 2020-2030: The second domestication of plants and animals, the disruption of the cow, and the collapse of industrial livestock farming. *Industrial Biotechnology (New Rochelle, N.Y.)*, *17*(2), 57–72. doi:10.1089/ind.2021.29240.ctu

Tuli, S., Mirhakimi, F., Pallewatta, S., Zawad, S., Casale, G., Javadi, B., Yan, F., Buyya, R., & Jennings, N. R. (2023). AI augmented Edge and Fog computing: Trends and challenges. *Journal of Network and Computer Applications*, *216*, 103648. doi:10.1016/j.jnca.2023.103648

Udawant, P., & Srinath, P. (2022). Cotton Leaf Disease Detection Using Instance Segmentation. *Journal of Cases on Information Technology*, *24*(4), 1–10.

Union, U. (2017). *World population projected to reach 9.8 billion in 2050, and 11.2 billion in 2100*. UN DESA.

Upadhyay, N. (2018). *CABology: Value of Cloud, Analytics and Big Data Trio Wave*. Springer. doi:10.1007/978-981-10-8675-5

US Environmental Protection Agency. (2005). *Pesticide fact sheet*. Retrieved from https://www3.epa.gov/pesticides/chem_search/reg_actions/registration/fs_PC-124871_11-Aug-05.pdf

Vats, K., Sharma, S., & Rathee, A. (2012). A Review of Cloud Computing and E-Governance. *International Journal of Advanced Research in Computer Science and Software Engineering*, *2*(2). http://www.ijarcsse.com/docs/papers/february2012/volume_2_issue/-V2I2074.pdf

Venkataramana, K., & Padmavathamma, M. (2012). A design of framework for agri-cloud. *IOSR Journal of Computer Engineering*, *4*(5), 1–6. doi:10.9790/0661-0450106

Vikaspedia. (n.d.). Available at https://vikaspedia.in/agriculture/crop-production/integrated-pest-managment/ipm-for-vegetables/ipm-strategies-for-tomato/tomato-diseases-and-symptoms

Vincent, D. R., Deepa, N., Elavarasan, D., Srinivasan, K., Chauhdary, S. H., & Iwendi, C. (2019). Sensors driven AI-based agriculture recommendation model for assessing land suitability. *Sensors (Basel), 19*(17), 3667. doi:10.339019173667 PMID:31450772

Vishnoi, V. K., Kumar, K., & Kumar, B. (2021). Crop Disease Classification Through Image Processing and Machine Learning Techniques Using Leaf Images. *2021 First International Conference on Advances in Computing and Future Communication Technologies (ICACFCT)*, 27–32. 10.1109/ICACFCT53978.2021.9837353

Viswanatha, V., Kumari, A., & Sathisha, B. M. (2022, October). Implementation of IoT in Agriculture: A Scientific Approach for Smart Irrigation. In *2022 IEEE 2nd Mysore Sub Section International Conference (MysuruCon)* (pp. 1-6). IEEE. 10.21203/rs.3.rs-2392461/v1

Vos, R., & Bellù, L. G. (2019). Global trends and challenges to food and agriculture into the 21st century. *Sustainable food and agriculture*, 11-30.

Waller, R., Kacira, M., Magadley, E., Teitel, M., & Yehia, I. (2021). Semi-transparent organic photovoltaics applied as greenhouse shade for spring and summer tomato production in arid climate. *Agronomy (Basel), 11*(6), 1152. doi:10.3390/agronomy11061152

Wang, H., Li, C., Zhang, J., Wang, J., Ma, Y., & Lian, Y. (2019). A new LSTM-based gene expression prediction model: L-GEPM. *Journal of Bioinformatics and Computational Biology, 17*(04), 1950022. doi:10.1142/S0219720019500227 PMID:31617459

Wang, L., Xu, L., Zheng, Z., Liu, S., Li, X., Cao, L., Li, J., & Sun, C. (2021). Smart contract-based agricultural food supply chain traceability. *IEEE Access : Practical Innovations, Open Solutions, 9*, 9296–9307. doi:10.1109/ACCESS.2021.3050112

Wang, Y. P. E., Lin, X., Adhikary, A., Grovlen, A., Sui, Y., Blankenship, Y., Bergman, J., & Razaghi, H. S. (2017). A primer on 3GPP narrowband Internet of Things. *IEEE Communications Magazine, 55*(3), 117–123. doi:10.1109/MCOM.2017.1600510CM

Weihua, J., & Shibing, S. (2013). Research on the Security Issues of Cloud Computing. In Z. Du (Ed.), *Intelligence Computation and Evolutionary Computation* (pp. 845–848). Springer. doi:10.1007/978-3-642-31656-2_115

Wei, Y., & Zhang, Y. (2018). Cloud Computing Data Security Protection Strategy. In X. Sun, Z. Pan, & E. Bertino (Eds.), *Cloud Computing and Security* (pp. 376–386). Springer International Publishing. doi:10.1007/978-3-030-00006-6_34

Wezel, A., Herren, B. G., Kerr, R. B., Barrios, E., Gonçalves, A. L. R., & Sinclair, F. (2020). Agroecological principles and elements and their implications for transitioning to sustainable food systems. A review. *Agronomy for Sustainable Development, 40*(6), 1–13. doi:10.100713593-020-00646-z

Williams, B. (2012). *The Economics of Cloud Computing*. Cisco Press.

Wolfert, S., Ge, L., Verdouw, C., & Bogaardt, M. J. (2017). Big data in smart farming–a review. *Agricultural Systems, 153*, 69–80. doi:10.1016/j.agsy.2017.01.023

Wordofa, M. G., & Sassi, M. (2020). Impact of agricultural interventions on food and nutrition security in Ethiopia: Uncovering pathways linking agriculture to improved nutrition. *Cogent Food & Agriculture, 6*(1), 1724386. doi:10.1080/23311932.2020.1724386

Workie, E., Mackolil, J., Nyika, J., & Ramadas, S. (2020). Deciphering the impact of COVID-19 pandemic on food security, agriculture, and livelihoods: A review of the evidence from developing countries. *Current Research in Environmental Sustainability, 2*, 100014. doi:10.1016/j.crsust.2020.100014 PMID:34977605

World Health Organization. (2020). *The state of food security and nutrition in the world 2020: transforming food systems for affordable healthy diets* (Vol. 2020). Food & Agriculture Org.

Wyld, D. C. (2010). The cloudy future of government IT: Cloud computing and the public sector around the world. *International Journal of Web & Semantic Technology, 1*(1), 1–20.

Xiaojing, Z., & Yuanguai, L. (2012, September). Zigbee implementation in intelligent agriculture based on internet of things. In *2nd International Conference on Electronic & Mechanical Engineering and Information Technology* (pp. 1842-1846). Atlantis Press.

Xiao, Z., & Chen, J. (2015). Cloud Computing Security Issues and Countermeasures. In W. E. Wong (Ed.), *Proceedings of the 4th International Conference on Computer Engineering and Networks* (pp. 731–737). Springer International Publishing. 10.1007/978-3-319-11104-9_85

Xia, P. (2020). Data Security Risk and Preventive Measures of Virtual Cloud Server Based on Cloud Computing. In Z. Xu, R. M. Parizi, M. Hammoudeh, & O. Loyola-González (Eds.), *Cyber Security Intelligence and Analytics* (pp. 40–45). Springer International Publishing. doi:10.1007/978-3-030-43306-2_6

Xie, S. (2019). *Does a dual reduction in chemical fertilizer and pesticides, improve nutrient loss and tea yield and quality? A pilot study in a green tea garden in Shaoxing.* Zhejiang Province. doi:10.100711356-018-3732-1

Xiong, H., Cheng, J., Wang, Y., Yang, X., & Wang, K. (2022). Research on Cloud Computing Security Technology of Power System. In Y. Tian, T. Ma, M. K. Khan, V. S. Sheng, & Z. Pan (Eds.), Big Data and Security (pp. 269–283). Springer. doi:10.1007/978-981-19-0852-1_21

Xu, L., Yuan, S., & Man, J. (2020). Changes in rice yield and yield stability in China during the past six decades. *Journal of the Science of Food and Agriculture, 100*(8), 3560–3569. doi:10.1002/jsfa.10385 PMID:32198766

Xu, Y., Zhang, X., Li, H., Zheng, H., Zhang, J., Olsen, M. S., Varshney, R. K., Prasanna, B. M., & Qian, Q. (2022). Smart breeding driven by big data, artificial intelligence and integrated genomic-enviromic prediction. *Molecular Plant, 15*(11), 1664–1695. doi:10.1016/j.molp.2022.09.001 PMID:36081348

Yadlapati, D., Siddhartha, N., Seelamneni, M., Nali, A. Y., Sangaraju, H. R., & Sridhar, P. S. V. S. (2023). Security Management Approaches Over the Cloud. *2023 International Conference on Sustainable Computing and Data Communication Systems (ICSCDS),* 1277–1282. 10.1109/ICSCDS56580.2023.10105026

Yakusheva, N.A. (2012). Calculation of economic efficiency of cloud computing. *Vestnik MGTU im. N.E. Baumann. Sir. "Priborostroenie", 5: Informatics and systems*, 224-235.

Yan, B., Fan, P., Lei, X., Liu, Z., & Yang, F. (2021). A real-time apple targets detection method for picking robot based on improved YOLOv5. *Remote Sensing, 13*(9), 1619.

Yanbykh, R., Saraikin, V., & Lerman, Z. (2020). Changes in Russia's agrarian structure: What can we learn from agricultural census? *Russian Journal of Economics, 6*(1), 26–41. doi:10.32609/j.ruje.6.49746

Yang, X., Shu, L., Chen, J., Ferrag, M. A., Wu, J., Nurellari, E., & Huang, K. (2021). A survey on smart agriculture: Development modes, technologies, and security and privacy challenges. *IEEE/CAA Journal of Automatica Sinica, 8*(2), 273-302.

Yan, L., Rong, C., & Zhao, G. (2009). Strengthen Cloud Computing Security with Federal Identity Management Using Hierarchical Identity-Based Cryptography. In M. G. Jaatun, G. Zhao, & C. Rong (Eds.), *Cloud Computing* (pp. 167–177). Springer. doi:10.1007/978-3-642-10665-1_15

Yan, X., Zhang, X., Chen, T., Zhao, H., & Li, X. (2012). The Research and Design of Cloud Computing Security Framework. In Y. Wu (Ed.), *Advances in Computer, Communication, Control and Automation* (pp. 757–763). Springer. doi:10.1007/978-3-642-25541-0_95

Yao, J., Qi, J., Zhang, J., Shao, H., Yang, J., & Li, X. (2021). A real-time detection algorithm for Kiwifruit defects based on YOLOv5. *Electronics (Basel)*, *10*(14), 1711.

Yatoo, A. A., & Sharma, A. (2021). A Novel Model for Automatic Crop Disease Detection. *2021 Sixth International Conference on Image Information Processing (ICIIP)*, *6*, 310–313. 10.1109/ICIIP53038.2021.9702553

Yeganeh, S., Babazadeh Sangar, A., & Azizi, S. (2023). A novel Q-learning-based hybrid algorithm for the optimal offloading and scheduling in mobile edge computing environments. *Journal of Network and Computer Applications*, *214*, 103617. doi:10.1016/j.jnca.2023.103617

Yeo, S.-S., & Park, J. H. (2013). Security Considerations in Cloud Computing Virtualization Environment. In Grid and Pervasive Computing (pp. 208–215). Springer. doi:10.1007/978-3-642-38027-3_22

Yoosefzadeh Najafabadi, M., Hesami, M., & Eskandari, M. (2023). Machine Learning-Assisted Approaches in Modernized Plant Breeding Programs. *Genes*, *14*(4), 777. doi:10.3390/genes14040777 PMID:37107535

Yu, S. (2018). Towards the Cloud Computing from Services to Securities and Challenges. In X. Sun, Z. Pan, & E. Bertino (Eds.), *Cloud Computing and Security* (pp. 334–343). Springer International Publishing. doi:10.1007/978-3-030-00009-7_31

Yusifov, F. & Farajova, A. (2021). Protection of personal data in electronic governance: "Data embassy" concept. *Information Society Problems, 2*, 119-130.

Zambare, R., Awati, C., Deshmukh, R., & Shirgave, S. (2022). A Theoretical Study on CNN based Cotton Crop Disease Detection. *2022 6th International Conference on Trends in Electronics and Informatics (ICOEI),* 1062–1066. 10.1109/ICOEI53556.2022.9777219

Zanella, A. (2020). *Security challenges to smart agriculture: Current state, key issues, and future directions*. Academic Press.

Zhang, T., Cai, Y., Zhuang, P., & Li, J. (2023). Remotely Sensed Crop Disease Monitoring by Machine Learning Algorithms: A Review. *Unmanned Systems*, 1–11. doi:10.1142/S2301385024500237

Zhang, D., & ... *Detection of systemic pesticide residues in tea products at trace level based on SERS and verified by GC – MS*. doi:10.100700216-019-02103-7

Zhang, H., Shukla, M. K., Larson, S., Rajendran, A. M., & Jiang, S. (2022). Molecular dynamics study of anisotropic shock responses in oriented α-quartz single crystal. *Journal of Materials Science*, *57*(12), 6688–6705. doi:10.100710853-022-07076-0

Zhang, N., Wang, M., & Wang, N. (2002). Precision agriculture—A worldwide overview. *Computers and Electronics in Agriculture*, *36*(2-3), 113–132. doi:10.1016/S0168-1699(02)00096-0

Zhang, S., Guo, Y., Li, S., Ke, Z., Zhao, H., Yang, J., Wang, Y., Li, D., Wang, L., Yang, W., & Zhang, Z. (2022). Investigation on environment monitoring system for a combination of hydroponics and aquaculture in greenhouse. *Information Processing in Agriculture*, *9*(1), 123–134. doi:10.1016/j.inpa.2021.06.006

Zhang, W. Y., Jing, T. Z., & Yan, S. C. (2017). Studies on prediction models of Den- drolimus superans occurrence area based on machine learning. *Journal of Beijing Forestry University*, *39*(1), 85–93.

Zhihua, S. (2016, August). Design of smart home system based on ZigBee. In *2016 International Conference on Robots & Intelligent System (ICRIS)* (pp. 167-170). IEEE. 10.1109/ICRIS.2016.35

Zhong, D. R. Y., Tan, P. K., & Bhaskaran, P. G. (2017). Data-driven food supply chain management and systems. *Industrial Management & Data Systems*, *117*(9), 1779–1781. doi:10.1108/IMDS-06-2017-0269

Zhong, Y., & Zhao, M. (2020). Research on deep learning in apple leaf disease recognition. *Computers and Electronics in Agriculture*, *168*, 105146.

Zhou, X., & Wang, S. (2019). Recent developments in radio frequency drying of food and agricultural products: A review. *Drying Technology*, *37*(3), 271–286. doi:10.1080/07373937.2018.1452255

Zhu, X., & Goldberg, A. B. (2022). *Introduction to Semi-Supervised Learning*. Springer Nature.

Zhu, Y., Song, J., & Dong, F. (2011). Applications of wireless sensor net- work in the agriculture environment monitoring. *Procedia Engineering*, *16*, 608–614. doi:10.1016/j.proeng.2011.08.1131

Ziani, A., & Medouri, A. (2019). Cloud Computing: Security and Privacy Issues. In M. Ben Ahmed, A. A. Boudhir, & A. Younes (Eds.), *Innovations in Smart Cities Applications* (2nd ed., pp. 668–677). Springer International Publishing. doi:10.1007/978-3-030-11196-0_55

Zichichi, A., & Ragaini, R. C. (1993). Innovative Technologies for Cleaning the Environment: Air, Water and Soil. World Scientific. doi:10.1142/9789812796974

Zwattendorfer, B., & Tauber, A. (2013). The Public Cloud for E-Government. *International Journal of Distributed Systems and Technologies*, *4*(4), 1–14. doi:10.4018/ijdst.2013100101

About the Contributors

Avinash Kumar Sharma currently works as Assistant Professor in Department of Computer Science & Engineering at ABES Institute of Technology Ghaziabad. Their current project is 'Optimization of service Model in Cloud Computing.'

Nitin Chanderwal is currently working as an Associate Professor Educator-Full Time in the Department of Electrical and Computer Engineering at University of Cincinnati. In the past I have worked as Associate Professor-Full Time of Information Systems and Analytics at IIM Shillong, Meghalaya, INDIA. During his tenure at IIM Shillong he also served as Chairperson for the Areas: {(Information Systems and Analytics) & (IT Services and Website Committee)}. During 2017-2018, He has worked as Professor Educator in the Department of EECS at University of Cincinnati, OH and during 2010-2011 as First Tier Bank Professor in the Peter Kiewit Institute at University of Nebraska at Omaha, NE, USA. In July 2001, he received B.Engg. in Computer Science & Engineering [Hons.] from Dr. B.R. Ambedkar University, Agra and M.Engg. in Software Engineering from Thapar University, erstwhile Thapar Institute of Engineering and Technology (Deemed University), Patiala, Punjab, INDIA in March 2003. In September 2008, he received Ph.D. in Computer Science & Engineering from Jaypee University of Information Technology, INDIA and University of Florida (UF), Gainesville, FL, USA under student exchange program, specifically he has completed 12 credits course work from UF. In May 2013, he received D.Sc. in Computer Science & Engineering from Uttarakhand Technical University, Dehradun, INDIA. I completed partial research work of D.Sc. at University of Nebraska at Omaha (UNO), NE, USA. He is a IBM certified engineer, a Life Member of IAENG, Senior Member of IEEE, ACM & IACSIT and Member of SIAM and ACIS and have published 200+ Research Papers in peer reviewed International Journals & Transactions, Book Chapters, Symposium, Conferences and Position. He has bagged more than 50 academic and research awards. My research interest includes Blockchain Technology, Cyber Physical Systems, Big Data Analytics, Social Networks especially Computer Mediated Communications & Flaming, Interconnection Networks & Architecture, Fault-tolerance & Reliability, NoCs, SoCs, and NiPs, Application of Stable Matching Problems, Stochastic Communication and Sensor Networks. He has received 2 Indian Patents and 1 Australian Patent during 2020-2021. he is also an Associate Editor of the International Journal of Parallel, Emergent and Distributed Systems, Taylor and Francis, UK and IEEE Access, IEEE, USA.

Rijwan Khan received his B. Tech Degree in Computer Science & Engineering from BIT, M. Tech in Computer Science & Engineering from IETE, and Ph.D.in Computer Engineering from Jamia Millia Islamia, New Delhi. He has 16 years of Academic & Research Experience in various areas of Computer

Science. He is currently working as Professor and Head of Department Computer Science and Engineering at ABES Institute of Technology, Ghaziabad U.P. He is author of three subject books. He published more than 30 research papers in different journals and conferences. He has been nominated in the board of reviewers of various peer-reviewed and refereed Journals. He is Editor of five research books. He is Editor of seven special issues of SCI and Scopus indexed journals. He was session chaired in three international conferences & keynote speakers in some the national and International conferences.

* * *

Prateek Chaurasia is a B.Tech final-year student at Ajay Kumar Garg Engineering College, he is majoring in Computer Science and Information Technology. He is proficient in full-stack web development. Prateek is also a winner of prestigious hackathons like the Smart India Hackathon and the UNESCO India Africa Hackathon.

Lakshmi D. is presently designated as a Senior Associate Professor in the School of Computing Science and Engineering (SCSE) & Assistant Director, at the Centre for Innovation in Teaching & Learning (CITL) at VIT Bhopal. She has 17 international conference presentations, and 21 international journal papers inclusive of SCOPUS & SCI (cumulative impact factor 31). 3 SCOPUS inee book chapters. A total of 24 patents are in various states and 18 patents have been granted at both national and international levels. One Edited book with Taylor & Francis (SCOPUS Indexed). She has won two Best Paper awards at international conferences, one at the IEEE conference and another one at EAMMIS 2021. She received two awards in the year 2022. She received two awards in the year 2022. She has addressed innumerable guest lectures, acted as a session chair, and was invited as a keynote speaker at several international conferences. She has conducted FDPs that cover approximately ~80,000 plus faculty members including JNTU, TEQIP, SERB, SWAYAM, DST, AICTE, MHRD, ATAL, ISTE, Madhya Pradesh Government-sponsored, and self-financed workshops across India on various titles.

Anupam Dwivedi is a B.Tech final-year student at Ajay Kumar Garg Engineering College, he is majoring in Computer Science and Information Technology.

Vishnuvarthanan Govindaraj, born in 1986, is currently working as an Associate Professor, for VIT-Bhopal University, Bhopal, Madhya Pradesh, in the Division of Data Science of School of Computing Science and Education. He has completed his B.E. in Instrumentation Control and Engineering by 2003, followed by M. Tech. In Electronics and Communication Engineering by 2009 and had been awarded a Doctoral Degree in the Department of Electronics and Communication Engineering for the thesis titled "Tumor Detection and Tissue Segmentation in Multimodal MR Brain Images Using Fuzzy and Optimization Techniques" by 2015. He has a total fourteen years of teaching experience in the three different reputed and premier engineering institutes of Tamil Nādu. With more than 100 journal paper publications, to till date he has published 31 international journals indexed in Science Citation Index Database with the highest impact factor publication of 17.560 and an average impact factor of 4.923. His significant research traits are the publications in IEEE Transactions on Fuzzy Systems (of Impact Factor: 12.253), IEEE Transactions on Industrial Informatics (of Impact Factor: 11.648) and Information Fusion, Elsevier. He has authored and co-authored 70+ conference papers indexed in the SCOPUS database. He has even published two Scopus indexed Book Chapters. He has been the reviewer for fourteen journals till now, and few are - IEEE Transactions on Fuzzy System, IEEE Transactions on Neural Networks and Learning Systems, Current Medical Imaging and Reviews, Neural Processing Letters. He has been a certified reviewer of three journals, namely Computers in Biology and Medicine, Applied Soft Computing and OPTIK. He has specialized in Medical Image Processing and pattern recognition.

Muskan Gupta is a B.Tech final-year student at Ajay Kumar Garg Engineering College, he is majoring in Computer Science and Information Technology.

Ishaan is a Bachelor of Technology (Computer Science and Engineering) student at ABES Institute of Technology, under Dr. A.P.J Abdul Kalam Technical University, passionate about computer science and engineering. He is skilled in various technologies and has participated in many hackathons and coding competitions. He is a curious and ambitious learner.

Mrignainy Kansal is currently pursuing her PhD in Computer Science and Engineering from Netaji Subhas University of Technology (NSUT), New Delhi. Alongside her studies, she holds the position of an Assistant Professor at Ajay Kumar Garg Engineering College, Ghaziabad. She completed her BTech Degree in Information Technology from AKTU and further pursued her MTech in Information Technology from Guru Gobind Singh Indraprastha University, Delhi. She has a good record of publications at various International Conferences. Additionally, she has qualified GATE-2016 and GATE-2023. Her expertise lies in diverse areas such as Machine Learning, Deep Learning, Natural Language Processing (NLP), Computer Vision, and Software Engineering.

Kaushal Kishor received his Ph.D. in Computer science and engineering from AKTU Lucknow, in the domain of Mobile Ad hoc Network. M.Tech & B.Tech in Computer Science & Engineering from UPTU Lucknow. Currently, he is working in ABES Institute of Technology, Ghaziabad as Associate Professor Information Technology. He has supervised more than 50 projects for graduate and post graduate students. He has more than 18 years of experience of teaching. He has worked for 6 years in IMS Engineering College, Ghaziabad as Assistant Professor of Computer Science & Engineering. He has worked for 4 years in Hi-Tech Engineering College, Ghaziabad as Assistant Professor of Computer Science & Engineering and He has worked for 3 years in G L BAJAJ Engineering College, Greater Noida as Associate Professor of Information technology. He is Gate Qualified 2003 score 94.5 percentile. He has book published and edited (1) "Cloud-based Intelligent Informative Engineering for Society 5.0" pp. 1-234. Chapman and Hall/CRC., 2023, eBook ISBN: 9781003213895 (2) Design and Analysis Algorithms (ISBN NO. 978-93-81695-20-3) (3) Computer Networks (ISBN NO. 978-93-81695-27-2) (4) Compiler Design (ISBN NO. 938169530 x and ISBN-13 9789381695302) (5) Design and Analysis of Algorithms: Techniques and Control Management (ISBN NO. 978-81-8220-516-1) (6) Computer Networks a System Approach (ISBN NO. 978-81-8220-516-3) (7) Compiler Design Principles, Techniques, and Tools (ISBN:978-81-8220-626-7) for various engineering field like B.Tech. and MCA student. He has published 20 papers in peer reviewed international/National journals and conferences. His research interest includes Artificial Intelligence, Computer Networks, Algorithm, Compiler Design Wireless and Sensor Networking. (1) Educational Qualification.

Indrajeet Kumar received his B Tech in Computer Science & Engineering from Sant Longowal Institute of Engineering and Technology, Punjab in 2009 and M Tech in Computer Engineering with specialization in Computer Network from YMCA University of Science & technology, Faridabad in 2013. He received his PhD on Analysis and Classification of Breast density classification using Mammographic images from GB Pant Institute of Engineering & Technology, Ghurdauri, Uttarakhand, India in 2019. During his PhD he worked on enhancing the potential of most commonly available Breast density classification for differential diagnosis between atypical cases of focal density cases. He served in academia in various reputed organizations like Galgotias Institutions, Greater Noida, U.P, India and graphic era hill University, Dehradun, India, for 6 years. His research interests include application of machine learning, Deep Learning and soft computing techniques for analysis of medical images.

Sabyasachi Pramanik is a professional IEEE member. He obtained a PhD in Computer Science and Engineering from Sri Satya Sai University of Technology and Medical Sciences, Bhopal, India. Presently, he is an Associate Professor, Department of Computer Science and Engineering, Haldia Institute of Technology, India. He has many publications in various reputed international conferences, journals, and book chapters (Indexed by SCIE, Scopus, ESCI, etc). He is doing research in the fields of Artificial Intelligence, Data Privacy, Cybersecurity, Network Security, and Machine Learning. He also serves on the editorial boards of several international journals. He is a reviewer of journal articles from IEEE, Springer, Elsevier, Inderscience, IET and IGI Global. He has reviewed many conference papers, has been a keynote speaker, session chair, and technical program committee member at many international conferences. He has authored a book on Wireless Sensor Network. He has edited 8 books from IGI Global, CRC Press, Springer and Wiley Publications.

Brijendra Singh received a PhD from VIT University, Vellore, Tamilnadu, India. He is an Assistant Professor (Senior) in the School of Information Technology and Engineering at VIT University Vellore, Tamilnadu, India. His research interests include Health Informatics, Soft Computing Techniques, Artificial Intelligence, Rough Set, Knowledge Representations and Deep Learning.

Manoj Kumar Singh is currently working as Assistant Professor in IMS Engineering College, Ghaziabad.

Pancham Singh currently working as an Assistant Professor in the Department of Information Technology, AKGEC, Ghaziabad Since 2007 to till date. He has more than 16+ years of teaching experience and 1.5 years of industry experience. He received B.Tech. degree in Computer Science & Engineering from Dr. A.P.J. Abdul Kalam Technical University, [formely known as a UPTU], Lucknow in 2005, Master's degree in Information Technology from RTU, Kota, Rajasthan in 2014 and Pursuing PhD from Netaji Subhas University of Technology (NSUT), Delhi since Jan' 2023. He has authored two books Operating Systems and Graph Theory. He has presented and published more than 10 papers in International Journals/ Conferences. He had reviewed papers for the International Conferences ICDT and ICISML 2023. He has published 6 National and International Patents. His area of interest are AI&ML, WSN, SE, Blockchain, IoT and Social Networking.

Mili Srivastava is currently pursuing her PhD in Computer Science and Engineering from Jaypee Institute of Information Technology, Noida. She is currently working as an Assistant Professor at Ajay Kumar Garg Engineering College, Ghaziabad.

Sanskar Varshney is a B.Tech final-year student at Ajay Kumar Garg Engineering College, he is majoring in Computer Science and Information Technology.

Farhad Yusifov received his Ph.D. degree in Information Technology in 2010 from the Institute of Information Technology in Baku, Azerbaijan. His primary research interests include various areas in e-government, public administration, e-governance, e-demography, e-voting, and social media, particularly in the area of data analytics, web usage mining, and multi-criteria decision-making. He is head of the education department of the Institute of Information Technology. He is the author of more than 70 papers.

Index

Recommended Reference Books

IGI Global's reference books are available in three unique pricing formats: Print Only, E-Book Only, or Print + E-Book.

Order direct through IGI Global's Online Bookstore at **www.igi-global.com** or through your preferred provider.

ISBN: 9781668445488
EISBN: 9781668445501
© 2022; 334 pp.
List Price: US$ **240**

ISBN: 9781668467503
EISBN: 9781668467527
© 2022; 296 pp.
List Price: US$ **240**

ISBN: 9781799889007
EISBN: 9781799889021
© 2022; 340 pp.
List Price: US$ **215**

ISBN: 9781799898511
EISBN: 9781799898535
© 2022; 276 pp.
List Price: US$ **250**

ISBN: 9781799896913
EISBN: 9781799896937
© 2022; 219 pp.
List Price: US$ **215**

ISBN: 9781668436868
EISBN: 9781668436875
© 2022; 2,064 pp.
List Price: US$ **2,085**

Do you want to stay current on the latest research trends, product announcements, news, and special offers?
Join IGI Global's mailing list to receive customized recommendations, exclusive discounts, and more.
Sign up at: **www.igi-global.com/newsletters.**

Publisher of Timely, Peer-Reviewed Inclusive Research Since 1988

IGI Global
PUBLISHER of TIMELY KNOWLEDGE

www.igi-global.com Sign up at www.igi-global.com/newsletters facebook.com/igiglobal twitter.com/igiglobal linkedin.com/igiglobal

www.ingramcontent.com/pod-product-compliance
Lightning Source LLC
LaVergne TN
LVHW062118300926
849161LV00010B/2484